Progress in Nonlinear Differential Equations and Their Applications
Volume 20

Editor
Haim Brezis
Université Pierre et Marie Curie
Paris
and
Rutgers University
New Brunswick, N.J.

Editorial Board
Antonio Ambrosetti, Scuola Normale Superiore, Pisa
A. Bahri, Rutgers University, New Brunswick
Luis Cafarelli, Institute for Advanced Study, Princeton
L. C. Evans, University of California, Berkeley
Mariano Giaquinta, University of Florence
David Kinderlehrer, Carnegie-Mellon University, Pittsburgh
S. Klainerman, Princeton University
Robert Kohn, New York University
P. L. Lions, University of Paris IX
Jean Mawhin, Université Catholique de Louvain
Louis Nirenberg, New York University
Lambertus Peletier, University of Leiden
Paul Rabinowitz, University of Wisconsin, Madison
John Toland, University of Bath

Topics in Geometry

In Memory of Joseph D'Atri

Simon Gindikin
Editor

Birkhäuser

Boston • Basel • Berlin

Simon Gindikin
Department of Mathematics
Rutgers University
New Brunswick, NJ 08903

Library of Congress In-Publication Data

Topics in geometry : in memory of Joseph D'Atri / Simon Gindikin, editor.
p. cm. -- (Progress in nonlinear differential equations and their applications ; v. 20)
Includes bibliographical references.
ISBN-13: 978-1-4612-7534-3 e-ISBN-13: 978-1-4612-2432-7
DOI: 10.1007/978-1-4612-2432-7

1. Geometry, Differential. I. D'Atri, J. E., 1938-1993 . II. Gindikin, S. G. (Semen Grigor'evich) III. Series.
QA641.T63 96-1703
516.3'6--dc20 CIP

Printed on acid-free paper

Birkhäuser

Typeset and reformatted by TeXniques, Boston, MA

9 8 7 6 5 4 3 2 1

Joseph D'Atri

In Memory of Joseph D'Atri, our Friend and Colleague

Contents

Preface

This collection of articles serves to commemorate the legacy of Joseph D'Atri, who passed away on April 29, 1993, a few days after his 55th birthday. Joe D'Atri is credited with several fundamental discoveries in geometry. In the beginning of his mathematical career, Joe was interested in the generalization of symmetrical spaces in the E. Cartan sense. Symmetric spaces, differentiated from other homogeneous manifolds by their geometrical richness, allows the development of a deep analysis. Geometers have been constantly interested and challenged by the problem of extending the class of symmetric spaces so as to preserve their geometrical and analytical abundance. The name of D'Atri is tied to one of the most successful generalizations: Riemann manifolds in which (local) geodesic symmetries are volume-preserving (up to sign). In time, it turned out that the majority of interesting generalizations of symmetrical spaces are D'Atri spaces: natural reductive homogeneous spaces, Riemann manifolds whose geodesics are orbits of one-parameter subgroups, etc.

The central place in D'Atri's research is occupied by homogeneous bounded domains in $\mathbb{C}^n$, which are not symmetric. Such domains were discovered by Piatetskii-Shapiro in 1959, and given Joe's strong interest in the generalization of symmetric spaces, it was very natural for him to direct his research along this path. For many years he moved towards the solution to the problem, which without doubt was principal for his work in this area: finding the geometrical characterization of symmetrical domains among homogeneous ones. The final result is surprisingly elegant: symmetric domains are characterized by the condition that its sectional curvature in its Bergmann metric is nonpositive.

An area of interest to Joe during the last years of his life were homogeneous pseudo-Kählerian manifolds. I was happy to have had an opportunity to collaborate with him on this project. We investigated the problem of the realization of pseudo-Hermitian symmetric spaces as nonconvex tube domains. We had a lot of other plans. Joe continued to work even though he was terminally ill. We reproduce here one of his last notes, which he sent to me from the hospital.

A significant part of this collection is composed of papers in geometrical areas, where D'Atri made outstanding contributions: Lie groups and homogeneous manifolds, first of all, D'Atri spaces, nonsymmetric homogeneous bounded domains, Siegel domains, homogeneous cones and different problems of symmetric spaces. The other part of this collection (which intersects with the first) is comprised of contributions of D'Atri's colleagues at Rutgers University. Joe D'Atri did a lot for the development of the Department of Mathematics. From 1985 to 1990 he was the Chair of the Department.

The topics of these papers reflects a broad view of the subject of geometry, which was inherent to him. We have here papers on differential equations of geometrical nature, automorphic functions, topology, Lorentz surfaces, etc. We also included biographical information and a summary of research prepared by D'Atri in 1992.

Joe D'Atri was an incredibly talented person. He had very broad interests not only in mathematics, but in other areas as well. He was distinguished by high professionalism in everything he did: research, teaching, administrative service, photography, cactus and orchid cultivation, model railroading, woodworking. Also he had a talent to be a devoted and generous husband, father, grandfather and friend. We felt that during his life and we feel it even more so now after his death.

Simon Gindikin
November 1995

Commentary

We reproduce on the next page one of the last computations which Joe D'Atri sent me from the hospital. During this time we discussed different interesting classes of pseudo Kählerian homogeneous manifolds. Joe wrote about an elementary version of nonconvex tube (following some more general constructions of R. Penney). He considers a nonconvex tube domain at $\mathbb{C}^3_{(z_1,z_2,z_3)}, z_1 = x_1 + iy_1, z_2 = x_2 + iy_2, w = u + iv$:

$$v > -y_1 y_2 + y_1^3/3 \qquad (*)$$

It is remarkable that this domain is affine homogeneous. The group of affine automorphisms is generated by all real translations, the dilatation

$$z_1 \mapsto \lambda z_1, \quad z_2 \mapsto \lambda^2 z_2, \quad w \mapsto \lambda^3 w, \qquad \lambda > 0$$

and by a 2-parameter unipotent subgroup

$$(z_1, z_2, w) \mapsto (z_1 + ia, z_2 + ib + az, w + i(-ab + a^3/3) - az_2 - bz_1)$$

It is evident that these transforms conserve (*) and we have a simple transitive group. Such transformation are similar to automorphisms of Hermitian Siegel domains but the presence of the cubic term in (*) is essentially a new phenomenon. Unfortunately it does not generalize to the matrix case.

Simon Gindikin

For Simon   Penney example 2 (nilball)   11/18/92

Domain is $\{(z_1, z_2, w) : \operatorname{Im} w > -x_1 x_2 + x_1^3/3\}$ where $z_1 = x_1 + iy_1$, $z_2 = x_2 + iy_2$

There is a simply-transitive action by a solvable group S (description later)

The Lie algebra $\mathfrak{s}$ has basis $X_1, Y_1, X_2, Y_2, Z_0, A$ with

$$[A, X_1] = \tfrac{1}{3}X_1, \quad [A, X_2] = \tfrac{2}{3}X_2, \quad [A, Z_0] = Z_0$$
$$[A, Y_1] = \tfrac{1}{3}Y_1, \quad [A, Y_2] = \tfrac{2}{3}Y_2$$
$$[X_1, Y_1] = Y_2, \quad [X_1, Y_2] = [X_2, Y_1] = Z_0$$

$\mathfrak{n} = [\mathfrak{s}, \mathfrak{s}]$ is nilpotent, $\mathfrak{s} = \mathfrak{a} \oplus \mathfrak{n}$ with $\mathfrak{a} = \mathbb{R}A$.

Let ε be the root on $\mathfrak{a}$ given by $\varepsilon(A) = 1$.

Then we have root spaces $\mathfrak{n}_\varepsilon = \mathbb{R}Z_0$, $\mathfrak{n}_{\frac{2}{3}\varepsilon} = \mathbb{R}X_2 \oplus \mathbb{R}Y_2$, $\mathfrak{n}_{\frac{1}{3}\varepsilon} = \mathbb{R}X_1 \oplus \mathbb{R}Y_1$

The induced j-structure is $jX_1 = Y_1$, $jX_2 = Y_2$, $jZ_0 = A$

The Koszul form is given by $\omega(Z_0) = 4$, $\omega(X_i) = \omega(Y_i) = \omega(A) = 0$

The associated inner product is $\langle X_1, X_2 \rangle = \langle Y_1, Y_2 \rangle = -4$

$$\langle Z_0, Z_0 \rangle = \langle A, A \rangle = 4$$

which is nondegenerate of type $(+,+,+,+,-,-)$ (in real coordinates)

The Nilpotent group $\mathcal{N}$ (corresponding to $\mathfrak{s}$) is diffeomorphic to $\mathbb{R}^5$.

Using coordinates $(a_1, b_1, a_2, b_2, x) \in \mathbb{R}^5 \propto \mathcal{N}$, $(z_1, z_2, w) \in \mathbb{C}^3$, we have

$$(a_1, b_1, a_2, b_2, x) : (z_1, z_2, w) \longmapsto (z_1 + a_1 + ia_2,\ z_2 + b_1 + ib_2 + a_1 z_1,\ w + x + i(-a_1 b_1 + a_1^3/3 - a_1 z_2 - b_1 z_1))$$

The multiplication in $\mathcal{N}$ is given by

$$(a_1', b_1', a_2', b_2', x')(a_1, b_1, a_2, b_2, x) = (a_1 + a_1',\ b_1 + b_1' + a_1 a_1',\ a_2 + a_2',\ b_2 + b_2' + a_1' a_2,$$
$$x + x' + a_1' b_2 + b_1' a_2)$$

(Yes, there is a more transparent way of writing this but that means going back through all of Penney's construction

X_1 corresponds to the 1-parameter group $(t, \tfrac{1}{2}t^2, 0, 0, 0)$ | X_2 corresponds to the 1-parameter group $(0, t, 0, 0, 0)$
Y_1 " " " " $(0, 0, t, 0, 0)$ | Y_2 " " " " " $(0, 0, 0, t, 0)$

and Z_0 to the 1-parameter group $(0, 0, 0, 0, t)$

$S = \mathcal{N} \rtimes \mathcal{A}$, where $\mathcal{A} = \exp tA$ and $\exp tA : (z_1, z_2, w) \longmapsto (e^{t/3} z_1, e^{2t/3} z_2, e^t w)$.

The base point of the domain is taken to be $(0, 0, i)$ (needed to compute j)

Summary of Research

Joseph D'Atri

Major emphasis of last 12 years: On the geometry of bounded homogeneous domains and similar homogeneous spaces

Fix some notation. D denotes a bounded homogeneous domain in some $\mathbb{C}^n$ or the equivalent homogeneous Siegel domain, G is the identity component of $\mathrm{Aut}(D)$, and S is the simply-transitive split-solvable subgroup of G whose Lie algebra $\mathfrak{s}$ is a normal j-algebra (with j coming from a $\mathbb{C}^n$). b is the Bergman metric on D which is, up to scalar multiple, the unique G invariant Kähler–Einstein metric on D.

The main results are:

(1) D is symmetric if the sectional curvature $(b) \leq 0$ (the converse is well-known). The final step is in [16] but requires the calculations of [12]–[14].

As far as I can determine, the study of the curvature properties of D has the following history. Kobayashi (*Transactions AMS*, Vol. 92, 1959) did some curvature calculations and stated, "It is very likely that most [of these] domains have negative sectional curvature." K. H. Look (Lu Chi-Keng) and Hsu I-Chau (Xu Yi Chao) (*Acta Math. Sinica* **11** (1961)) gave the first example of D where the Bergman metric has some positive curvature (even positivie holomorphic curvature) but the example is very computational. [12] gives a whole class of examples depending only on a simple criterion in terms of root spaces of $\mathfrak{s}$, namely, if $\exists$ roots $1/2\ \varepsilon m$, $1/2\ \varepsilon n$, $2(\varepsilon m - \varepsilon n)$ with $\dim \mathfrak{n}_{\frac{1}{2}(\varepsilon m - \varepsilon n)} = 1$ (later improved to any odd dimension [14]), then D has some positive sectional curvature in the Bergman metric (as well as for any S-invariant admissible Kähler metric, see (2)).

(2) There are conditions, expressed in terms of the root spaces of $\mathfrak{s}$, which say that for some domaines D, there is no S-invariant admissible Kähler metric on D with (i) sectional curvature ≤ 0; (ii) holomorphic sectional curvature ≤ 0. These results apply in particular to the Bergman metric.

Examples are also given of nonsymmetric domains which do admit S-invariant admissible Kähler metrics with either sectional curvature or holomorphic sectional curvature ≤ 0. All of this is in [12]–[14].

The S-invariant admissible Kähler metrics are those defined by the inner product $< X, Y >= \omega[jX, Y]$, $X, Y \in \mathfrak{s}$ for any linear form ω on $(\mathfrak{s}, j)$ which satisfies the normal j-algebra axioms. These can be parameterized by an r-dimensional cone where $r = \operatorname{rank} D = \dim \mathfrak{a}(\mathfrak{s} = \mathfrak{a} \oplus [\mathfrak{s}, \mathfrak{s}])$ although each ray in the cone describes metrics differing only by multiplicative constant. The Bergman metric is determined by a formula giving the values of ω on a canonical basis of the fundamental root spaces in terms of the dimensions of all root spaces.

(3) The following conditions are equivalent:

(a) D is symmetric,
(b) the almost complex structure map on the tangent space T_pD (p is a canonical base point for the Siegel domain D) is in the image of the isotropy representation,
(c) there exist no nontrivial G-invariant vector fields,
(d) the algebra of G-invariant differential operators on D is commutative,
(e) the isotropy acts transitively on the Šilov boundary.

These are all proved in [18]. Properties (b) and (e) are important in the usual development of harmonic analysis on symmetric domains and are an indication that such analysis on nonsymmetric domains will be substantially harder.

These results mostly depend on (4) below.

(4) Let D be given the Bergman metric. There is a Levi decomposition $\mathfrak{S} = \mathfrak{S}_{SS} \oplus \mathfrak{r}$ ($\mathfrak{r}$ = the radical) so that if $\mathfrak{h}_{SS}$ is the sum of the noncompact ideals of $\mathfrak{S}_{SS}$, then $\mathfrak{s}$ is the orthogonal direct sum of $\mathfrak{s}_1$ and $\mathfrak{s}_2$ where $\mathfrak{s}_1 = \mathfrak{s} \cap \mathfrak{h}_{ss}, \mathfrak{s}_2 = \mathfrak{s} \cap \mathfrak{r}$. Further, one has $\mathfrak{s}_\iota = \mathfrak{a}_\iota \oplus \mathfrak{n}_\iota$ where $\mathfrak{a}_\iota = \mathfrak{a} \cap \mathfrak{s}_\iota, \mathfrak{n}_\iota = [\mathfrak{s}_\iota, \mathfrak{s}_\iota] = \mathfrak{n} \cap \mathfrak{s}_\iota\ \iota = 1, 2$. In this decomposition one has:

(a) The image of the isotropy representation is spanned by the set of covariant derivatives ∇_X, $X \in \mathfrak{n}_1$, together with the algebra of skew symmetric derivations of $\mathfrak{s}$.
(b) The orbit of exp $\mathfrak{s}$ (at the appropriate base point) is a totally geodesic Riemannian locally symmetric submanifold D, and D is symmetric if and only if $\mathfrak{a}_2 = 0$.
(c) $\mathfrak{n}_1$ is the direct sum of the root spaces $\mathfrak{n}_\alpha$ for which $\alpha|\mathfrak{a}_2 = 0$ and $\mathfrak{a}_2$ is the maximal subspace of $\mathfrak{a}$ with this property. Similary, $\mathfrak{n}_2$ is the direct sum of the root spaces $\mathfrak{n}_\alpha$ for which $\alpha|\mathfrak{a}_2 \neq 0$.
(d) There are conditions on roots α, β which imply $\mathfrak{n}_\alpha \subset \mathfrak{n}_1 \Rightarrow \mathfrak{n}_\beta \subset \mathfrak{n}_1$.

This theorem is proved in [18], using much of the algebraic work of Dorfmeister. Another proof, giving most but not all of the above results, is given in [19], using only differential geometric operators (covariant derivatives, curvature) and results on the isotropy algebra going back to Ambrose–Singer and simplified by Azencott–Wilson. The second proof may also have extensions to the more general class of NC-algebras.

(5) There exist domains D with geodesics which have focal points in the Bergman metric [17].

This is interesting because of the implications D symmetric $\Leftrightarrow$ sectional curvature $(D, b) \leq 0 \Rightarrow D$ has no focal points $\Rightarrow D$ has an "ideal" boundary consisting of classes of asymptotic geodesics.

(6) An irreducible Siegel domain D is quasi-symmetric if and only if the metric induced by the Bergman metric of D on the tube domain $D' \subset D$ is symmetric [21].

In [16] we are given another characterization of irreducible quasi-symmetric domains in terms of dimensions of root spaces; although useful, this was probably already known.

Satake introduced the notion of a quasi-symmetric Siegel domain and an equivalent formulation was later given by Dorfmeister. Essentially these definitions say that D is quasi-symmetric if and only if one of the following equivalent conditions holds:

(a) The cone of D is self dual with respect to a canonically defined inner product.

(b) The connectedness (or connection) algebra of D (as defined by Vinberg) is a Jordan algebra.

Part of the work in [13, 14, 16] involves showing that Dorfmeister's product agrees with Vinberg's product and that both may be computed by a simple formula involving the covariant derivative and j.

Note that quasi-symmetric, as applied to bounded homogeneous domains, is *not* the same as quasi-symmetric in [6].

(7) (a) For $X \in \mathfrak{s}$, the induced vector $X^*|p \in T_pD$ is tangent to a canonical flat if and only if $X \in \mathfrak{a} \oplus \mathfrak{n}_1$. In particular, D is symmetric if and only if every vector is tangent to a canonical flat [21].

Here a canonical flat means a flat totally-geodesic submanifold representable as an orbit $(\exp \mathfrak{a}) \cdot p$ for the abelian component $\mathfrak{a}$ of a normal j-algebra $\mathfrak{s} = \mathfrak{a} \oplus [\mathfrak{s}, \mathfrak{s}]$, $\mathfrak{s}$ corresponding to any simply-transitive split-solvable $S \subset G$. Also $\mathfrak{n}_1$ is as in (4), p is any point in D and $X^*|_p = \frac{d}{dt}|_{t=0}(\exp tX) \cdot p$.

(b) For D irreducible quasi-symmetric and $X \in \mathfrak{s}, X^*|_p$ is tangent to a canonical flat if and only if $\nabla_x R = 0$ [21].

Note that not every flat totally-geodesic submanifold of D is contained in a canonical flat when D is nonsymmetric (examples in [21]). Further, [17] shows that geodesics tangent to a canonical flat behave essentially like geodesics in symmetric D.

(8) The holomorphic bisectional curvature of a quasi-symmetric D is nonpositive [22].

This depends on a formula due to Zelow who gave a case-by-case proof and applied it only to the holomorphic curvature. [22] gives a classification free proof of Zelow's formula and, in the process, gives numerous useful results on covariant derivatives in quasi-symmetric domains (these have been extended and made more systematic by Azumaya).

(9) [24] studies geodesics in a class of left-invariant metrics on Lie groups S whose Lie algebra $\mathfrak{s}$ has a structure like $\mathfrak{s} = \mathfrak{a} \oplus [\mathfrak{s}, \mathfrak{s}]$. The result shows that geodesics are "asymptotically tangent" to the S-invariant distribution induced by $\mathfrak{a}$.

(10) [25] studies the affine symmetric spaces X_0 which are pseudo-Hermitian. For classical groups, we find all cases where there exists an open dense subdomain $D \subset X_0$ which is a generalized Siegel domain of tube type and give explicit constructions. D is generalized in the sense that the cone V defining D is no longer convex and we find interesting classes of new nonconvex cones.

Older work

[11] gives a general construction for finding nonsymmetric naturally reductive metrics, primarily given as left-invariant metrics on a compact semi-simple Lie group. [11] also gives general constructions for finding homogeneous Einstein metrics. This gives a large number of examples of "exotic" Einstein metrics on, for example, spheres.

[3,7,9] introduce a study of spaces whose geodesic symmetries are volume preserving (called for a while, e.g., in Besse's Einstein spaces, "D'Atri spaces"). These are interesting because of the connections with naturally reductive spaces, commutative spaces, and harmonic spaces. These connections were greatly expanded and developed by, for example, Vanhecke and Szabo.

This work is not totally separate. For example, I have long thought that examples of noncompact nonsymmetric harmonic spaces might come from some of the algebras akin to n in a normal j-algebra. Recently examples have been found by Damek and Ricci on extensions of generalized Heisenberg groups.

Publications

[1] *Homomorphisms of continuous pseudo groups* (Joseph E. D'Atri), Nagoya Math. Jour., **25** (1965), 143-163

[2] *The existence of special orthonormal frames* (J. E. D'Atri and H. K.Nickerson), J. Diff.Geom., **2** (1968), 393-409.

[3] *Divergence-preserving geodesic symmetries* (J. E. D'Atri and H. K. Nickerson), Jour. Diff. Geom.,**3** (1969), 467-476.

[4] *Examples of Riemmanian manifolds having divergence-free orthonormal frames* (J. E. D'Atri and H.K. Nickerson), Tensor, N.S. **22** (1971), 360-362.

[5] *Connections and symmetry structures* (J.E.D'Atri), Tensor, N.S., **25** (1972), 448-450.

[6] *Quasi-symmetric is locally symmetric* (J.E.D'Atri), J. Diff. Geom., **9** (1974), 275-277.

[7] *Geodesic symmetries in spaces with special curvature tensors* (J.E. D'Atri and H.K. Nickerson), Jour. Diff. Geom., **9** (1974), 251-262

[8] *Goedesic conformal transformations and symmetric spaces* (J.E. D'Atri), Kodai Math. Sem. Rep., **26** (March 1975), 201-203.

[9] *Geodesic spheres and symmetries in naturally reductive spaces* (J.E. D'Atri), Mich. Math. Jour., **22** (1975), 71-76.

[10] *Certain isoparametric families of hypersurfaces in symmetric spaces* (J.E. D'Atri) Jour. Diff. Geom., **14** (1979) 21-40.

[11] *Naturally reductive metrics and Einstein metrics on compact Lie groups* (J.E. D'Atri and W. Ziller), Memoirs of Amer. Math. Soc., **18** (215) (1979), 72 pages.

[12] *The curvature of homogenous Siegel domains* (J.E. D'Atri), J. Diff. Goemetry, **15** (1980), 61-70.

[13] *Holomorphic sectional curvatures of bounded homogenous domains and related questions* (J.E. D'Atri), Trans. A.M.S., **256** (1979), 405-413.

[14] *Sectional curvatures and quasisymmetric domains* (J.E. D'Atri), J. Diff. Geometry, **16** (1981), 11-18.

[15] *Selected papers on Geometry, Vol.* **4** (J.E. D'Atri), Editor (With A.K Stehey, T.K. Milnor, T.F. Banchoff), (1979), MAA.

[16] *A characterization of bounded symmetric domains by curvature* (J.E. D'Atri and I. Dotti de Miatello), Trans. A.M.S., **276** (1983), 531-540.

[17] *Geodesics and Jacobi fields in bounded homogeneous domains* (J.E. D'Atri and Zhao Yan Da), Proc. A.M.S., **89** (1983), 55-62.

[18] *The isotropy representation for homogenous Siegel domains* (J.E. D'Atri and J. Dorfmeister and Zhao Yan Da), Pac, J. Math., **120** (1985), 295-326.

[19] *A refined structure for normal j-algebras* (J.E. D'Atri and Zhao Yan da), Algebra, Groups and Geometries, **5** (1988), 33-60.

[20] *Codazzi tensors and harmonic curvature for left invariant metrics* (J.E. D'Atri), Geometriae Dedicata, **19** (1985), 229-236.

[21] *Flat totally geodesic submanifolds of quasisymmetric Siegel domains* (J.E. D'Atri and J. Dorfmeister), Geometriae Dedicata, **28** (1988), 321-336.

[22] *The biholomorphic curvature of quasisymmetric Siegel domains* (J.E. D'Atri and J. Dorfmeister), J. Differential Geometry, **31** (1990), 73-100.

[23] *Eigenvalues of the curvature operator of certain homogenous manifolds* (J.E. DAtri and I. Dotti. Miatello), Canad. J. Math, **XLII** (1990), 981-999.

[24] *The long-time behavior of geodesics in certain left-invariant metrics* (J.E. D'Atri), Proc. Amer. Math. Soc., **116**:3, (1992), 813-818.

[25] *Siegel Domain realization of pseudo-Hermitian symmetric manifolds* (J.E. D'Atri and S. Gindikin), Geometriae Dedicata, **46** (1993), 91-126.

Curriculum Vitae

Birthdate: April 20, 1938

Education: A.B. Columbia College, 1959
Ph.D., Princeton University, 1964

Positions: Instructor, Columbia College summer session, 1959/60/61
Instructor, Princeton University, 1962/63
Lecturer, Rutgers University, 1963/64
Assistant Professor, Rutgers University, 1964/69
Associate Professor, Rutgers University, 1969/75
Professor, Rutgers University, 1975...
Visitor, IAS, 1976/77
Member, IAS, 1990/91

Talks: *Homogeneous Siegel domains,* Invited series of 4 talks, University of Virginia, June, 1981,

Homogeneous Kähler manifolds, Invited survey, AMS Summer Research Conference in Applications of Lie Groups in Differential Geometry, Arcata, CA, July, 1985 (cf. Abstracts AMS, Issue 41, Vol. 6, No. 6, 1985)

Geometry of homogeneous bounded domains, Invited talk at special session on Topics in Differential Geometry, AMS, Issue 40, Vol. 6, No. 5, 1985

Nonsymmetric homogeneous Siegel domains, Columbia University Geometry and Analysis Seminar, April, 1986

Geometry of bounded homogeneous domains, Institute for Advanced Study, Princeton, January, 1991

Professional Activities

Member, MAA Editorial Committee for R. W. Brink Selected Papers Volume 4 (selected papers on geometry), 1978/79
Member, AMS Organizing Committee for summer research conference in Applications of Lie Groups in Differential Geometry, 7/84-7/85
Reviewer for National Science Foundation (approx. 2 papers per yr)
Referee for Annals, Transations of AMS, Proceedings of AMS, Geom. Dedicata, Math. Zeitschrift, Illinois J.,
Representative of MAA to Rutgers Department of Mathematics, 1983...

Rutgers Activities

Member, Rutgers Speakers Bureau, June, 1987-1992
Member, Coordinating Council of the (New Brunswick) Office of Minority Undergraduate Science Programs, Sept., 1988-1992
Fellow, Douglass College, since inception
Member, Mc Nair Post-Baccalaureate Achievement Program Faculty Advisory Committee, Oct. 1990-

Some Notable Past University Activity

Chair, Dept. of Mathematics, Douglass College, 1973-1976

Member, Ad-hoc Appointments and Promotion Committee for Cook College, July 1973

Representative from Douglass College to University Senate

Chair, Senate Committee on Equal Opportunity

Member, FAS Curriculum Committee, Sept. 1982-June-1985

Scheduling Officer, Dept. of Mathematics, January-1984-Feb. 1985

Acting Vice-Chair for Undergraduate Affairs, Math. Dept., September 1984-February 1985

Member, University Ad-hoc Committee on International Faculty and Student Services, Spring 1987

Member, University Ad-hoc Committee on Warsaw Exchange, Spring 1987

Member, Provost's Committee on Faculty and Student Services, February-June 1988

Member, Provost's Ad-hoc Committee on Faculty Development, February-June 1988

Member, Douglass College Committee on Bunting-Cobb awards, 1988-89

Chair, Dept of Mathematics, February, 1985-July, 1990

Chair, Hill Center Space Committee, February, 1985-July, 1990

Member, Executive Committee of FAS chairs, February, 1985-March 1989

Member, CAIP Building Committee, August, 1985-June 1989

Member, CAIP Academic Advisory Committee, December,1987-June, 1989

Chair, University Mathematics Section, February, 1985-July, 1990

Member, Provost's Committee on Faculty Governance, Sept, 1988-Feb, 1989

Member, President's Coordinating Council on International Programs, Feb, 1990- June, 1990

Member, President's Ad-hoc Affirmitive Action Advisory Committee, Sept, 1989-June, 1990

Delegate from Douglass College to New Brunswick Council of Fellows, July, 1990-June, 1991

3/27/92

Topics in Geometry

In Memory of Joseph D'Atri

Non-Linear Elliptic Equations on Riemannian Manifolds with the Sobolev Critical Exponent

A. Bahri and H. Brezis

Introduction

Given (M^n, g) a Riemannian compact manifold, without boundary of dimension n and a function $q \in L^\infty(M, \mathbb{R})$, we study by variational methods the following class of equations:

$$\begin{cases} (-\Delta + q)u = u^{\frac{n+2}{n-2}} \\ u > 0 \text{ on } M^n \end{cases} .$$

Extending a method devised in [7] and [10], which enables to find solutions to such equations although the corresponding variational problem fails the Palais-Smale condition, we show that this method works also in this new case, where the back-ground metric is not flat. Our result provides, in particular, a new proof of the existence of a solution to the Yamabe conjecture in dimensions 3, 4, and 5. For the locally conformally flat case ($n \geq 6$), this method also works as has been established elsewhere ([11]).

In dimensions 3, 4, and 5, we do not require any condition on M^n, besides being a closed manifold without boundary. The only assumption is that the operator $-\Delta + q$ has to be coercive — this is also a necessary condition — which can be viewed as a condition on q.

We chose a presentation which starts with the case $M^n = S^n$, equipped with its standard metric, which we study thoroughly in the first four sections of this paper. We extend, in these four sections, the technical tools of [7] and [10] to this more general framework. We then move to the case of a general compact manifold without boundary. In dimension $n \geq 6$, besides the positivity condition on $-\Delta + q$, we have to introduce other conditions on the first, second, or third homology group of M with $\mathbb{Z}_2$-coefficients. It would be interesting to understand how much such conditions are necessary. It woud also be interesting to remove the hypothesis $q \in L^\infty(M, \mathbb{R})$ and replace it by some other L^p ($p = n/2$?) condition on q.

The results of this paper were announced without proofs in [12]. Earlier work concerning the same type of model problems on domains—which immediately sets the framework for a functional analysis and variational attack on such equations, beyond the minimization techniques used in more geometric methods—has been done in [13].

It is with deep feelings of recollection and sorrow that we have dedicated this paper to the memory of Joseph D'Atri, our late friend. May these mathematics testify to our profound respect and to our deep regret for having lost such a good colleague and friend.

1. Notations

Let

$$L = \Delta_{\mathbb{R}^{n+1}} - k_n \quad , \quad \text{where} \quad k_n = \frac{n(n-2)}{4}$$

be the conformal Laplacian of S^n.

We introduce, for a given function $q \in L^\infty(M, \mathbb{R})$, the problem:

$$\begin{cases} -Lu + qu = u^{\frac{n+2}{n-2}} \\ u > 0 \end{cases} \quad \text{on } S^n \tag{1}$$

We will assume, throughout this paper, that the operator $-L + q$ is coercive, i.e:

There exists $\delta > 0$ such that, for any $\varphi \in H^1(\mathbb{R}^n)$, we have:

$$\int |\nabla\varphi|^2 + \int (k_n + q)\varphi^2 \geq \delta \|\varphi\|^2_{H^1} . \tag{2}$$

We introduce, for $\lambda > 0$ given and $a \in \mathbb{R}^n$, the family of functions:

$$U_{\lambda,a} = \frac{\lambda^{\frac{n-2}{2}}}{(1 + \lambda^2|x-a|^2)^{\frac{n-2}{2}}} \,, \quad x \in \mathbb{R}^n . \tag{3}$$

This family satisfies the equation:

$$-\Delta U = n(n-2)U^{\frac{n+2}{n-2}} . \tag{4}$$

Let u be a solution of (1) on S^n. Let $\prod^{-1}$ be the stereographic projection from S^n into $\mathbb{R}^n$ (Π^{-1} is analytically defined below). For x in $\mathbb{R}^n$, we introduce:

$$v(x) = \left(\frac{2}{1+r^2}\right)^{\frac{n-2}{2}} u(\Pi(x)) \text{ , where } r = |x| . \tag{5}$$

Then, v satisfies:

$$-\Delta v + a(x)v = v^{\frac{n+2}{n-2}} \tag{6}$$

where

$$a(x) = \frac{4}{(1+r^2)^2} q\big(\Pi(x)\big) \ , \quad x \in \mathbb{R}^n \ , \ r = |x| \ . \tag{7}$$

Indeed, as we prove in the appendix, the transformation defined by (5) yields:

$$\int_{S^n} (-L+q)u \cdot u dv_{S^n} = \int_{\mathbb{R}^n} \Big(\left|\nabla_{\mathbb{R}^n} v\right|^2 dx + \tag{8}$$

$$+ \int_{\mathbb{R}^n} \frac{\left(1+r^2\right)^2}{4} a(x) \left(\frac{1+r^2}{2}\right)^{n-2} v^2(x) \Big) \, dy$$

where $y = \Pi(x)$.

Since

$$dy = \left(\frac{2}{1+r^2}\right)^n dr \ ,$$

we have:

$$\int_{S^n} (-L+q)u \cdot u dv_{S^n} = \int_{\mathbb{R}^n} \left(\left|\nabla_{\mathbb{R}^n} v\right|^2 + a(x)v^2 \right) dx \ . \tag{9}$$

We also have:

$$\int_{S^n} |u|^{\frac{2n}{n-2}}(y) dy = \int_{\mathbb{R}^n} |v(x)|^{\frac{2n}{n-2}} dx \ . \tag{10}$$

(9) and (10) imply that the variational formulation for (1) transforms, through (5), into the variational formulation for (6). Thus, using the appropriate variational spaces, we can solve (1) or (6), indifferently. In S^n, $H^1(S^n, \mathbb{R})$ is the appropriate variational space. The corresponding space in $\mathbb{R}^n$ is the space

$$H = \left\{ v \ \text{ such that } v \in L^{\frac{2n}{n-2}}\left(\mathbb{R}^n\right) \ \text{ and } \ \nabla v \in L^2\left(\mathbb{R}^n\right) \right\} \ . \tag{11}$$

We conclude this section with some analytical formulae for Π and U. Let $y \in S^n$ and let S be the south pole of S^n:

$$S = \begin{pmatrix} 0 \\ \vdots \\ 0 \\ -1 \end{pmatrix} . \tag{12}$$

Let

$$\delta_{S,\lambda}(y) = \left(\frac{2}{1+r^2}\right)^{-\frac{n-2}{2}} U_{0,\lambda}\left(\Pi^{-1}(y)\right) \quad , \quad r = \left|\Pi^{-1}(y)\right| . \tag{13}$$

Geometrically, Π^{-1} is represented as follows:

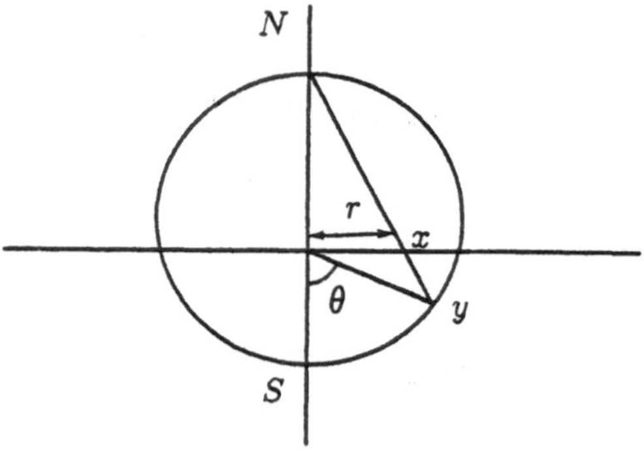

(14)

The last coordinate of y is

$$\frac{r^2-1}{r^2+1} . \tag{15}$$

Thus,

$$\cos\vartheta = \frac{1-r^2}{1+r^2} \quad , \quad 1-\cos\vartheta = \frac{2r^2}{1+r^2} \tag{16}$$

where $\vartheta =$ dist (y, S) on S^n. Then,

$$\begin{aligned} \delta_{S,\lambda}(y) &= \left(\frac{2}{1+r^2}\right)^{-\frac{n-2}{2}} \frac{\lambda^{\frac{n-2}{2}}}{(1+\lambda^2 r^2)^{\frac{n-2}{2}}} = \\ &= \frac{2^{-\frac{n-2}{2}} \lambda^{\frac{n-2}{2}}}{\left(1+\frac{\lambda^2-1}{2}(1-\cos\vartheta)\right)^{\frac{n-2}{2}}} . \end{aligned} \tag{17}$$

Setting:

$$\mu^2 = \frac{\lambda^2 - 1}{2}\,, \tag{18}$$

we have

$$\delta_{S,\mu}(y) = 2^{-\frac{n-2}{2}} \frac{\left(1+2\mu^2\right)^{\frac{n-2}{4}}}{\left(1+\mu^2(1-\cos\vartheta)\right)^{\frac{n-2}{2}}}\,. \tag{19}$$

$\delta_{S,\mu}$ satisfies (this is readily derived from the equation satisfied by U):

$$-L\delta_{S,\mu} = n(n-2)\delta_{S,\mu}^{\frac{n+2}{n-2}}\,. \tag{20}$$

Given $b \in S^n$, we set:

$$\delta_{b,\mu}(y) = 2^{-\frac{n-2}{2}} \frac{(1+2\mu^2)^{\frac{n-2}{4}}}{\left(1+\mu^2(1-\cos\vartheta)\right)^{\frac{n-2}{2}}} \tag{21}$$

where

$$\vartheta = \text{ dist } (b,y) \qquad \text{on } S^n\,. \tag{22}$$

One can easily see that there exists two positive constants α and β, which do not depend on b and μ, such that:

$$\frac{\alpha\mu^{\frac{n-2}{2}}}{(1+\mu^2 d^2)^{\frac{n-2}{2}}} \le \delta_{b,\mu}(y) \le \beta \frac{\mu^{\frac{n-2}{2}}}{(1+\mu^2 d^2)^{\frac{n-2}{2}}}\,, \tag{23}$$

where

$$d = \vartheta = \text{ dist } (b,y)\,.$$

This concludes the first section of this chapter.

2. Self–interaction estimates

Let b be given, for example b is the south pole S of S^n. In this section, we denote δ_λ the function $\delta_{b,\lambda}$. We introduce the function φ_λ, defined by the equation:

$$(-L+q)\varphi_\lambda = -L\delta_\lambda \quad \text{ on } S^n\,. \tag{24}$$

Since $-L+q$ is positive, (24) has a unique solution φ_λ and this function is positive. Let

$$H_\lambda = \lambda^{\frac{n-2}{2}}(\varphi_\lambda - \delta_\lambda)\,. \tag{25}$$

We assume that λ is larger than or equal to 1. The following estimates then hold:

Proposition 1. *There exists a constant C such that, for any $x \in \mathbb{R}^n$ and $\lambda \geq 1$*

i) $|H_\lambda(x)| \leq C$ *if* $n = 3$

ii) $|H_\lambda(x)| \leq C\left(1 + \left|\log\left(\frac{1}{\lambda^2} + d(x)^2\right)\right|\right)$ *if* $n = 4$

iii) $|H_\lambda(x)| \leq C\lambda^{\frac{n-4}{2}}\delta_\lambda^{\frac{n-4}{n-2}}(x)$ *if* $n > 4$

where $d = d(x) = dist\ (x, S)$.

Proof of Proposition 1. Let C and C' denote general constants. H_λ satisfies:

$$(-L+q)\left(\frac{H_\lambda}{\lambda^{\frac{n-2}{2}}} + \delta_\lambda\right) = -L\delta_\lambda\,. \tag{26}$$

Thus

$$(-L+q)H_\lambda = -q\delta_\lambda\lambda^{\frac{n-2}{2}}\,. \tag{27}$$

Observe that

$$\frac{\lambda^2 d^2}{1+\lambda^2(1-\cos d)} \leq \frac{\lambda^2 d^2}{\lambda^2(1-\cos d)} \leq \frac{d^2}{1-\cos d} \leq C'\,. \tag{28}$$

Thus

$$\left|q\lambda^{\frac{n-2}{2}}\delta_\lambda\right| \leq \frac{C\lambda^{n-2}}{(1+\lambda^2(1-\cos d))^{\frac{n-2}{2}}} \leq \frac{C'}{d^{n-2}}\,. \tag{29}$$

We then introduce the function W defined by the equation:

$$(-L+q)W = \frac{1}{d^{n-2}(x)} \qquad \text{on } S^n\,. \tag{30}$$

Using the maximum principle, we derive that, for a suitable constant C', we have:

$$|H_\lambda(x)| \leq C'W(x)\,. \tag{31}$$

Standard estimates on W readily imply then:

$$|W| \leq C \qquad \text{if } n=3 \quad \left(\text{since } \frac{1}{d} \in L^2\right) \tag{32}$$

$$|W(x)| \leq C(1+|\log d|) \qquad \text{if } n=4 \tag{33}$$

$$|W(x)| \leq \frac{C}{d(x)^{n-4}} \qquad \text{if } n>4\,. \tag{34}$$

(32)–(34) imply Proposition 1 if $d(x) \geq \frac{1}{\lambda}$. Indeed, we then have

$$|H_\lambda(x)) \leq C\big(1+|\log d|\big) \leq C'\left(1+\left|\log\left(\frac{1}{\lambda^2}+d^2\right)\right|\right) \quad \text{for } n=4 \tag{35}$$

$$|H_\lambda(x)| \leq \frac{C}{d(x)^{n-4}} \leq C'\lambda^{\frac{n-4}{2}}\delta_\lambda(x)^{\frac{n-4}{n-2}} \quad \text{for } n>4 \tag{36}$$

since

$$\lambda d(x)^2 \delta_\lambda^{\frac{2}{n-2}}(x) = \frac{\lambda^2 d(x)^2}{1+\lambda^2(1-\cos d)} \geq \frac{1}{C'} > 0 \tag{37}$$

if $\lambda d(x) \geq 1$.

Denoting:

$$B_{1/\lambda} = \big\{x \in S^n \text{ such that } d(x) < \frac{1}{\lambda}\big\} \tag{38}$$

we have:

$$\begin{cases} (-L+q)H_\lambda \leq f \text{ in } B_{1/\lambda} \\ H_\lambda \leq g \text{ on } \partial B_{1/\lambda}\,, \end{cases} \tag{39}$$

where

$$f(x) = \frac{C\lambda^{n-2}}{\big(1+\lambda^2(1-\cos d)\big)^{\frac{n-2}{2}}} \leq C\lambda^{n-2} \tag{40}$$

$$|g(x)| = |H_\lambda(x)| \leq \begin{cases} C\big(1+|\log\lambda|\big) & \text{for } n=4, \text{ on } \partial B_{1/\lambda} \\ C\lambda^{n-4} & \text{for } n>4, \text{ on } \partial B_{1/\lambda}\,. \end{cases} \tag{41}$$

We then use the following lemma:

Lemma 1. *There exists a constant C independent of ϱ such that if u satisfies:*

$$-Lu + qu \leq f \text{ in } B_\varrho$$
$$u \leq g \text{ on } \partial B_\varrho$$

then,

$$||u||_{L^\infty(B_\varrho)} \leq C\left(||f||_{L^\infty(B_\varrho)}\varrho^2 + ||g||_{L^\infty(\partial B_\varrho)}\right) .$$

Using lemma 1, we derive:

$$||H_\lambda||_{L^\infty(B_{1/\lambda})} \leq C \begin{cases} 1 + |\log \lambda| & \text{for } n = 4 \\ \lambda^{n-4} & \text{for } n > 4 . \end{cases} \tag{42}$$

This implies ii) of Proposition 1, for $n = 4$.

Observing that, on $B_{1/\lambda}$, we have:

$$C\lambda \leq \delta_\lambda^{\frac{2}{n-2}} = \frac{\lambda}{1 + \lambda^2(1 - \cos d)} \tag{43}$$

we derive from (42):

$$|H_\lambda(x)| \leq C\lambda^{n-4} \leq C\lambda^{\frac{n-4}{2}} \delta_\lambda^{\frac{n-4}{n-2}}(x) \quad \text{for } n > 4 . \tag{44}$$

iii) of Proposition 1 is thereby proven.

We prove now Lemma 1. Let

$$v(x) = u(x) - ||g||_{L^\infty(\partial B_\varrho)} . \tag{45}$$

Then v satisfies:

$$\begin{cases} -Lv + qv \leq f + C||g||_{L^\infty(\partial B_\varrho)} & \text{in } B_\varrho \\ v \leq 0 & \text{on } \partial B_\varrho . \end{cases} \tag{46}$$

Using the fact that $-L + q$ is coercive and the maximum principle, Lemma 1 rests upon the following result. Let w satisfy:

$$\begin{cases} -Lw + qw = 1 & \text{in } B_\varrho \\ w = 0 & \text{on } \partial B_\varrho . \end{cases} \tag{47}$$

Then

$$w \leq C\varrho^2 . \tag{48}$$

Let us first observe that w is upperbounded by a constant C, independent of ϱ. Indeed, w is upperbounded by the function $\bar{w}$ which solves:

$$\begin{cases} -L\bar{w} + q\bar{w} = 1 & \text{on } S^n \\ \bar{w} > 0 \end{cases} \tag{49}$$

and $\bar{w} \in L^\infty(S^n)$. We thus need to establish (48) only for ϱ small enough. By the results of Hartman–Stampacchia [1], the solution ω of

$$\begin{cases} -L\omega = 1 & \text{in } B_\varrho \\ \omega = 0 & \text{on } \partial B_\varrho \end{cases} \tag{50}$$

satisfies

$$||\omega||_{L^\infty(B_\varrho)} \le C\varrho^2 \,, \tag{51}$$

where C is independent of ϱ and depends only on the ellipticity constant of the operator L. Thus

$$|w|_{L^\infty(B_\varrho)} \le C\varrho^2 \left(1 + ||q||_{L^\infty} ||w||_{L^\infty(B_\varrho)}\right) \tag{52}$$

which implies (47) for ϱ small enough.

3. Interaction between two different bubbles

The main result of this section is described in the following theorem:

Theorem 1. *There exists a constant C positive such that, for any a and b belonging to S^n and any $\lambda \ge 2$, we have:*

i) $\int \varphi^5_{a,\lambda}\varphi_{b,\lambda} \ge \left(1 - \frac{C}{\lambda}\right) \int \delta^5_{a,\lambda}\varphi_{b,\lambda}$ *for* $n = 3$

ii) $\int \varphi^3_{a,\lambda}\varphi_{b,\lambda} \ge \left(1 - \frac{C}{\lambda^2}|\log\lambda|^2\right) \int \delta^3_{a,\lambda}\varphi_{b,\lambda}$ *for* $n = 4$

iii) $\int \varphi^{\frac{n+2}{n-2}}_{a,\lambda}\varphi_{b,\lambda} \ge \left(1 - \frac{C}{\lambda^2}|\log\lambda|\right) \int \delta^{\frac{n+2}{n-2}}_{a,\lambda}\varphi_{b,\lambda}\,.$

The proof of Theorem 1 requires the following lemmas:

Lemma 2. *There exists $\alpha > 0$ such that:*

$$\varphi_{a,\lambda} \ge \frac{\alpha}{\lambda^{\frac{n-2}{2}}} \qquad \forall a \in S^n, \;\; \forall \lambda \ge 1\,.$$

Proof of Lemma 2. Let $G(x,y)$ be the Green's function of the positive operator $-L+q$. One can find a positive constant α such that

$$G(x,y) \geq \alpha > 0 \qquad \forall x,\ \forall y \in S^n \,. \tag{53}$$

Then, using (24), we have:

$$\begin{aligned}\varphi_{a,\lambda}(x) = \int_{S^n} G(x,y)\,(-L\delta_{a,\lambda})\,(y)dy = n(n-2)\int_{S^n} G(x,y)\delta_{a,\lambda}^{\frac{n+2}{n-2}}(y)dy \geq \\ \geq n(n-2)\alpha \int_{S^n} \delta_{a,\lambda}^{\frac{n+2}{n-2}}(y)dy\,. \end{aligned}\tag{54}$$

On the other hand, we have, through the change of variables $x = \Pi^{-1}(y)$:

$$\begin{aligned}\int_{S^n} \delta_{a,\lambda}^{\frac{n+2}{n-2}}(y)dy &= \int_{S^n} \delta_{\delta,\lambda}^{\frac{n+2}{n-2}}(y)dy \\ &= \int_{\mathbf{R}^n} (1+|x|^2)^{\frac{n+2}{2}} \frac{\lambda^{\frac{n+2}{2}}}{(1+\lambda^2|x|^2)^{\frac{n+2}{2}}}\,|J_{ac}\Pi|\,dx \\ &= \int_{\mathbf{R}^n} \frac{1}{(1+|x|^2)^{\frac{n-2}{2}}} \frac{\lambda^{\frac{n+2}{2}}}{(1+\lambda^2|x|^2)^{\frac{n+2}{2}}}dx \\ &\geq c\int_0^1 \frac{\lambda^{\frac{n+2}{2}}}{(1+\lambda^2r^2)^{\frac{n+2}{2}}} r^{n-1}dr \geq c'\lambda^{1-\frac{n}{2}}\,. \end{aligned}\tag{55}$$

Lemma 2 follows from (54) and (55).

We now have:

Lemma 3. *There exist positive constants α,β such that:*

$$\alpha\delta_{a,\lambda}(x) \leq \varphi_{a,\lambda}(x) \leq \beta\delta_{a,\lambda}(x) \qquad \forall a \in S^n,\ \ \forall\lambda \geq 1$$

Proof of Lemma 3. Since $\varphi_\lambda = \delta_\lambda + \frac{H_\lambda}{\lambda^{\frac{n-2}{2}}}$ and since, for $n=3$, H_λ is upperbounded by (i) of Proposition 1, we have:

$$\varphi_\lambda \leq \delta_\lambda + \frac{C}{\sqrt{\lambda}} \leq \beta\delta_\lambda \qquad \text{for } n=3\,. \tag{56}$$

We used in (56) the following inequality, which holds for any $n \geq 3$:

$$\delta_\lambda \geq \frac{\alpha}{\lambda^{\frac{n-2}{2}}}\,. \tag{57}$$

In order to prove that $\varphi_\lambda \le \beta\delta_\lambda$ for $n = 4$, we use (ii) of Proposition 1. We have:

$$\varphi_\lambda \le \delta_\lambda + \frac{C}{\lambda}\left(1 + \left|\log\left(\frac{1}{\lambda^2} + d^2\right)\right|\right), \tag{58}$$

and we claim that

$$1 + \left|\log\left(\frac{1}{\lambda^2} + d^2\right)\right| \le C\frac{\lambda^2}{1 + \lambda^2(1 - \cos d)}. \tag{59}$$

Indeed, with appropriate constants c and c' and c'':

$$\begin{aligned}\frac{C\lambda^2}{1 + \lambda^2(1 - \cos d)} &\ge \frac{C\lambda^2}{1 + c\lambda^2 d^2} = \\ &= \frac{C}{\frac{1}{\lambda^2} + cd^2} \ge \frac{c'}{\frac{1}{\lambda^2} + d^2} \ge c''\left(1 + \left|\log\left(\frac{1}{\lambda^2} + d^2\right)\right|\right).\end{aligned} \tag{60}$$

We used, for (60), the fact that $1 + |\log t|$ is upperbounded by $\frac{C}{t}$ for $t \le 1 + C\Pi^2$. (58) and (59) imply that φ_λ is upperbounded by $\beta\delta_\lambda$ for $n = 4$.

For $n > 4$, we know by (iii) of Proposition 1 that:

$$H_\lambda \le C\lambda^{\frac{n-4}{2}}\delta_\lambda^{\frac{n-4}{n-2}}. \tag{61}$$

We want to prove that

$$\frac{1}{\lambda^{\frac{n-2}{2}}}H_\lambda \le C\frac{1}{\lambda^{\frac{n-2}{2}}}\lambda^{\frac{n-4}{2}}\delta_\lambda^{\frac{n-4}{n-2}} \le C'\delta_\lambda \tag{62}$$

(62) will yield the upperbound inequality of Lemma 3 in the case $n > 4$; (62) is equivalent to:

$$\frac{C''\lambda^2}{1 + \lambda^2(1 - \cos d)} \ge 1 \tag{63}$$

which holds obviously with an appropriate constant C''.

In order to prove the reverse inequality, we start with:

$$\varphi_\lambda = \delta_\lambda + \frac{1}{\lambda^{\frac{n-2}{2}}}H_\lambda \ge \delta_\lambda - \frac{|H_\lambda|}{\lambda^{\frac{n-2}{2}}}. \tag{64}$$

By Lemma 2, we know that:

$$\varphi_\lambda \ge \frac{\alpha}{\lambda^{\frac{n-2}{2}}}. \tag{65}$$

For $n = 3$, (64) yields:

$$\varphi_\lambda \geq \delta_\lambda - \frac{C}{\sqrt{\lambda}} . \tag{66}$$

Observe that, if $\gamma > 0$ is chosen small enough, the sphere S^3 can be described as the union of two sets (I) and (II), defined as follows:

$$S^3 = (I) \cup (II) \quad (I) = \left\{ x \in S^3 \Big/ \delta_\lambda(x) \geq \frac{C}{(1-\gamma)\sqrt{\lambda}} \right\} ,$$
$$(II) = \left\{ x \in S^3 \Big/ \delta_\lambda(x) \leq \frac{\alpha}{\gamma\sqrt{\lambda}} \right\} . \tag{67}$$

(65), (66) and (67) imply that:

$$\varphi_\lambda \geq \gamma \delta_\lambda \qquad \text{for } n = 3 . \tag{68}$$

The proof of (68) for $n \geq 4$ uses the same ideas, although the technical details may differ for $n = 4$ and $n \geq 5$. We provide the proof for $n \geq 5$. We then have, by iii) of Proposition 1 and by Lemma 2:

$$\varphi_\lambda \geq \delta_\lambda - \frac{C\lambda^{\frac{n-4}{2}}}{\lambda(1+\lambda^2 d^2)^{\frac{n-4}{2}}} , \quad \varphi_\lambda \geq \frac{\alpha}{\lambda^{\frac{n-2}{2}}} . \tag{69}$$

Observe that, for $\gamma > 0$ small enough, we can write S^n as the union of two sets (I) and (II) where:

$$(I) = \left\{ x \in S^3 \Big/ 1 + \lambda^2 d(x)^2 \leq \frac{1-\gamma}{C} \lambda^2 \right\}$$
$$(II) = \left\{ x \in S^3 \quad \text{s.t} \quad 1 + \lambda^2 d^2 \geq \left(\frac{\gamma}{\alpha}\right)^{2/n-2} \lambda^2 \right\} . \tag{70}$$

In (I), the following inequality holds:

$$\delta_\lambda \geq \frac{C\lambda^{\frac{n-4}{2}}}{(1-\gamma)\lambda(1+\lambda^2 d^2)^{\frac{n-4}{2}}} . \tag{71}$$

In (II), (72) holds:

$$\delta_\lambda \leq \frac{\alpha}{\gamma \lambda^{\frac{n-2}{2}}} . \tag{72}$$

(69), (71) and (72) yield that:

$$\varphi_\lambda \geq \gamma\delta_\lambda \qquad \text{on all of } S^3 . \tag{73}$$

The proof of Lemma 3 is thereby complete.

Proof of Theorem 1. Since $\delta_{a,\lambda} = \varphi_{a,\lambda} + \frac{1}{\lambda^{\frac{n-2}{2}}} H_{a,\lambda}$, we have, with an appropriate constant C:

$$\delta_{a,\lambda}^{\frac{n+2}{n-2}} \leq \varphi_{a,\lambda}^{\frac{n+2}{n-2}} + C\left(\varphi_{a,\lambda}^{\frac{4}{n-2}} \frac{|H_{a,\lambda}|}{\lambda^{\frac{n-2}{2}}} + \frac{|H_{a,\lambda}|^{\frac{n+2}{n-2}}}{\lambda^{\frac{n+2}{2}}}\right) . \tag{74}$$

By Lemma 3, we know that:

$$\varphi_{a,\lambda}^{\frac{4}{n-2}} \frac{|H_{a,\lambda}|}{\lambda^{\frac{n-2}{2}}} \leq C \frac{\delta_{a,\lambda}^{\frac{4}{n-2}} |H_{a,\lambda}|}{\lambda^{\frac{n-2}{2}}} . \tag{75}$$

On the other hand, also by Lemma 3, we know that:

$$\frac{1}{\lambda^{\frac{n-2}{2}}} |H_{a,\lambda}| = |\varphi_{a,\lambda} - \delta_{a,\lambda}| \leq C\delta_{a,\lambda} . \tag{76}$$

Thus

$$\frac{1}{\lambda^2} |H_{a,\lambda}|^{\frac{4}{n-2}} \leq C' \delta_{a,\lambda}^{\frac{4}{n-2}} , \tag{77}$$

and

$$\frac{|H_{a,\lambda}|^{\frac{n+2}{n-2}}}{\lambda^{\frac{n+2}{2}}} \leq \frac{C'}{\lambda^{\frac{n-2}{2}}} \delta_{a,\lambda}^{\frac{4}{n-2}} |H_{a,\lambda}| . \tag{78}$$

(74), (75) and (78) yield:

$$\delta_{a,\lambda}^{\frac{n+2}{n-2}} \leq \varphi_{a,\lambda}^{\frac{n+2}{n-2}} + C \frac{\delta_{a,\lambda}^{\frac{4}{n-2}} |H_{a,\lambda}|}{\lambda^{\frac{n-2}{2}}} . \tag{79}$$

Thus,

$$\int \varphi_{b,\lambda} \delta_{a,\lambda}^{\frac{n+2}{n-2}} \leq \varphi_{b,\lambda} \varphi_{a,\lambda}^{\frac{n+2}{n-2}} + \underbrace{\frac{C}{\lambda^{\frac{n-2}{2}}} \int \delta_{b,\lambda} \delta_{a,\lambda}^{\frac{4}{n-2}} \; |H_{a,\lambda}|}_{R} . \tag{80}$$

In order to estimate R, we need the following two lemmas:

Lemma 4.

$$\int_{S^n} \delta_{a,\lambda}^{\frac{n+2}{n-2}} \delta_{b,\lambda} = F\left(\frac{\lambda^2 \sin\left(\frac{1}{2}d(a,b)\right)}{\sqrt{1+\lambda^2}}\right)$$

where $F(0) > 0$, $\underset{t\to+\infty}{F(t)} \sim \frac{K}{t^{n-2}}$, $K = \int \frac{dx}{(1+|x|^2)^{\frac{n+2}{2}}}$.

Proof. Let us consider a geodesic curve from a to b on S^n and let N, the north pole of S^n, be the middle point on this geodesic.

Using the stereographic projection Π^{-1}, whose formulae are provided in Appendix 1, we have:

$$\int_{S^n} \delta_{a,\lambda}^{\frac{n+2}{n-2}} \delta_{b,\lambda} = \frac{1}{n(n-2)} \int_{S^n} -L\delta_{a,\lambda}\delta_{b,\lambda} = \frac{1}{n(n-2)} \int_{\mathbb{R}^n} \nabla U_{\hat{a},\nu} \nabla U_{\hat{b},\nu} \tag{81}$$

where, denoting $\text{Proj}_{\mathbb{R}^n}$ the orthogonal projection onto $\mathbb{R}^n$:

$$\begin{cases} \hat{a} = \frac{\lambda^2 \text{Proj}_{\mathbb{R}^n} a}{1+\lambda^2\left(1-\cos\vartheta_0\right)} & \hat{b} = \frac{\lambda^2 \text{Proj}_{\mathbb{R}^n} b}{1+\lambda^2\left(1-\cos\vartheta_0\right)} \\ \nu = \frac{1+\lambda^2(1-\cos\vartheta_0)}{\sqrt{1+2\lambda^2}} \quad , & \vartheta_0 = \text{dist}\,(b, N)\,. \end{cases} \tag{82}$$

Thus, using the fact that $\int_{\mathbb{R}^n} U_{\hat{a},\nu}^{\frac{n+2}{n-2}} U_{\hat{b},\nu}$ is translation–invariant:

$$\int_{S^n} \delta_{a,\lambda}^{\frac{n+2}{n-2}} \delta_{b,\lambda} = \frac{1}{n(n-2)} \int_{\mathbb{R}^n} U_{\hat{a},\nu}^{\frac{n+2}{n-2}} U_{\hat{b},\nu} = F\left(\nu\left|\hat{a}-\hat{b}\right|\right) = \tag{83}$$

$$= F\left(\frac{2\lambda^2 \sin\vartheta_0}{\sqrt{1+2\lambda^2}}\right)\,.$$

We then have:

Lemma 5. $\frac{1}{n(n-2)} \int_{\mathbb{R}^n} \frac{dx}{(1+|x|^2)^{\frac{n+2}{2}} (1+|x-c|^2)^{\frac{n-2}{2}}} = F(|c|)$ *where* $F(t)$ *has the properties described in Lemma 4.*

Lemma 4 follows readily from Lemma 5.

Observe now that:

$$n(n-2)F(|c|) = I_1 + I_2 + I_3 = \int_{|y-c|<\frac{|c|}{2}} + \int_{\substack{|y-c|>\frac{|c|}{2} \\ |y|<2|c|}} + \int_{|y|>2|c|} \tag{84}$$

where the integrands are the same than the one in the statement of the lemma. We then have, when $|c| \to +\infty$:

$$I_1 \sim \frac{1}{|c|^{n+2}} \int_{|y|<\frac{|c|}{2}} \frac{dy}{(1+|y|^2)^{\frac{n-2}{2}}} \sim \frac{1}{|c|^n} \tag{85}$$

$$I_2 \sim \frac{1}{|c|^{n-2}} \left(\int_0^{2|c|} \frac{dy}{(1+|y|^2)^{\frac{n+2}{2}}} - \frac{|c|^n}{|c|^{n+2}} \right) =$$

$$= \frac{1}{|c|^{n-2}} \left(\int_0^{2|c|} \frac{r^{n-1}dr}{r^{n+2}} - \frac{1}{|c|^2} \right) \sim \frac{1}{|c|^{n-2}} \int \frac{dx}{(1+|x|^2)^{\frac{n+2}{2}}} \tag{86}$$

$$I_3 \sim \int_{2|c|}^{+\infty} \frac{r^{n-1}dr}{r^{2n}} \sim \frac{1}{|c|^n} \, . \tag{87}$$

Thus

$$F(|c|) \sim \frac{1}{|c|^{n-2}} \int \frac{dx}{(1+|x|^2)^{\frac{n+2}{2}}} \qquad \text{when } |c| \to +\infty \tag{88}$$

as claimed.

We now proceed with the estimates on R.

$n = 3$: Then

$$R \leq \frac{1}{\sqrt{\lambda}} \int_{S^3} \delta_{a,\lambda}^4 \delta_{b,\lambda} \, . \tag{89}$$

Using the results of Appendix 1, we have:

$$R \leq \frac{1}{\sqrt{\lambda}} \int_{\mathbf{R}^3} U_{\hat{a},\nu}^4(x) U_{-\hat{a},\nu}(x) dx =$$

$$\frac{1}{\sqrt{\lambda}} \nu^{\frac{5}{2}} \int_{\mathbf{R}^3} \frac{1}{(1+\nu^2|x-\hat{a}|^2)^2} \, \frac{1}{(1+\nu^2|x+\hat{a}|^2)^{\frac{1}{2}}} dx \, . \tag{90}$$

Setting

$$\xi = \nu(x-\hat{a}) \, , \tag{91}$$

we then have

$$R \leq \frac{1}{\sqrt{\lambda}} \nu^{-1/2} \int_{\mathbf{R}^3} \frac{1}{\left(1+|\xi|^2\right)^2} \, \frac{1}{\left(1+|\xi+2\hat{a}\nu|^2\right)^{1/2}} dx =$$

$$= \frac{1}{\sqrt{\lambda}} \nu^{-1/2} G(2|\hat{a}|\nu) = \frac{1}{\sqrt{\lambda}} \nu^{-1/2} G\left(\frac{2\lambda^2 \sin \vartheta_0}{\sqrt{1+2\lambda^2}} \right) \tag{92}$$

where $G(t)$ is equivalent to C/t when t tends to $+\infty$.

Thus

$$G(t) \le C\ F(t)\,. \tag{93}$$

Hence,

$$R \le \frac{C}{\sqrt{\lambda\nu}} \int_{S^3} \delta_{a,\lambda}^5 \delta_{b,\lambda} \le \frac{C'}{\lambda} \int_{S^3} \delta_{a,\lambda}^5 \delta_{b,\lambda} \tag{94}$$

which provides in with the estimate on R we need.

We now verify that $G(t)$ is equivalent to C/t when t tends to $+\infty$. We have:

$$\begin{aligned} G(c) &= \int_{\mathbf{R}^3} \frac{dy}{(1+|y|^2)^2 (1+|y-c|^2)^{1/2}} = \\ &= \int_{|y-c|<\frac{|c|}{2}} + \int_{\substack{|y-c|>\frac{|c|}{2} \\ |y|<2|c|}} + \int_{|y|>2|c|} = I_1 + I_2 + I_3 \end{aligned} \tag{95}$$

where the integrands in I_1, I_2 and I_3 are the same than the one used in $\int_{\mathbf{R}^3}$. We then find

$$I_1 \sim \frac{1}{|c|^4} \int_0^{|c|/2} \frac{r^2 dr}{(1+r^2)^{\frac{1}{2}}} \sim \frac{1}{|c|^2} \tag{96}$$

$$I_2 \sim \frac{1}{|c|} \int_0^{2|c|} \frac{r^2 dr}{(1+r^2)^2} - \frac{1}{|c|^2} \sim \frac{1}{|c|} \tag{97}$$

$$I_3 \sim \frac{1}{|c|} \int_{2|c|}^{+\infty} \frac{r^2 dr}{(1+r^2)^2} \sim \frac{1}{|c|^2}\,. \tag{98}$$

The result follows.

$n = 4$: Then,

$$R \le \frac{C}{\lambda} \int_{S^4} \delta_{b,\lambda} \delta_{a,\lambda}^2 \left(1 + \left|\log\left(\frac{1}{\lambda^2} + d^2(x)\right)\right|\right) \tag{99}$$

where $d(x) = \text{dist}(x,a)$. Observe that

$$\left|\log\left(\frac{1}{\lambda^2} + d^2\right)\right| \le C\,|\log(\lambda\delta_{a,\lambda})|\,. \tag{100}$$

We choose the south pole in the middle of (a,b). We then have:

$$\int_{S^4} \delta_{b,\lambda}\delta_{a,\lambda}^2 (1 + |\log(\lambda\delta_{a,\lambda})|)\, dy \le C \int_{\mathbf{R}^4} U_{\hat{b},\nu}(x) U_{\hat{a},\nu}^2(x) \frac{1}{(1+|x|^2)^{-3}} \times$$

$$\times \left(1 + \left|\log \lambda \left(1 + |x|^2\right) U_{\hat{a},\nu}(x)\right|\right) \left(1 + |x|^2\right)^{-4} =$$

$$= C \int_{\mathbf{R}^4} U_{\hat{b},\nu} U_{\hat{a},\nu}^2 \frac{1}{1 + |x|^2} \left(1 + |\log \lambda U_{\hat{a},\nu}| + \log\left(1 + |x|^2\right)\right) \leq$$

$$\leq C \int_{\mathbf{R}^4} U_{\hat{b},\nu} U_{\hat{a},\nu}^2 \left(1 + |\log \lambda U_{\hat{a},\nu}|\right) \leq$$

$$\leq C \int_{\mathbf{R}^4} U_{\hat{b},\nu} U_{\hat{a},\nu}^2 \left(1 + |\log \nu U_{\hat{a},\nu}|\right) \tag{101}$$

since

$$\nu = \frac{1 + \lambda^2(1 + \cos \vartheta_0)}{\sqrt{1 + 2\lambda^2}} \qquad 0 \leq \vartheta_0 \leq \pi/2 \tag{102}$$

hence ν is equivalent to λ. Then

$$R \leq \frac{C}{\lambda^2} |\log \lambda| G_1(|c|) + \frac{C}{\lambda^2} G_2(|c|) \tag{103}$$

where

$$|c| = \frac{2\lambda^2 \sin \vartheta_0}{\sqrt{1 + 2\lambda^2}} \qquad \vartheta_0 = \pi - \frac{1}{2} d(a, b) \tag{104}$$

and

$$G_1(|c|) \leq C \frac{\log |c|}{|c|^2} \tag{105}$$

for $|c|$ large

$$G_2(|c|) \leq C \frac{(\log |c|)^2}{|c|^2} \,. \tag{106}$$

Therefore, for every c, we have:

$$G_1(|c|) \leq C \log(2 + |c|) F(|c|) \tag{107}$$

$$G_2(|c|) \leq C \left(\log(2 + |c|)\right)^2 F(|c|) \,. \tag{108}$$

Since $|c|$ is less than or equal to $|\lambda|$, we have:

$$R \leq \frac{C}{\lambda^2} (\log \lambda)^2 F(|c|) = \frac{C}{\lambda^2} (\log \lambda)^2 \int_{S^4} \delta_{b,\lambda} \delta_{a,\lambda}^3 \leq$$

$$\leq C' \frac{(\log \lambda)^2}{\lambda^2} \int_{S^4} \varphi_{b,\lambda} \delta_{a,\lambda}^3 \,. \tag{109}$$

$n > 4$: We have

$$R \leq \frac{C}{\lambda} \int \delta_{b,\lambda} \delta_{a,\lambda}^{\frac{n}{n-2}} . \tag{110}$$

Using the Appendix 1, we derive:

$$R \leq \frac{C}{\lambda} \int U_{\hat{a},\nu}^{\frac{n}{n-2}} U_{\hat{b},\nu} =$$
$$\frac{C}{\lambda} \nu^{n-1} \int \frac{1}{(1+\nu^2|x-\hat{a}|^2)^{\frac{n}{2}}} \times \frac{1}{\left(1+\nu^2|x-\hat{b}|^2\right)^{\frac{n-2}{2}}} \leq \frac{C}{\lambda^2} G\left(\nu|\hat{a}-\hat{b}|\right) \tag{111}$$

where

$$G(c) = \int \frac{dy}{(1+|y|^2)^{\frac{n}{2}} (1+|y-c|^2)^{\frac{n-2}{2}}} . \tag{112}$$

We split $G(c)$, as usual, in three integrals:

$$G(c) = \int_{|y-c|<\frac{|c|}{2}} + \int_{\substack{|y-c|>\frac{|c|}{2} \\ |y|<2|c|}} + \int_{|y|>2|c|} = I_1 + I_2 + I_3 . \tag{113}$$

We then have:

$$I_1 \sim \frac{1}{|c|^n} \int_0^{|c|/2} \frac{r^{n-1} dr}{(1+r^2)^{\frac{n-2}{2}}} \sim \frac{1}{|c|^{n-2}} \tag{114}$$

$$I_2 \sim \frac{1}{|c|^{n-2}} \left(\int_0^{2|c|} \frac{r^{n-1} dr}{(1+r^2)^{n/2}} - 1 \right) \sim \frac{\log |c|}{|c|^{n-2}} \tag{115}$$

$$I_3 \sim \int_{2|c|}^{+\infty} \frac{r^{n-1} dr}{(1+r^2)^{n-1}} \sim \frac{1}{|c|^{n-2}} . \tag{116}$$

Therefore, there exists $c > 0$ such that

$$G(|c|) \leq C \log(2+|c|) \qquad F(|c|) \tag{117}$$

for any c. Since $|c| \leq C\lambda$, we obtain:

$$R \leq \frac{C}{\lambda^2} \log \lambda F(|c|) = \frac{C}{\lambda^2} \log \lambda \int_{S^n} \delta_{a,\lambda}^{\frac{n+2}{n-2}} \delta_{b,\lambda} . \tag{118}$$

Proposition 2.

$$\int(-L+q)\varphi_{a,\lambda}\varphi_{a,\lambda} = S + \begin{cases} O\left(\frac{1}{\lambda}\right) & \text{for } n=3 \\ O\left(\frac{\log\lambda}{\lambda^2}\right) & \text{for } n=4 \\ O\left(\frac{1}{\lambda^2}\right) & \text{for } n>4 \end{cases}$$

where $S = n(n-2)\int_{\mathbf{R}^n} \frac{dx}{(1+|x|^2)^n}$.

Proof. Using the definition of $\varphi_{a,\lambda}$, we have:

$$\int(-L+q)\varphi_{a,\lambda}\varphi_{a,\lambda} = -\int L\delta_{a,\lambda}\varphi_{a,\lambda} = n(n-2)\int_{S^n}\delta_{a,\lambda}^{\frac{2n}{n-2}} + R \tag{119}$$

where

$$R = \frac{n(n-2)}{\lambda^{\frac{n-2}{2}}}\int_{S^n}\delta_{a,\lambda}^{\frac{n+2}{n-2}}H_{a,\lambda}\,. \tag{120}$$

We thus can use the estimates that we completed earlier, with $c=0$.

We can derive:

$$|R| \le \frac{C}{\sqrt{\lambda\nu}}G(0) \qquad \text{, where } \nu = \sqrt{1+\lambda^2} \qquad \text{for } n=3 \tag{121}$$

$$|R| \le C\frac{|\log\lambda|}{\lambda^2}G_1(0) + \frac{C}{\lambda^2}G_2(0) \tag{122}$$

$$|R| \le C\frac{C}{\lambda^2}G(0)\,. \tag{123}$$

Hence the result.

We now have the last result of this section:

Lemma 6.

$$\int \varphi_{a,\lambda}^{\frac{2n}{n-2}} = S + \begin{cases} O\left(\frac{1}{\lambda}\right) & \text{for } n=3 \\ O\left(\frac{\log\lambda}{\lambda^2}\right) & \text{for } n=4 \\ O\left(\frac{1}{\lambda^2}\right) & \text{for } n>5\,. \end{cases}$$

The proof follows from the same arguments.

3. Expansion of the functional at infinity

We denote:

$$J(u) = \frac{1}{n(n-2)} \frac{\int(-L+q)uu\,dx}{\|u\|^2_{2n/n-2}} = \frac{N}{D} \,. \tag{124}$$

Let

$$I = \frac{1}{p}\sum_{i\neq j}\int \varphi_{a_i,\lambda}^{\frac{n+2}{n-2}}\varphi_{a_j,\lambda} \tag{125}$$

and let O_n be a quantity of the type:

$$O_n = \begin{cases} O\left(\frac{1}{\lambda}\right) & \text{for } n=3 \\ O\left(\frac{\log\lambda}{\lambda^2}\right) & \text{for } n=4 \\ O\left(\frac{1}{\lambda^2}\right) & \text{for } n\geq 5\,. \end{cases} \tag{126}$$

Let

$$\|\alpha\|_q = \left(\sum_{i=1}^{p}\alpha_i^q\right)^{1/q} . \tag{127}$$

Let

$$\omega_n(p) = \begin{cases} p & n=3 \\ \sqrt{p}\log(1+p) & n=4 \\ \sqrt{p\log(1+p)} & n\geq 5 \end{cases} \tag{128}$$

and let

$$O'_n = \begin{cases} O\left(\frac{1}{\lambda}\right) & \text{for } n=3 \\ O\left(\left(\frac{\log\lambda}{\lambda}\right)^2\right) & \text{for } n=4 \\ O\left(\frac{\log\lambda}{\lambda^2}\right) & \text{for } n\geq 5\,. \end{cases} \tag{129}$$

Let

$$\theta_n(p) = \begin{cases} p^3 & n=3 \\ p^{3/2}\log(1+p) & n=4 \\ (p\log(1+p))^{3/2} & n\geq 5\,. \end{cases} \tag{130}$$

We then have:

Proposition 3. *There exists $c_n > 0$ such that for any p, $\alpha_1, \ldots, \alpha_p$, for any $\lambda \geq c_n w_n(p)$,*

$$J\left(\sum_{i=1}^{p} \alpha_i \varphi_{a_i,\lambda}\right) < \left((p+1)\, S\right)^{2/n} .$$

Remark. The aim of section 3 is to improve the estimate provided by Proposition 3 so that, under some condition on p, we will derive that $J\left(\sum_{i=1}^{p} \alpha_i \varphi_{a_i,\lambda}\right)$ is upperbounded by $(pS)^{2/n}$.

Proof of Proposition 3. N is the numerator of $J\left(\sum_{i=1}^{p} \alpha_i \varphi_{a_i,\lambda}\right)$. We have:

$$\begin{aligned} N &= \int \left(\sum \alpha_i \delta_{a_i,\lambda}^{\frac{n+2}{n-2}}\right)\left(\sum \alpha_j \varphi_{a_j,\lambda}\right) = \int \left(\sum \alpha_i \varphi_{a_i,\lambda}^{\frac{n+2}{n-2}}\right)\left(\sum \alpha_j \varphi_{a_j,\lambda}\right) + \\ &+ \int \sum \alpha_i \left(\delta_{a_i,\lambda}^{\frac{n+2}{n-2}} - \varphi_{a_i,\lambda}^{\frac{n+2}{n-2}}\right) \sum \alpha_j \varphi_{a_j,\lambda} = \\ &= \left(\int \left(\sum \alpha_i \varphi_{a_i,\lambda}^{\frac{n+2}{n-2}}\right)\left(\sum \alpha_j \varphi_{a_j,\lambda}\right)\right)(1+R) \end{aligned} \tag{131}$$

where R satisfies:

$$\begin{aligned} |R| &\leq \frac{\int \left(\sum \alpha_i \left|\delta_{a_i,\lambda}^{\frac{n+2}{n-2}} - \varphi_{a_i,\lambda}^{\frac{n+2}{n-2}}\right|\right) \sum \alpha_j \varphi_{a_j,\lambda}}{\sum \alpha_i^2 \int \varphi_{a_i,\lambda}^{2n/n-2} + \sum_{i\neq j} \alpha_i \alpha_j \int \varphi_{a_i,\lambda}^{\frac{n+2}{n-2}} \varphi_{a_j,\lambda}} \leq \\ &\leq \frac{C}{\lambda^{\frac{n-2}{2}}} \frac{\int \left(\sum \alpha_i \delta_{a_i,\lambda}^{\frac{4}{n-2}} H_{a_i,\lambda}\right)\left(\sum \alpha_j \delta_{a_j,\lambda}\right)}{(1+I_1)\left(\sum \alpha_i^2\right)} , \end{aligned} \tag{132}$$

where

$$I_1 = \frac{\sum_{i\neq j} \alpha_i \alpha_j \int \varphi_{a_i,\lambda}^{\frac{n+2}{n-2}} \varphi_{a_i,\lambda}}{\sum \alpha_i^2} . \tag{133}$$

From Proposition 1 of section 2, we derive that:

$$\sum_i \alpha_i^2 \int \delta_{a_i,\lambda}^{\frac{n+2}{n-2}} \frac{1}{\lambda^{\frac{n-2}{2}}} H_{a_i,\lambda} \leq O_n \left(\sum \alpha_i^2\right) . \tag{134}$$

The estimates of section 2 also imply:

$$\sum_{i\neq j} \alpha_i \alpha_j \int \delta_{a_i,\lambda}^{4/n-2} \delta_{a_j,\lambda} \frac{H_{a_i,\lambda}}{\lambda^{\frac{n-2}{2}}} \leq (I_1 O_n') \left(\sum \alpha_i^2\right) . \tag{135}$$

I_1 will be the estimated later on in this section.

We thus have, using (134) and (135):

$$N \leq \int \left(\sum \alpha_i \varphi_{a_i,\lambda}^{\frac{n+2}{n-2}}\right)\left(\sum \alpha_j \varphi_{a_j,\lambda}\right)\left(1+\frac{O_n}{1+I_1}+\frac{I_1 O'_n}{1+I_1}\right). \tag{136}$$

Thus

$$J\left(\sum_{i=1}^{p} \alpha_i \varphi_{a_i,\lambda}\right) \leq \frac{\int \left(\sum \alpha_i \varphi_{a_i,\lambda}^{\frac{n+2}{n-2}}\right)\left(\sum \alpha_j \varphi_{a_j,\lambda}\right)}{\|\sum \alpha_i \varphi_{a_i,\lambda}\|^2_{2n/n-2}}\left(1+\frac{O_n}{1+I_1}+\frac{I_1 O'_n}{1+I_1}\right). \tag{137}$$

Since O_n is upperbounded by O'_n, we have:

$$\left(1+\frac{O_n}{1+I_1}+\frac{I_1 O'_n}{1+I_1}\right) \leq 1+O'_n. \tag{138}$$

Using the results of Appendix 2, we have:

$$\frac{\int \left(\sum \alpha_i \varphi_{a_i,\lambda}^{\frac{n+2}{n-2}}\right)\left(\sum \alpha_j \varphi_{a_j,\lambda}\right)}{\|\sum \alpha_i \varphi_{a_i,\lambda}\|^2_{2n/n-2}} \leq \left(\sum_i \int \frac{\alpha_i \varphi_{a_i,\lambda}}{\sum \alpha_k \varphi_{a_k,\lambda}} \varphi_{a_i,\lambda}^{2n/n-2}\right)^{2/n}. \tag{139}$$

We now have:

$$\sum_i \int \frac{\alpha_i \varphi_{a_i,\lambda}}{\sum \alpha_k \varphi_{a_k,\lambda}} \varphi_{a_i,\lambda}^{2n/n-2} \leq \sum_i \int \varphi_{a_i,\lambda}^{2n/n-2} = p\,(S+O_n) = pS\,(1+O_n). \tag{140}$$

Thus

$$J\left(\sum_{i=1}^{p} \alpha_i \varphi_{a_i,\lambda}\right) \leq (pS)^{2/n}(1+O_n)(1+O'_n) = (pS)^{2/n}(1+O'_n). \tag{141}$$

The conditions on λ imply the existence of a constant d_n such that

$$O'_n \leq \frac{d_n}{p} \tag{142}.$$

d_n can be chosen small, at our convenience, provided c_n is large enough. Proposition 3 follows.

We now prove.

Proposition 4. *There exist* $\vartheta_0 \in (0,1)$, $\gamma_0 > 0$, $\beta > 0$ *such that for any* p, *for any* $\alpha_i > 0$ *such that* $\alpha_i/\alpha_j \in (1-\vartheta_0, 1+\vartheta_0)$, *for any* $\lambda \geq 1$, *for any* $a_1, \neg a_p \in S^n$ *satisfying:*

$$I < \beta$$

then

$$J\left(\sum_{i=1}^{p} \alpha_i \varphi_{a_i,\lambda}\right) \leq \frac{\sum \alpha_i^2}{\left(\sum \alpha_i^{2n/n-2}\right)^{\frac{n-2}{n}}} \times S^{2/n} \times$$

$$\times \left(1 + \begin{cases} O\left(\frac{1}{\lambda}\right) & \text{for } n=3 \\ O\left(\frac{\log \lambda}{\lambda^2}\right) & \text{for } n=4 \\ O\left(\frac{\log \lambda}{\lambda^2}\right) & \text{for } n \geq 5 \end{cases} - \gamma_0 \frac{I}{S}\left(1 + O\left(\tfrac{1}{\lambda}\right)\right)\right) =$$

$$= \frac{\sum \alpha_1^2}{\left(\sum \alpha_i^{2n/n-2}\right)^{\frac{n-2}{n}}} \times S^{2/n}\left(1 + O_n - \gamma_0 \frac{I}{S}\left(1 + O\left(\frac{1}{\lambda}\right)\right)\right).$$

Proof. N is the numerator of $J\left(\sum_{i=1}^{p} \alpha_i \varphi_{a_i,\lambda}\right)$. It is equal to:

$$N = \int \left(\sum \alpha_i \delta_{a_i,\lambda}^{\frac{n+2}{n-2}}\right)\left(\sum a_j \varphi_{a_j,\lambda}\right) =$$

$$= \sum \alpha_i^2 \int \delta_{a_i,\lambda}^{\frac{n+2}{n-2}} \varphi_{a_i,\lambda} + \sum_{i \neq j} \alpha_i \alpha_j \int \delta_{a_i,\lambda}^{\frac{n+2}{n-2}} \varphi_{a_j,\lambda}.$$

Using the estimates of section 2, we estimate the second term of the right hand side:

$$\sum_{i\neq j} \alpha_i \alpha_j \int \delta_{a_i,\lambda}^{\frac{n+2}{n-2}} \varphi_{a_j,\lambda} \leq \sum_{i \neq j} \alpha_i \alpha_j \left(1 + O\left(\frac{1}{\lambda}\right)\right) \int \varphi_{a_i,\lambda}^{\frac{n+2}{n-2}} \varphi_{a_j,\lambda}. \tag{144}$$

We also know, from Proposition 1 of section 2, that:

$$\int \delta_{a_i,\lambda}^{\frac{n+2}{n-2}} \varphi_{a_i,\lambda} = S + O_n. \tag{145}$$

Therefore,

$$N \leq \left(\sum \alpha_i^2\right) S\left(1 + O_n\right)\left(1 + \frac{\sum_{i\neq j} \alpha_i \alpha_j \left(1 + O\left(\frac{1}{\lambda}\right)\right) \int \varphi_{a_i,\lambda}^{\frac{n+2}{n-2}} \varphi_{a_j,\lambda}}{S \sum \alpha_i^2}\right) \tag{146}$$

Since $\alpha_i/\alpha_j \in (1-\vartheta_0, 1+\vartheta_O)$,

$$\frac{\alpha_i}{\sqrt{\sum \alpha_i^2}} \le \frac{1+\vartheta_0}{\sqrt{p}} \,. \tag{147}$$

Thus

$$N \le \left(\sum \alpha_i^2\right) S\,(1+O_n)\left(1+\frac{1+O\left(\frac{1}{\lambda}\right)}{S}\,(1+\vartheta_0)^2\, I\right) . \tag{148}$$

We, now, turn to estimate the denominator D. We have:

$$D = \left\|\sum \alpha_i \varphi_{a_i,\lambda}\right\|^2_{\frac{2n}{n-2}} \;;\quad D^{n/n-2} = \int \left(\sum \alpha_i \varphi_{a_i,\lambda}\right)^{2n/n-2} . \tag{149}$$

Using Lemma A2 of the Appendix 2, we derive:

$$D^{n/n-2} \ge \int \sum \alpha_i^{2n/n-2} \varphi_{a_i,\lambda}^{2n/n-2} + \frac{\gamma_n}{n-2} \sum_{i\neq j} \alpha_i^{\frac{n+2}{n-2}} \alpha_j \varphi_{a_i,\lambda}^{\frac{n+2}{n-2}} \varphi_{a_j,\lambda} \,. \tag{150}$$

Using again the estimates of section 2, we know that $\int \varphi_{a_i,\lambda}^{2n/n-2} = S + O_n$.
Therefore:

$$D^{n/n-2} \ge \left(\sum \alpha_i^{2n/n-2}\right)(S+O_n)\left(1+\frac{\gamma_n}{n-2}\,(1-\vartheta_0)^{2n/n-2}\,\frac{I\,(1+O_n)}{S}\right) \tag{151}$$

since

$$\frac{\alpha_i}{\left(\sum \alpha_i^{2n/n-2}\right)^{\frac{n-2}{2n}}} \ge \frac{1-\vartheta_0}{p^{\frac{n-2}{2n}}} \,. \tag{152}$$

Therefore,

$$D \ge \|\alpha\|^2_{2n/n-2} S^{\frac{n-2}{n}}\,(1+O_n)\left(1+\frac{\gamma_n}{n-2}\,(1-\vartheta_0)^{2n/n-2}\,\frac{I\,(1+O_n)}{S}\right)^{\frac{n-2}{n}} \tag{153}$$

and

$$J \le \frac{\|\alpha\|_2^2}{\|\alpha\|^2_{2n/n-2}}\; S^{2/n}\,(1+O_n) \times \\ \times \frac{\left(1+\left(\frac{1+O\left(\frac{1}{\lambda}\right)}{S}\right)(1+\vartheta_0)^2\, I\right)}{\left(1+\frac{\gamma_n}{n-2}\,(1-\vartheta_0)^{2n/n-2}\,\frac{I}{S}\left(1+O\left(\frac{1}{\lambda}\right)\right)\right)^{\frac{n-2}{n}}} \tag{154}$$

since $O_n \leq O\left(\frac{1}{\lambda}\right)$.

If u is a small positive number, then:

$$\frac{1}{(1+u)^{\frac{n-2}{n}}} \leq 1 - \frac{n-2}{n} u . \tag{155}$$

Therefore, if $I \leq \beta$ is small enough, we have:

$$J \leq \frac{\|\alpha\|_2^2}{\|\alpha\|_{2n/n-2}^2} S^{2/n} (1 + O_n) \times$$
$$\times \left(1 + \left(1 + O\left(\frac{1}{\lambda}\right)\right) \left((1+\vartheta_0)^2 - \gamma (1-\vartheta_0)^{2n/n-2}\right) \frac{I}{S}\right) . \tag{156}$$

Since γ is shortly larger than 1, we choose ϑ_0 small enough so that:

$$\gamma_0 = (1+\vartheta_0)^2 - \gamma (1-\vartheta_0)^{2n/n-2} < 0 . \tag{157}$$

The proof of Proposition 4 is thereby complete.

We now have:

Proposition 5. *For any $\vartheta_0 \in (0,1)$, there exists $\gamma_1' > 0$ such that, for any p and for any $\alpha_i > 0$ such that $\alpha_i/\alpha_j \in (1-\vartheta_0, 1+\vartheta_0)$, for any $\lambda \geq 1$, for any a_1, a_2, ..., a_p,*

$$J\left(\sum_{i=1}^{p} \alpha_i \varphi_{a_i,\lambda}\right) \leq (pS)^{2/n} \left(1 + O_n - \gamma_1' \frac{I}{p}\right)^{2/n} (1 + O_n') .$$

Proof. We start the proof as for Proposition 4. We have:

$$J \leq \left(\sum_i \int \frac{\alpha_i \varphi_{a_i,\lambda}}{\sum \alpha_k \varphi_{a_k,\lambda}} \varphi_{a_i,\lambda}^{2n/n-2} \right)^{2/n} (1 + O_n') \tag{158}$$

$$\sum_i \int \frac{\alpha_i \varphi_{a_i,\lambda}}{\sum \alpha_k \varphi_{a_k,\lambda}} \varphi_{a_i,\lambda}^{2n/n-2} \leq \sum_i \int \varphi_{a_i,\lambda}^{2n/n-2} +$$
$$- \sum_i \int \frac{\sum_{j\neq i} \alpha_j \varphi_{a_j,\lambda}}{\sum \alpha_k \varphi_{a_k,\lambda}} \varphi_{a_i,\lambda}^{2n/n-2} = pS(1 + O_n) - Q . \tag{159}$$

Since $\alpha_i/\alpha_j \in (1-\vartheta_0, 1+\vartheta_0)$, we have:

$$Q \geq (1-\vartheta_0)\sum_i \int \frac{\sum_{j\neq i}\varphi_{a_j,\lambda}}{\sum \varphi_{a_k,\lambda}}\varphi_{a_i,\lambda}^{2n/n-2} =$$

$$= (1-\vartheta_0)\sum_i\sum_{j\neq i}\int \frac{\varphi_{a_i,\lambda}}{\sum_j \varphi_{a_k,\lambda}}\varphi_{a_i,\lambda}^{\frac{n+2}{n-2}}\varphi_{a_j,\lambda}\,. \tag{160}$$

We wish to lowerbound

$$T = \sum_i\sum_{j\neq i}\int \frac{\varphi_{a_i,\lambda}}{\sum \varphi_{a_k,\lambda}}\varphi_{a_i,\lambda}^{\frac{n+2}{n-2}}\varphi_{a_j,\lambda} \tag{161}$$

by $\gamma_1 I$.

Let $\varepsilon_0 < 1$ be given. Then,

$$T \geq \varepsilon_0/p\sum_{j\neq i}\int_{E_i}\varphi_i^{\frac{n+2}{n-2}}\varphi_j \qquad \text{where } E_i = \left\{x/\varphi_i \geq \frac{\varepsilon_o}{p}\sum_k \varphi_k\right\}\,. \tag{162}$$

We, then, split I as follows:

$$pI = \sum_{i\neq j}\int \varphi_i^{\frac{n+2}{n-2}}\varphi_j = \sum_{i\neq j}\int_{E_i} + \sum\int_{{}^cE_i} \leq \frac{pT}{\varepsilon_0} + \sum_{i\neq j}\int_{{}^cE_i}\varphi_i^{\frac{n+2}{n-2}}\varphi_j\,. \tag{163}$$

We now observe that, for any j:

$${}^cE_i \subset \left\{x/\varphi_i(x) < \frac{2\varepsilon_0}{p}\varphi_j\right\} \cup \left\{x/\varphi_i(x) < \frac{2\varepsilon_0}{p}\sum_{k\neq j}\varphi_k(x)\right\}\,. \tag{164}$$

Thus

$$\int_{{}^cE_i}\varphi_i^{\frac{n+2}{n-2}}\varphi_j \leq \left(\frac{2\varepsilon_0}{p}\right)^{2/n-2}\int \varphi_i^{n/n-2}\varphi_j^{n/n-2} + \int\left(\frac{2\varepsilon_0}{p}\sum_{k\neq j}\varphi_k\right)^{\frac{n+2}{n-2}}\varphi_j\,. \tag{165}$$

Observe now that:

$$\int \varphi_i^{n/n-2}\varphi_j^{n/n-2} \leq \int \varphi_i^{\frac{n+2}{n-2}}\varphi_j + \int \varphi_j^{\frac{n+2}{n-2}}\varphi_i\,. \tag{166}$$

Using the convexity of the function $x \longrightarrow |x|^{\frac{n+2}{n-2}}$, we have:

$$\left(\frac{1}{p}\sum_{k\neq j}\varphi_k\right)^{\frac{n+2}{n-2}} \leq \left(\frac{1}{p-1}\sum_{k\neq j}\varphi_k\right)^{\frac{n+2}{n-2}} \leq \frac{1}{p-1}\sum_{k\neq j}\varphi_k^{\frac{n+2}{n-2}}\,. \tag{167}$$

Therefore,

$$pI \leq \frac{pT}{\varepsilon_0} + \left(\frac{2\varepsilon_0}{p}\right)^{2/n-2} \times 2 \sum_{i \neq j} \int \varphi_i^{\frac{n+2}{n-2}} \varphi_j + (2\varepsilon_0)^{\frac{n+2}{n-2}} \sum_{i \neq j} \int \varphi_i^{\frac{n+2}{n-2}} \varphi_j \, . \quad (168)$$

Thus

$$pI \left(1 - 2\left(\frac{2\varepsilon_0}{p}\right)^{2/n-2} - (2\varepsilon_0)^{\frac{n+2}{n-2}}\right) \leq pT/\varepsilon_0 \, . \quad (169)$$

The result follows, with $\gamma_1 = \varepsilon_0 \left(1 - 2\,(2\varepsilon_0)^{2/n-2} - (2\varepsilon_0)^{\frac{n+2}{n-2}}\right)$.

We now have:

Proposition 6. *Let $\vartheta_0 \in (0,1)$ be given. There exists a constant $C'_n(\vartheta_0)$ such that, for any p, if $O'_n \leq C'_n(\vartheta_0)/p$, (this can be reformulated in: $\exists c_n(\vartheta_0)$ s.t if $\lambda \geq c_n(\vartheta_0)p$ if $n = 3$, $\lambda \geq c_n(\vartheta_0)\sqrt{p}\log(1+p)$ for $n = 4$, $\lambda \geq c_n(\vartheta_0)\sqrt{p\log(1+p)}$ for $n \geq 5$), then, for any $(\alpha_1, \ldots, \alpha_p)$ such that $\alpha_i/\alpha_j \in (1-\vartheta_0, 1+\vartheta_0)$ and $\alpha_{i_0}/\alpha_{i_0} \notin (1-\vartheta_0/2, 1+\vartheta_0/2)$ for a couple (i_0, j_0), we have:*

$$J\left(\sum \alpha_i \varphi_{a_i,\lambda}\right) < (pS)^{2/n} \, .$$

The proof of Proposition 6 uses the following lemma, which is established in Appendix 2:

Lemma 6. *Given $\vartheta_0 \in (0,1)$, there exists $C(\vartheta_0) > 0$ such that, for any $(\alpha_i, \ldots, \alpha_p)$ such that $\alpha_{i_0}/\alpha_{j_0} \notin (1-\vartheta_0/2, 1+\vartheta_0/2)$, for a couple (i_0, j_0), then*

$$\frac{\sum_{i=1}^p \alpha_i^2}{\left(\sum_{i=1}^p \alpha_i^{2n/n-2}\right)^{\frac{n-2}{n}}} \leq p^{2/n}\left(1 - \frac{C(\vartheta_0)}{p}\right) \, .$$

Proof of Proposition 6. The proof splits into two cases:

1st case: $I < \beta$.

We then use Proposition 4 and Lemma 6. We derive that

$$J \leq (pS)^{2/n}\left(1 - \frac{C(\vartheta_0)}{p}\right)(1 + O_n) \, . \quad (170)$$

Since O_n is less than $O'_n = o\left(\frac{1}{p}\right)$, the proof is complete.

2nd case: $I \geq \beta$.

We then apply Proposition 5, we thus have:

$$J \leq (pS)^{2/n}\left(1 + O_n - \gamma'_1 \frac{\beta}{p}\right)^{2/n}(1 + O'_n)\ . \tag{171}$$

We conclude using the same argument. The proof is thereby complete.

We have defined I_1 in (133). Let

$$\varphi_i = \varphi_{a_i,\lambda}\ . \tag{172}$$

We then have the two following lemmas:

Lemma 7. *There exists $\gamma_1 > 0$ such that, for any $\alpha_i \geq 0$, for any p, for any $\lambda \geq 1$,*

$$J\left(\sum_{i=1}^{p} \alpha_i \varphi_i\right) \leq \frac{\sum \alpha_i^2}{\left(\sum \alpha_i^{2n/n-2}\right)^{\frac{n-2}{n}}} S^{2/n}(1 + O_n + \gamma_1 I_1)\ .$$

Lemma 8. *There exists $\gamma'_1 \geq 0$ such that, for any $\alpha_i \geq 0$,*

$$J\left(\sum_{i=1}^{p} \alpha_i \varphi_i\right) \leq (pS)^{2/n}(1 + O'_n)\left(1 - \frac{\gamma'_1}{p^2} I_1\right)^{2/n}\ .$$

Observe that, in contrast with the statements of Proposition 5 and 6, these lemmas do not impose any restriction on the α_i's. These two lemmas imply the following Proposition 7:

Proposition 7. *Let $\vartheta_0 \in (0,1)$ be given. There exists $C''_n(\vartheta_0)$ such that, for any p, for any $(\alpha_1, \ldots, \alpha_p)$ such that $\alpha_{i_0}/\alpha_{j_0} \notin (1 - \vartheta_0, 1 + \vartheta_0)$ for one couple of indices (i_0, j_0),*

$$J\left(\sum_{i=1}^{p} \alpha_i \varphi_i\right) < (pS)^{2/n}$$

if $O'_n \leq \frac{C'_n(\vartheta_0)}{p^3}$.

The above condition on O'_n translates on λ as follows:

$$\lambda \geq c_n(\vartheta_0)\,\theta_n(p)\,. \tag{173}$$

Proof of Proposition 7. We use Lemma 8. Proposition 7 follows if I_1 is larger than $C_n'''(\vartheta_0)/p$, where $C_n'''(\vartheta_0)$ can be chosen as small as we wish, if $C_n''(\vartheta_0)$ is chosen appropriately small.

We thus assume, in the sequel, that:

$$I_1 < \frac{C_n'''(\vartheta_0)}{p}\,. \tag{174}$$

We apply Lemmas 6 and 7. We then derive:

$$J\left(\sum_{i=1}^{p}\alpha_i\varphi_i\right) \leq (pS)^{2/n}\left(1-\frac{C(\vartheta_0)}{p}\right)\left(1+O_n+\frac{\gamma_1 C_n'''(\vartheta_0)}{p}\right) \tag{175}$$

where $C(\vartheta_0)$ is the constant of Lemma 6. We choose $C_n'''(\vartheta_0)$ small enough so that:

$$O_n+\frac{\gamma_1 C_n'''(\vartheta_0)}{p} < \frac{C(\vartheta_0)}{2p}\,. \tag{176}$$

Proposition 7 follows.

Proof of Lemma 7. Coming back to the proof of Proposition 4, we have:

$$N \leq \|\alpha_{j,\lambda}\|_2^2 S\,(1+O_n)\left(1+\left(1+O\left(\frac{1}{\lambda}\right)\right)\frac{I_1}{S}\right)\,. \tag{177}$$

We lowerbound the denominator D as follows:

$$\begin{aligned} D^{n/n-2} &\geq \int\sum\alpha_i^{2n/n-2}\varphi_i^{2n/n-2} \geq (S+O_n)\left(\sum\alpha_i^{2n/n-2}\right) = \\ &= (S+O_n)\left(\sum\alpha_i^{2n/n-2}\right) = S\,(1+O_n)\,\|\alpha\|_{2n/n-2}^{2n/n-2}\,. \end{aligned} \tag{178}$$

Thus

$$\frac{N}{D} \leq \frac{\|\alpha\|_2^2}{\|\alpha\|_{2n/n-2}^2}S^{2/n}\,(1+O_n+\gamma_1 I_1)\,. \tag{179}$$

The proof of Lemma 7 is thereby complete.

Proof of Lemma 8. Coming back to the proof of Propositions 4 and 5, we have:

$$J\left(\sum_{i=1}^{p}\alpha_i\varphi_i\right) \leq \left(\sum_i \int \frac{\alpha_i\varphi_i}{\sum \alpha_k\varphi_k}\varphi_i^{2n/n-2}\right)^{2/n}(1+O'_n)\,. \tag{180}$$

Let (i_0, j_0) be such that:

$$\int \varphi_{i_0}^{\frac{n+2}{n-2}}\varphi_{j_0} = \operatorname*{Sup}_{i\neq j}\int \varphi_i^{\frac{n+2}{n-2}}\varphi_j\,. \tag{181}$$

Using the results of section 2, there exists a constant $C > 0$ such that

$$\int \varphi_{j_0}^{\frac{n+2}{n-2}}\varphi_{i_0} \geq c\int \varphi_{i_0}^{\frac{n+2}{n-2}}\varphi_{j_0}\,. \tag{182}$$

We assume that $\alpha_{i_0} \geq \alpha_{j_0}$.

We then observe that:

$$\frac{\alpha_i\varphi_i}{\sum \alpha_k\varphi_k} \leq 1 \tag{183}$$

$$\int \varphi_i^{2n/n-2} = S + O_n\,. \tag{184}$$

Thus

$$J\left(\sum_{i=1}^{p}\alpha_i\varphi_i\right) \leq \left((p-1)(S+O_n) + \left(\int \frac{\alpha_{j_0}\varphi_{j_0}}{\sum \alpha_k\varphi_k}\varphi_{j_0}^{2n/n-2}\right)\right)^{2/n}(1+O'_n)\,. \tag{185}$$

Let L be:

$$L = \int \frac{\alpha_{j_0}\varphi_{j_0}}{\sum \alpha_k\varphi_k}\varphi_{j_0}^{2n/n-2}\,. \tag{186}$$

Then

$$L \leq \int \frac{\alpha_{j_0}\varphi_{j_0}}{\alpha_{i_0}\varphi_{i_0}+\alpha_{j_0}\varphi_{j_0}}\varphi_{j_0}^{2n/n-2} \leq \int \frac{\varphi_{j_0}}{\varphi_{i_0}+\varphi_{j_0}}\varphi_{j_0}^{2n/n-2} = \\ = S + O_n - \int \frac{\varphi_{j_0}}{\varphi_{i_0}+\varphi_{j_0}}\varphi_{j_0}^{\frac{n+2}{n-2}}\varphi_{i_0} = S + O_n - A\,. \tag{187}$$

Let $0 < \varepsilon_0 < 1$ be given and let

$$E_O = \{x/\varphi_{j_0}(x) \geq \varepsilon_0 (\varphi_{i_0}(x) + \varphi_{j_0}(x))\} \ . \tag{188}$$

Then

$$\begin{aligned} A &\geq \varepsilon_0 \int_{E_0} \varphi_{j_0}^{\frac{n+2}{n-2}} \varphi_{i_0} = \varepsilon_0 \left(\int \varphi_{j_0}^{\frac{n+2}{n-2}} \varphi_{i_0} - \int_{^cE_0} \varphi_{j_0}^{\frac{n+2}{n-2}} \varphi_{i_0} \right) \geq \\ &\geq \varepsilon_0 \left(\int \varphi_{j_0}^{\frac{n+2}{n-2}} \varphi_{i_0} - \left(\frac{\varepsilon_0}{1-\varepsilon_0} \right)^{2/n-2} \int \varphi_{j_0}^{n/n-2} \varphi_{i_0}^{n/n-2} \right) \ . \end{aligned} \tag{189}$$

Since

$$\begin{aligned} \int \varphi_{i_0}^{\frac{n}{n-2}} \varphi_{j_0}^{\frac{n}{n-2}} &\leq \int \varphi_{i_0}^{\frac{n+2}{n-2}} \varphi_{j_0} + \int \varphi_{j_0}^{\frac{n+2}{n-2}} \varphi_{i_0} \leq \\ &\leq \left(\frac{1}{C} + 1 \right) \int \varphi_{j_0}^{\frac{n+2}{n-2}} \varphi_{i_0} \ , \end{aligned} \tag{190}$$

we have:

$$A \geq \varepsilon_0 \int \varphi_{j_0}^{\frac{n+2}{n-2}} \varphi_{i_0} \left(1 - \left(\frac{1}{C} + 1 \right) \left(\frac{\varepsilon_0}{1-\varepsilon_0} \right)^{2/n-2} \right) \ . \tag{191}$$

Therefore, for ε_0 small enough:

$$A \geq \frac{\varepsilon_0}{2} \int \varphi_{j_0}^{\frac{n+2}{n-2}} \varphi_{i_0} \geq \frac{C\varepsilon_0}{2} \int \varphi_{i_0}^{\frac{n+2}{n-2}} \varphi_{j_0} \tag{192}$$

and

$$L \leq S + O_n - \frac{c\varepsilon_0}{2} \int \varphi_{i_0}^{\frac{n+2}{n-2}} \varphi_{j_0} \ . \tag{193}$$

This implies:

$$J\left(\sum_{i=1}^{p} \alpha_i \varphi_i \right) \leq (pS)^{2/n} (1 + O'_n) \left(1 - \frac{C'}{p} \operatorname*{Sup}_{i \neq j} \int \varphi_i^{\frac{n+2}{n-2}} \varphi_j \right) \ . \tag{194}$$

Since

$$\operatorname*{Sup}_{i \neq j} \int \varphi_i^{\frac{n+2}{n-2}} \varphi_j \geq \frac{I_1}{p} \ , \tag{195}$$

which follows from the fact that

$$\frac{\sum_{i,j}\alpha_i\alpha_j}{\sum_{i=1}^{p}\alpha_i^2}\le p\,, \tag{196}$$

we derive:

$$J\le (pS)^{2/n}\,(1+O'_n)\left(1-\frac{C'I_1}{p}\right)^{2/n}\,. \tag{197}$$

The proof of Lemma 8 is thereby complete.

Lemma 9. *There exists a constant $\overline{c}_n$ positive such that, for any $p\in\mathbb{N}^*$, for any A such that $A\ge \overline{c}_n\omega_n(p)$, there exists $\varepsilon_0(p,A)>0$ such that, for any μ such that $\overline{c}_n\omega_n(p)\le\mu\le A$, for any $(\alpha_1,\dots,\alpha_p)$, we have:*

$$J\left(\sum_{i=1}^{p}\alpha_i\varphi_{a_i,\mu}\right)<(pS)^{2/n}$$

provided

$$\underset{i\neq j}{\mathrm{Min}}\, d\,(a_i,a_j)<\varepsilon_0\,(p,A)\,.$$

Proof. Using (194) with $\varphi_i=\varphi_{a_i,\mu}$ and recalling from Lemma 3 of section 3 that:

$$\int\varphi_{a_i,\mu}^{\frac{n+2}{n-2}}\varphi_{a_j,\mu}\ge\alpha\int\delta_{a_i,\mu}^{\frac{n+2}{n-2}}\delta_{a_j,\mu}=\alpha F\left(\frac{\mu^2\sin\left(\frac{1}{2}d\,(a_i,a_j)\right)}{\sqrt{1+\mu^2}}\right) \tag{198}$$

where F is a continuous and strictly positive function, thus, if $\mathrm{Min}_{i\neq j}\, d(a_i,a_j)<\varepsilon_0\,(p,A)$, $\varepsilon_0\,(p,A)$ small enough:

$$\underset{i\neq j}{\mathrm{Sup}}\int\delta_{a_i,\mu}^{\frac{n+2}{n-2}}\delta_{a_j,\mu}\ge\frac{\alpha F(0)}{2}=\alpha_1\ , \tag{199}$$

α_1 independent of A, we derive

$$J\left(\sum_{i=1}^{p}\alpha_i\varphi_{a_i,\mu}\right)\le(pS)^{2/n}\,(1+O'_n)\left(1-\frac{C'\alpha_1}{p}\right)^{2/n}\,. \tag{200}$$

Hence, Lemma 9. Observe that the preceding argument uses only the fact that $\underset{d(a_i,a_j)\to 0}{\underline{\lim}}\int\varphi_{a_i,\mu}^{\frac{n+2}{n-2}}\varphi_{a_j,\mu}=S+O_n$ is larger than $S/2$, if $\mu\ge\overline{c}_n\omega_n(\rho)$, for an appropriate constant $\overline{c}_n$.

4. The map $f_p(\lambda)$ is homologically zero for $n = 3, 4, 5$ and p large

Notations:

Let ϑ_0 be given by Proposition 4, in section 3. In this section, we will make use of this ϑ_0 exclusively, in particular when applying Proposition 5, 6 and 7 of section 3. The function $\vartheta_n(p)$ upperbounds the function $\omega_n(p)$ for all p.

Lemma 5 and Proposition 4 provide us with constants $\overline{c}_n$ and c_n. Proposition 6 and 7 provide us with constants $c_n(\vartheta_0)$ and $c'_n(\vartheta_0)$. Let C be larger than $\max(c_n, \overline{c}_n, c_n(\vartheta_0), c'_n(\vartheta_0))$. Then if λ is larger than, or equal to $C\vartheta_n(p)$, all propositions and all lemmas of the third section hold, and if λ is larger than or equal to $C\omega_n(p)$, Proposition 3 and 6 hold.

Let

$$\Delta_{p-1} = \left\{(\alpha_1, \ldots, \alpha_p) \quad ; \quad \alpha_i \geq 0 \ ; \ \sum_{i=1}^{p} \alpha_i = 1\right\} . \tag{201}$$

Let M be a compact manifold. Let

$$B_p(M) = \left\{\sum_{i=1}^{p} \alpha_i \delta_{a_i} \quad ; \quad \alpha_i \geq 0 \ ; \ \sum_{i=1}^{p} \alpha_i = 1 \ ; \ a_i \in M\right\} \tag{202}$$

where δ_{a_i} is the Dirac mass at a_i. Let $B_0(M) = \phi$. Let

$$F_p(M) = \{(a_1, \ldots, a_p) \in M^p / \exists i \neq j \text{ with } a_i = a_j\} .$$

Let σ_p be the symmetric group of order p. σ_p acts on F_p.

Let T_p be a σ_p-equivariant tubular neighborhood of F_p, in M_p. (The existence of T_p is derived in the book by G. Bredon [2]). Some further requirements on T_p will be made later. Let

$$V_p = \overline{M^p - T_p}\,, \quad \delta V_p \text{ the boundary of } V_p . \tag{203}$$

Let

$$\Sigma^+ = \left\{u \geq 0 \Big/ \|u\|_{-L+q} = 1\right\} . \tag{204}$$

Let

$$W_p = \left\{u \in \Sigma^+ \Big/ J(u) < b_p = \left((p+1)S\right)^{2/n}\right\} . \tag{205}$$

Let, for $\lambda > 0$ and $p \in \mathbb{N} - \{0\}$, $f_p(\lambda)$ denote:

$$\begin{aligned} f_p(\lambda) \ : \ B_p(M) &\longrightarrow \Sigma^+ \\ \sum_{i=1}^{p} \alpha_i \delta_{a_i} &\longrightarrow \frac{\sum_{i=1}^{p} \alpha_i \varphi_{a_i,\lambda}}{\|\sum_{i=1}^{p} \alpha_i \varphi_{a_i,\lambda}\|_{-L+q}} \,. \end{aligned} \tag{206}$$

By Proposition 3 of the third section, we know that, if $\lambda \geq C\omega_n(p)$, then $f_p(\lambda)$ maps $B_p(M)$ into W_p, hence $\big(B_p(M),\ B_{p-1}(M)\big)$ into (W_p, W_{p-1}). We then have:

Proposition 8. *There exists $p_1 \in \mathbb{N} - \{0\}$ such that, for any $p \geq p_1$, and for any $\lambda \geq C\vartheta_n(p)$, the map:*

$$f_p(\lambda) : \quad \big(B_p(M), B_{p-1}(M)\big) \longrightarrow (W_p, W_{p-1})$$

is homotopic to a map valued in (W_{p-1}, W_{p-1}) and is therefore homologically trivial.

Proof of Proposition 8. We construct, for p large enough and for $\lambda \geq C\vartheta_n(p)$, a homotopy $U(t,\cdot)$ such that:

$$\begin{cases} U(t,\cdot) : & [0,1] \times B_p(M) \longrightarrow W_p \qquad U \text{ continuous} \\ U(t, B_{p-1}) & \subset W_{p-1} \forall t; \quad U(0,\cdot) = f_p(\lambda)(\cdot) \\ U(1, B_p) & \subset W_{p-1} \,. \end{cases} \tag{207}$$

Let $0 < \vartheta < 1$ be given. We denote:

$$\begin{aligned} \Delta^{\vartheta}_{p-1} = \big\{(\alpha_1, \ldots, \alpha_p) \in \Delta_{p-1} \text{ such that } \alpha_i/\alpha_j \in [1-\vartheta, 1+\vartheta], \\ \forall i, \forall j\big\} \ ; \delta\Delta^{\vartheta}_{p-1} \text{ is the boundary of } \Delta^{\vartheta}_{p-1} \,. \end{aligned} \tag{208}$$

Let T_p and T_p^1 be two σ_p–equivariant tubular neighborhoods of F_p, such that

$$\overline{T_p} \subset \overset{0}{T}{}^1_p \,. \tag{209}$$

T_p and T_p^1 are assumed to be small enough. Let

$$V_p^1 = \overline{M^p - T_p^1} \,. \tag{210}$$

We single out, in $B_p(M)$, three sets A, B and C, defined as follows:

$$A = \left\{ \sum_{i=1}^{p} \alpha_i \delta_{a_i} \text{ such that } (a_1, \ldots, a_p, \alpha_1, \ldots, \alpha_p) \in V_p \times \Delta^{\vartheta_0}_{p-1} \right\} \tag{211}$$

$$^cA = B = \left\{ \sum_{i=1}^{p} \alpha_i \delta_{a_i} \text{ such that any corresponding } (a_1, ..., a_p, \alpha_1, ..., \alpha_p) \right.$$

$$\left. \text{is either in } T_p \times \Delta_{p-1} \text{or in } V_p \times \left(\Delta_{p-1} - \Delta_{p-1}^{\vartheta_0}\right) \right\} \tag{212}$$

$$C \subset \overset{0}{A}; C = \left\{ \sum_{i=1}^{p} \alpha_i \delta_{a_i} \text{ such that } (a_1, ..., a_p, \alpha_1, ..., \alpha_p) \in V_p^1 \times \Delta_{p-1}^{\vartheta_0/2} \right\} . \tag{213}$$

Let us observe that $B_{p-1}(M)$ is contained in B. Let

$$\lambda_p = C_1 p\,. \tag{214}$$

The value of C_1 will be chosen later.

Let, for $(a, \alpha) = (a_1, \ldots, a_p, \alpha_1, \ldots, \alpha_p) \in M_\times^p \Delta_{p-1}, \Psi(a, \alpha)$ be a continuous function, valued in $[0, 1]$, equal to 1 on C and to zero on ^cA. $U(t, \cdot)$ is defined as follows:

1) If $\sum_{i=1}^p \alpha_i \delta_{a_i} \in C$, we define $U\left(t, \sum_{i=1}^p \alpha_i \delta_{a_i}\right)$ to be:

$$U\left(t, \sum_{i=1}^{p} \alpha_i \delta_{a_i}\right) = \frac{\sum_{i=1}^p \alpha_i \varphi_{a_i,(1-t)\lambda + t\lambda_p}}{\left\|\sum_{i=1}^p \alpha_i \varphi_{a_i,(1-t)\lambda + t\lambda_p}\right\|_{-L+q}} .$$

2) If $\sum_{i=1}^p \alpha_i \delta_{a_i} \in B$, we define $U\left(t, \sum_{i=1}^p \alpha_i \delta_{a_i}\right)$ to be:

$$U\left(t, \sum_{i=1}^{p} \alpha_i \delta_{a_i}\right) = \frac{\sum_{i=1}^p \alpha_i \varphi_{a_i,\lambda}}{\left\|\sum_{i=1}^p \alpha_i \varphi_{a_i,\lambda}\right\|_{-L+q}} = f_p(\lambda) \left(\sum_{i=1}^{p} \alpha_i \delta_{a_i,\lambda}\right) .$$

3) If $\sum_{i=1}^p \alpha_i \delta_{a_i} \in A - C$, we define $U\left(t, \sum_{i=1}^p \alpha_i \delta_{a_i}\right)$ to be:

$$U\left(t, \sum_{i=1}^{p} \alpha_i \delta_{a_i}\right) = \frac{\sum_{i=1}^p \alpha_i \varphi_{a_i,(1-t\psi(a,\alpha))\lambda + t\psi(a,\alpha)\lambda_p}}{\left\|\sum_{i=1}^p \alpha_i \varphi_{a_i,(1-t\psi(a,\alpha))\lambda + t\psi(a,\alpha)\lambda_p}\right\|_{-L+q}} .$$

Since λ_p is larger than or equal to $C\omega_n(p)$, any barycenter of λ and λ_p is lowerbounded by $C\omega_n(p)$ and $U(t, \cdot)$ is, by Proposition 3 of the third section, valued in W_p.

$B_{p-1}(M)$ is contained in B and $U(t, \cdot) = f_p(\lambda)(\cdot)$ on B. Thus

$$U(t, B_{p-1}) = f_p(\lambda)\left(B_{p-1}\right) \subset W_{p-1}\,. \tag{215}$$

$U(t, \cdot)$ therefore maps, for any t, (B_p, B_{p-1}) into (W_p, W_{p-1}).

At $t = 0$, $U(0,\cdot)$ is equal to $f_p(\lambda)$ and U is continuous on $[0,1]\times B_p(M)$. In order to complete the proof of Proposition 8, we need only to check that $U(1,\cdot)$ is valued in W_{p-1}.

In order to check this fact, we distinguish three cases:

1) $\sum_{i=1}^p \alpha_i\delta_{a_i} \in A - C$ i.e $(a_1,\dots,a_p,\alpha_1,\dots,\alpha_p) \in V_{p\times}\Delta_{p-1}^{\vartheta_0}$ and either

$$(a_1,\dots,a_p) \in T_p^1 \text{ or } (\alpha_1,\dots,\alpha_p) \notin \Delta_{p-1}^{\vartheta_0/2} .$$

Let us observe that any barycenter of λ and λ_p is upperbounded by $A = \max(\lambda,\lambda_p)$. W choose T_p^1 so that $\mathrm{Min}_{i\neq j}\, d(a_i,a_j) < \varepsilon_0(p,A)$ if $(a_1,\dots,a_p) \in T_p^1$. $\varepsilon_o(p,A)$ has been defined in Lemma 9 of the third section.

Since λ and λ_p ar lowerbounded by $C\omega_n(p)$, A is also lowerbounded by $C\omega_n(p)$ and Lemma 9 of section 3 holds. We thus have:

$$J\left(\sum_{i=1}^p \alpha_i\varphi_{a_i,\mu}\right), (pS)^{2/n} \quad \text{for any } C\omega_n(p) \leq \mu \leq A \tag{216}$$

which we use with $\mu = (1 - t\psi(a,\alpha))\,\lambda + t\psi(a,\alpha)\lambda_p$.

The result follows in case $(a_1,\dots,a_p) \in T_p^1$.

If $(\alpha_1,\dots,\alpha_p) \notin \Delta_{p-1}^{\vartheta_0/2}$, we apply Proposition 6 of the third section, with

$$\lambda = \mu = (1 - t\psi(a,\alpha))\,\lambda + t\psi(a,\alpha)\lambda_p \geq C\omega_n(p) . \tag{217}$$

Thus

$$J\left(\sum \alpha_i\varphi_{a_i,\mu}\right) < (pS)^{2/n} . \tag{218}$$

The result follows also in this case.

2) $\sum_{i=1}^p \alpha_i\delta_{a_i} \in C$. Thus $(a_1,\dots,a_p,\alpha_1,\dots,\alpha_p) \in V_p^1 \times \Delta_{p-1}^{\vartheta_0/2}$

$$U\left(1,\sum_{i=1}^p \alpha_i\delta_{a_i}\right) \text{ is equal to } \frac{\sum_{i=1}^p \alpha_i\varphi_{a_i,\lambda_p}}{\left\|\sum_{i=1}^p \alpha_i\varphi_{a_i,\lambda_p}\right\|_{-L+q}} .$$

We distinguish two subcases:

2.1) either

$$I = \frac{1}{p}\sum_{i\neq j}\int \varphi_{a_i,\lambda_p}^{\frac{n+2}{n-2}}\varphi_{a_j,\lambda_p} < \beta$$

where β has been defined in Proposition 4.

2.2) or

$$I \geq \beta .$$

In case 2.1 holds, we have, by Proposition 4:

$$J\left(\sum_{i=1}^{p} \alpha_i \varphi_{a_i,\lambda_p}\right) \leq \frac{\sum \alpha_i^2}{\left(\sum \alpha_i^{2n/n-2}\right)^{n-2/2}} \times S^{2/n} \times$$
$$\times \left(1 + O_n(\lambda_p) - \frac{\gamma_0 I}{S}\left(1 + o\left(\frac{1}{\lambda}\right)\right)\right) . \tag{219}$$

Observe that

$$I \geq \frac{(p-1)c}{\lambda_p^{n-2}}$$

since $\int \varphi_{a_i,\lambda_p}^{\frac{n+2}{n-2}} \varphi_{a_j,\lambda_p} \geq \frac{c}{\lambda_p^{n-2}}$ by lemmas 3 and 4 of section 3. Furthermore,

$$\frac{\sum_{i=1}^{p} \alpha_i^2}{\left(\sum_{i=1}^{p} \alpha_i^{2n/n-2}\right)^{\frac{n-2}{n}}} \leq p^{2/n} . \tag{220}$$

Thus,

$$J\left(\sum_{i=1}^{p} \alpha_i \varphi_{a_i,\lambda_p}\right) \leq (pS)^{2/n}\left(1 + O_n(\lambda_p) - \frac{\gamma_0 (p-1) c}{\lambda_p^{n-2}}\left(1 + O\left(\frac{1}{\lambda_p}\right)\right)\right) . \tag{221}$$

We would like to have:

$$O_n(\lambda_p) - \frac{\gamma_0 (p-1) c}{\lambda_p^{n-2}}\left(1 + O\left(\frac{1}{\lambda_p}\right)\right) < 0 \tag{222}$$

for p large enough; λ_p must also be larger than or equal to $C\omega_n(p)$. This yields:

For $n = 3$:

$$\begin{cases} O\left(\frac{1}{\lambda_p}\right) & < \frac{\gamma_0 (p-1) c}{\lambda_p^{n-2}}\left(1 + O\left(\frac{1}{\lambda_p}\right)\right) \\ \lambda_p & \geq Cp \end{cases} \tag{223}$$

which can be restated as:

$$\begin{cases} O(1) & < \gamma_0 (p-1) c \left(1 + O\left(\frac{1}{\lambda_p}\right)\right) \\ \lambda_p & \geq Cp\,. \end{cases} \tag{224}$$

Clearly λ_p, as defined in (214), satisfies such requirements.

For $n = 4$:

$$\begin{cases} O\left(\frac{\log \lambda_p}{\lambda_p^2}\right) & < \frac{\gamma_0(p-1)c}{\lambda_p^2} \left(1 + O\left(\frac{1}{\lambda_p}\right)\right) \\ \lambda_p & \geq C\sqrt{p} \log(1+p) \end{cases} \tag{225}$$

which is equivalent to

$$\begin{cases} O(\log \lambda_p) & < (p-1) c \left(1 + O\left(\frac{1}{\lambda_p}\right)\right) \\ \lambda_p & \geq C\sqrt{p} \log(1+p)\,. \end{cases} \tag{226}$$

Again, λ_p, as defined in (214), satisfies those requirements.

For $n = 5$:

$$\begin{cases} O\left(\frac{1}{\lambda_p^2}\right) & < \frac{\gamma_0(p-1)c}{\lambda_p^3} \left(1 + O\left(\frac{1}{\lambda_p}\right)\right) \\ \lambda_p & \geq C\sqrt{p \log(1+p)} \end{cases} \tag{227}$$

which is equivalent to

$$\begin{cases} O(\lambda_p) & < \gamma_0 (p-1) c \left(1 + O\left(\frac{1}{\lambda_p}\right)\right) \\ \lambda_p & \geq C\sqrt{p \log(1+p)}\,. \end{cases} \tag{228}$$

If C_1 is chosen small enough, λ_p, as defined in (214), again satisfies those requirements. This argument cannot work anymore for $n \geq 6$, since we would have to satisfy:

$$\begin{cases} O(\lambda_p^{n-4}) & < \gamma_0 (p-1) C \left(1 + O\left(\frac{1}{\lambda_p}\right)\right) \\ \lambda_p & \geq C\sqrt{p \log(1+p)}\,. \end{cases} \tag{229}$$

Since $n - 4 \geq 2$, those two statements cannot be easily satisfied at the same time. Thus, if $n \leq 5$ and if 2.1 holds, $U(1, \cdot)$ is valued in W_{p-1}. In case 2.2 holds and I is lowerbounded by β, we apply Proposition 5.

$\alpha_i/\alpha_j \in (1-\vartheta_0/2, 1+\vartheta_0/2)$. Thus

$$J\left(\sum_{i=1}^{p} \alpha_i \varphi_{a_i,\lambda_p}\right) \leq (pS)^{2/n}\left(1+O_n(\lambda_p)-\gamma_0'\beta/p\right)^{2/n}\left(1+O_n'(\lambda_p)\right). \tag{230}$$

For $n = 3$:

$$\begin{cases} O_n(\lambda_p) = \frac{1}{C_1} O_3\left(\frac{1}{p}\right) \\ O_n'(\lambda_p) = \frac{1}{C_1} O_3'\left(\frac{1}{p}\right). \end{cases} \tag{231}$$

Thus, if C_1 is large enough,

$$\left(1+\frac{1}{C_1}O_3'\left(\frac{1}{p}\right)-\gamma_1'\frac{\beta}{p}\right)^{2/3}\left(1+\frac{1}{C_1}O_3'\left(\frac{1}{p}\right)\right)<1 \tag{232}$$

for p large enough and $U(1,\cdot)$ is again valued in W_{p-1} in this case.

In case $n = 4$,

$$\begin{cases} O_n(\lambda_p) = O\left(\frac{\log\lambda_p}{\lambda_p^2}\right) = O\left(\frac{\log p}{p^2}\right) \\ O_n'(\lambda_p) = O\left(\frac{(\log\lambda_p)^2}{\lambda_p^2}\right) = O\left(\frac{\log^2 p}{p^2}\right). \end{cases} \tag{233}$$

Again, for p large enough,

$$\left(1+O\left(\frac{\log p}{p^2}\right)-\frac{\gamma_1'\beta}{p}\right)^{1/2}\left(1+O\left(\frac{\log^2 p}{p^2}\right)\right)<1 \tag{234}$$

from which the result follows.

In case $n = 5$:

$$\begin{cases} O_n(\lambda_p) = O\left(\frac{1}{\lambda_p^2}\right) = O\left(\frac{1}{p^2}\right) \\ O_n'(\lambda_p) = O\left(\frac{\log\lambda_p}{\lambda_p^2}\right) = O\left(\frac{\log p}{p^2}\right). \end{cases} \tag{235}$$

Again, for p large enough,

$$\left(1+O\left(\frac{1}{p^2}\right)-\frac{\gamma_1'\beta}{p}\right)^{2/5}\left(1+O\left(\frac{\log p}{p^2}\right)\right)<1, \tag{236}$$

hence the result.

We consider now the third and last case.

3) $\sum_{i=1}^{p} \alpha_i \delta_{a_i} \in B$. We then have two subcases:

3.1) $(a_1, \ldots, a_p, \alpha_1, \ldots, \alpha_p) \in T_p \times \Delta_{p-1}$

We then use Lemma 9 with $\mu = \lambda$ and $A = \lambda$. Observe that

$$\lambda \geq \overline{c}_n \omega_n(p) . \tag{237}$$

Also, $\overline{T}_p$ is contained in T_p^1. Therefore, if T_p^1 is chosen small enough, then

$$\operatorname*{Min}_{i \neq j} d(a_i, a_j) < \varepsilon_0(p, \lambda) \quad \forall (a_1, \ldots, a_p) \in \overline{T}_p . \tag{238}$$

Thus, by Lemma 9,

$$J\left(\sum_{i=1}^{p} \alpha_i \varphi_{a_i,\lambda}\right) < (pS)^{2/n} \tag{239}$$

from which the result follows.

3.2) $(a_1, \ldots, a_p, \alpha_1, \ldots, \alpha_p) \in V_p \times \left(\Delta_{p-1} - \Delta_{p-1}^{\vartheta_0}\right)$

There are then two indices α_{i_0} and α_{j_0} such that

$$\alpha_{i_0}/\alpha_{j_0} \notin (1 - \vartheta_0, 1 + \vartheta_0) . \tag{240}$$

$U\left(1, \sum_{i=1}^{p} \alpha_i \delta_{a_i}\right)$ is equal to $\frac{\sum_{i=1}^{p} \alpha_i \varphi_{a_i,\lambda}}{\left\|\sum_{i=1}^{p} \alpha_i \varphi_{a_i,\lambda}\right\|_{-L+q}}$. Since λ is larger than or equal to $C\vartheta_n(p)$, with $C \geq c_n(\vartheta_0)$, we can use Proposition 7 of section 3. We thus have

$$J\left(\sum_{i=1}^{p} \alpha_i \varphi_{a_i,\lambda}\right) < (pS)^{2/n} . \tag{241}$$

This concludes the proof of Proposition 8.

Let us observe that, in the three dimensional case, there is no need to complete a homotopy of $f_p(\lambda)$, which is valued in W_{p-1} for p large enough. Indeed, if $n = 3$, using Proposition 4, we have, for $\alpha_i/\alpha_j \in (1 - \vartheta_0, 1 + \vartheta_0)$, $\forall i \neq j$:

$$J\left(\sum_{i=1}^{p} \alpha_i \varphi_{a_i,\lambda}\right) \leq \frac{\sum \alpha_i^2}{\left(\sum \alpha_i^6\right)^{1/3}} \times$$
$$\times S^{2/3}\left(1 + O_3\left(\frac{1}{\lambda}\right) - \frac{\gamma_0 I}{S}\left(1 + O\left(\frac{1}{\lambda}\right)\right)\right) \tag{242}$$

if $I < \beta$.

I is lowerbounded by $\frac{C(p-1)}{\lambda}$ and this implies that

$$J\left(\sum_{i=1}^{p}\alpha_i\varphi_{a_i,\lambda}\right) \leq$$

$$\leq (pS)^{2/3}\left(1+O_3\left(\frac{1}{\lambda}\right)\right.$$

$$\left. - \frac{c(p-1)}{\lambda S}\left(1+O\left(\frac{1}{\lambda}\right)\right)\right) < (pS)^{2/3}\,. \tag{243}$$

If I is larger than or equal to β, we apply Proposition 5, which implies that $J\left(\sum_{i=1}^{p}\alpha_i\varphi_{a_i,\lambda}\right)$ is strictly less than $(pS)^{2/3}$, if λ/p is large enough. Finally, if $\alpha_{i_0}/\alpha_{j_0} \notin (1-\vartheta_0, 1+\vartheta_0)$, we apply Proposition 7 with λ_p/p^3 large enough.

Summing up, if λ_p/p^3 is large enough, $f_p(\lambda)$ is valued in W_{p-1}, for p large enough. This fact had already been observed and used in [10]. In dimensions 4 and 5, $f_p(\lambda)$ does not necessarily have this property, anymore. However, $f_p(\lambda)$ is homotopic to a map valued in W_{p-1}, which suffices for our purposes.

5. The case of a general, compact and closed, Riemannian manifold (M^n, g)

Let Δ_g be the Laplace–Beltrami operator on (M^n, g). We consider the equation:

$$\begin{cases} (-\Delta_g + q)\,u = u^{\frac{n+2}{n-2}} \\ u > 0\,. \end{cases} \quad \text{on } M\ ;\ q \in L^\infty(M) \tag{244}$$

We assume that the operator $-\Delta_g + q$ is coercive, hence that there exists a positive constant δ such that

$$\int\left(|\nabla_g\varphi|^2 + q\varphi^2\right)dv_g \geq \delta\,\|\varphi\|^2_{H^1(M)}\,. \tag{245}$$

We prove, in this section, the following theorem:

Theorem 2. *If* $3 \leq \dim M \leq 5$ *and if* M *is compact and closed, problem (244) has, at least, one solution.*

The proof is similar to the one completed when $M = S^n$. It involves several modifications, which are detailed in the sequel.

Definition of the functions $\hat{\delta}_{a,\lambda}$.

Given a point in M, the exponential map at a, $\exp_a$, identifies a neighborhood of zero in the tangent space to a neighborhood of a in M.

At any point a of M, the tangent space is equipped with a scalar product, defined by the metric g. We denote, for $x \in T_a(M)$, $|x|_a$ the norm $\sqrt{g_a(x,x)}$. Since M is compact, there exists $\rho > 0$ such that the ball $B_\rho(0)$ of $T_a(M)$ is contained in a neighborhood where the exponential map is defined. This map is differentiable in the variables a and x.

We pick up a C^∞ function $\chi : \mathbb{R}^+ \to [0,1]$ which is equal to 1 if $t \leq \rho/4$ and is equal to zero if $t \geq \rho/2$. We introduce the function:

$$\delta_{a,\lambda}(x) = \frac{\lambda^{\frac{n-2}{2}}}{(1+\lambda^2|x|_a^2)^{\frac{n-2}{2}}} \,. \tag{246}$$

This function satisfies the equation:

$$\sum g^{ij}(a)\frac{\partial^2}{\partial x_i x_j}\delta_{a,\lambda} + n(n-2)\delta_{a,\lambda}^{\frac{n+2}{n-2}} = 0 \,. \tag{247}$$

Let

$$\hat{\delta}_{a,\lambda}(y) = \chi\left(|x|_a\right)\delta_{a,\lambda}(x) \tag{248}$$

where

$$\begin{cases} x = \exp_a^{-1}(y) & \text{if } x \in B_\rho \\ \hat{\delta}_{a,\lambda}(y) = 0 & \text{otherwise.} \end{cases} \tag{249}$$

Definition of $\varphi_{a,\lambda}$.

We define the family of functions $\varphi_{a,\lambda}$, for $a \in M$ and $\lambda > 0$, as follows:

$$(-\Delta_g + q)\,\varphi_{a,\lambda} = n(n-2)\hat{\delta}_{a,\lambda}^{\frac{n+2}{n-2}} \,. \tag{250}$$

$H_{a,\lambda}$ is, as before:

$$H_{a,\lambda} = \lambda^{\frac{n-2}{2}}\left(\varphi_{a,\lambda} - \hat{\delta}_{a,\lambda}\right) \,. \tag{251}$$

We now indicate the modifications to sections 1, 2 and 3 in order to prove Theorem 2. The new sections are sections 1′, 2′ and 3′.

Section 1′.

Proposition 1′. *There exists $C > 0$ independent of a and λ such that*

i) $|H_{a,\lambda}(y)| \le C$ *for* $n = 3$

ii) $|H_{a,\lambda}(y)| \le C\left(1 + \left|\log\left(\frac{1}{\lambda^2} + \text{ dist } (a,x)^2\right)\right|\right)$ *for* $n = 4$

iii) $|H_{a,\lambda}(y)| \le C\left(1 + \lambda^{\frac{n-4}{2}} \hat{\delta}_{a,\lambda}^{\frac{n-4}{n-2}}\right)$ *for* $n \ge 5$.

Proof. Let $H_\lambda = H_{a,\lambda}$ H_λ satisfies:

$$(-\Delta_g + q)\, H_\lambda = \lambda^{\frac{n-2}{2}} \left(n(n-2)\hat{\delta}_{a,\lambda}^{\frac{n+2}{n-2}} + \Delta_g \hat{\delta}_{a,\lambda} + q\hat{\delta}_{a,\lambda}\right) . \tag{252}$$

In geodesic normal coordinates around a (see T. Aubrin [3], for example), we have:

$$\Delta_g = \sum \frac{\partial}{\partial x_j}\left((\delta_i^j + O\left(|x|_a^2\right))\frac{\partial}{\partial x_i}\right) , \tag{253}$$

which means that, in a suitable chart defined by the exponential map,

$$g^{i,j}(x) = \delta_i^j + O\left(|x|_a^2\right) . \tag{254}$$

$\chi\left(|x|_a\right)$ is identically equal to 1 in $B_{\rho/2}$, thus

$$\delta_{a,\lambda}(x) = \hat{\delta}_{a,\lambda}(y) \qquad \text{for } x \in B_{\rho/2} . \tag{255}$$

Thus

$$\Delta_g \hat{\delta}_{a,\lambda}(x) + n(n-2)\hat{\delta}_{a,\lambda}^{\frac{n+2}{n-2}} = \sum \frac{\partial}{\partial x_j}\left(O\left(|x|_a^2\right) \frac{\partial}{\partial x_i}\hat{\delta}_{a,\lambda}(x)\right) . \tag{256}$$

This last quantity is easily seen to be upperbounded by $C\hat{\delta}_{a,\lambda}$. Therefore, on $B_{\rho/4}$, we have:

$$|(-\Delta_g + q)\, H_\lambda| \le C\lambda^{\frac{n-2}{2}} \hat{\delta}_{a,\lambda} . \tag{257}$$

On ${}^C B_{\rho/4}$, we have:

$$|(-\Delta_g + q)\, H_\lambda| \le C . \tag{258}$$

Indeed, since $|x|_a \geq \rho/2$, the following estimates hold:

$$\begin{cases} \hat{\delta}_{a,\lambda} \leq \frac{C(\rho)}{\lambda^{\frac{n-2}{2}}} \\ |\Delta_g \hat{\delta}_{a,\lambda}| \leq \frac{C(\rho)}{\lambda^{\frac{n-2}{2}}} \end{cases} \tag{259}$$

from which (258) is easily derived. Thus, for any $y \in M$,

$$|(-\Delta_g + q)\, H_\lambda| \leq \frac{C}{(\text{dist }(a,y))^{n-2}}\,.$$

The argument, starting from this estimate, proceeds as in section 1.

Section 2′. Let $\eta > 0$ be given.
Let $\tilde{\delta}_{a,\lambda} = \hat{\delta}_{a,\lambda} + \frac{\eta}{\lambda^{\frac{n-2}{2}}}$ We then have:

Theorem 1′. *For any $\eta > 0$, there exists a constant $C(\eta)$ such that, for any a and b in M, for any $\lambda \geq 2$, we have:*

1) $\int \varphi^5_{a,\lambda}\varphi_{b,\lambda} \geq \left(1 - \frac{C(\eta)}{\lambda}\right) \int \tilde{\delta}^5_{a,\lambda}\varphi_{b,\lambda}$ *for $n = 3$*

2) $\int \varphi^3_{a,\lambda}\varphi_{b,\lambda} \geq \left(1 - \frac{C(\eta)}{\lambda^2}(\log\lambda)^2\right) \int \tilde{\delta}^3_{a,\lambda}\varphi_{b,\lambda}$ *for $n = 4$*

3) $\int \varphi^{\frac{n+2}{n-2}}_{a,\lambda}\varphi_{b,\lambda} \geq \left(1 - C(\eta)\frac{|\log\lambda|}{\lambda^2}\right) \int \tilde{\delta}^{\frac{n+2}{n-2}}_{a,\lambda}\varphi_{b,\lambda}$ *for $n \geq 5$.*

Proof of the Theorem 1′. Lemma 1′'s statement is identical to the one of the Lemma 1. The only modification required in its proof is:

$$\varphi_{a,\lambda} = \left(\int G(x,y)\hat{\delta}^{\frac{n+2}{n-2}}_{a,\lambda}\right) n(n-2) \geq n(n-2)\alpha \int_{\exp_a(B_{\rho/2})} \hat{\delta}^{\frac{n+2}{n-2}}_{a,\lambda}\, dv \geq$$
$$\geq C \int_{|x|\leq\rho/4} \frac{\lambda^{\frac{n+2}{2}}}{(1+\lambda^2|x|^2)^{\frac{n+2}{2}}}\, dx \geq \frac{C(\rho)}{\lambda^{\frac{n-2}{2}}}\,. \tag{261}$$

We then have:

Lemma 2′. *There exist $\alpha_{j,\lambda} > 0$ and $\beta > 0$ such that*

$$\alpha_{j,\lambda}\tilde{\delta}_{a,\lambda} \leq \varphi_{a,\lambda} \leq \beta\tilde{\delta}_{a,\lambda} \qquad \forall a \in M, \forall\lambda \geq 1\,.$$

Proof. Let

$$\tilde{H}_\lambda = H_\lambda - \eta\,. \tag{262}$$

We then have

$$\varphi = \tilde{\delta} + \frac{\tilde{H}_\lambda}{\lambda^{\frac{n-2}{2}}} . \tag{263}$$

$\tilde{H}_\lambda$ satisfies Proposition 1 with C replaced by $C(\eta)$.

In a first step, we would like to prove that

$$\varphi_{a,\lambda} \geq \alpha \tilde{\delta}_{a,\lambda} . \tag{264}$$

Observe that, if $d(x)$ is $\operatorname{dist}(a,x)$ on M, we have:

$$\frac{C_1' \lambda^{\frac{n-2}{2}}}{(1+\lambda^2 d^2)^{\frac{n-2}{2}}} \leq \tilde{\delta}_{a,\lambda}(x) \leq \frac{C_1 \lambda^{\frac{n-2}{2}}}{(1+\lambda^2 d^2)^{\frac{n-2}{2}}} \tag{265}$$

starting from these inequalities, the proof of (264), for $n = 3$, is unchanged.

For $n = 4$, the proof rested on the inequality:

$$\frac{1}{\lambda}\left(1 + \left|\log\left(\frac{1}{\lambda^2} + d^2\right)\right|\right) \leq \frac{C\lambda}{1+\lambda^2 d^2} . \tag{266}$$

Since $\tilde{H}_\lambda$ is upperbounded by $C\left(1 + \left|\log\left(\frac{1}{\lambda^2} + d^2\right)\right|\right)$ and since $\tilde{\delta}_\lambda$ upperbounds $C_1' \frac{\lambda}{1+\lambda^2 d^2}$, the former proof extends to the new framework. When $n > 4$, H_λ satisfies, by (iii) of Proposition 1:

$$|H_\lambda| \leq C\left(\lambda^{\frac{n-4}{2}} \tilde{\delta}_{a,\lambda}^{\frac{n-4}{n-2}} + 1\right) . \tag{267}$$

Since $\tilde{\delta}_{a,\lambda} = \hat{\delta}_{a,\lambda} + \frac{\eta}{\lambda^{\frac{n-2}{2}}}$, $\eta > 0$, we have:

$$|H_\lambda| \leq C(\eta) \lambda^{\frac{n-4}{2}} \tilde{\delta}_\lambda^{\frac{n-4}{n-2}} . \tag{268}$$

Starting from this inequality, the proof proceeds as in the S^n–case.

The reverse inequality follows the same lines, once $\delta_{a,\lambda}$ and H_λ are replaced by $\tilde{\delta}_{a,\lambda}$ and $\tilde{H}_\lambda$.

Proof of Theorem 1′. It rests upon the following lemma.

Lemma 4′. *There exists a continuous function* $F_1 : \mathbb{R}^+ \to \mathbb{R}^+$, $F_1(t) > 0$ $\forall t$, $F_1(t) \geq C/t^{n-2}$ *when* $t \to +\infty$ *and such that*

$$\int_M \tilde{\delta}_{a,\lambda}^{\frac{n+2}{n-2}} \tilde{\delta}_{b,\lambda} \geq F_1(\lambda d(a,b)) \quad , \forall a, \forall b, \forall \lambda \geq 1 .$$

Proof. There are two cases

1st case:

$$d(a,b) \le \rho/8 .$$

Let

$$b_1 = \exp_a^{-1}(b) . \tag{269}$$

We have:

$$\begin{aligned}
&\int_M \tilde\delta_{a,\lambda}^{\frac{n+2}{n-2}} \tilde\delta_{b,\lambda} \ge \int_{\exp_a(B_{\rho/8})} \tilde\delta_{a,\lambda}^{\frac{n+2}{n-2}} \tilde\delta_{b,\lambda} \ge \\
&\ge c_1 \int_{|x|\le\rho/8} \frac{\lambda^{\frac{n+2}{2}}}{(1+\lambda^2|x|^2)^{\frac{n+2}{2}}} \times \frac{\lambda^{\frac{n-2}{2}}}{(1+\lambda^2|x-b_1|^2)^{\frac{n-2}{2}}} dx = \\
&= C_1 \left(F(\lambda|b_1|) - \int_{|x|\ge\rho/8} \frac{\lambda^n}{(1+\lambda^2|x|^2)^{\frac{n+2}{2}} (1+\lambda^2|x-b_1|^2)^{\frac{n-2}{2}}} dx \right) .
\end{aligned} \tag{270}$$

On the other hand, we have:

$$\begin{aligned}
&\int_{|x|\ge\rho/8} \frac{\lambda^n}{(1+\lambda^2|x|^2)^{\frac{n+2}{2}} (1+\lambda^2|x-b_1|^2)^{\frac{n-2}{2}}} dx \le \\
&\quad \le \int_{\substack{|x-b_1|\le\rho \\ |x|\ge\rho/8}} + \int_{\substack{|x-b_1|\ge\rho \\ |x|\ge\rho/8}} \le \qquad (271) \\
&\quad \le \frac{C(\rho)}{\lambda^{n+2}} \int_{r\le\lambda\rho} \frac{r^{n-1}dr}{(1+r^2)^{\frac{n-2}{2}}} + \frac{C(\rho)}{\lambda^{n-2}} \int_{r\ge\lambda\rho/8} \frac{r^{n-1}dr}{(1+r^2)^{\frac{n+2}{2}}} = O\left(\frac{1}{\lambda^n}\right) .
\end{aligned}$$

Thus,

$$\begin{aligned}
&\int_M \tilde\delta_a^{\frac{n+2}{n-2}} \tilde\delta_{b,\lambda} \ge c_1 \left(F(\lambda|b_1|) - O\left(\frac{1}{\lambda^n}\right)\right) \ge \\
&\quad \ge \frac{C}{(\lambda|b_1|)^{n-2}} - \frac{C}{\lambda^n} \ge \frac{C}{2(\lambda|b_1|)^{n-2}} \ge \frac{C'}{(\lambda d(a,b))^{n-2}}
\end{aligned} \tag{272}$$

if ρ is small enough.

2nd case: $d(a,b) \ge \rho/8$. We then have:

$$\int_M \tilde\delta_{a,\lambda}^{\frac{n+2}{n-2}} \tilde\delta_{b,\lambda} \ge \frac{C}{\lambda^{\frac{n-2}{2}}} \int_M \tilde\delta_{a,\lambda}^{\frac{n+2}{n-2}} \ge \frac{C}{\lambda^{\frac{n-2}{2}}} \int_{\exp_a(B_{\rho/8})} \tilde\delta_{a,\lambda}^{\frac{n+2}{n-2}} \ge$$

$$\geq \frac{C'}{\lambda^{\frac{n-2}{2}}} \int_{|x|,\rho/8} \frac{\lambda^{\frac{n+2}{2}}}{(1+\lambda^2|x|^2)^{\frac{n+2}{2}}} dx = \frac{C''}{\lambda^{n-2}} \,. \tag{273}$$

The proof of Lemma 4′ is thereby complete.

We now proceed with the proof of Theorem 1′. When compared to the proof of Theorem 1, the first part of the proof remains unchanged, provided that we replace $\delta_{a,\lambda}$ by $\tilde{\delta}_{a,\lambda}$ and $H_{a,\lambda}$ by $\tilde{H}_{a,\lambda}$. We then derive:

$$\int \varphi_{b,\lambda} \tilde{\delta}_{a,\lambda}^{\frac{n+2}{n-2}} \leq \int \varphi_{b,\lambda} \varphi_{a,\lambda}^{\frac{n+2}{n-2}} + C(\eta) R \tag{274}$$

where

$$R = \frac{1}{\lambda^{\frac{n-2}{4}}} \int \tilde{\delta}_{b,\lambda} \tilde{\delta}_{a,\lambda}^{4/n-2} \left|\tilde{H}_{a,\lambda}\right| \,. \tag{275}$$

Next, we have:

$$\begin{aligned}
\lambda^{\frac{n-2}{2}} R \leq & \int_{\exp_b(B_\rho)} \left(\tilde{\delta}_{b,\lambda} + \frac{\eta}{\lambda^{\frac{n-2}{2}}}\right) \tilde{\delta}_{a,\lambda}^{4/n-2} \left|\tilde{H}_{a,\lambda}\right| + \\
& + \frac{\eta}{\lambda^{\frac{n-2}{2}}} \int_M \tilde{\delta}_{a,\lambda}^{4/n-2} \left|\tilde{H}_{a,\lambda}\right| \leq \\
& \leq C(\eta) \Bigg| \int_{\exp_b(B_\rho)} \tilde{\delta}_{b,\lambda} + \tilde{\delta}_{a,\lambda}^{4/n-2} \left|\tilde{H}_{a,\lambda}\right| \\
& + \frac{1}{\lambda^{\frac{n-2}{2}}} \int_{\exp_a(B_\rho)} \tilde{\delta}_{a,\lambda}^{4/n-2} \left|\tilde{H}_{a,\lambda}\right| + \\
& + \frac{1}{\lambda^{\frac{n+2}{2}}} \int_M \left|\tilde{H}_{a,\lambda}\right| \Bigg| = (1) + (2) + (3) \,.
\end{aligned} \tag{276}$$

We estimate, in a first step, (1):

1st case: $d(a,b) \geq \rho$.

Then,

$$(1) \leq \frac{C(\eta)}{\lambda^2} \int_{\exp_b(B_\rho)} \tilde{\delta}_{b,\lambda} \left|\tilde{H}_{a,\lambda}\right| \,. \tag{277}$$

Since $d(a,b) \geq \rho$ we have on $\exp_b(B_\rho)$, by Proposition 1′:

$$\left|\tilde{H}_{a,\lambda}\right| \leq c \,. \tag{278}$$

Thus

$$(1) \le \frac{C(\eta)}{\lambda^2} \int_{|x|\le\rho} \delta_\lambda dx = \frac{C(\eta)}{\lambda^2} \times \frac{1}{\lambda^{\frac{n+2}{2}}} \int_{|x|\le\lambda\rho} \frac{dx}{(1+|x|^2)^{\frac{n-2}{2}}} = \frac{C(\eta)}{\lambda^{\frac{n+2}{2}}} . \tag{279}$$

2nd case: $d(a,b) \le \rho$.

Let $a_1 = \exp_b^{-1}(a)$. Observe that $\exp_b^{-1}(b) = 0$. Clearly:

$$(1) \le C(\eta) \int_{|x|\le\rho} \frac{\lambda^{\frac{n-2}{2}}}{(1+\lambda^2|x|^2)^{\frac{n-2}{2}}} \times \times \left(\frac{\lambda^{\frac{n-2}{2}}}{\left(1+\lambda^2|x-a_1|^2\right)^{\frac{n-2}{2}}} + \frac{C}{\lambda^{\frac{n-2}{2}}} \right)^{4/n-2} \left|H'_{a_1,\lambda}\right| \tag{280}$$

where $H'_{a_1,\lambda}$ is a new function, replacing $H_{a_1,\lambda}$ and satisfying Proposition 1 with $d = \text{dist}(x, a_1)$ in $\mathbb{R}^n$ (in fact, in $B_\rho(0)$).

Observe that

$$\frac{C}{\lambda^{\frac{n-2}{2}}} \le \frac{\lambda^{\frac{n-2}{2}}}{\left(1+\lambda^2|x-a_1|^2\right)^{\frac{n-2}{2}}} . \tag{281}$$

Therefore,

$$(1) \le C(\eta) \int_{\mathbb{R}^n} \delta_{0,\lambda} \delta_{a_1,\lambda}^{4/n-2} \left|H'_{a_1,\lambda}\right| . \tag{282}$$

In section 2, we provided estimates for the right hand side and we proved that we could upperbound this right hand side by:

$$\begin{cases} \frac{C}{\sqrt{\lambda}} \int \delta_{0,\lambda}\delta_{a_1,\lambda}^5 & \text{for } n=3 \\ \frac{C}{\lambda} (\log\lambda)^2 \int \delta_{0,\lambda}\delta_{a_1,\lambda}^3 & \text{for } n=4 \\ \lambda^{\frac{n-2}{2}} \times \frac{C}{\lambda^2} \log\lambda \int \delta_{0,\lambda}\delta_{a_1,\lambda}^{\frac{n+2}{n-2}} & \text{for } n \ge 5 . \end{cases} \tag{283}$$

We now estimate (2). Using the exponential map at a, we have:

$$(2) \le \frac{1}{\lambda^{\frac{n-2}{2}}} \int_{|x|\le\rho} \frac{\lambda^{\frac{n-2}{2}}}{\left(1+\lambda^2|x|^2\right)^2} |H'_\lambda|\, dx \tag{284}$$

where H'_λ satisfies Proposition 1 with $d(x) = |x|$. Thus, we have:

$$\begin{cases} (2) \leq \frac{C}{\lambda^{3/2}} & \text{for } n = 3 \\ (2) \leq \frac{C(\log \lambda)^2}{\lambda^3} & \text{for } n = 4 \\ (2) \leq \frac{C \log \lambda}{\lambda^{\frac{n+2}{2}}} & \text{for } n \geq 5\,. \end{cases} \tag{285}$$

We now estimate (3). Using Proposition 1, we easily derive that, with an appropriate constant C, we have:

$$\forall a\,,\ \forall \lambda \geq 1 \quad \int_M \left|\tilde{H}_{a,\lambda}\right| \leq C\,,$$

hence,

$$(3) \leq \frac{C}{\lambda^{\frac{n+2}{2}}}\,. \tag{286}$$

Summing up, we have, if $d(a,b) < \rho$:

$$\begin{cases} \sqrt{\lambda} R \leq \frac{C}{\lambda\sqrt{\lambda}} + \frac{C}{\sqrt{\lambda}} \int_{\mathbf{R}^3} \delta_{0,\lambda} \delta^5_{a_1,\lambda} & \text{for } n = 3 \\ \lambda R \leq C \frac{(\log\lambda)^2}{\lambda^3} + C \frac{(\log\lambda)^2}{\lambda} \int_{\mathbf{R}^4} \delta_{0,\lambda} \delta^3_{a_1,\lambda} & \text{for } n = 4 \\ \lambda^{\frac{n-2}{2}} R \leq C \frac{\log\lambda}{\lambda^{\frac{n+2}{2}}} + \lambda^{\frac{n-2}{2}} \frac{C \log\lambda}{\lambda^2} \int_{\mathbf{R}^n} \delta_{0,\lambda} \delta^{\frac{n+2}{n-2}}_{a_1,\lambda} & \text{for } n \geq 5\,. \end{cases} \tag{287}$$

If $d(a,b) \geq \rho$, we can cancel the second term in the upperbound.

We conclude now the proof of Theorem 1′. Since M is compact, Lemma 4′ implies:

$$\int_M \tilde{\delta}^{\frac{n+2}{n-2}}_{a,\lambda} \tilde{\delta}_{b,\lambda} \geq \frac{C}{\lambda^{n-2}} \qquad \forall \lambda \geq 1\,. \tag{288}$$

Furthermore, Lemma 4 and 4′ imply that there exists a constant $C > 0$ such that for any a and b in M, with $d(a,b) \leq \rho$ and for any $\lambda \geq 1$, we have:

$$\int_M \tilde{\delta}^{\frac{n+2}{n-2}}_{a,\lambda} \tilde{\delta}_{b,\lambda} \geq C \int_{\mathbf{R}^n} \delta^{\frac{n+2}{n-2}}_{a_1,\lambda} \delta_{0,\lambda} \tag{289}$$

where $\quad a_1 = \exp_b^{-1}(a)$.

We also know that:

$$\tilde{\delta}_{b,\lambda} \leq C(\eta)\varphi_{b,\lambda}\,. \tag{290}$$

Therefore:

$$\begin{cases} R \le \frac{C}{\lambda}\int_M \tilde\delta^5_{a,\lambda}\varphi_{b,\lambda} & \text{for } n=3 \\ R \le \frac{C(\log\lambda)^2}{\lambda^2}\int_M \tilde\delta^3_{a,\lambda}\varphi_{b,\lambda} & \text{for } n=4 \\ R \le C\frac{\log\lambda}{\lambda^2}\int_M \tilde\delta^{\frac{n+2}{n-2}}_{a,\lambda}\varphi_{b,\lambda} & \text{for } n\ge 5\,. \end{cases} \tag{291}$$

Theorem 1′ follows.

Proposition 2 and Lemma 5 extend also. We need, in order to complete this extension, only to observe that

$$\int_{\mathbb{R}^n} \delta^{2n/n-2}_{a,\lambda} = \int_{B_{\rho/8}} \delta^{2n/n-2}_{a,\lambda} + O\left(\frac{1}{\lambda^n}\right)\,. \tag{292}$$

Section 3: Proposition 3's statement is unchanged. Some details have to be changed in the proof. In particular, N should be taken to be equal to:

$$N = \int \left(\sum \alpha_i \hat\delta^{\frac{n+2}{n-2}}_{a_i,\lambda_i}\right) \times \left(\sum \alpha_j \varphi_{a_j,\lambda}\right)\,. \tag{293}$$

We also modify the upperbound in (132) as follows:

$$\left|\tilde\delta^{\frac{n+2}{n-2}}_{a_i,\lambda} - \varphi^{\frac{n+2}{n-2}}_{a_i,\lambda}\right| \le C\tilde\delta^{4/n-2}_{a_i,\lambda}\left|\tilde H_{a_i,\lambda}\right|\,. \tag{294}$$

The remainder of the proof remains unchanged, if we replace $\delta_{a_i,\lambda}$ by $\tilde\delta_{a_i,\lambda}$ and $H_{a_i,\lambda}$ by $\tilde H_{a_i,\lambda}$.

Proposition 4's statement is unchanged. $\tilde\delta_{a_i,\lambda}$ should be replaced by $\tilde\delta_{a_i,\lambda}$, in its proof. This change should be completed in the remainder of the section, which then extends readily.

6. Continuous Selection of points

Notations. Let λ_i and λ_j , positive, b given. Let a_i and $a_j \in M$.

We denote:

$$\varepsilon_{ij} = \left(\frac{\lambda_i}{\lambda_j} + \frac{\lambda_j}{\lambda_i} + \lambda_i\lambda_j d\left(a_i,a_j\right)^2\right)^{-\frac{n-2}{2}}\,. \tag{295}$$

Let μ_i, μ_j, positive, be given. Let b_i and $b_j \in M$. We denote:

$$\varepsilon'_{ij} = \left(\frac{\mu_i}{\mu_j} + \frac{\mu_j}{\mu_i} + \mu_i\mu_j d\left(b_i,b_j\right)^2\right)^{-\frac{n-2}{2}}\,. \tag{296}$$

We prove, in Appendix 3 of this Chapter, that:

$$c_1\varepsilon_{ij} \leq \int \varphi_{a_i,\lambda_i}^{\frac{n+2}{n-2}}\varphi_{a_j,\lambda_j} \leq c_2\varepsilon_{ij} \,. \tag{297}$$

We will use this result in this section.

We denote:

$$\begin{cases} \beta_j = \sum_{\substack{i=1 \\ j\neq i}}^{p} \varepsilon_{ij} \\ \beta'_j = \sum_{\substack{i=1 \\ j\neq i}}^{p} \varepsilon'_{ij} \,. \end{cases} \tag{298}$$

We assume, throughout this section, that λ_i and μ_i are larger than A, where A is a given constant, independent of p and which will be chosen large enough.

Let, for $u \in \Sigma^+, \gamma_u$ be the function:

$$\gamma_u\left(a_1,\ldots,a_p,\alpha_1,\ldots,\alpha_p,\lambda_1,\ldots,\lambda_p\right) = \\ = \gamma_a\left(a,\alpha,\lambda\right) = \left|u - \sum_{i=1}^{p}\alpha_i\varphi_{a_i,\lambda_i}\right|^2_{-L+q} \,.$$

All concentrations λ_i are assumed to be larger than A so that all lemmas of Appendix 3 apply. We also assume that $\alpha_i \leq 2$. We then have:

Lemma 10. *For any $0, \vartheta_0, 1$, there exist $\varepsilon(\vartheta_0)$ and $c_1(\vartheta_0)$ which depend only on ϑ_0, such that, assuming $\alpha_i/\alpha_j \in (1-\vartheta_0, 1+\vartheta_0)$ for any i and j, assuming $\sum_{\substack{i=1 \\ j\neq i}}^{p} \varepsilon_{ij} < c_1(\vartheta_0)$ and assuming that $\gamma_u(a,\alpha,\lambda) < \varepsilon^2(\vartheta_0)$, then the function γ_u is strictly convex in a neighborhood of the point (a,α,λ).*

Proof of Lemma 10. We compute the second derivative of γ_u, at $(a,\alpha_{j,\lambda},\lambda)$, with respect to a variation $(\delta a_1,\ldots,\delta a_p,\delta\alpha_1,\ldots,\delta\alpha_p,\delta\lambda_1,\ldots,$ $\delta\lambda_p)$. We use the more convenient coordinates:

$$h = \left(\lambda_1\delta a_1,\ldots,\lambda_p\delta a_p,\delta\alpha_1,\ldots,\delta\alpha_p,\frac{\delta\lambda_1}{\lambda_1},\ldots,\frac{\delta\lambda_p}{\lambda_p}\right) \tag{299}$$

The second derivative of γ_u is:

$$D^2\gamma_u\left(a,\alpha,\lambda\right).h.h = \\ = 2\left|\sum_{i=1}^{p}\left(\delta\alpha_i\varphi_{a_i,\lambda_i} + \frac{\alpha_i}{\lambda_i}\frac{\partial\varphi_{a_i,\lambda_i}}{\partial a_i}\left(\lambda_i\delta a_i\right) + \alpha_i\lambda_i\frac{\partial\varphi_{a_i,\lambda_i}}{\partial\lambda_i}\left(\frac{\delta\lambda_i}{\lambda_i}\right)\right)\right|^2_{-L+q} + \\ +2\left(u - \sum_{i=1}^{p}\alpha_i\varphi_{a_i,\lambda_i}, \sum_{i=1}^{p}\left(\delta\alpha_i\left(\lambda_i\frac{\partial\varphi_{a_i,\lambda_i}}{\partial\lambda_i}\left(\frac{\delta\lambda_i}{\lambda_i}\right) + \frac{1}{\lambda_i}\frac{\partial\varphi_{a_i,\lambda_i}}{\partial a_i}\left(\lambda_i\delta a_i\right)\right)+\right.\right.$$

$$+ \quad \alpha_i \lambda_i^2 \frac{\partial^2 \varphi_{a_i,\lambda_i}}{\partial \lambda_i^2} \left(\frac{\delta \lambda_i}{\lambda_i} \right)^2 + \alpha_i \frac{\partial^2 \varphi_{a_i,\lambda_i}}{\partial \lambda_i \partial a_i} \left(\lambda_i \delta_{a_i} \right) \frac{\delta \lambda_i}{\lambda_i} +$$

$$+ \quad \frac{\alpha_i}{\lambda_i^2} \frac{\partial^2 \varphi_{a_i,\lambda_i}}{\partial a_i^2} \left(\lambda_i \delta a_i, \lambda_i \delta a_i \right) \Bigg) \Bigg) . \tag{300}$$

We derive then, from Lemma A9 of Appendix 3:

$$D^2 \gamma_u (a, \alpha, \lambda) .h.h \geq C_1' \left(\sum \delta \alpha_i^2 + \sum |\lambda_i \delta a_i|^2 + \sum \left(\frac{\delta \lambda_i}{\lambda_i} \right)^2 \right) +$$
$$- C_2 \sum_{\substack{i,j \\ i \neq j}}^{p} \varepsilon_{ij} \left(\lambda_i |\delta a_i| |\delta \alpha_j| + |\delta \alpha_j| \frac{|\delta \lambda_i|}{\lambda} + |\delta a_i| |\delta \lambda_j| \right) +$$
$$- C \varepsilon (\vartheta_0) \left(\sum \delta \alpha_i^2 + \sum (\lambda_i \delta a_i)^2 + \sum \left(\frac{\delta \lambda_i}{\lambda_i} \right)^2 \right) . \tag{301}$$

C_1' depends only on ϑ_0 and on the various constants of Lemmas of Appendix 3. The matrix $D^2 \gamma_u$ is therefore lowerbounded by a $3p \times 3p$ matrix whose diagonal terms are $C_1' - \varepsilon(\vartheta_0)$ and the non diagonal terms a_{ij} are $O(\varepsilon_{ij})$. Lemma 10 follows.

We now have:

Lemma 11. *Let* $(a_1^k, \ldots, a_p^k)$, $(\lambda_1^k, \ldots, \lambda_p^k)$, $(b_1^k, \ldots, b_p^k)$, $(\mu_1^k, \ldots, \mu_p^k)$ *be sequences such that:*

$$\left| \sum_{i=1}^{p} \varphi_{a_i^k, \lambda_i^k} - \sum_{i=1}^{p} \alpha_i^k \varphi_{b_i^k, \mu_i^k} \right|_{-L+q} = o(1) \underset{k \to +\infty}{\longrightarrow} 0$$

and such that

$$\beta_j^k = \sum_{\substack{i=1 \\ i \neq j}}^{p} \varepsilon_{ij}^k = o(1) \underset{k \to +\infty}{\longrightarrow} 0 \quad \forall j .$$

Then,

$$\beta_j'^k = \sum_{\substack{i=1 \\ i \neq j}}^{k} \varepsilon'^k_{ij} = o(1) \underset{k \to +\infty}{\longrightarrow} 0 \quad \forall j$$

$$\alpha_i^k = 1 + o(1) \underset{k \to +\infty}{\longrightarrow} 1 \quad \forall j$$

and

$$\forall i \ , \quad \exists j^k(i) = j(i) \ \textit{unique such that}$$

$$\left|\varphi_{b_i^k,\mu_i^k} - \varphi_{a^k_{j(i)},\lambda^k_{j(i)}}\right|_{-L+q} = o(1) \ . \underset{k\to+\infty}{\longrightarrow} 0$$

Proof. Under the hypotheses of the lemma, the sequences $\left(\alpha_i^k\right)$ are bounded independently of k. Indeed, we have:

$$C\sum_{i=1}^{p}\alpha_i^2 \leq \left|\sum_{i=1}^{p}\alpha_i^k\varphi_{b_i^k,\mu_i^k}\right|^2_{-L+q} \leq \left|\sum_{i=1}^{p}\alpha_i^k\varphi_{a_i^k,\lambda_i^k}\right|^2_{-L+q} + o(1) < pC_1 \ . \quad (302)$$

On the other hand, there exists a universal constant C such that

$$\varepsilon_{ij}\varepsilon_{jr} < C\varepsilon_{ir} \qquad \forall i,j,r \text{ positive distinct.}$$

Indeed, it is easy to check that:

$$\frac{\lambda_1}{\lambda_2}+\frac{\lambda_2}{\lambda_1}+\lambda_1\lambda_2 d\left(a_1,a_2\right)^2 \leq$$
$$\leq C^{\frac{n-2}{2}}\left(\frac{\lambda_1}{\lambda_3}+\frac{\lambda_3}{\lambda_1}+\lambda_1\lambda_3 d\left(a_1,a_3\right)^2\right)\left(\frac{\lambda_2}{\lambda_3}+\frac{\lambda_3}{\lambda_2}+\lambda_2\lambda_3 d\left(a_2,a_3\right)^2\right) .$$

This inequality implies that, for any $i \in (1,\dots,p)$, there is at most one $j^k(i) = j(i)$ such that

$$\left(\varphi_{b_i^k,\mu_i^k},\varphi_{a^k_{j(i)},\lambda^k_{j(i)}}\right)_{-L+q} \neq o(1) \ . \quad (303)$$

Indeed, arguing by contradiction, we would have for two distinct indices $j(i) \neq j_1(i)$:

$$\left(\varphi_{a^k_{j(i)},\lambda^k_{j(i)}},\varphi_{b_i^k,\mu_i^k}\right)_{-L+q}\left(\varphi_{b_i^k,\mu_i^k},\varphi_{a^k_{j_1(i)},\lambda^k_{j_1(i)}}\right)_{-L+q} \leq$$
$$\leq C\left(\frac{\lambda^k_{j(i)}}{\mu_i^k}+\frac{\mu_i^k}{\lambda^k_{j(i)}}+\lambda^k_{j(i)}\mu_i^k d\left(a^k_{j(i)},b_i^k\right)^2\right)^{-\frac{n-2}{2}} \times$$
$$\times\left(\frac{\lambda^k_{j_1(i)}}{\mu_i^k}+\frac{\mu_i^k}{\lambda^k_{j_1(i)}}+\lambda^k_{j_1(i)}\mu_i^k d\left(a^k_{j_1(i)},b_i^k\right)^2\right)^{-\frac{n-2}{2}} \leq$$

$$\leq C\left(\frac{\lambda^k_{j(i)}}{\lambda^k_{j_1(i)}}+\frac{\lambda^k_{j_1(i)}}{\lambda^k_{j(i)}}+\lambda^k_{j_1(i)}\lambda^k_{j(i)}d\left(a^k_{j_1(i)},a^k_{j(i)}\right)^2\right)^{-\frac{n-2}{2}}=$$
$$=C'\left(\varphi_{a^k_{j(i)},\lambda^k_{j(i)}},\varphi_{a^k_{j_1(i)},\lambda^k_{j_1(i)}}\right)_{-L+q}\underset{k\to+\infty}{\longrightarrow}0\,. \tag{304}$$

Hence, the fact that $j(i)$ is unique.

We now prove the existence of $j^k(i)=j(i)$. Indeed, arguing by contradiction, there exists one index i such that, on a subsequence which we do not index, we have:

$$\left(\varphi_{b^k_i,\mu^k_i},\varphi_{a^k_j,\lambda^k_j}\right)_{-L+q}=o(1)\underset{k\to+\infty}{\longrightarrow}0 \quad \forall j\in\{1,\ldots,p\}\,. \tag{305}$$

Then, there is an index $r_k=r$, such that

$$\left(\varphi_{a^k_r,\lambda^k_r},\varphi_{b^k_j,\mu^k_j}\right)_{-L+q}=o(1)\underset{k\to+\infty}{\longrightarrow}0 \quad \forall j\in\{1,\ldots,p\}\,. \tag{306}$$

This implies that:

$$o(1)=\left|\sum_{i=1}^{p}\varphi_{a^k_i,\lambda^k_i}-\sum_{i=1}^{p}\alpha^k_i\varphi_{b^k_i,\mu^k_i}\right|^2_{-L+q}\geq$$
$$\geq\left|\varphi_{a^k_r,\lambda^k_r}\right|^2_{-L+q}-2\sum_{i=1}^{p}\alpha^k_i\left(\varphi_{a^k_r,\lambda^k_r},\varphi_{b^k_i,\mu^k_i}\right)^2_{-L+q}\geq$$
$$\geq\left|\varphi_{a^k_r,\lambda^k_r}\right|^2_{-L+q}+o(1)\,. \tag{307}$$

This yields a contradiction, since $\left|\varphi_{a^k_r,\lambda^k_r}\right|^2_{-L+q}$ is lowerbounded by a positive constant C, independent of k. Thus, $j(i)$ exists and is unique. We can reorder the indices so that $j(i)=i$. We do not mention anymore, for the sake of simplifying the notations, the index k. We then have:

$$o(1)=\left|\sum_{i=1}^{p}\left(\varphi_{a_i,\lambda_i}-\alpha_i\varphi_{b_i,\mu_i}\right)\right|^2_{-L+q}\geq$$
$$\geq pS+o(1)-2\sum_{i=1}^{p}\alpha_i\left(\varphi_{a_i,\lambda_i},\varphi_{b_i,\mu_i}\right)_{-L+q}+$$
$$+\sum_{i=1}^{p}\alpha_i^2\left|\varphi_{b_i,\mu_i}\right|^2_{-L+q}+$$
$$+2\sum_{i\neq j}^{p}\alpha_i\alpha_j\left(\varphi_{b_i,\mu_i},\varphi_{b_j,\mu_j}\right)_{-L+q}\,. \tag{308}$$

Furthermore, we have for any i:

$$\begin{aligned} 2\alpha_i\left(\varphi_{a_i,\lambda_i},\varphi_{b_i,\mu_i}\right)_{-L+q} &\le \\ &\le |\varphi_{a_i,\lambda_i}|^2_{-L+q} + \alpha_i^2\,|\varphi_{b_i,\mu_i}|^2_{-L+q} \le \\ &\le S + o(1) + \alpha_i^2\,|\varphi_{b_i,\mu_i}|^2_{-L+q}\ . \end{aligned} \tag{309}$$

From (308) and (309), we derive that:

$$\sum_{i\neq j}\alpha_i\alpha_j\left(\varphi_{b_i,\mu_i},\varphi_{b_j,\mu_j}\right) = o(1)\underset{k\to+\infty}{\longrightarrow} 0 \tag{310}$$

and

$$|\varphi_{a_i,\lambda_i}|^2_{-L+q} + \alpha_i^2\,|\varphi_{b_i,\mu_i}|^2_{-L+q} - 2\alpha_i\left(\varphi_{a_i,\lambda_i},\varphi_{b_i,\mu_i}\right)^2_{-L+q} = o(1)\underset{k\to+\infty}{\longrightarrow} 0\ . \tag{311}$$

Thus,

$$|\varphi_{a_i,\lambda_i} - \alpha_i\varphi_{b_i,\mu_i}|^2_{-L+q} = o(1)\ . \tag{312}$$

Hence,

$$\mu_i \to +\infty \quad \text{and} \quad \alpha_i = 1 + o(1)\underset{k\to+\infty}{\longrightarrow} 0\ . \tag{313}$$

Coming back to (310), we derive:

$$o(1) = \sum_{i\neq j}\left(\varphi_{b_i,\mu_i},\varphi_{b_j,\mu_j}\right)_{-L+q} \ge c\sum_{i\neq j}\varepsilon'^{k}_{ij}\ . \tag{314}$$

Lemma 11 follows.

From Lemma 10 and 11, we derive the following propositions.

Proposition 9. *For any $p \in \mathbb{N} - \{0\}$, there exists $\varepsilon_0(p) > 0$ such that, for any $u \in \sum^+$ satisfying:*

$$\left|u - \frac{1}{\sqrt{pS}}\sum_{i=1}^{p}\varphi_{a_i,\lambda_i}\right|_{-L+q} < \varepsilon_0(q) \quad ,$$

where the φ_{a_i,λ_i}'s satisfy $\sum_{i\neq j}\varepsilon_{ij}, \varepsilon_0(p)$, then the minimization problem:

$$\operatorname*{Min}_{\alpha_i,b_i,\mu_i}\left|u - \sum_{i=1}^{p}\alpha_i\varphi_{b_i,\mu_i}\right|^2_{-L+q}$$

has a unique solution, up to a permutation on the set of indices $1,\ldots,p$.

Proof. From lemma 10, we derive the existence of $\varepsilon_0 > 0$ such that for any $0 < \varepsilon < \varepsilon_0$, for any $(a_1, \ldots, a_p)$, $(\lambda_1, \ldots, \lambda_p)$, $(b_1, \ldots, b_p)$, $(\mu_1, \ldots, \mu_p)$, $(\alpha_1, \ldots, \alpha_p)$ satisfying:

$$\left|\sum_{i=1}^{p} \varphi_{a_i,\lambda_i} - \sum_{i=1}^{p} \alpha_i \varphi_{b_i,\mu_i}\right|_{-L+q} < \varepsilon < \varepsilon_0 \tag{315}$$

and

$$\sum_{i \neq j} \varepsilon_{ij} < \varepsilon < \varepsilon_0 , \tag{316}$$

then

$$\sum_{i \neq j} \varepsilon'_{ij} < \varepsilon_1(\varepsilon) \quad , \quad |\alpha_i - 1| < \varepsilon_1(\varepsilon) \quad \forall i \tag{317}$$

with

$$\lim_{\varepsilon \to 0} \varepsilon_1(\varepsilon) = 0 .$$

This implies the existence of a solution to the minimization problem when $\varepsilon_0(p)$ is small enough ($\varepsilon_0(p) < \overline{\varepsilon}_0$). Indeed, if $\varepsilon_0(p)$ is small enough ($\varepsilon_0(p) < \overline{\varepsilon}_0$), any minimizing sequence $\left(\alpha_i^k, b_i^k, \mu_i^k\right)$ will satisfy, for k large enough,

$$\left|\sum_{i=1}^{p} \alpha_i^k \varphi_{b_i^k,\mu_i^k} - \sum_{i=1}^{p} \frac{1}{\sqrt{pS}} \varphi_{a_i,\lambda_i}\right|_{-L+q} < 2\varepsilon_0(p) . \tag{318}$$

This implies that

$$\left|\left(\alpha_i^k - \frac{1}{\sqrt{pS}}\right)\right| < \frac{1}{2\sqrt{pS}} . \tag{319}$$

On the other hand, any sequence $\left(\mu_i^k\right)$ is upperbounded, since:

$$\begin{aligned}\left(\alpha_i^k\right)^2 \left|\varphi_{b_i^k,\mu_i^k}\right|_{-L+q}^2 - \frac{2\alpha_i^k}{\sqrt{pS}} \left(\varphi_{b_i^k,\mu_i^k}, \sum_{i=1}^{p} \varphi_{a_i,\lambda_i}\right)_{-L+q} &\leq \\ \leq \left|\sum_{i=1}^{p} \alpha_i^k \varphi_{b_i^k,\mu_i^k} - \sum_{i=1}^{p} \frac{1}{\sqrt{pS}} \varphi_{a_i,\lambda_i}\right|_{-L+q}^2 &< 4\varepsilon_0(p)^2 .\end{aligned} \tag{320}$$

Assuming μ_i^k tend to $+\infty$, we would have, for k large enough:

$$\frac{1}{4pS}(S+o(1)) < 4\varepsilon_0^2(p)\,, \tag{321}$$

which yields a contradiction if $\varepsilon_0(p)$ is small enough. The existence of a minimum follows, provided $\varepsilon_0(p)$ is small enough.

Lemma 11 implies that any minimum will satisfy the requirements of Lemma 10, if $\varepsilon_0(p)$ is small enough. The strict convexity of γ_u at this minimum, then implies that it is locally unique. Let, then,

$$\omega = \frac{\sum_{i=1}^p \varphi_{a_i,\lambda_i}}{|\sum_{i=1}^p \varphi_{a_i,\lambda_i}|_{-L+q}}\,. \tag{322}$$

Let

$$\begin{cases} v_t = (1-t)u + t\dfrac{\sum_{i=1}^p \varphi_{a_i,\lambda_i}}{|\sum_{i=1}^p \varphi_{a_i,\lambda_i}|_{-L+q}} \\ \omega_t = \dfrac{v_t}{|v_t|_{-L+q}}\,. \end{cases} \tag{323}$$

ω_t is then a continuous path from u to ω, in $\sum^+$, which satisfies:

$$\operatorname*{Sup}_{t\in[0,1]} \left|\omega_t - \frac{1}{\sqrt{pS}}\sum_{i=1}^p \varphi_{a_i,\lambda_i}\right|_{-L+q} \longrightarrow 0 \tag{324}$$

when $\varepsilon_0(p) \to 0$.

If we require

$$\operatorname*{Sup}_{t\in[0,1]} \left|\omega_t - \frac{1}{\sqrt{pS}}\sum_{i=1}^p \varphi_{a_i,\lambda_i}\right|_{-L+q} < \bar{\varepsilon}_0 \quad , \tag{325}$$

the minimization problem associated to ω_t, for any $t \in [0,1]$, will have solutions which are strict minima. The number of solutions will therefore be the same at $t=0$ and at $t=1$, i.e. at ω and at u. At ω, the solution (α_i, b_i, μ_i) of the minimization problem are such that $\sum_{i=1}^p \alpha_i\varphi_{b_i,\mu_i} = \omega$.

The number of these solutions does not depend on $\omega = \sum_{i=1}^p \alpha_i\varphi_{b_i,\mu_i}$ provided $\alpha_i = \frac{1}{\sqrt{pS}}(1+o(1))$ and $\sum_{i\neq j}\varepsilon_{ij}, \varepsilon_0(p)$, since the set of such ω's is arcwise connected. Indeed, it is the image through the map

$$(\alpha_i, b_i, \mu_i) \longrightarrow \sum_{i=1}^p \alpha_i\varphi_{b_i,\mu_i} \tag{326}$$

of the set $\left\{(\alpha, b, \mu) \text{ such that } \left|\alpha_i - \frac{1}{\sqrt{pS}}\right|, \bar{\varepsilon}_1, \mu_i > \frac{1}{\bar{\varepsilon}_1}, \varepsilon_{ij} < \bar{\varepsilon}_1, \forall i \neq j\right\}$ and one can prove, directly, that this set is arcwise connected.

We choose then $\omega = \sum_{i=1}^{p} \frac{1}{\sqrt{pS}} \varphi_{b_i,\mu_i}$ with

$$d(b_1, b_j) \geq 2\rho \qquad \text{for } j \geq 2\,. \tag{327}$$

Such a choice is possible if 2ρ is less than half the diameter of M, which we will assume.

Assume, then, that we have another solution of the minimization problem for ω

$$\omega = \sum_{i=1}^{p} \overline{\alpha}_i \varphi_{\overline{b}_i,\overline{\mu}_i}\,. \tag{328}$$

Applying then Lemma 10, for any i, there exists a unique $j(i)$, which we can assume, after reordering, to be equal to i, such that

$$\left|\varphi_{\overline{b}_i,\overline{\mu}_i} - \varphi_{b_i,\mu_i}\right|_{-L+q} = o(1) \tag{329}$$

where $o(1) \to 0$ when $\bar{\varepsilon}_0 \to 0$. Thus, for $\bar{\varepsilon}_0$ small enough, we have:

$$d\left(b_i, \overline{b}_i\right) < \frac{\varrho}{16}\,. \tag{330}$$

Therefore,

$$d\left(b_1, \overline{b}_i\right) \geq \frac{3\varrho}{2} \qquad \forall i \neq 1\,. \tag{331}$$

Applying then the operator $-\Delta_g + q$ to the identity

$$\sum_{i=1}^{p} \frac{1}{\sqrt{pS}} \varphi_{b_i,\mu_i} = \sum_{i=1}^{p} \overline{\alpha}_i \varphi_{\overline{b}_i,\overline{\mu}_i} \tag{332}$$

and using the fact that the support of χ, as defined in section 5, is contained in $B(0,\varrho)$, we derive from (332) and (250) and from the fact that $\chi \equiv 1$ on $B(0,\varrho/4)$ that:

$$\frac{1}{\sqrt{pS}} \hat{\delta}_{b_1,\mu_1}^{\frac{n+2}{n-2}} = \alpha_1 \hat{\delta}_{\overline{b}_1,\overline{\mu}_1}^{\frac{n+2}{n-2}} \qquad \text{on} \qquad B(b_1, \varrho/16)\,. \tag{333}$$

The above arguments depend very little on the metric and we can continuously deform the metric g among a compact, connected, set of metrics without changing the number of solutions of one minimization problem (provided $\bar{\varepsilon}_0$ is uniform and small enough). We can therefore assume that g is euclidian on $B(b_1, \varrho/16)$.

(333) becomes then an identity in a neighborhood of 0 in $\mathbb{R}^n$ and it implies that:

$$\alpha_1 = \frac{1}{\sqrt{pS}} \quad , \bar{b}_1 = b_1, \bar{\mu}_1 = \mu_1 \, . \tag{334}$$

Thus, using this particular ω, with this particular metric, the two first bubbles identify and (333) becomes an identity between $(p-1)$ elements. In fact, the above argument shows that there are as many solutions for p bubbles as there are for $p-1$.

But, we can repeat the above argument for $p = p - 1$. Arguing in this way, we end up with the case where $p = 1$. The conclusion then follows and Proposition 9 is thereby established.

7. Existence arguments in higher dimensions, under other hypotheses

Theorem 3. *Let (M^n, g) be a compact n–dimensional, Riemannian manifold, without boundary. Assuming that*

1) *either n is arbitrary and $H_1(M, \mathbb{Z}_2)$ or $H_2(M, \mathbb{Z}_2)$ is non zero.*
2) *or $n = 6, 7$ and $H_3(M, \mathbb{Z}_2)$ is non zero.*

Then, equation (1) has a solution.

Remark. There are various examples of manifolds having a non zero H_1 with $\mathbb{Z}_2$–coefficients, such as the tori T^n, the projective space $P\mathbb{R}^n$ and, more generally, any manifold with a commutative and infinite fundamental group.

$H_1(M, \mathbb{Z})$ is the abelianized group of the fundamental group and $H_1(M, \mathbb{Z}_2)$ can be derived from $H_1(M, \mathbb{Z})$, using the universal coefficients formula (see A. Dold [4], for example).

Let us also point out that, if (M^n, g) is a connected Riemannian manifold equipped with a fixed point free involution T, which is an isometry and such that $q(Tx) = q(x)$, then equation (1) has a solution u such that $u(Tx) = u(x)$, as we can derive from Theorem applied to $(M^n/T, g)$.

Indeed, M^n/T has then a non zero H_1 with $\mathbb{Z}_2$–coefficients. Let us observe that $H_i(M, \mathbb{Z}_2)$ is non zero if and only if $H^i(M, \mathbb{Z}_2)$ (or $H_{n-i}(M, \mathbb{Z}_2)$) is non zero, by Poincaré duality. Let $\omega \in H^i(M, \mathbb{Z}_2), \omega \neq 0$. Then, by the results of R. Thom [5], there exists a compact manifold V and a continuous map

$$g: \quad V \longrightarrow M^n \quad \text{such that} \quad g^*(\omega) = 0_V,$$

where 0_V designates the $\mathbb{Z}_2$–orientation class of V. Thus V is of dimension i. Since we can assume that

$$2i \leq n \tag{335}$$

using Poincaré duality, g can be taken to be a C^∞ immersion (see M. Hirsch [6], for the density of immersions in $C^0(V, M)$).

In the sequel, we consider a C^∞–immersion g of a q–dimensional compact manifold V into M^n. We define, following section 4,

$$B_p(V) = \left\{ \sum_{i=1}^{p} \alpha_i \delta_{v_i} ; \alpha_i \geq 0 ; \sum_{i=1}^{p} \alpha_i = 1 \; ; \; v_i \in V \right\} \tag{336}$$

where δ_{v_i} denotes the Dirac mass at v_i.

Using the map g, we define:

$$\begin{aligned} g_p \; &: B_p(V) \longrightarrow B_p(M) \\ &\sum_{i=1}^{p} \alpha_i \delta_{v_i} \longrightarrow \sum \alpha_i \delta_{g(v_i)} \, . \end{aligned} \tag{337}$$

In section 4, we had defined the map:

$$\begin{aligned} f_p(\lambda) \; &: B_p(M) \longrightarrow \Sigma^+ \\ &\sum_{i=1}^{p} \alpha_i \delta_{a_i} \longrightarrow \frac{\sum_{i=1}^{p} \alpha_i \varphi_{a_i, \lambda}}{\|\sum_{i=1}^{p} \alpha_i \varphi_{a_i, \lambda}\|_{-L+q}} \, . \end{aligned} \tag{338}$$

By composition, we have thus defined a map:

$$h_p(\lambda) \; : B_p(V) \longrightarrow \Sigma^+$$

$$h_p(\lambda) = f_p(\lambda) \, o g_p \, . \tag{339}$$

If λ is larger than $C\omega_n(p)$, then $h_p(\lambda)$ maps $\left(B_p(V), B_{p-1}(V)\right)$ into (W_p, W_{p-1}). We then have the following Proposition.

Proposition 10. *Under the conditions of Theorem 3, there exists $p_1 \in \mathbb{N}^*$ such that for any $p \geq p_1$ and for any $\lambda \geq C\omega_n(p)$, the map $h_p(\lambda)$ is homotopic to a map valued in (W_{p-1}, W_{p-1}) and is therefore homologically trivial.*

The proof of Proposition 10 requires the following lemma:

Lemma 12. *Let V b a compact, q–dimensional manifold.*

There exists a constant $C > 0$ such that, for any $p \in \mathbb{N}^*$, for any $(v_1, \dots, v_p)$, points of V; for any $\lambda \geq 1$, we have:

$$C inf\left(1, \frac{1}{\lambda^{n-2}} p^{\frac{n-2}{q}}\right) \leq \frac{1}{p} \sum_{i \neq j} \frac{1}{\left(1+\lambda^2 d_{ij}^2\right)^{\frac{n-2}{2}}} .$$

Proof. In order to prove the lemma, we can assume that V is the unit cube of $\mathbb{R}^q$. For a given λ, we split the p points in two packs:

$$\begin{cases} A_1 = \{i / \exists j \neq i \quad \text{and} \quad \lambda d_{ij} \leq 1\} \\ A_2 = {}^c A_1 . \end{cases} \tag{340}$$

Lemma 12 holds if the cardinal of A_1 is larger than $2\left(Cp2^{\frac{n-2}{2}}\right)$.

Since C can be taken arbitrarily small, we can assume that the cardinal of A_2 is of order p. Therefore, we need to prove the result when

$$A_2 = \{1, \dots, p\} .$$

(The result for the other cases will follow). Under such an assumption, we have:

$$\frac{1}{p} \sum_{i \neq j} \frac{1}{\left(1+\lambda^2 d_{ij}^2\right)^{\frac{n-2}{2}}} \geq \frac{1}{\lambda^{n-2}} \left(\frac{1}{2^{\frac{n-2}{2}}} \frac{1}{p} \sum_{i \neq j} \frac{1}{d_{ij}^{n-2}} \right) . \tag{341}$$

We want to prove that:

$$\frac{1}{p} \sum_{i \neq j} \frac{1}{d_{ij}^{n-2}} \geq C p^{\frac{n-2}{q}} . \tag{342}$$

Let then:

$$B_1 = \left\{ i / \forall j \neq i \quad d_{ij} \geq 2p^{-1/q} \right\} \quad ; B_2 = {}^c B_1 . \tag{343}$$

Using a volume argument, we derive that the cardinal of B_1 is at most $p/2$ since:

$$\operatorname{vol}(V) = 1 \geq (\text{card } B_1) \frac{2^q}{p} . \tag{344}$$

The cardinal of B_2 is thus lowerbounded by $p/2$. Therefore,

$$\sum_{i=1}^{p} \frac{1}{d_{ij}^{n-2}} \geq \sum_{\substack{i \neq j \\ (i,j) \in B_2^2}} \frac{1}{d_{ij}^{n-2}} \,. \tag{345}$$

For any i in B_2, there exists an index j such that

$$d_{ij} \leq 2p^{-1/q} \,. \tag{346}$$

We therefore have at least $[p/4]$ distinct couples such that

$$d_{ij} < 2p^{-1/q} \,. \tag{347}$$

Thus

$$\sum_{i \neq j} \frac{1}{d_{ij}^{n-2}} \geq \left(\frac{1}{2^{n-2}} p^{\frac{n-2}{q}} \right) \left[\frac{p}{4} \right] \,. \tag{348}$$

Lemma 12 follows.

Proof of Proposition 10. From Proposition 4 of section 3 and Proposition 4′ of section 5, we derive:

Assuming:

$$\alpha_i/\alpha_j \in (1 - \vartheta_0, 1 + \vartheta_0) \text{ and } I = \frac{1}{p} \sum_{i \neq j} \int \varphi_{a_i,\lambda}^{\frac{n+2}{n-2}} \varphi_{a_j,\lambda} < \beta \ , \tag{349}$$

then

$$J\left(\sum_{i=1}^{p} \alpha_i \varphi_{a_i,\lambda} \right) \geq \frac{\sum \alpha_i^2}{\left(\sum \alpha_i^{2n/n-2} \right)^{\frac{n-2}{2}}} \cdot \times S^{2/n} \times \left(1 + O\left(\frac{1}{\lambda^2} \right) - \frac{\gamma_0' I}{S} \right) \tag{350}$$

Since the map we are considering is $h_p(\lambda)$, the points a_i are equal to $g(v_i)$, $v_i \in V$ and we thus have:

$$\frac{1}{p} \sum_{i \neq j} \int \varphi_{a_i,\lambda}^{\frac{n+2}{n-2}} \varphi_{a_j,\lambda} \geq \frac{C}{p} \sum_{i \neq j} \frac{1}{\left(1 + \lambda^2 d(a_i, a_j)^2 \right)^{\frac{n-2}{2}}} \geq$$

$$\geq \frac{C}{p} \sum_{i \neq j} \frac{1}{\left(1 + \lambda^2 d(g(v_i), g(v_j))^2 \right)^{\frac{n-2}{2}}} \geq \frac{C'}{p} \sum_{i \neq j} \frac{1}{\left(1 + \lambda^2 d(v_i, v_j)^2 \right)^{\frac{n-2}{2}}} \,. \tag{351}$$

In particular, for λ large enough, we either have:

$$J\left(\sum_{i=1}^{p} \alpha_i \varphi_{g(v_i)<\lambda}\right) < (pS)^{2/n} \tag{352}$$

or

$$J\left(\sum_{i=1}^{p} \alpha_i \varphi_{g(v_i),\lambda}\right) \leq \frac{\sum \alpha_i^2}{\left(\sum \alpha_i^{2n/n-2}\right)^{\frac{n-2}{2}}} S^{2/n}\left(1 + O\left(\frac{1}{\lambda^2}\right) - C\frac{p^{\frac{n-2}{q}}}{\lambda^{n-2}}\right). \tag{353}$$

We now observe that

$$\operatorname*{Max}_{\lambda \geq 1}\left(\frac{1}{\lambda^2} - \frac{Cp^{\frac{n-2}{q}}}{\lambda^{n-2}}\right) = 0\left(\frac{1}{p^2 q^{\frac{n-2}{n-4}}}\right) = 0\left(\frac{1}{p}\right) \tag{354}$$

under the conditions of Theorem 3. ($q = 1, 2$ if $n > 4$; $q = 3$ for $n = 6, 7$). Furthermore, the function $\frac{1}{\lambda^2} = C\frac{p^{\frac{n-2}{q}}}{\lambda^{n-2}}$ vanishes for $\lambda = \lambda_p = C^{\frac{1}{n-4}} p^{\frac{n-2}{q(n-4)}}$ and is therefore negative for $1 \leq \lambda \leq \lambda_p$. We then copy the proof of Proposition 8 of section 4. It suffices then to verify that

$$\lambda_p \geq C\sqrt{p \log(1+p)} \tag{355}$$

which is equivalent to

$$\frac{n-2}{q(n-4)} > \frac{1}{2}. \tag{356}$$

Under the conditions of the theorem, this inequality holds. The proof of Proposition 10 is thereby complete.

8. The topological argument

The topological argument follows from Lemma 17 of section 9, Proposition 9 of section 6 and from the fact, which we established in sections 4 and 7, that the map $f_p(\lambda)$ is homologically zero, if p is sufficiently large.

We recall here those Lemmas and Propositions. We denote:

$$(1) \quad \begin{cases} (-\Delta + q)\, u = u^{\frac{n+2}{n-2}} \\ u > 0\,. \end{cases}$$

Throughout, we assume, arguing by contradiction, that (1) has no solution.

Let $\varepsilon > 0$ be given. We introduce the sets $V(p,\varepsilon)$ (compare with the $V'(p,\varepsilon)$ of section 9).

$$V(p,\varepsilon) = \Big\{ u \in \Sigma^+ \text{ such that } \exists (a_1,\dots,a_p) \in M^p\,,\ \exists (\lambda_1,\dots,\lambda_p)\,, \lambda_i > \frac{1}{\varepsilon},$$

$$\varepsilon_{ij} < \varepsilon \text{ and } \left| u - \sum_{i=1}^{p} \frac{1}{\sqrt{pS}} \varphi_{a_i,\lambda} \right|_{-\Delta+q} < \varepsilon \Big\}. \tag{357}$$

We then have:

Lemma 13. *(Lemma 17 of section 9). If $\varepsilon > 0$ is small enough, W_p retracts by deformation onto $W_{p-1} \cup A$, where $A \subset V(p,\varepsilon)$.*

Lemma 14. *(Proposition 9 of section 6). For any $p \in \mathbb{N}^*$, there exists $\varepsilon_0(p) > 0$ such that, for any $u \in V(p,\varepsilon)$, with $\varepsilon < \varepsilon_0(p)$, the minimization problem*

$$\operatorname*{Min}_{\alpha_i, a_i, \lambda_i} \left| u - \sum_{i=1}^{p} \alpha_i \varphi_{a_i,\lambda_i} \right|_{-\Delta+q}$$

has a unique solution up to permutation.

Lemma 15. *(Proposition 8 of section 4, 10 of section 7 and 13 of section 9). Assuming that the topological conditions of sections 4 and 7 hold, there exists $p_0(M)$ such that $f_p(\lambda)$ is homologically trivial for $p \geq p_0(M)$.*

Lemmas 13, 14 and 15 imply the following Proposition. In what follows, the homology and cohomology are taken with $\mathbb{Z}_2$–coefficients.

Proposition 11. *The homology of (W_p, W_{p-1}), $H_*(W_p, W_{p-1})$ is a module (through the cap–product) on the cohomology of M^p/σ_p, $H^*(M^p/\sigma_p)$.*

Before giving the proof of Proposition 11, we observe that the homology of (B_p, B_{p-1}), $H_*(B_p, B_{p-1})$, is also a module (also by cap–product) over the cohomology of M^p/σ_p, $H^*(M^p/\sigma_p)$. This will be explained in details later on, in this paper.

The key point in the argument is provided in the following Proposition:

Proposition 12. *The homology homomorphism $f_{p*}(\lambda)$ is $H^*(M^p/\sigma_p)$–linear.*

The existence argument follows from a contradiction between Lemma 15 and Proposition 12.

We will provide the proof for $3 \leq n \leq 5$. The generalization to the other cases is straightforward.

Proof of theorem 3.

The following diagram commutes:

$$\begin{array}{ccc}
(B_p,B_{p-1}) & \xrightarrow[f_p(\lambda)]{} & (W_p,W_{p-1}) \\
\downarrow & & \downarrow \\
(B_{p-1},B_{p-2}) & \xrightarrow[f_{p-1}(\lambda)]{} & (W_{p-1},W_{p-2})
\end{array} \tag{358}$$

Passing to homology, we derive that the following diagram commutes:

$$\begin{array}{ccc}
H_*(B_{p-1},B_{p-2}) & \xrightarrow{f_{p-1*}(\lambda)} & H_*(W_{p-1},W_{p-2}) \\
\downarrow & & \uparrow \\
H_*(B_p,B_{p-1}) & \xrightarrow{f_{p*}(\lambda)} & H_*(W_p,W_{p-1}) \\
\partial_1\downarrow & & \partial_2\downarrow \\
H_{*-1}(B_{p-1},B_{p-2}) & \xrightarrow{f_{p-1*}(\lambda)} & H_{*-1}(W_{p-1},W_{p-1})
\end{array} \tag{359}$$

∂_1 and ∂_2 are the connecting homomorphisms of the triads (B_p,B_{p-1},B_{p-2}) and (W_p,W_{p-1},W_{p-2}) (see Dold [4]). We will only use the lower part of the above diagram, i.e the formula: Let

$$\mathcal{M}_p = V_p \underset{\sigma_p}{\times} \Delta^{1-\varepsilon}_{p-1} \tag{361}$$

where $V_p = \overline{M^p - T_p}$ and Δ^{ϑ}_{p-1} have been defined in section 4.

$\mathcal{M}_p$ is a manifold, which can be viewed as a subset of B_p. The pair $(B_p, {}^c\mathcal{M}_p)$ retracts by deformation onto (B_p, B_{p-1}). We thus have:

$$H_*(B_p,B_{p-1}) = H_*(B_p, {}^c\mathcal{M}_p)\ . \tag{362}$$

Thus, using excision

$$H_*(B_p,B_{p-1}) = H_*(\mathcal{M}_p, \partial\mathcal{M}_p)\ . \tag{363}$$

Since any manifold is orientable (modulo its boundary), with $\mathbb{Z}_2$–coefficients, we have a non zero orientation class in $H_*(B_p,B_{p-1})$, which we denote:

$$\omega_p \in H_{np+p-1}(B_p,B_{p-1})\ . \tag{364}$$

The following lemma relates ω_p and ω_{p-1}, via the cap–product denoted $\cap$, by an element of $H^*(M^p/\sigma_p)$.

Lemma 16. *There exists* $o_p^* \in H^n(M^p/\sigma_p)$ *such that:*

$$(*) \qquad \partial_1\left(o_p^* \cap \omega_p\right) = \omega_{p-1}$$

The existence argument proceeds then as follows. We first observe that:

$$f_{1*}(\lambda)(\omega_1) \neq 0 \qquad \text{(see Proposition 14 of section 9).} \tag{365}$$

We then start an induction and we assume that:

$$f_{p-1*}(\lambda)(\omega_p) \neq 0\,. \tag{366}$$

We then derive:

$$f_{p-1*}(\lambda)\left(\partial_1\left(o_p^* \cap \omega_p\right)\right) = f_{p-1*}(\lambda)(\omega_{p-1}) \neq 0\,. \tag{367}$$

By (359), we have:

$$f_{p-1*}(\lambda)\,\partial_1 = \partial_2 f_{p*}(\lambda)\ . \tag{368}$$

Therefore,

$$\partial_2 f_{p*}(\lambda)\left(o_p^* \cap \omega_p\right) \neq 0\,. \tag{369}$$

$f_{p*}(\lambda)$ is $H^*(M^p/\sigma_p)$–linear, by Proposition 12. Thus,

$$f_{p*}(\lambda)\left(o_p^* \cap \omega_p\right) = o_p^* \cap f_{p*}(\lambda)(\omega_p)\ . \tag{370}$$

This implies that:

$$o_p^* \cap f_{p*}(\lambda)(\omega_p) \neq 0\,. \tag{371}$$

Thus

$$f_{p*}(\lambda)(\omega_p) \neq 0\,. \tag{372}$$

Since this should hold for any p, it contradicts Lemma 15.

The existence of solution for (1) follows.

Proof of Proposition 11. Lemma 13 implies that

$$H_*(W_p, W_{p-1}) = H_*(W_{p-1} \cup A, W_{p-1}) \ . \tag{373}$$

Since $A \subset V(p,\varepsilon)$, which is an open set, we have, by excision (see Dold [4]):

$$H_*(W_p, W_{p-1}) = H_*((W_{p-1} \cup A) \cap V(p,\varepsilon), W_{p-1} \cap V(p,\varepsilon)) \ . \tag{374}$$

Given any pair of spaces (X,Y) $(Y \subset X)$, the cohomology of X acts by cap-product on the homology of (X,Y) and this action is linear. Therefore, the homology of (W_p, W_{p-1}), $H_*(W_p, W_{p-1})$, is a module over the cohomology of $(W_{p-1} \cup A) \cap V(p,\varepsilon)$.

Using the results of section 6, we can define a continuous map s_p, as follows:

$$s_p \ : \ V(p,\varepsilon) \longrightarrow (M^p/\sigma_p) \ . \tag{375}$$

To any $u \in V(p,\varepsilon)$, we assign the p-uple $(a_1, \ldots, a_p)$, which solves the minimization problem of Lemma 14. We thus have:

$$H^*(M^p/\sigma_p) \overset{j_p^* o s_p^*}{\longrightarrow} H^*((W_{p-1} \cup A) \cap V(p,\varepsilon)) \tag{376}$$

where j_p is the injection of $(W_{p-1} \cup A) \cap V(p,\varepsilon)$ into $V(p,\varepsilon)$. Since s_p^* is linear, $H_*(W_p, W_{p-1})$ is a module over the cohomology of (M^p/σ_p). q.e.d.

Proof of Proposition 12. Observe that $f_p(\lambda)$ maps, for λ large enough and ϑ small enough, the set

$$\mathcal{M}_p(\vartheta) = V_p \underset{\sigma_p}{\times} \Delta_{p-1}^{\vartheta} \quad \text{into } V(p,\varepsilon) \ .$$

We already pointed out that $H_*(B_p, B_{p-1})$ was equal to $H_*(\mathcal{M}_p(\vartheta), \partial\mathcal{M}_p(\vartheta))$. $\mathcal{M}_p(\vartheta)$ retracts by deformation onto

$$V_p \underset{\sigma_p}{\times} \{1/p, \ldots, 1/p\} \tag{377}$$

which can be mapped, in the obvious way, into (M^p/σ_p).

The homology of $(\mathcal{M}_p(\vartheta), \partial\mathcal{M}_p(\vartheta))$ is therefore, via cap-product, a module over the cohomology of (M^p/σ_p). Therefore, the homology of (B_p, B_{p-1}) is also a module over the same cohomology.

Considering the diagram:

$$\begin{array}{ccc} \mathcal{M}_p(\vartheta) & \xrightarrow{f_p(\lambda)} & V(p,\varepsilon) \\ \downarrow r_p & & \downarrow s_p \\ V_p/\sigma_p & \xrightarrow{i_p} & M^p/\sigma_p \end{array} \tag{378}$$

where r_p is the projection map $V_p \underset{\sigma_p}{\times} \Delta^{\vartheta}_{p-1} \longrightarrow V_p/\sigma_p$, and i_p is the injection of V_p/σ_p into (M^p/σ_p), we observe that $H_*(B_p, B_{p-1})$ is a module over $H^*(M^p/\sigma_p)$ via $r_p^* o i_p^*$. $H_*(W_p, W_{p-1})$ is a module over $H^*(M^p/\sigma_p)$ via s_p^*. Since the diagram is commutative, $f_{p*}(\lambda)$ is $H^*(M^p/\sigma_p)$–linear.

Proof of Lemma 16. We first define o_p^* and then prove the formula:

$$\partial_1\left(o_p^* \cap \omega_p\right) = \omega_{p-1} \tag{379}$$

(M^p/σ_p) can be viewed as a gradient space of the space $M_\times M^{p-1}/\sigma_{p-1}$. Indeed, denoting $\sigma_1 \times \sigma_{p-1} = \{1\} \times \sigma_{p-1}$, the subgroup of σ_p consisting of elements which map 1 to 1, we have:

$$M^p/\sigma_p = M_\times M^{p-1}/\sigma_{p-1} \Big/ \left(\sigma_p/\sigma_1 \times \sigma_{p-1}\right) \tag{380}$$

Let

$$q \;:\; M_\times M^{p-1}/\sigma_{p-1} \longrightarrow M^p/\sigma_p \tag{381}$$

be the projection map. In the complement of the degenerate set for the action of $\sigma_p/\sigma_1 \times \sigma_{p-1}$ on $M_\times M^{p-1}/\sigma_{p-1}$, q defines a covering with associated group $G = \sigma_p/\sigma_1 \times \sigma_{p-1}$. We then have (see Bredon [2]) a well defined transfer map:

$$\mathrm{tr}^* \;:\; H^*\left(M_\times M^{p-1}/\sigma_{p-1}\right) \longrightarrow H^*\left(M^p/\sigma_p\right). \tag{382}$$

Let

$$\vartheta \in H^n(M; Z_2) \tag{383}$$

be the cohomology orientation class of M. ϑ can be seen, under the form $\vartheta \otimes 1$, in the cohomology of order n of $M_\times M^{p-1}/\sigma_{p-1}$. Indeed, $H^*\left(M_\times M^{p-1}/\sigma_{p-1}\right) = \oplus_{i=0}^n H^i(M) \otimes H^{n-i}\left(M^{p-1}/\sigma_{p-1}\right)$ contains

$H^n(M) \otimes H^0\left(M^{p-1}/\sigma_{p-1}\right)$, to which $\vartheta \otimes 1$ belongs. We thus have:

$$\begin{array}{ccccc} M & \longrightarrow & M_{\times} M^{p-1}/\sigma_{p-1} & \longrightarrow & M \\ m & \longrightarrow & (m,z) & \longrightarrow & m\,. \end{array} \tag{384}$$

Thus, $H^n(M)$ is a direct factor in $H^n\left(M_{\times} M^{p-1}/\sigma_{p-1}\right)$.

We then define o_p^* to be:

$$o_p^* = \mathrm{tr}^*\,(\vartheta \otimes 1)\ . \tag{385}$$

We would like to prove (*), which can be rewritten under the form:

$$\partial_1\left(\mathrm{tr}^*\,(\vartheta \otimes 1) \cap \omega_p\right) = \omega_{p-1}\ . \tag{386}$$

This identity can be traced back to a known identity about transfer maps (see A.Dold [4] p.314).

Assuming that we are given a covering of a (non necessarily compact) manifold $\mathcal{M}$ by $\tilde{\mathcal{M}}$:

$$\begin{array}{ccc} \tilde{\mathcal{M}} & \xrightarrow{\ q\ } & \mathcal{M} \\ \cup & & \cup \\ \tilde{\mathcal{N}} & \xrightarrow{\ q\ } & \mathcal{N} \end{array} \tag{387}$$

where $\tilde{\mathcal{N}}$ is a submanifold of $\tilde{\mathcal{M}}$ and $\mathcal{N}$ is a submanifold of $\mathcal{M}$, and given $y \in H^*\left(\tilde{\mathcal{M}}, \mathcal{Z}_2\right)$ and $\xi \in H_*\left(\mathcal{M}, \mathcal{N}, \mathcal{Z}_2\right)$, the following formula holds:

$$\mathrm{tr}^* y \cap \xi = q_*\left(y \cap \mathrm{tr}_* \xi\right) \tag{388}$$

where tr^* is the cohomological transfer and tr_* is the homological one.

In order to use this formula, we first relate ω_p and ω_{p-1} to other spaces, which are manifolds and we find a covering map: ω_p can be seen as the orientation class of $(\mathcal{M}_p, \partial\mathcal{M}_p)$ and ω_{p-1} as the orientation class of $(\mathcal{M}_{p-1}, \partial\mathcal{M}_{p-1})$. Both are manifolds with boundary.

$\mathcal{M}_p$ retracts by deformation onto V_p/σ_p, which injects into M^p/σ_p.

$\sigma_1 \times \sigma_{p-1}$, as well as σ_p, act on V_p and we have the following commutative diagrams:

$$\begin{array}{ccc} V_p/\sigma_1 \times \sigma_{p-1} & \xrightarrow{\ \tilde{k}\ } & M_{\times} M^{p-1}/\sigma_{p-1} \\ \downarrow & & \downarrow \\ V_p/\sigma_p & \xrightarrow{\ k\ } & M^p/\sigma_p \end{array} \tag{389}$$

$$\begin{array}{ccc} V_{p_{\sigma_1 \times \sigma_{p-1}}} \overset{\times}{} \Delta_{p-1}^{1-\varepsilon} & \longleftarrow & \partial\left(V_{p_{\sigma_1 \times \sigma_{p-1}}} \overset{\times}{} \Delta_{p-1}^{1-\varepsilon}\right) \\ \Big\downarrow q_1 & & \Big\downarrow q_1 \\ \mathcal{M}_p & \longleftarrow & \partial \mathcal{M}_p \end{array} \tag{390}$$

q_1 is a covering map.

Denoting ω_p' (it is ω_p viewed after excision) the orientation class of $(\mathcal{M}_p, \partial \mathcal{M}_p)$, the cap–product of ω_p' with an element of the cohomology of V_p/σ_p (which is a retract by deformation of $\mathcal{M}_p$) is well defined. We pick up this element to be:

$$k^* \left(\mathrm{tr}^* \left(\vartheta \otimes 1\right)\right) \tag{391}$$

which can be identified, using the naturality of the transfer map (see Bredon [2]) with:

$$\mathrm{tr}^* \; o \; \tilde{k}^* \left(\vartheta \otimes 1\right) . \tag{392}$$

We then introduce the element:

$$\mathrm{tr}^* \; o \; \tilde{k}^* \left(\vartheta \otimes 1\right) \cap \omega_p' . \tag{393}$$

We can also define:

$$\tilde{k}^* \left(\vartheta \otimes 1\right) \cap \mathrm{tr}_* \left(\omega_p'\right) . \tag{394}$$

Indeed, $\tilde{k}^* (\vartheta \otimes 1)$ is in the cohomology of $H^* (V_p/\sigma_1 \times \sigma_{p-1})$ and $V_{p_{\sigma_1 \times \sigma_{p-1}}} \overset{\times}{} \Delta_{p-1}^{1-\varepsilon}$ retracts by deformation onto $V_p/\sigma_1 \times \sigma_{p-1}$.

Using (388), we derive:

$$\begin{aligned} q_{1*} &= \left(\tilde{k}^* \left(\vartheta \otimes 1\right) \cap \mathrm{tr}_* \left(\omega_p'\right)\right) = \\ &= \mathrm{tr}^* \; o \; \tilde{k}^* \left(\vartheta \otimes 1\right) \cap \omega_p' . \end{aligned} \tag{395}$$

This, once we rewrite it using the spaces (B_p, B_{p-1}) and (B_{p-1}, B_{p-2}) and once we apply the connecting homomorphism ∂_1, yields our result: the homomorphism tr_*, since it is related to the action of $\sigma_1 \times \sigma_{p-1}$, which leaves the first factor invariant, maps ω_p' onto the tensor product of the orientation class of M with a suspension of order 1 of the orientation class of $\left(M^{p-1}\right)^* \underset{\sigma_p}{\times} \Delta_{p-2}^{1-\varepsilon}$ or $(\mathcal{M}_{p-1}, \partial \mathcal{M}_{p-1})$. Taking the cap–product with

$\tilde{h}^* (\vartheta \otimes 1)$, we cancel the orientation class of M. Applying the connecting homomorphism, we cancel the suspension and find $(*)$.

The details of the above argument can also be found in [7] and [10]. q.e.d.

9. Deformation Lemmas and non triviality of $f_1(\lambda)_*$

All the results that are described in the paper work for a more general class of equation than (1). This class include equations of the type

$$(-\Delta_g + q)\, u = u^{\frac{n+2}{n-2}} + h(x,u) \tag{396}$$

where $h(x,s) = o(s)$ when s tends to zero and $h(x,s) = o\left(s^{\frac{n+2}{n-2}}\right)$ when s tends to $+\infty$.

For the sake of simplicity, we only discuss the case where $h = 0$. We keep all the notations of the previous section. Let

$$J_0(u) = \frac{\int \left(|\nabla_g u|^2 + \frac{n-2}{4(n-1)} R_g u^2\right) dv}{\|u\|^2_{2n/n-2}} \tag{397}$$

where u belongs to $H^1(M) - \{o\}$ and R_g is the scalar–curvature of (M^n, g). Let

$$F(u) = \frac{1}{2}\left(\int |\nabla_g u|^2 + qu^2\right) dv - \frac{n-2}{2n}\int u^{+2n/n-2}\,. \tag{398}$$

Let

$$\hat{F}(u) = \operatorname*{Max}_{\lambda \geq 0} F(\lambda u) \qquad , \; u \in H^1(M) - \{o\}\,. \tag{399}$$

We have:

$$\hat{F}(u) = \frac{1}{n} J(u)^{n/2} \tag{400}$$

where

$$J(u) = \frac{\int |\nabla_g u|^2 + qu^2}{\|u^+\|^2_{2n/n-2}}\,. \tag{401}$$

We denote

$$Z_p = \left\{u \in H^1(M) \;\text{ s.t }\; F(u) < b'_p\right\}, \tag{402}$$

where

$$b'_p = \frac{1}{n}(p+1)\,S\,. \tag{403}$$

We denote, for $p > 0$

$$B'_p(M) = \left\{ \sum_{i=1}^{p} \alpha_i \delta_{a_i}, \alpha_i \geq 0, \sum_{i=1}^{p} \alpha_i \neq 0\ ,\ a_i \in M \right\}. \tag{404}$$

δ_{a_i} is the Dirac mass at a_i.

$$B'_0(M) = \left\{ \alpha\delta_a; a \in M; \alpha \in \left[\frac{1}{2}, \frac{3}{4}\right] \cup \left[\frac{5}{4}, \frac{3}{2}\right] \right\}. \tag{405}$$

We normalize, in this section, $\delta(a,\lambda)(x)$ on $\mathbb{R}^n$ so that:

$$-\Delta\delta = \delta^{\frac{n+2}{n-2}}\,. \tag{406}$$

The definition of the function $\varphi_{a,\lambda}$ is kept unchanged.

Let, for $\varepsilon > 0$ given:

$$V'(p,\varepsilon) = \left\{ u \in H^1(M) \text{ s.t } \left| u - \sum_{i=1}^{p} \varphi_{a_i,\lambda_i} \right|_{H^1} < \varepsilon; \lambda_i \geq \frac{1}{\varepsilon}; \frac{1}{\varepsilon_{ij}} \geq \frac{1}{\varepsilon} \right\} \tag{407}$$

where

$$\varepsilon_{ij} = \left(\frac{\lambda_i}{\lambda_j} + \frac{\lambda_j}{\lambda_i} + \lambda_i\lambda_j d\,(a_i,a_j)^2 \right)^{-\frac{n-2}{2}}. \tag{408}$$

Assuming that F has no critical point but zero, then the functional F satisfies the Palais–Smale condition at any level different from b'_p, $p \in \mathbb{N}$ (see [7],[8],[9] and [10]. The framework is not identical to the present one, since the manifolds involved in [7]-[10] are not general compact Riemannian manifolds. However, the extension is straightforward).

On another hand, if a sequence (u_k) satisfies

$$F\,(u_k) \to b'_p \qquad F'\,(u_k) \to \infty \tag{409}$$

then, given $\varepsilon > 0$, the sequence lives in $V'(p,\varepsilon)$ for p large enough (same extension). We then have the following deformation result:

Lemma 17. *For any $\varepsilon > 0$ and any $p \geq 1$, Z_p retracts by deformation onto $Z_{p-1} \cup A$, where A is contained in $V'(p,\varepsilon)$ and contains some $V'(p,\delta)$.*

Before proving Lemma 17, we introduce the map $h_p(\lambda)$:

$$\begin{array}{ccc} B'_p(M) & \xrightarrow{h_p(\lambda)} & H^1(M) - \{0\} \\ \sum_{i=1}^p \alpha_i \delta_{a_i} & \xrightarrow{h_p(\lambda)} & \sum_{i=1}^p \alpha_i \varphi_{a_i,\lambda} \,. \end{array} \tag{410}$$

Proof of Lemma 17. Given ϑ and $\bar{\varepsilon} > 0$, we pick up a function $\mu : [0, +\infty[\to \mathbb{R}$, which behaves as follows: $\vartheta\bar{\varepsilon}$ will be taken to be small. We define

$$G(u) = F(u) - \mu\left(|F'(u)|^2\right) \tag{411}$$

$\bar{\varepsilon}$ as well as $\vartheta\bar{\varepsilon}$ can be chosen small enough so that

$$G'(u) \cdot F'(u) \geq 0 \qquad \forall u \text{ such that } F(u) \leq b'_p \,. \tag{412}$$

Z_p can then be deformed, using the flow of $-F'(u)$ onto:

$$G_{p-1} = \left\{u/G(u) \leq b'_{p-1}\right\} \,. \tag{413}$$

Clearly

$$G_{p-1} = Z_{p-1} \cup A_1 \tag{414}$$

where $A_1 = \left\{u \in G_{p-1}/F(u) > b'_{p-1}\right\}$.

If $u \in A_1$, the $\mu\left(|F'(u)|^2\right)$ is positive. Thus,

$$|F'(u)| < \vartheta\bar{\varepsilon} \tag{415}$$

and therefore, if ϑ is small enough, u belongs to $V'(p, \varepsilon)$, thus A_1 is contained in $V'(p, \varepsilon)$ as claimed.

When δ tends to zero, $\mathrm{Sup}_{u \in V'(p,\delta)} |F'(u)|^2$ tends to zero. Therefore,

$$\overline{\lim_{\substack{\delta \to 0 \\ u \in V'(p,\delta)}}} \ \mathrm{Sup}\, G(u) = b_{p-1} - \bar{\varepsilon}, b_{p-1} \,. \tag{416}$$

Thus

$$A = V'(p, \delta) \cup A_1 \ \subset \ G_{p-1} \tag{417}$$

for δ small enough, as claimed.

We now define:

$$f_p(\lambda)\ :\ B_p(M) \longrightarrow \Sigma^+ = \{u \in H^1(M) - \{0\}, u \geq 0, |u|_{-\Delta_g + q} = 1\}$$
$$\sum_{i=1}^p \alpha_i \delta_{a_i} \longrightarrow \frac{\sum_{i=1}^p \alpha_i \varphi_{a_i,\lambda}}{\|\sum_{i=1}^p \alpha_i \varphi_{a_i,\lambda}\|_{-\Delta_g+q}}\,. \tag{418}$$

We then have the following proposition:

Proposition 13.

1) $h_p(\lambda)\left(B'_p(M)\right) \subset Z_p$ *if and only if* $f_p(\lambda)\left(B_p(M)\right) \subset W_p$
2) *There exists* $\lambda_p \geq 1$ *such that, for* $\lambda \geq \lambda_p$,
$h_p(\lambda)\left(B'_p(M) B'_{p-1}(M)\right) \subset (Z_p, Z_{p-1})$
3) *If* $\lambda \geq \lambda_p$,
$h_p(\lambda)_* : H_*\left(B'_p(M) B'_{p-1}(M)\right) \longrightarrow H_*\left(Z_p Z_{p-1}\right)$ *is zero if*
$f_p(\lambda)_* : H_*\left(B_p(M) B_{p-1}(M)\right) \longrightarrow H_*\left(W_p W_{p-1}\right)$ *is zero.*

Remark. Since we took $h(x,s) = 0$, λ_p is given by section 3 and is equal to $C\omega_n(p)$.

Proof of Proposition 13.

1) and 2) are straightforward.

Proof of 3). Let $\mu > 0$ be given. Let

$$\begin{array}{rccc} r\ : & \left(B'_p(M), B'_{p-1}(M)\right) & \longrightarrow & \left(B_p(M), B_{p-1}(M)\right) \\ & \sum_{i=1}^p \alpha_i \delta_{a_i} & \longrightarrow & \sum_{i=1}^p \frac{\alpha_i}{\sum_{j=1}^p \alpha_j} \delta_{a_i}\,. \end{array} \tag{419}$$

Let

$$\begin{array}{rccc} \gamma_\mu\ : & (W_p, W_{p-1}) & \longrightarrow & (Z_p, Z_{p-1}) \\ & u & \longrightarrow & \mu u\,. \end{array} \tag{420}$$

Let

$$\begin{aligned} &k_p(\lambda) = \gamma_\mu \circ f_p(\lambda) \circ r \\ &k_p(\lambda) : \left(B'_p(M), B'_{p-1}(M)\right) \longrightarrow \left(Z_p, Z_{p-1}\right)\,. \end{aligned} \tag{421}$$

$k_p(\lambda)$ and $h_p(\lambda)$ are homotopic maps. The homotopy is provided by:

$$U_t\left(\sum_{i=1}^p \alpha_i \delta_{a_i}\right) = \left(\frac{(1-t)\,\mu}{\|\sum_{i=1}^p \alpha_i \varphi_{a_i,\lambda}\|_{-\Delta_g+q}} + t\right) \sum_{i=1}^p \alpha_i \varphi_{a_i,\lambda}\,. \tag{422}$$

Thus, $h_p(\lambda)_* = \gamma_{\mu *} \circ f_p(\lambda)_* \circ r_*$

The proof of Proposition 13 is thereby complete.

The results of section 6 imply readily the following lemma:

Lemma 18. *For any $p \geq 1$, there exists $\varepsilon_0(p)$ such that, for any $0 < \varepsilon < \varepsilon_0(p)$, the minimization problem:*

$$\underset{\substack{\alpha_i \geq 0 \\ a_i \in M \\ \lambda_i \geq 0}}{\operatorname{Min}} \left| u - \sum_{i=1}^{p} \alpha_i \varphi_{a_i, \lambda_i} \right|_{-\Delta_g + q}$$

has a unique solution, up to permutation, if $u \in V'(p, \varepsilon)$.

We now have:

Proposition 14. *There exists $\lambda_1 \geq 1$ such that, for any $\lambda \geq \lambda_1$, $h_1(\lambda)$ maps $(B_1'(M), B_0'(M))$ into (Z_1, Z_0).*

Furthermore, $h_1(\lambda)_*$ is injective in homology, if F has no other critical point than zero.

Proof of Proposition 14. Let (μ_k) be a sequence of positive numbers; Let (a_k) be a sequence of M; Let (λ_k) be a sequence tending to $+\infty$. Then, $\hat{F}(\mu_k \varphi_{a_k, \lambda_k})$ tends to S if and only if μ_k tends to 1. In such a case, $J(\varphi_{a_k, \lambda_k})$ tends to S and $(J - J_0)(\varphi_{a_k, \lambda_k})$ tends to zero. Therefore, for $\varepsilon > 0$ small enough, the set $J_{S+\varepsilon} = \left\{u; |u|_{-\Delta g+q} = 1; J(u) < S^{2/n} + \varepsilon\right\}$ retracts on M.

Indeed, let

$$\begin{aligned} r_\lambda \;:\; M &\longrightarrow \textstyle\sum \\ a &\longrightarrow \frac{\varphi_{a,\lambda}}{\|\varphi_{a,\lambda}\|_{-\Delta g+q}} \,. \end{aligned} \tag{423}$$

Let $\varepsilon_0 > 0$ be given. Since $\lim_{\lambda \to +\infty} J_0(\varphi_{a,\lambda})$ is equal to S, the map r_λ is valued in $J_{0\ S+\varepsilon_0}$ for λ large enough. On another hand, given $\varepsilon_1 > 0$, on can choose $\varepsilon_0 > 0$ small enough so that:

$$J_{0,S+\varepsilon_0} \subset V(1, \varepsilon_1)\,. \tag{424}$$

The results of section 6 imply then that the minimization problem:

$$\underset{\alpha_j, \lambda, a, \lambda}{\operatorname{Min}} |u - \alpha \varphi_{a,\lambda}|_{-\Delta g+q}$$

has a unique solution $\left(\overline{\alpha}(u), \overline{a}(u), \overline{\lambda}(u)\right)$, up to permutation. Clearly,

$$\overline{a}(r_\lambda(a)) = a\,. \tag{425}$$

Thus, $J_{0,S+\varepsilon_0}$ retracts on M, for ε_0 small enough.

Observe now that, given $\varepsilon_2 > 0$, there exists $\varepsilon_0 > 0$ such that

$$J_{0,S+\varepsilon_0} \subset J_{S+\varepsilon_2} \,. \tag{426}$$

Also, for any $\varepsilon_3 > 0$, there exists $\varepsilon_2 > 0$ such that

$$J_{S+\varepsilon_2} \subset J_{0,S+\varepsilon_3} \,. \tag{427}$$

We then have the following sequence of maps:

$$\underbrace{M \xrightarrow{r_\lambda} J_{0,S+\varepsilon_0} \hookrightarrow J_{S+\varepsilon_2} \hookrightarrow J_{0,S+\varepsilon_3} \xrightarrow{\bar{a}} M}_{Id_M} \tag{428}$$

which readily implies that $J_{S+\varepsilon_2}$ retracts on M.

Let now A be given, by Lemma 17, such that:

$$V'(p,\varepsilon_5) \subset A \subset V'(p,\varepsilon_4) \tag{429}$$

and

$$(Z_1, Z_0) \text{ retracts by deformation onto } (Z_0 \cup A, Z_0) \,. \tag{430}$$

Let ε_6 and $\alpha_{j,\lambda} > 0$ be given, small enough so that, for any $u \in J_{S+\varepsilon_6}$, for any $\mu \in [1-\alpha, 1+\alpha]$, μu is in $V'(p,\varepsilon_5)$.

Taking ε_6 smaller if necessary, we can assume that:

$$F(\mu u), S \text{ if } \mu \notin (1-\alpha, 1+\alpha), \forall u \in J_{S+\varepsilon_6} \,. \tag{431}$$

We thus have a well defined map k:

$$\begin{gathered}(J_{S+\varepsilon_6} \times [1-\alpha, 1+\alpha], J_{S+\varepsilon_6} \times \{1-\alpha, 1+\alpha\}) \longrightarrow (Z_1, Z_0) \\ (u,\mu) \xrightarrow{k} \mu u \,.\end{gathered} \tag{432}$$

k maps $J_{S+\varepsilon_6} \times [1-\alpha_{j,\lambda}, 1+\alpha_{j,\lambda}]$ into $V'(p,\varepsilon_5) \subset A$.

The set $F_S = \{u \in H^1(M) \text{ s.t } F(u) \leq S\}$ is not connected. Indeed, for any non zero u, $F(\lambda u)$ is strictly less than S if $\lambda > 0$ is small or $\lambda > 0$ is large. These two types of elements of F_S cannot be connected unless, for some u_0, $J(u_0)$ is less than or equal to S.

A mountain–pass argument (observe that the Palais–Smale condition is satisfied below the level S, <u>S included</u>) shows, then, that F has a strictly positive critical value, contradicting our assumption. Therefore, there exists a continuous function

$$\sigma : F_S \longrightarrow \{0,1\} \tag{433}$$

which is zero on λu, for $\lambda > 0$ small enough and is 1 on λu for $\lambda > 0$ large.

ε_2 being given, one can define, for ε_4 small enough:

$$\tau : (A, A \cap Z_0) \longrightarrow (J_{S+\varepsilon_2} \times [1/2, 3/2], J_{S+\varepsilon_2} \times [1/2, 1) \cup (1, 3/2]) \quad (434)$$

using the formula:

$$\tau(u) = \left(\frac{u}{\|u\|_{-\Delta_g+q}}, \vartheta(u) \right) , \quad (435)$$

where $\vartheta(u) : A \longrightarrow [1/2, 3/2]$ is a continuous extension of

$$\begin{cases} \vartheta_1(u) \ : \ A \cap Z_0 \longrightarrow [1/2, 1) \cup (1, 3/2) \\ \vartheta_1(u) \ = \ \begin{cases} F(u)/J_0(u) & \text{if } \sigma(u) = 0 \\ J_0(u)/F(u) & \text{if } \sigma(u) = 1 \end{cases} \end{cases} \quad (436)$$

(observe that, since σ has been defined on F_S, it makes sense on $A \cap Z_0$).

Clearly, since A is contained in $V'(p, \varepsilon_4)$, $\frac{u}{\|u\|_{-\Delta_g+q}}$ belongs to $J_{S+\varepsilon_2}$, if u is in A. On the other hand, $F(u)$ is strictly less than S if u belongs to Z_0, whil $J_0(u)$ is larger than or equal to S. τ maps, therefore, $A \cap Z_0$ into $J_{S+\varepsilon_2} \times [1/2, 1) \cup (1, 3/2]$. Since k maps $J_{S+\varepsilon_6} \times [1-\alpha, 1+\alpha]$ into $V'(p, \varepsilon_5) \subset A$, $\tau_0 k$ is well defined from

$$(J_{S+\varepsilon_6} \times [1-\alpha, 1+\alpha], J_{S+\varepsilon_6} \times \{1-\alpha, 1+\alpha\})$$

into

$$(J_{S+\varepsilon_2} \times [1/2, 3/2], J_{S+\varepsilon_2} \times [1/2, 1) \cup (1, 3/2]) \ .$$

We thus have:

$$\tau_0 k\,(u, \mu) = (u, \vartheta\,(\mu u)) \ . \quad (437)$$

Since $J\,(\mu u)$ is strictly less than S, for any $u \in J_{S+\varepsilon_6}$ and $0 \le \mu \le 1 - \alpha$ or $\mu \ge 1 + \alpha$, we have:

$$\vartheta\,((1-\alpha)\,u) = \frac{F\,((1-\alpha)\,u)}{J_0(u)} < 1 \quad (438)$$

$$\vartheta((1+\alpha)u) = \frac{J_0(u)}{F((1+\alpha)u)} > 1\,. \tag{439}$$

The first component of the map $\tau_0 k$ maps therefore $J_{S+\varepsilon_6}$ into $J_{S+\varepsilon_2}$, through the natural injection. $\tau_0 k$ also maps

$$J_{S+\varepsilon_6} \times \{1-\alpha\} \qquad \text{into} \qquad J_{S+\varepsilon_2} \times [\frac{1}{2}, 1)$$

and

$$J_{S+\varepsilon_6} \times \{1+\alpha\} \qquad \text{into} \qquad J_{S+\varepsilon_2} \times (1, 3/2]\ .$$

Therefore, for ε_2 small enough, $(\tau_0 k)_*$ injects

$$(J_{S+\varepsilon_6} \times [1-\alpha, 1+\alpha], J_{S+\varepsilon_6} \times \{1-\alpha, 1+\alpha\})$$

into

$$H_*\left(J_{S+\varepsilon_2} \times [1/2, 3/2], J_{S+\varepsilon_2} \times [1/2, 1) \cup (1, \frac{3}{2}]\right)\ .$$

Furthermore, $(\tau_0 k)_*$ factorizes through k_*, which is valued in $H_*(Z_1, Z_0)$. $H_*(J_{S+\varepsilon_6} \times [1-\alpha, 1+\alpha], J_{S+\varepsilon_6} \times \{1-\alpha, 1+\alpha\})$ is therefore a direct factor in the homology of (Z_1, Z_0). Since $J_{S+\varepsilon_6}$ retracts onto M, the homology of

$$(M_\times [1-\alpha, 1+\alpha], M_\times \{1-\alpha, 1+\alpha\})$$

is a direct factor in $H_*(Z_1, Z_0)$.

Observe now that

$$H_*(M_\times [1-\alpha, 1+\alpha], M_\times \{1-\alpha, 1+\alpha\})$$

is a direct factor in the homology of (Z_1, Z_0) through the map:

$$(a, \mu) \xrightarrow{kor_\lambda} \mu\varphi_{a,\lambda} \tag{440}$$

for λ large enough. This map can be easily identified with $h_1(\lambda)$, if we forget the fact that $f_1(\lambda)$ is defined on

$$(B_1'(M), B_0'(M)) = (M_\times \mathbb{R}^+ - \{0\}, M_\times ([1/2, 3/4] \cup [5/4, 3/2]))$$

Written under this form, this set is homotopy equivalent to

$$(M_\times [1-\alpha, 1+\alpha], M_\times \{1-\alpha, 1+\alpha\})$$

Therefore, $(k \circ r_\lambda)_*$ and $h_1(\lambda)_*$ are equal, up to composition with an isomorphism. Since $(k \circ r_\lambda)_*$ is injective, so is $h_1(\lambda)_*$. Proposition 14 is thereby established.

Appendix 1

The formulae for the (reverse) stereographic projecton from $\mathbb{R}^n$ into S^n are:

$$\Pi(x) = \frac{2}{1+r^2}\begin{pmatrix} x_1 \\ \vdots \\ x_n \\ \frac{r^2-1}{2} \end{pmatrix} \qquad x \in \mathbb{R}^n, r = |x| \tag{A1}$$

where

$$|Jac\Pi| = \det(\varphi, \varphi_{x_1}, \ldots, \varphi_{x_n}) = \left(\frac{2}{1+r^2}\right)^{n+1} \begin{vmatrix} x_1 & 1 & & & 0 \\ \vdots & \vdots & \ddots & & \\ \vdots & \vdots & & \ddots & \\ \vdots & \vdots & 0 & & \ddots \\ x_n & \vdots & \cdots & \cdots & 1 \\ \frac{r^2-1}{2} & x_1 & \cdots & \cdots & x_n \end{vmatrix} = \left(\frac{2}{1+r^2}\right)^n \tag{A2}$$

Let

$$\varphi : S^n \longrightarrow \mathbb{R}. \tag{A3}$$

We define then:

$$\psi : \mathbb{R}^n \longrightarrow \mathbb{R} \quad \psi(x) = \left(\frac{2}{1+r^2}\right)^{\frac{n-2}{2}} \varphi(\Pi(x)). \tag{A4}$$

We claim that

$$\int_{\mathbb{R}^n} \left(|\nabla_{\mathbb{R}^n}\psi|^2 + a\psi^2\right) dx = \int_{S^n} \left(|\nabla_{S^n}\varphi|^2 + \left(\frac{n(n-2)}{4} + q\right)\varphi^2\right) dv_{S^n}. \tag{A5}$$

Indeed,

$$\nabla\psi = \nabla\left(\frac{2}{1+r^2}\right)^{\frac{n-2}{2}} \varphi(\Pi(x)) + \left(\frac{2}{1+r^2}\right)^{\frac{n-2}{2}} \nabla\Big(\varphi(\Pi(x))\Big) \tag{A6}$$

hence,

$$\begin{aligned}\int_{\mathbf{R}^n} |\nabla\psi|^2 = \int_{\mathbf{R}^n} \left|\nabla\left(\frac{2}{1+r^2}\right)^{\frac{n-2}{2}}\right|^2 \varphi^2(\Pi(x))\, dx + \\ + \int_{\mathbf{R}^n} \left(\frac{2}{1+r^2}\right)^{n-2} \left|\nabla\Big(\varphi(\Pi(x))\Big)\right|^2 + \\ - \frac{1}{2}\int_{\mathbf{R}^n} \varphi^2(\Pi(x))\, \Delta\left(\frac{2}{1+r^2}\right)^{n-2} .\end{aligned} \tag{A7}$$

Observe now that:

$$-\frac{1}{2}\delta\left(\frac{2}{1+r^2}\right)^{n-2} \left|\nabla\left(\frac{2}{1+r^2}\right)^{\frac{n-2}{2}}\right|^2 = \frac{n(n-2)}{4}\left(\frac{2}{1+r^2}\right)^{n} . \tag{A8}$$

On the other hand, setting

$$u = \frac{1}{(1+r^2)^{n-2}} \tag{A9}$$

we derive

$$\Delta u = \frac{n(n-2)}{4} Jac\pi . \tag{A10}$$

Thus,

$$\begin{aligned}\int_{\mathbf{R}^n} |\nabla\psi|^2 = \int_{\mathbf{R}^n} \varphi^2(\Pi(x))\, n\frac{(n-2)}{4}(Jac\pi) + \\ + \int_{\mathbf{R}^n} \left(\frac{2}{1+r^2}\right)^{n-2} \left|\nabla\Big(\varphi(\Pi(x))\Big)\right|^2 dx = \\ = \int_{S^n} \varphi^2(y)\frac{n(n-2)}{4} dy + \int_{S^n} |\nabla_{S^n}\varphi|^2\, dy .\end{aligned} \tag{A11}$$

Indeed, since Π is a conformal map, any infinitesimal length is multiplied, through $d\Pi$ by a factor of $(Jac\pi)^{1/n}$. Thus

$$|\nabla_{\mathbf{R}^n}(\varphi(\Pi(x)))|^2 = |\nabla_{S^n}\varphi|^2(\Pi(x))\,(Jac\pi)^{2/n} . \tag{A12}$$

Hence

$$\int_{\mathbb{R}^n}\left(\frac{2}{1+r^2}\right)^{n-2}\left|\nabla_{\mathbb{R}^n}\Big(\varphi(\Pi(x))\Big)\right|^2=$$

$$=\int_{\mathbb{R}^n}|\nabla_{S^n}\varphi|^2\,(\Pi(x))\,Jac\pi=\int_{S^n}|\nabla_{S^n}\varphi|^2\,(y)\,dy \tag{A13}$$

which yields the result.

Let, now, for $x\in\mathbb{R}^n$ and $b\in S^n$,

$$v(x)=\left(\frac{2}{1+|x|^2}\right)^{\frac{n-2}{2}}\delta_{b,\mu}\left(\Pi(x)\right)\ . \tag{A14}$$

Then, v satisfies:

$$-\Delta v=n(n-2)v^{\frac{n+2}{n-2}}\ . \tag{A15}$$

Thus,

$$v(x)=U_{a,\nu}(x)=\frac{\nu^{\frac{n-2}{2}}}{(1+\nu^2|x-a|^2)^{\frac{n-2}{2}}}\quad,\ \forall x \tag{A16}$$

where ν and a depend on b and μ.

ν and a are given by:

$$\nu=\frac{1+\mu^2\left(1-\cos\vartheta_0\right)}{\sqrt{1+2\mu^2}} \tag{A17}$$

$$a=\frac{\mu^2\mathrm{Proj}_{\mathbb{R}^n}b}{1+\mu^2\left(1-\cos\vartheta_0\right)} \tag{A18}$$

where $\cos\vartheta_0=be\cdot e_{n+1}=\cos\ \mathrm{dist}\ (b,N)$ and N is the north pole. Indeed, completing an expansion, where $|x|$ tends to infinity, we have:

$$U_{a,\nu}(x)=\frac{\nu^{-\frac{n-2}{2}}}{|x|^{n-2}}\left(1+(n-2)\,\frac{x\cdot a}{|x|^2}+O\left(\frac{1}{|x|^2}\right)\right)\ . \tag{A19}$$

On the other hand,

$$\left(\frac{2}{1+|x|^2}\right)^{\frac{n-2}{2}}=\frac{2^{\frac{n-2}{2}}}{|x|^{n-2}}\left(1+O\left(\frac{1}{|x|^2}\right)\right) \tag{A20}$$

$$\delta_{b,\mu}\left(\Pi(x)\right) = 2^{-\frac{n-2}{2}} \frac{\left(1+2\mu^2\right)^{\frac{n-2}{4}}}{\left(1+\mu^2\left(1-\cos\vartheta\right)\right)^{\frac{n-2}{2}}} \tag{A21}$$

where $\vartheta = \text{dist}\ (b, \Pi(x))$. Thus,

$$\cos\vartheta = b\cdot\Pi(x)\ . \tag{A22}$$

Coming back to the formula of $\Pi(x)$, we have, when $|x|$ tends to infinity.

$$\Pi(x) = e_{n+1} + \frac{2x}{|x|^2} + O\left(\frac{1}{|x|^2}\right)\ . \tag{A23}$$

Thus

$$\cos\vartheta = b\cdot e_{n+1} + 2\frac{x\cdot b}{|x|^2} + O\left(\frac{1}{|x|^2}\right)\ . \tag{A24}$$

Thus

$$\begin{aligned} v(x) = \frac{1}{|x|^{n-2}} & \frac{\left(1+2\mu^2\right)^{\frac{n-2}{4}}}{\left(1+\mu^2\left(1-\cos\vartheta_0\right)\right)^{\frac{n-2}{2}}}\times \\ & \times\left(1+\frac{(n-2)\,\mu^2 x\cdot b}{|x|^2\left(1+\mu^2\left(1-\cos\vartheta_0\right)\right)} + O\left(\frac{1}{|x|^2}\right)\right) \end{aligned} \tag{A25}$$

which implies the result by identifying the terms in the expansion provided by (A19) and (A25).

Given, now, two points a and b on the sphere and choosing the south pole at the middle point on the shortest geodesic from a to b, we have the following identity, where $y = \Pi(x)$.

$$\int_{S^n} \delta^p_{a,\lambda}(y)\delta_{b,\lambda}(y)dy = \int_{\mathbf{R}^n} U^p_{\hat{a},\nu}(x)U^p_{\hat{b},\nu}(x)\left(\frac{1+|x|^2}{2}\right)^{\frac{n-2}{2}(p+1)-n} dx \tag{A26}$$

where

$$-\hat{b} = \hat{a} = \frac{\lambda^2\sin\vartheta_0}{1+\lambda^2\left(1-\cos\vartheta_0\right)} \qquad \vartheta_0 = \pi - \frac{1}{2}d(a,b) \tag{A27}$$

$$\nu = \frac{1+\lambda^2\left(1-\cos\vartheta_0\right)}{\sqrt{1+2\lambda^2}}\ . \tag{A28}$$

If $p \le \frac{n+2}{n-2}$, we thus have:

$$\int_{S^n} \delta^p_{a,\lambda}(y)\delta_{b,\lambda}(y)dy \le \int_{\mathbf{R}^n} U^p_{\hat{a},\nu}(x)U^p_{\hat{b},\nu}(x)dx \tag{A29}$$

and the equality holds if $p = \frac{n+2}{n-2}$.

Appendix 2

Lemma A2. *Let $q > 2$ be given; there exists $\gamma(q) > 1$ such that for any $a_1, \ldots, a_p \geq 0$, we have:*

$$\left(\sum_{i=1}^{p} a_i\right)^q \geq \sum a_i^q + \frac{\gamma q}{2} \sum_{i \neq j} a_i^{q-1} a_j$$

Proof. We can consider the case $p = 2$. The general case follows by induction, keeping the same γ, from the fact that $q - 1 \geq 1$ and $(a+b)^{q-1} \geq a^{q-1} + b^{q-1}$ $(a, b \geq 0)$. We thus would like to prove that:

$$(a+b)^q - a^q - b^q \geq \frac{\gamma q}{2} \left(a^{q-1} b + b^{q-1} a\right) . \tag{A30}$$

The function

$$t \longrightarrow (t+1)^q - t^q - 1 = \varphi(t) \tag{A31}$$

is convex, since $q > 2$. We thus have:

$$\varphi(t) - \varphi(0) \geq \varphi'(0) t = qt \tag{A32}$$

or

$$(t+1)^q - t^q - 1 \geq qt . \tag{A33}$$

For $t > 0$, the function is strictly convex, therefore:

$$(t+1)^q - t^q - 1 > qt . \tag{A34}$$

Changing t into $1/t$, we derive:

$$(t+1)^q - t^q - 1 > qt^{q-1} . \tag{A35}$$

Adding up, we derive:

$$(t+1)^q - t^q - 1 > \frac{q}{2} \left(t + t^{q-1}\right) . \tag{A36}$$

Observe now that

$$\lim_{t\to 0}\frac{(t+1)^q-t^q-1}{t+t^{q-1}}=q>\frac{q}{2} \tag{A37}$$

and, changing t into $1/t$,

$$\lim_{t\to \infty}\frac{(t+1)^q-t^q-1}{t+t^{q-1}}=q>\frac{q}{2}\,. \tag{A38}$$

Therefore, there exists $\gamma>1$ such that

$$(t+1)^q-t^q-1\geq\frac{\gamma q}{2}\left(t+t^{q-1}\right)\quad \forall t\geq 0 \tag{A39}$$

i.e.

$$(a+b)^q-a^q-b^q\geq\frac{\gamma q}{2}\left(a^{q-1}b+ab^{q-1}\right)\,. \tag{A40}$$

The result follows.

Lemma A3. *(see also [10]).*

$$\frac{N}{D}=\frac{\int\left(\sum\alpha_i\varphi_{a_i,\lambda}^{\frac{n+2}{n-2}}\right)\left(\sum\alpha_i\varphi_{a_i,\lambda}\right)}{\|\sum\alpha_i\varphi_{a_i,\lambda}\|_{2n/n-2}^2}\leq\left(\sum_i\int\frac{\alpha_i\varphi_{a_i,\lambda}}{\sum\alpha_k\varphi_{a_k,\lambda}}\varphi_{a_i,\lambda}^{2n/n-2}\right)^{2/n}$$

Proof. The function $x\longrightarrow|x|^{n/2}$ is convex. Denoting $\varphi_i=\varphi_{a_i,\lambda}$, we thus have:

$$\left|\frac{\sum\alpha_i\varphi_i}{\sum\alpha_k\varphi_k}\varphi_i^{4/n-2}\right|^{n/2}\leq\frac{\sum\alpha_i\varphi_i}{\sum\alpha_k\varphi_k}\varphi_i^{2n/n-2} \tag{A41}$$

i.e.

$$\left(\sum\alpha_i\varphi_i^{\frac{n+2}{n-2}}\right)^{n/2}\leq\left(\sum\alpha_k\varphi_k\right)^{n/2}\left(\frac{\sum\alpha_i\varphi_i}{\sum\alpha_k\varphi_k}\varphi_i^{2n/n-2}\right)\,. \tag{A42}$$

Thus

$$N\leq\left\|\sum\alpha_i\varphi_i^{\frac{n+2}{n-2}}\right\|_{\frac{2n}{n+2}}\left\|\sum\alpha_i\varphi_i\right\|_{\frac{2n}{n+2}}\,. \tag{A43}$$

From (A42), we derive:

$$\left(\sum \alpha_i \varphi_i^{\frac{n+2}{n-2}}\right)^{\frac{2n}{n+2}} \leq \left(\sum \alpha_k \varphi_k\right)^{\frac{2n}{n+2}} \left(\frac{\sum \alpha_i \varphi_i}{\sum \alpha_k \varphi_k} \varphi_i^{2n/n-2}\right)^{\frac{4}{n+2}} . \qquad \text{(A44)}$$

Thus

$$\left\| \sum \alpha_i \varphi_i^{\frac{n+2}{n-2}} \right\|_{\frac{2n}{n+2}}^{\frac{2n}{n+2}} \leq \left\| \sum \alpha_k \varphi_k \right\|_{\frac{2n}{n-2}}^{\frac{2n}{n+2}} \left(\int \frac{\sum \alpha_i \varphi_i}{\sum \alpha_k \varphi_k} \varphi_i^{2n/n-2} \right)^{\frac{4}{n+2}} . \qquad \text{(A45)}$$

Now

$$N \leq \left\| \sum \alpha_k \varphi_k \right\|_{\frac{2n}{n-2}}^{2} \left(\int \frac{\sum \alpha_i \varphi_i}{\sum \alpha_k \varphi_k} \varphi_i^{2n/n-2} \right)^{2/n} . \qquad \text{(A46)}$$

Hence

$$\frac{N}{D} \leq \left(\int \frac{\sum \alpha_i \varphi_i}{\sum \alpha_k \varphi_k} \varphi_i^{2n/n-2} \right)^{\frac{2}{n+2}}$$

q.e.d.

Lemma A4. *For any $\vartheta_0 \in (0,1)$, there exists $C(\vartheta_0) > 0$ such that, for any $\alpha_1, \ldots, \alpha_p$ having the property that there exists a couple of indices (i_0, j_0) such that $\alpha_{i_0}/\alpha_{j_0} \notin \left(1 - \frac{\vartheta_0}{2}, 1 + \frac{\vartheta_0}{2}\right)$, then*

$$\frac{\|\alpha\|_2^2}{\|\alpha\|_{2n/n-2}^2} \leq p^{2/n} \left(1 - \frac{C(\vartheta_0)}{p}\right) \qquad \forall p \geq 2 .$$

Proof. Let $q = \frac{n}{n-2} > 1$. Let

$$u_i = \frac{p\alpha_i^2}{\sum \alpha_i^2} . \qquad \text{(A47)}$$

Then

$$\sum u_i = p \qquad \text{(A48)}$$

and

$$\frac{u_{i_0}}{u_{j_0}} \notin \left(\left(1 - \frac{\vartheta_0}{2}\right)^2, \left(1 + \frac{\vartheta_0}{2}\right)^2 \right) . \qquad \text{(A49)}$$

We would like to prove that

$$\sum u_i^q \geq \frac{p}{1 - \frac{C(\vartheta_0)}{p}} \tag{A50}$$

which can be rewritten as:

$$\sum (u_i^q - 1) \geq p \left(\frac{1}{1 - \frac{C(\vartheta_0)}{p}} - 1 \right) \quad \sim C(\vartheta_0) \,. \tag{A51}$$

We then have:

$$\sum u_i^q = \sum_{\substack{i \neq i_0 \\ i \neq j_0}} u_i^q + u_{i_0}^p + u_{j_0}^p \geq (p-2) \left(\frac{p - u_{i_0} - u_{j_0}}{p-2} \right)^q + u_{i_0}^q + u_{j_0}^q \tag{A52}$$

where we used the convexity of the function $x \longrightarrow |x|^q$ as follows:

$$\left(\frac{1}{p-2} \sum_{\substack{i \neq i_0 \\ i \neq j_0}} u_i \right)^q \left(\frac{1}{p-2} (p - u_{i_0} - u_{j_0}) \right)^q \leq \frac{1}{p-2} \sum_{\substack{i \neq i_0 \\ i \neq j_0}} u_i^q \,. \tag{A53}$$

Let

$$a = \frac{u_{i_0}}{p} \qquad b = \frac{u_{j_0}}{p} \,. \tag{A54}$$

Then

$$a + b \leq 1 \quad ; \quad a/b \notin \left(\left(1 - \frac{\vartheta_0}{2}\right)^2, \left(1 + \frac{\vartheta_0}{2}\right)^2 \right) \,. \tag{A55}$$

We would like to check that:

$$p^q \frac{(p-2)}{(p-2)^q} (1 - a - b)^q + p^q (a^q + b^q) - p \geq C(\vartheta_0) \tag{A56}$$

which is equivalent to:

$$p^q \left(\frac{(1 - a - b)^q}{(p-2)^{q-1}} + a^q + b^q \right) \geq p + C(\vartheta_0) \,. \tag{A57}$$

Setting $a + b = k \leq 1$, prescribed, we consider the function:

$$F(a) = a^q + (k - a)^q \qquad 0 \leq a \leq k \,. \tag{A58}$$

We choose $0 \leq \vartheta'_0 \leq 1$ so that

$$(\vartheta'_0, 1/\vartheta'_0) \subset \left[\left(1 - \frac{\vartheta_0}{2}\right)^2, \left(1 + \frac{\vartheta_0}{2}\right)^2 \right] . \tag{A59}$$

We can then assume that

$$a/b \notin [\vartheta'_0, 1/\vartheta'_0] . \tag{A60}$$

Thus, after replacing a by b if necessary:

$$\frac{a}{k-a} \leq \vartheta'_0 \qquad \text{i.e} \quad a \leq \frac{\vartheta'_0 k}{1+\vartheta'_0} . \tag{A61}$$

When a runs in $\left[0, \frac{\vartheta'_0 k}{1+\vartheta'_0}\right]$, $F(a)$ can be lowerbounded by:

$$F(a) = a^q + (k-a)^q \geq \left(\frac{\vartheta'_0 k}{1+\vartheta'_0}\right)^q + \frac{k^q}{(1+\vartheta'_0)^q} = k^q \frac{1+(\vartheta'_0)^q}{(1+\vartheta'_0)^q} . \tag{A62}$$

We need thus to estimate

$$p^q \left(\frac{(1-k)^q}{(p-2)^{q-1}} + k^q \frac{1+(\vartheta'_0)^q}{(1+\vartheta'_0)^q} \right)$$

when k runs in $[0,1]$.

If $p = 2$, then k is equal to 1 and we observe that, since $\vartheta'_0 < 1$,

$$2^q \frac{1+(\vartheta'_0)^q}{(1+\vartheta'_0)^q} > 2 + C(\vartheta_0) . \tag{A63}$$

If $p \geq 3$, we let

$$A^{q-1} = \frac{1+(\vartheta'_0)^q}{(1+\vartheta'_0)^q} . \tag{A64}$$

A is strictly larger than $1/2$ since ϑ'_0 is strictly less than 1.

The function

$$G(k) = p^q \left(\frac{(1-k)^q}{(p-2)^{q-1}} + k^q A^{q-1} \right) \tag{A65}$$

achieves it minimum on $[0,1]$, at $k = k_0 = \frac{1}{1+(p-2)A}$.

We compute $G(k_0)$:

$$G(k_0) = \frac{p^q A^{q-1}}{(1+(p-2)A)^{q-1}} = p\left(\frac{1}{1+(\frac{1}{A}-2)\frac{1}{p}}\right)^{q-1} \geq p\left(1+\frac{c(\vartheta_0)}{p}\right). \tag{A66}$$

The proof of Lemma A4 is thereby complete.

Appendix 3

We use again the notations of section 5, in particular for the functions $\hat{\delta}_{a,\lambda}$ and $\tilde{\delta}_{a,\lambda} = \hat{\delta}_{a,\lambda} + \frac{\eta}{\lambda^{\frac{n-2}{2}}}$, also for φ. We have:

Lemma A5. *There exists a positive constant C such that, for any $\lambda \geq 1$, for any $a \in M$, for any $y \in M$, we have:*

$$\begin{array}{llll}
\left|\frac{\partial\tilde{\delta}_{a,\lambda}}{\partial a}\right|(y) \leq C\lambda\tilde{\delta}_{a,\lambda}(y) & & \left|\frac{\partial^2\tilde{\delta}_{a,\lambda}}{\partial a^2}\right|(y) \leq C\lambda^2\tilde{\delta}_{a,\lambda}(y) \\
\left|\frac{\partial^2\tilde{\delta}_{a,\lambda}}{\partial\lambda\partial a}\right|(y) \leq C\tilde{\delta}_{a,\lambda}(y) & & \left|\frac{\partial\tilde{\delta}_{a,\lambda}}{\partial \lambda}\right|(y) \leq \frac{C}{\lambda}\tilde{\delta}_{a,\lambda}(y) \\
\left|\frac{\partial^2\tilde{\delta}_{a,\lambda}}{\partial\lambda^2}\right|(y) \leq \frac{C}{\lambda^2}\tilde{\delta}_{a,\lambda}(y)\,. & &
\end{array}$$

The estimate on the second derivative is local and can be completed in geodesic normal coordinates. We also have:

Lemma A6. *There exist C_1 and $C_2 > 0$ such that, for any $\lambda \geq 1$, for any $a \in M$, we have:*

$$C_1 \leq \frac{1}{\lambda}\left|\frac{\partial\tilde{\delta}_{a,\lambda}}{\partial a}\right|_{H^1} \leq C_2 \quad ; \quad C_1 \leq \lambda\left|\frac{\partial\tilde{\delta}_{a,\lambda}}{\partial \lambda}\right|_{H^1} \leq C_2$$

Lemma A7. *There exist $C > 0$ such that, for any $\lambda \geq 1$ and for any $a \in M$, we have:*

$$\left|\frac{\partial^2\tilde{\delta}_{a,\lambda}}{\partial a^2}\right|_{H^1} \leq C\lambda^2\,; \quad \left|\frac{\partial^2\tilde{\delta}_{a,\lambda}}{\partial a\partial\lambda}\right|_{H^1} \leq C\,; \quad \left|\frac{\partial^2\tilde{\delta}_{a,\lambda}}{\partial \lambda^2}\right|_{H^1} \leq \frac{C}{\lambda^2}\,.$$

Here again, the estimate on $\frac{\partial^2\tilde{\delta}}{\partial a^2}$ is of local nature ($\frac{\partial\tilde{\delta}}{\partial a}$ is zero outside of an order ρ–neighborhood of a. The coordinates are geodesic normal coordinates). Let

$$\varepsilon_{ij} = \left(\frac{\lambda_i}{\lambda_j} + \frac{\lambda_j}{\lambda_i} + \lambda_i\lambda_j d(a_i,a_j)^2\right)^{-\frac{n-2}{2}}. \tag{A67}$$

Lemma A8. *There exists C_1, C_2 and A positive such that, for any $\lambda_1, \lambda_2 \geq A$, for any $a_1, a_2 \in M$, we have:*

$$C_1\varepsilon_{12} \leq \int \varphi_{a_1,\lambda_1}^{\frac{n+2}{n-2}} \varphi_{a_2,\lambda_2} \leq C_2\varepsilon_{12} \,.$$

We recall that the function $\varphi_{a,\lambda}$ satisfy:

$$(-\Delta + q)\,\varphi_{a,\lambda} = n\,(n-2)\,\hat{\delta}_{a,\lambda}^{\frac{n+2}{n-2}} \,. \tag{A68}$$

We also recall that:

$$\tilde{\delta}_{a,\lambda}(y) = \chi\left(\left|\exp_a^{-1}(y)\right|_a^2\right) \frac{\lambda^{\frac{n-2}{2}}}{\left(+\lambda^2\left|\exp_a^{-1}(y)\right|_a^2\right)^{\frac{n-2}{2}}} + \frac{\eta}{\lambda^{\frac{n-2}{2}}} \tag{A69}$$

where $\quad |x|_a^2 = \sum g_{ij}(a)x_ix_j$. Clearly,

$$\left|\frac{\partial}{\partial a}\left|\exp_a^{-1}(y)\right|_a^2\right| \leq C\left|\exp_a^{-1}(y)\right|_a \tag{A70}$$

where C is independent of y and a.

If ρ is chosen small enough, we also have the reverse inequality; there exists $c > 0$ such that, for any a, for any y, such that

$$d\,(a,y) \leq \rho \tag{A71}$$

we have:

$$\left|\frac{\partial}{\partial a}\left|\exp_a^{-1}(y)\right|_a^2\right| \geq c\left|\exp_a^{-1}(y)\right|_a \,. \tag{A72}$$

Indeed, denoting δa a variation of a, we have:

$$\begin{aligned}\left(\frac{\partial}{\partial a}\left|\exp_a^{-1}(y)\right|_a^2\right)(\delta a) =& 2\left(\exp_a^{-1}(y), \frac{\partial}{\partial a}\exp_a^{-1}(y)\,(\delta a)\right) + \\ &+ |\delta a|\, O\left(\left|\exp_a^{-1}(y)\right|_a^2\right) \,.\end{aligned} \tag{A73}$$

Since $\exp_a^{-1}(y)$ is zero, we have:

$$\frac{\partial}{\partial y}\left(\exp_a^{-1}(\cdot)\right)(a) + \frac{\partial}{\partial y}\exp_a^{-1}(a) = 0 \,. \tag{A74}$$

Therefore:

$$\frac{\partial}{\partial y}\exp_a^{-1}(y) = -\frac{\partial}{\partial y}\left(\exp_a^{-1}\right)(a) + \rho O(1)\,. \tag{A75}$$

Since $\exp_a^{-1}$ is diffeomorphism, $\frac{\partial}{\partial y}\left(\exp_a^{-1}\right)(a)$ is invertible and its inverse is bounded independently of a; therefore:

$$\begin{aligned}\left|\left(\frac{\partial}{\partial a}\left|\exp_a^{-1}(y)\right|_a^2\right)(\delta a)\right| &\geq c\left|\exp_a^{-1}(y)\right|_a|\delta a| + \\ &- C_\rho\left|\exp_a^{-1}(y)\right|_a|\delta a| - |\delta a|\, O\left(\left|\exp_a^{-1}(y)\right|_a^2\right)\end{aligned} \tag{A76}$$

hence, (A72) for ρ small enough.

We denote, in the sequel,

$$\exp_a^{-1}(y) = x(a)\,. \tag{A77}$$

Proof of Lemma A5. χ' is bounded above and

$$\frac{\chi'\lambda^{\frac{n-2}{2}}}{\left(1+\lambda^2|x(a)|_a^2\right)^{\frac{n-2}{2}}} \leq \frac{C(\rho)}{\lambda^{\frac{n-2}{2}}}\,. \tag{A78}$$

Indeed, χ' is zero if $|x(a)|_a^2 \leq \rho/8$ or $|x(a)|_a^2 \geq \rho$. Computing $\frac{\partial\tilde{\delta}}{\partial a}$, we derive:

$$\begin{aligned}\left|\frac{\partial\tilde{\delta}}{\partial a}\right| \leq C&\left|\chi'\left(|x(a)|_a^2\right)\right|\left|\frac{\partial}{\partial a}\left(|x(a)|_a^2\right)\right| \times \\ &\times \frac{\lambda^{\frac{n-2}{2}}}{\left(1+\lambda^2|x(a)|_a^2\right)^{\frac{n-2}{2}}} + \frac{(n-2)\,\lambda^2\tilde{\delta}\left|\frac{\partial}{\partial a}\left(|x(a)|_a^2\right)\right|}{1+\lambda^2|x(a)|_a^2}\,.\end{aligned} \tag{A79}$$

Thus

$$\left|\frac{\partial\tilde{\delta}}{\partial a}\right| \leq C\lambda\tilde{\delta}\,. \tag{A80}$$

Similarly, using the fact that $\frac{\partial^2}{\partial a^2}|x(a)|_a^2$ is bounded independently of a and of y in M, we derive:

$$\left|\frac{\partial^2\tilde{\delta}}{\partial a^2}\right| \leq C\lambda^2\tilde{\delta}\,. \tag{A81}$$

Finally, using the fact that:

$$\left|\frac{\partial}{\partial\lambda}\frac{1}{1+\lambda^2|x(a)|_a^2}\right| = \left|\frac{2\lambda|x(a)|_a^2}{\left(1+\lambda^2|x(a)|_a^2\right)^2}\right| \le \frac{2}{\lambda}\frac{\rho^2}{1+\lambda^2|x(a)|_a^2} \tag{A82}$$

we derive that

$$\left|\frac{\partial^2\tilde{\delta}}{\partial a\partial\lambda}\right| \le C\tilde{\delta}\,. \tag{A83}$$

The other estimates of Lemma A5 are immediate.

Proof of Lemma A6.
(1) We want to prove that

$$\frac{1}{\lambda}\left|\frac{\partial\tilde{\delta}_{a,\lambda}}{\partial a}\right|_{H^1} \ge C\,. \tag{A84}$$

We have:

$$\begin{aligned}\frac{1}{\lambda}\left|\frac{\partial\tilde{\delta}_{a,\lambda}}{\partial a}\right|_{H^1} &\ge \frac{c}{\lambda}\left|\frac{\partial\tilde{\delta}_{a,\lambda}}{\partial a}\right|_{L^{2n/n-2}} \ge c\left|\frac{\lambda\hat{\delta}\left|\frac{\partial}{\partial a}\left(|x(a)|_a^2\right)\right|}{1+\lambda^2|x(a)|_a^2}\right|_{L^{2n/n-2}} + \\ &- \frac{C}{\lambda}\left(\int_{\rho/8\le|x(a)|_a\le\rho}\frac{\lambda^n}{\left(1+\lambda^2|x(a)|_a^2\right)^n}dv\right)^{\frac{n-2}{2n}} \ge \\ &\ge c|\omega|_{L^{2n/n-2}} - \frac{c}{\lambda^{n/2}}\,.\end{aligned} \tag{A85}$$

Using the fact that $\left|\frac{\partial}{\partial a}|x(a)|_a^2\right| \ge c|x(a)|_a$, we derive:

$$\begin{aligned}&|\omega|_{L^{2n/n-2}} \\ &\ge c\left(\int_{|x(a)|_a\le\rho/8}\frac{\lambda^{2n/n-2}}{\left(1+\lambda^2|x(a)|_a^2\right)^{n+\frac{2n}{n-2}}}\times\lambda^n|x(a)|_a^{2n/n-2}dv\right)^{\frac{n-2}{2n}} \ge \\ &\ge c'\int_{r=0}^{r=\frac{\lambda\rho}{8}}\frac{r^{2n/n-2}\times r^{n-1}dr}{(1+r^2)^{n+\frac{2n}{n-2}}} \ge C'' > 0\end{aligned} \tag{A86}$$

hence the lower bound of $\frac{1}{\lambda}\left|\frac{\partial}{\partial a}\tilde{\delta}_{a,\lambda}\right|_{H^1}$.

We now want to prove:

(2) $\lambda\left|\frac{\partial\tilde{\delta}_{a,\lambda}}{\partial\lambda}\right|_{H^1} \geq c_1$. We have:

$$\begin{aligned}
\lambda\left|\frac{\partial}{\partial\lambda}\tilde{\delta}_{a,\lambda}\right|_{H^1} &\geq c\lambda\left|\frac{\partial\tilde{\delta}_{a,\lambda}}{\partial\lambda}\right|_{2n/n-2} \geq \\
&\geq c\lambda\left|\frac{2\lambda\,|x(a)|_a^2\,\lambda^{\frac{n-2}{2}}}{\left(1+\lambda^2\,|x(a)|_a^2\right)^n}\chi\right|_{2n/n-2} - \frac{C\lambda}{\lambda^{n/2}} \geq \\
&\geq c\lambda\left(\int_{r=0}^{\lambda\rho/8}\frac{\lambda^{-2n/n-2}r^{4n/n-2}}{(1+r^2)^n}r^{n-1}dr\right)^{\frac{n-2}{2n}} - \frac{C}{\lambda^{\frac{n-2}{2}}} \geq \\
&\geq c' - \frac{C}{\lambda^{\frac{n-2}{2}}} \geq c''
\end{aligned} \tag{A87}$$

hence the estimate.

The upper–estimates of Lemmas A6 and A7 are derived as follows. We recall, from section 5, that:

$$-\Delta_g\hat{\delta} = n(n-2)\hat{\delta}^{\frac{n+2}{n-2}} + \sum\frac{\partial}{\partial x_j}\left(O\left(|x|_a^2\right)\frac{\partial}{\partial x_i}\hat{\delta}\right). \tag{A88}$$

We differentiate this relation with respect to a and λ. We denote

$$R = -\sum\frac{\partial}{\partial x_j}\left(O\left(|x|_a^2\right)\right)\frac{\partial}{\partial x_i}\hat{\delta}. \tag{A89}$$

We easily verify that:

$$\begin{array}{llllll}
\left|\frac{\partial R}{\partial a}\right| & \leq C\lambda\tilde{\delta} & ; & \left|\frac{\partial R}{\partial\lambda}\right| & \leq \frac{C\tilde{\delta}}{\lambda} & ; \quad \left|\frac{\partial^2 R}{\partial\lambda^2}\right| \leq \frac{C\tilde{\delta}}{\lambda^2} \\
\left|\frac{\partial^2 R}{\partial a^2}\right| & \leq C\lambda^2\tilde{\delta} & ; & \left|\frac{\partial^2 R}{\partial\lambda\partial a}\right| & \leq C\tilde{\delta}\,. &
\end{array} \tag{A90}$$

Using then the estimates of Lemma A5, we derive these upperbounds. Let us, for example, show that

$$\left|\frac{\partial\tilde{\delta}_{a,\lambda}}{\partial a}\right|_{H^1} \leq C_2\lambda\,. \tag{A91}$$

Since

$$\left|-\Delta_g\frac{\partial\hat{\delta}}{\partial a}\right| \leq C\hat{\delta}^{4/n-2}\left|\frac{\partial\hat{\delta}}{\partial a}\right| \tag{A92}$$

we derive, using Lemma A5:

$$\left|-\Delta_g \frac{\partial \hat{\delta}}{\partial a}\right| \leq C\lambda \tilde{\delta}^{\frac{n+2}{n-2}} . \tag{A93}$$

Thus

$$\left|\frac{\partial \hat{\delta}}{\partial a}\right|_{H^1} \leq C\lambda \int \tilde{\delta}^{\frac{n+2}{n-2}} \left|\frac{\partial \hat{\delta}}{\partial a}\right| + C \int \left|\frac{\partial \hat{\delta}}{\partial a}\right|^2 \leq$$
$$\leq C\lambda^2 \left(\int_M \tilde{\delta}^{2n/n-2} + \int_M \tilde{\delta}^2\right) \leq C'\lambda^2 . \tag{A94}$$

The result follows.

Proof of Lemma A8. We recall (see [10]) that:

$$\int_{\mathbf{R}^n} \delta_{a,\lambda}^{\frac{n+2}{n-2}} \delta_{b,\mu} = c\left(\frac{\lambda}{\mu} + \frac{\mu}{\lambda} + \lambda\mu\,|a-b|^2\right)^{-\frac{n-2}{2}} (1+o(1)) \tag{A95}$$

where c is an appropriate constant.

There are two cases:

1st case: $d(a_1, a_2) \geq \rho/16$.

2nd case: $d(a_1, a_2) \leq \rho/16$.

In first case, we have:

$$\int \varphi_{a_1,\lambda_1}^{\frac{n+2}{n-2}} \varphi_{a_2,\lambda_2} \leq c \int \tilde{\delta}_{a_1,\lambda_1}^{\frac{n+2}{n-2}} \tilde{\delta}_{a_2,\lambda_2} \leq \frac{c'}{\lambda_2^{\frac{n-2}{2}}} \int_{B(a_1,\rho/30)} \tilde{\delta}_{a_1,\lambda_1}^{\frac{n+2}{n-2}} +$$
$$+ \frac{c'}{\lambda_1^{\frac{n-2}{2}}} \int_{B(a_2,\rho/32)} \tilde{\delta}_{a_2,\lambda_2} + \frac{c'}{\lambda_1^{\frac{n-2}{2}}} \times \frac{1}{\lambda_2^{\frac{n-2}{2}}}$$
$$\leq \frac{c}{(\lambda_1\lambda_2)^{\frac{n-2}{2}}} \leq C\varepsilon_{12} . \tag{A96}$$

For the lowerbound, we have:

$$\int \varphi_{a_1,\lambda_1}^{\frac{n+2}{n-2}} \varphi_{a_2,\lambda_2} \geq \frac{c}{\lambda_2^{\frac{n-2}{2}}} \int_M \varphi_{a_1,\lambda_1}^{\frac{n+2}{n-2}} \geq \frac{c}{\lambda_2^{\frac{n-2}{2}}} \int_{B(a_1,\rho/8)} \tilde{\delta}_{a_1,\lambda_1}^{\frac{n+2}{n-2}} \geq$$
$$\geq \frac{c}{(\lambda_1\lambda_2)^{\frac{n-2}{2}}} \geq C'\varepsilon_{12} . \tag{A97}$$

The result follows, in this case.

In the second case, we use the exponential map at a_1, for example.

Let

$$b = \exp_{a_1}^{-1} a_2 . \tag{A98}$$

Observe that, for ρ small enough,

$$\left|\exp_{a_1}^{-1} \exp_{a_2} x\right| \sim \left|x - \exp_{a_1}^{-1} a_2\right|_{a_1} . \tag{A99}$$

Let us also observe that

$$|b|_{a_1} \sim d(a_1, a_2) . \tag{A100}$$

We have:

$$\int \tilde{\delta}_{a_1,\lambda_1}^{\frac{n+2}{n-2}} \tilde{\delta}_{a_2,\lambda_2} \leq \int_M \varphi_{a_1,\lambda_1}^{\frac{n+2}{n-2}} \varphi_{a_2,\lambda_2} \leq c_1 \int_M \tilde{\delta}_{a_1,\lambda_1}^{\frac{n+2}{n-2}} \tilde{\delta}_{a_2,\lambda_2} . \tag{A101}$$

We split:

$$\int_M \tilde{\delta}_{a_1,\lambda_1}^{\frac{n+2}{n-2}} \tilde{\delta}_{a_2,\lambda_2} = \int_{B(a_1,\rho/8)} \tilde{\delta}_{a_1,\lambda_1}^{\frac{n+2}{n-2}} \tilde{\delta}_{a_2,\lambda_2} + O\left(\frac{1}{\lambda_1^{\frac{n+2}{2}} \lambda_2^{\frac{n-2}{2}}}\right) . \tag{A102}$$

$\int_{B(a_1,\rho/8)} \tilde{\delta}_{a_1,\lambda_1}^{\frac{n+2}{n-2}} \tilde{\delta}_{a_2,\lambda_2}$ is of the same order as

$$\int_{|x|\leq\rho/8} \delta_{0,\lambda_1}^{\frac{n+2}{n-2}} \delta_{b,\lambda_2} + O\left(\frac{1}{(\lambda_1\lambda_2)^{\frac{n-2}{2}}}\right) .$$

Thus

$$\int_M \tilde{\delta}_{a_1,\lambda_1}^{\frac{n+2}{n-2}} \tilde{\delta}_{a_2,\lambda_2} \tag{A103}$$

is equivalent to $\int_{\mathbf{R}^n} \delta_{0,\lambda_1}^{\frac{n+2}{n-2}} \delta_{b,\lambda_2} - \int_{|x|\geq\rho/8} \delta_{0,\lambda_1}^{\frac{n+2}{n-2}} \delta_{b,\lambda_2} + O\left(\frac{1}{(\lambda_1\lambda_2)^{\frac{n-2}{2}}}\right)$ where

$$|b| \leq \rho/16 . \tag{A104}$$

One can easily check that

$$\int_{|x|\geq\rho/8} \delta_{0,\lambda_1}^{\frac{n+2}{n-2}} \delta_{b,\lambda_2} = O\left(\frac{1}{\lambda_1^{\frac{n+2}{2}}} \times \frac{1}{\lambda_2^{\frac{n-2}{2}}}\right) . \tag{A105}$$

We thus derive that $\int \varphi_{a_1,\lambda_1}^{\frac{n+2}{n-2}} \varphi_{a_2,\lambda_2}$ is equivalent to

$$\int_{\mathbb{R}^n} \delta_{0,\lambda_1}^{\frac{n+2}{n-2}} \delta_{b,\lambda_2} + O\left(\frac{1}{(\lambda_1\lambda_2)^{\frac{n-2}{2}}}\right) = c\varepsilon_{12} + O\left(\frac{1}{(\lambda_1\lambda_2)^{\frac{n-2}{2}}}\right).$$

If ρ is small enough and $\lambda_1\lambda_2$ are large enough, then:

$$O\left(\frac{1}{(\lambda_1\lambda_2)^{\frac{n-2}{2}}}\right) \leq \frac{c\varepsilon_{12}}{2}. \tag{A106}$$

Thus

$$c_1\varepsilon_{12} \leq \int \varphi_{a_1,\lambda_1}^{\frac{n+2}{n-2}} \varphi_{a_2,\lambda_2} \leq c_2\varepsilon_{12} \tag{A107}$$

as claimed. The result, in all cases, follows.

Lemma A9. *There exist constant C_1, C_2 and A positive such that if λ_1 and λ_2 are larger than or equal to A, we have:*

$$\left|(\varphi_{a_1,\lambda_1}, \varphi_{a_2,\lambda_2})_{-\Delta_g+q}\right| \leq C_2\varepsilon_{12} \tag{1}$$

$$C_1 \leq |\varphi_{a_1,\lambda_1}|^2_{-\Delta_g+q} \leq C_2 \tag{2}$$

$$\left|\left(\varphi_{a_1,\lambda_1}, \frac{\partial}{\partial\lambda_2}\varphi_{a_2,\lambda_2}\right)_{-\Delta_g+q}\right| \leq C_2\frac{\varepsilon_{12}}{\lambda_2} \tag{3}$$

$$\left|\left(\varphi_{a_1,\lambda_1}, \frac{\partial}{\partial a_2}\varphi_{a_2,\lambda_2}\right)_{-\Delta_g+q}\right| \leq C_2\lambda_2\varepsilon_{12} \tag{4}$$

$$C_1\lambda_1^2 \leq \left|\frac{\partial}{\partial a_1}\varphi_{a_1,\lambda_1}\right|^2_{-\Delta_g+q} \leq C_2\lambda_1^2 \tag{5}$$

$$\left|\left(\frac{\partial}{\partial\lambda_1}\varphi_{a_1,\lambda_1}, \frac{\partial}{\partial\lambda_2}\varphi_{a_2,\lambda_2}\right)_{-\Delta_g+q}\right| \leq C_2\lambda_1\lambda_2\varepsilon_{12} \tag{6}$$

$$\left|\left(\frac{\partial}{\partial a_1}\varphi_{a_1,\lambda_1}, \frac{\partial}{\partial a_2}\varphi_{a_2,\lambda_2}\right)_{-\Delta_g+q}\right| \le C_2\frac{\lambda_1}{\lambda_2}\varepsilon_{12} \tag{7}$$

$$\frac{C_1}{\lambda_1^2} \le \left|\frac{\partial}{\partial a_1}\varphi_{a_1,\lambda_1}\right|^2_{-\Delta_g+q} \le \frac{C_2}{\lambda_1^2} \tag{8}$$

$$\left|\left(\frac{\partial}{\partial \lambda_1}\varphi_{a_1,\lambda_1}, \frac{\partial}{\partial \lambda_2}\varphi_{a_2,\lambda_2}\right)_{-\Delta_g+q}\right| \le \frac{C_2}{\lambda_1\lambda_2}\varepsilon_{12} \tag{9}$$

$$\left|\frac{\partial^2}{\partial a_1^2}\varphi_{a_1,\lambda_1}\right|_{-\Delta_g+q} \le C_2\lambda_1^2 \tag{10}$$

$$\left|\frac{\partial^2}{\partial \lambda_1^2}\varphi_{a_1,\lambda_1}\right|_{-\Delta_g+q} \le \frac{C_2}{\lambda_1^2} \tag{11}$$

$$\left|\frac{\partial^2}{\partial a_1\partial \lambda_1}\varphi_{a_1,\lambda_1}\right|_{-\Delta_g+q} \le C_2\,. \tag{12}$$

Proof.

(1) reads:

$$\int \hat{\delta}_{a_1,\lambda_1}^{\frac{n+2}{n-2}}\varphi_{a_2,\lambda_2} < c_2\varepsilon_{12} \tag{A108}$$

which follows from Lemma A8.

(2) reads:

$$c_1 \le \int \hat{\delta}_{a_1,\lambda_1}^{\frac{n+2}{n-2}}\varphi_{a_1,\lambda_1} < c_2 \tag{A109}$$

$\int \hat{\delta}_{a_1,\lambda_1}^{\frac{n+2}{n-2}}\varphi_{a_1,\lambda_1}$ is upperbounded by c_2 and lowerbounded by $c\int \hat{\delta}_{a_1,\lambda_1}^{2n/n-2}$. The result follows.

(3) reads:

$$\left|\int \varphi_{a_1,\lambda_1} \frac{\partial}{\partial\lambda_2} \hat{\delta}_{a_2,\lambda_2}^{\frac{n+2}{n-2}}\right| \le c_2 \frac{\varepsilon_{12}}{\lambda_2} . \tag{A110}$$

By Lemma A5,

$$\left|\frac{\partial}{\partial\lambda_2} \hat{\delta}_{a_2,\lambda_2}\right| \le \frac{C}{\lambda_2} \tilde{\delta}_{a_2,\lambda_2} . \tag{A111}$$

Thus

$$\left|\left(\varphi_{a_1,\lambda_1}, \frac{\partial}{\partial\lambda_2}\varphi_{a_2,\lambda_2}\right)_{-\Delta_g+q}\right| \le \frac{C}{\lambda_2} \int \varphi_{a_1,\lambda_1} \varphi_{a_2,\lambda_2}^{\frac{n+2}{n-2}} \le \frac{C\varepsilon_{12}}{\lambda_2} . \tag{A112}$$

Lemma A9 follows.

(4) is derived in the same way, since, by Lemma A5,

$$\left|\frac{\partial}{\partial a_2} \hat{\delta}_{a_2,\lambda_2}\right| \le C\lambda_2 \tilde{\delta}_{a_2,\lambda_2} . \tag{A113}$$

R has been defined in (A89) and we have:

$$(-\Delta_g + q)\, H_{a,\lambda} = \lambda^{\frac{n-2}{2}} \left(R + q\hat{\delta}_{a,\lambda}\right) . \tag{A114}$$

Therefore, using the estimates on R and Lemma A5, we derive the following inequalities, which will be useful in the sequel:

$$\begin{aligned}
\left|(-\Delta_g + q)\, \frac{\partial H_{a,\lambda}}{\partial a}\right| &\le C\lambda^{\frac{n-2}{2}} \left(\lambda\tilde{\delta}_{a,\lambda}\right) \\
\left|(-\Delta_g + q)\, \frac{\partial H_{a,\lambda}}{\partial a}\right| &\le C\lambda^{\frac{n-2}{2}} \left(\frac{1}{\lambda}\tilde{\delta}_{a,\lambda}\right) \\
\left|(-\Delta_g + q)\, \frac{\partial^2 H_{a,\lambda}}{\partial a^2}\right| &\le C\lambda^{\frac{n-2}{2}} \left(\lambda^2\tilde{\delta}_{a,\lambda}\right) \\
\left|(-\Delta_g + q)\, \frac{\partial^2 H_{a,\lambda}}{\partial \lambda^2}\right| &\le C\lambda^{\frac{n-2}{2}} \frac{\tilde{\delta}_{a,\lambda}}{\lambda^2} \\
\left|(-\Delta_g + q)\, \frac{\partial^2 H_{a,\lambda}}{\partial \lambda \partial a}\right| &\le C\lambda^{\frac{n-2}{2}} \tilde{\delta}_{a,\lambda} .
\end{aligned} \tag{A115}$$

We then have the following lemma, which we will prove later.

Lemma A10. *There exists a constant $c > 0$ such that, for any a and for any $\lambda \geq 1$, for any ψ satisfying $|(-\Delta_g + q)\,\psi| \leq \tilde{\delta}_{a,\lambda}$, we have:*

$$|\psi| \leq C\tilde{\delta}_{a,\lambda}\,.$$

From Proposition 9′, Lemma A5 and Lemma A10, we derive that:

$$\frac{1}{\lambda}\left|\frac{\partial}{\partial a}\varphi_{a,\lambda}\right|(x) + \lambda\left|\frac{\partial}{\partial a}\varphi_{a,\lambda}\right|(x) + \frac{1}{\lambda^2}\left|\frac{\partial^2}{\partial a^2}\varphi_{a,\lambda}\right|(x) + \\ + \lambda^2\left|\frac{\partial^2}{\partial \lambda^2}\varphi_{a,\lambda}\right|(x) + \left|\frac{\partial^2}{\partial\lambda^2\partial a^2}\varphi_{a,\lambda}\right|(x) \leq C\tilde{\delta}_{a,\lambda}\,. \tag{A116}$$

(5) to (12) are derived from (A116), with a line of proof close to the one of (1) to (4).

The lowerbound estimates (5) and (8) are derived in the same way, which we describe for (5). Setting $a_1 = a$ and $\lambda_1 = \lambda$, we have:

$$\left|\frac{\partial}{\partial a}\varphi_{a,\lambda}\right|^2_{-\Delta_g+q} = \int \frac{\partial}{\partial a}\hat{\delta}_{a,\lambda}^{\frac{n+2}{n-2}}\frac{\partial}{\partial a}\left(\hat{\delta}_{a,\lambda} - \frac{1}{\lambda^{\frac{n-2}{2}}}H_{a,\lambda}\right) = \\ = \int \frac{\partial}{\partial a}\hat{\delta}_{a,\lambda}^{\frac{n+2}{n-2}}\frac{\partial}{\partial a}\hat{\delta}_{a,\lambda} - \int \frac{\partial}{\partial a}\varphi_{a,\lambda}\,(-\Delta_g + q)\left(\frac{1}{\lambda^{\frac{n-2}{2}}}\frac{\partial}{\partial a}H_{a,\lambda}\right)\,. \tag{A117}$$

Using then (A116), we derive:

$$\left|\frac{\partial}{\partial a}\varphi_{a,\lambda}\right|^2_{-\Delta_g+q} \geq \int \frac{\partial}{\partial a}\hat{\delta}_{a,\lambda}^{\frac{n+2}{n-2}}\frac{\partial}{\partial a}\hat{\delta}_{a,\lambda} - \lambda^2\int_M \tilde{\delta}^2_{a,\lambda} \geq \\ \geq \int \frac{\partial}{\partial a}\hat{\delta}_{a,\lambda}^{\frac{n+2}{n-2}}\frac{\partial}{\partial a}\hat{\delta}_{a,\lambda} - \begin{cases} n=3 & C\lambda \\ n=4 & C\log\lambda \\ n\geq 5 & C\,. \end{cases} \tag{A118}$$

Using a chart, we thus estimate:

$$\int_{|k-a|\leq\rho/16}\left|\frac{\partial}{\partial a}\delta_{a,\lambda}\right|^2\delta_{a,\lambda}^{4/n-2} \tag{A119}$$

which is lowerbounded by $C\lambda^2$. We thus have:

$$\left|\frac{\partial}{\partial a}\varphi_{a,\lambda}\right|^2_{-\Delta_g+q} \geq c\lambda^2 - \begin{cases} C\lambda & \text{for } n=3 \\ C\log\lambda & \text{for } n=4 \\ C & \text{for } n\geq 5 \end{cases} \tag{A120}$$

The result follows.

Proof of Lemma A10. Propositions 1 and 1′ apply to $\lambda^{\frac{n-2}{2}}\psi$, since we only used, in the proof of Proposition 1, the fact that:

$$\frac{1}{\lambda^{\frac{n-2}{2}}}\left|(-\Delta_g + q)\, H_{a,\lambda}\right| \leq C\tilde{\delta}_{a,\lambda}\,. \tag{A121}$$

Therefore,

$$\begin{aligned} |\psi(x)| &\leq \tfrac{C}{\sqrt{\lambda}} && \text{for } n = 3 \\ |\psi(x)| &\leq \tfrac{C}{\lambda}\left(1 + \left|\log \lambda\tilde{\delta}_{a,\lambda}\right|\right) && \text{for } n = 4 \\ |\psi(x)| &\leq \frac{C}{\lambda^{\frac{n-2}{2}}}\lambda^{\frac{n-4}{2}}\tilde{\delta}_{a,\lambda}^{\frac{n-4}{n-2}} = \tfrac{C}{\lambda}\tilde{\delta}_{a,\lambda}^{\frac{n-4}{n-2}} && \text{for } n \geq 5\,. \end{aligned} \tag{A122}$$

In any dimension, the above estimates imply that $|\psi(x)|$ is upperbounded by $C\tilde{\delta}_{a,\lambda}$, as claimed. The proof of Lemma A10 is thereby complete.

References

[1] P. Hartman and G. Stampacchia, On some linear elliptic differential equations, *Acta. Math.* **115** (1966), 271–310.

[2] G. Bredon, *Introduction to Compact Transformation Groups*, Academic Press, New York (1972).

[3] T. Aubin, *Nonlinear Analysis on Manifolds, Monge-Ampère Equations*, Grundlehren der Mathematischen Wissenchaft **252**, Springer-Berlin.

[4] A. Dold, *Lectures on Algebraic Topology*, Springer-Verlag, Berlin-New York (1972).

[5] R. Thom, Sous-variétés et classes d'homologie de variétés differentiables. II Résultats et applications, *C.R. Acad. Sci. Paris* **236** (1953), 453–455.

[6] M. Hirsch, *Differential Topology*, Springer-Verlag, New York (1972).

[7] A. Bahri, *Critical Points at Infinity in Some Variational Problems*, Pitman Research Notes in Mathematics **182**, Longman, London.

[8] M. Struwe, A global compactness result for elliptic boundary value problems involving non-linearities, *Math. Z.* **187** (1984), 511–519.

[9] H. Brezis and J. M. Coron, Convergence of solutions of H-systems or how to blow bubbles, *Arch. Rational Mech. Anal.* **89** (1985), 21–56.

[10] A. Bahri and J. M. Coron, On a nonlinear elliptic equation involving the critical Sobolev exponent: The effect of the topology of the domain, *Comm. Pure Appl. Math.* **41** (1988), 253–294.

[11] A. Bahri, Proof of the Yamabe conjecture, without the positive mass theorem, for locally conformally flat manifolds, in *Einstein Metrics and Yang-Mills Connections*, Proceedings of the 27th Tanigachi International Symposium, T. Mabuchi and S. Mukaï, eds.

[12] A Bahri and H. Brezis, Equations elliptiques nonlinéaires sur des variétés avec exposant de Sobolev critique, *C. R. Acad. Sci. Paris* **307** (1988), 573–576.

[13] H. Brezis and L. Nirenberg, Positive solutions of nonlinear elliptic equations involving critical Sobolev exponents, *Comm. Pure Appl. Math.* **36** (1988), 437–477.

A. Bahri
Department of Mathematics
Rutgers University
New Brunswick, NJ 08903
and
H. Brezis
Analyse Numérique
Université Pierre et Marie Curie
4, pl. Jussieu
75252 Paris Cedex 5

Department of Mathematics
Rutgers University
New Brunswick, NJ 08903

Received April 1995

Symmetric Cones*

*Josef Dorfmeister***

Introduction

Joe D'Arti was an ever great source for open problems. One of the questions he enjoyed was the following:

Let K denote an open convex cone in $\mathbb{R}^n$, with vertex at the origin, that does not contain any straight line. Then $K^* = \{y \in \mathbb{R}^n : \langle y, x\rangle > 0 \text{ for all } x \in \overline{K} \setminus \{0\}\}$ denotes the dual cone for K, where $\langle\cdot,\cdot\rangle$ is the standard inner product on $\mathbb{R}^n$ and $\overline{K}$ denotes the closure in $\mathbb{R}^n$ of K. Consider the function

$$\eta(x) = \int_{K^*} \mathrm{e}^{-\langle y,x\rangle} \mathrm{d}y.$$

This function is defined for all $x \in K$ and real analytic [2, §3]. Moreover η is "logarithmically convex", i.e.,

$$g_x(u,v) = (\mathrm{d}_x^2 \ln \eta)(u,v)$$

is, for every $x \in K$, a positive definite symmetric bilinear form on $\mathbb{R}^n$ [2, §3.5]. In other words, $(g_x)_{x\in K}$ defines a Riemannian metric on K.

Assume now that (K,g) is a Riemannian symmetric space. Then the identity component $G = I(K)^0$ of all isometries $I(K)$ of (K,g) is a reductive connected Lie group that acts transitively on K [3, Chapter IV, Lemma 5.4, and Chapter V, Theorem 1.1]. In view of the fact that "domains of positivity" [2, §7] are examples for this situation, the question is, whether G acts necessarily linearly on $K \subset \mathbb{R}^n$.

In this note we answer this question, up to a diffeomorphism, affirmatively.

* Supported by NSF Grant DMS-9205293 and the Deutsche Forschungsgemeinschaft.

** Permanent address: Department of Mathematics, University of Kansas, Lawrence, KS 66045, USA.

This paper was written during a visit at the TU-München. The author would like to thank the TU-München for its hospitality. He would also like to thank Jost Eschenburg for his interest and constructive discussions.

1. Basic Differential Geometric Facts

1.1. First we collect some general results on η. From the definition of η it follows

$$\eta(tx) = t^{-n}\eta(x) \quad \text{for all } x \in K,\ t > 0. \tag{1.1.1}$$

Hence

$$\mathrm{d}_x\eta(x) = -n\eta(x) \quad \text{for all } x \in K, \tag{1.1.2}$$

whence

$$\mathrm{d}_x^2\eta(x,u) = -(n+1)\mathrm{d}_x\eta(u) \quad \text{for all } x \in K,\ u \in \mathbb{R}^n. \tag{1.1.3}$$

In what follows, of special interest will be

$$\mathcal{M} = \{x \in K : \eta(x) = 1\}. \tag{1.1.4}$$

We will see below that $\mathcal{M}$ is an $(n-1)$-dimensional submanifold of K. For $u \in \mathrm{T}_x\mathcal{M}$, the tangent space to $\mathcal{M}$ at $x \in \mathcal{M}$, we have

$$\mathrm{d}_x\eta(u) = 0 \quad \text{for all } x \in \mathcal{M},\ u \in \mathrm{T}_x\mathcal{M}. \tag{1.1.5}$$

In particular

$$\mathrm{d}_x^2\eta(x,u) = 0 \quad \text{for all } x \in \mathcal{M},\ u \in \mathrm{T}_x\mathcal{M}. \tag{1.1.6}$$

For later use we differentiate (1.1.1) for u

$$\mathrm{d}_{tx} \ln \eta(tu) = \mathrm{d}_x \ln \eta(u), \tag{1.1.7}$$

$$\mathrm{d}_{tx}^2 \ln \eta(tu, tv) = \mathrm{d}_x^2 \ln \eta(u,v), \tag{1.1.8}$$

$$g_{tx}(u,v) = t^{-2}g_x(u,v) \quad \text{for all } x \in K,\ t > 0,\ u,v \in \mathbb{R}^n. \tag{1.1.9}$$

1.2. A straightforward computation yields

$$g_x(u,v) = \{\mathrm{d}_x^2\eta(u,v)\cdot\eta(x) - \mathrm{d}_x\eta(u)\cdot\mathrm{d}_x\eta(v)\}\cdot\eta(x)^{-2} \tag{1.2.1}$$

for all $x \in K$, $u, v \in \mathbb{R}^n$.

For the special case $u = x$ we know $\mathrm{d}_x\eta(x) = -n\eta(x)$ from (1.1.2) and $\mathrm{d}_x^2\eta(x,v) = -(n+1)\mathrm{d}_x\eta(v)$ from (1.1.3). Hence

$$g_x(x,v) = \{-(n+1)\mathrm{d}_x\eta(v)\cdot\eta(x) + n\eta(x)\cdot\mathrm{d}_x\eta(v)\}\cdot eta(x)^{-2},$$

$$g_x(x,v) = -\mathrm{d}_x\eta(v)\cdot\eta(x)^{-1} \quad \text{for } x \in K,\ v \in \mathbb{R}^n. \tag{1.2.2}$$

In view of (1.1.5) we obtain

$$g_x(x,v) = 0, \quad \text{for } x \in \mathcal{M},\ v \in \mathrm{T}_x\mathcal{M}. \tag{1.2.3}$$

As a consequence,

(1.2.4) *Let $x \in \mathcal{M}$,*
then x is perpendicular to $\mathcal{M}$ at $x \in \mathcal{M}$ relative to (K, g).

Specializing also $v = x$ in (1.2.2), we obtain, because of (1.1.2),

$$g_x(x,x) = n \quad \text{for } x \in K. \tag{1.2.5}$$

1.3. In this section we investigate the map

$$\psi : \mathbb{R}^+ \times \mathcal{M} \to K \quad (\alpha, x) \mapsto \alpha x, \tag{1.3.1}$$

where $\mathbb{R}^+ = \{x \in \mathbb{R} : x > 0\}$. It is straightforward to verify

$$\mathrm{d}_{(\alpha,x)}\psi(r,u) = rx + \alpha u \quad \text{for } \alpha > 0,\ x \in \mathcal{M},\ r \in \mathbb{R},\ u \in \mathrm{T}_x\mathcal{M}. \tag{1.3.2}$$

For the pull back $\tilde{g}$ of g to $\mathbb{R}^+ \times \mathcal{M}$ we obtain for $\alpha > 0$, $x \in \mathcal{M}$, $r, s \in \mathbb{R}$, $u, v \in \mathrm{T}_x\mathcal{M}$

$$\tilde{g}_{(\alpha,x)}((r,u),(s,v)) = g_{\alpha x}(rx + \alpha u, sx + \alpha v). \tag{1.3.3}$$

In view of (1.2.3) we have $g_{\alpha x}(rx+\alpha u, sx+\alpha v) = rs g_{\alpha x}(x,x) + \alpha^2 g_{\alpha x}(u,v)$. Using also (1.2.5)

$$\tilde{g}_{(\alpha,x)}((r,u),(s,v)) = \alpha^{-2}\cdot n\cdot r\cdot s + \alpha^2 g_{\alpha x}(u,v) \tag{1.3.4}$$

for $\alpha > 0$, $x \in \mathcal{M}$, $r, s \in \mathbb{R}$, $u, v \in T_x\mathcal{M}$. Applying (1.1.8) we finally obtain

$$\tilde{g}_{(\alpha,x)}((r,u),(s,v)) = \alpha^{-2} \cdot n \cdot r \cdot s + g_x(u,v) \tag{1.3.5}$$

for all $\alpha > 0$, $x \in \mathcal{M}$, $r, s \in \mathbb{R}$, $u, v \in T_x\mathcal{M}$.

2. The main goal of this section is to investigate the effect of the translation $x \mapsto tx$ of K.

2.1. Let T_t denote the map

$$T_t x = tx, \quad x \in K, \ t > 0. \tag{2.1.1}$$

Then the map $\tilde{T}_t = \psi^{-1} T_t \psi$ is given by

$$\tilde{T}_t(\alpha, x) = (t\alpha, x), \quad t, \alpha > 0, \ x \in \mathcal{M}. \tag{2.1.2}$$

By $\mathfrak{g}$ we denote the Lie algebra of the connected Lie group G introduced in the Introduction. Then $\mathfrak{g}$ can be realized on K by vector fields, i.e., C^∞-maps from K to $\mathbb{R}^n$. Via ψ we can also realize $\mathfrak{g}$ by vector fields on $\mathbb{R}^+ \times \mathcal{M}$. While we denote the vector fields on K by X, Y etc., we will denote the corresponding vector fields on $\mathbb{R}^+ \times \mathcal{M}$ by $\tilde{X}, \tilde{Y}$ etc.

In this section we consider the vector fields introduced from T_t and $\tilde{T}_t$ respectively

$$\tau(x) = \left.\frac{\mathrm{d}}{\mathrm{d}t}\right|_{t=1} T_t x = x. \tag{2.1.3}$$

For $\tilde{\tau}$ we obtain from (2.1.2) and (1.3.2)

$$\tilde{\tau}(\alpha, x) = (\alpha, 0). \tag{2.1.4}$$

Next we choose an arbitrary vector field $\tilde{X}$ on $\mathbb{R}^+ \times \mathcal{M}$. Then

$$[\tilde{\tau}, \tilde{X}](\alpha, x) = \left.\frac{\mathrm{d}}{\mathrm{d}t}\right|_{t=1} \mathrm{d}_{\alpha,x} \tilde{T}_t \tilde{X} \tilde{T}_t^{-1}(\alpha, x) =$$

$$\left.\frac{\mathrm{d}}{\mathrm{d}t}\right|_{t=1} (t\tilde{X}_0(t^{-1}\alpha, x), \hat{X}(t^{-1}\alpha, x)),$$

where $\tilde{X} = (\tilde{X}_0, \hat{X})$ is the natural decomposition of $\tilde{X}$ on $\mathbb{R}^+ \times \mathcal{M}$,

$$\tilde{X}_0 : \mathbb{R}^+ \times \mathcal{M} \to \mathbb{R} \quad \text{and} \quad \hat{X}(\alpha, x) \in T_x\mathcal{M}.$$

As a consequence we obtain

(2.1.5) $$[\tilde{\tau}, \tilde{X}](\alpha, x) = (\tilde{X}_0(\alpha, x) - \alpha\partial_\alpha \tilde{X}_0(\alpha, x), -\alpha\partial_\alpha \hat{X}(\alpha, x)).$$

2.2. In Section 1.3 we have computed the pull back metric $\tilde{g}$. In matrix form it reads

(2.2.1) $$\tilde{g}_{(\alpha,x)}((r,u),(s,v)) = (r,u)J_{(\alpha,x)}(s,v)^{\mathrm{T}},$$

where $J_{(\alpha,x)} = \mathrm{diag}(\alpha^{-2}n, g_x(\cdot,\cdot))$. Then a diffeomorphism $\tilde{\varphi}$ of $\mathbb{R}^+ \times \mathcal{M}$ is an isometry iff

(2.2.2) $$(\mathrm{d}_{(\alpha,x)}\tilde{\varphi})^{\mathrm{T}} J_{\tilde{\varphi}(\alpha,x)} \mathrm{d}_{(\alpha,x)}\tilde{\varphi} = J_{(\alpha,x)} \quad \text{for all } \alpha > 0,\ x \in \mathcal{M}.$$

If $\tilde{\varphi} = \tilde{\varphi}_\varepsilon$ and $\tilde{\varphi}_0 = I$, then by differentiation we obtain for the corresponding vector field $\tilde{X}$

(2.2.3) $$(\mathrm{d}_{(\alpha,x)}\tilde{X})^{\mathrm{T}} J_{(\alpha,x)} + J_{(\alpha,x)}\mathrm{d}_{(\alpha,x)}\tilde{X} + \mathrm{d}_{(\alpha,x)}J(\tilde{X}(\alpha,x)) = 0.$$

We will evaluate (2.2.3) in matrix form. For this we set

(2.2.4) $$\mathrm{d}_{(\alpha,x)}\tilde{X}\begin{pmatrix} \chi & c^{\mathrm{T}} \\ b & A \end{pmatrix} = \begin{pmatrix} \partial_\alpha \tilde{X}_0 & \partial_x \tilde{X}_0 \\ \partial_\alpha \hat{X} & \partial_x \hat{X} \end{pmatrix}$$

and obtain from (2.2.3) the set of equations

(2.2.5a) $$2n\alpha^{-2}\chi - 2\alpha^{-3}n\tilde{X}_0 = 0.$$

(2.2.5b) $$g_x \cdot b + \alpha^{-2}nc = 0.$$

(2.2.5c) $$A^{\mathrm{T}} \cdot g_x + g_x \cdot A + \partial_x g_x(A) = 0.$$

2.3. Since $\chi = \partial_\alpha \tilde{X}_0$ the equation (2.2.3) yields

$$\partial_\alpha \tilde{X}_0 = \alpha^{-1}\tilde{X}_0.$$

This implies

$$\tilde{X}_0(\alpha, x) = \alpha C(x), \quad \text{for all } \alpha > 0,\ x \in \mathcal{M}.$$

Substituting this into (2.1.5) yields

(2.3.2) $$[\tilde{\tau}, \tilde{X}](\alpha, x) = (0, -\alpha\partial\hat{X}(\alpha, x))$$

for all $\alpha > 0$, $x \in \mathcal{M}$ and all $\tilde{X} \in \mathfrak{g}$. Let $\tilde{Y} = (0, \hat{Y}(\alpha, x)) \in \mathfrak{g}$. Then in (2.2.4) we have $c = 0$. Therefore, from (2.2.5b) we obtain $b = 0$, since g_x is positive definite. But this shows $\partial_\alpha \hat{Y} = 0$.

As a consequence we get

(2.3.3) $$\text{If } \tilde{Y} = (0, \hat{Y}) \in \mathfrak{g}, \text{ then } \partial_\alpha \hat{Y} = 0.$$

As a corollary to this we obtain in view of (2.3.2)

(2.3.4) $$(\operatorname{ad} \tilde{\tau})^2 = 0.$$

2.4. Since K is a symmetric space, the Lie algebra is reductive, i.e.,

(2.4.1) $$\mathfrak{g} = \mathfrak{c} + \hat{\mathfrak{g}}$$

where $\hat{\mathfrak{g}} = [\mathfrak{g}, \mathfrak{g}]$ is semisimple and $\mathfrak{c}$ is the center of $\mathfrak{g}$. Splitting $\tau = \tau_{\mathfrak{c}} + \hat{\tau}$ we obtain from (2.3.4) that $(\operatorname{ad} \hat{\tau})^2 = 0$ holds.

Theorem. $\operatorname{ad} \tau = 0$.

Proof. It clearly suffices to show $\operatorname{ad} \hat{\tau} = 0$. Suppose that $\operatorname{ad} \hat{\tau} \neq 0$, then by the Jacobson-Morozov Theorem [1, §11.2] there exist $\hat{\sigma}, \hat{\chi} \in \hat{\mathfrak{g}}$ such that $\mathbb{R}\hat{\sigma} + \mathbb{R}\hat{\chi} + \mathbb{R}\hat{\tau} \cong \mathrm{sl}(2, \mathbb{R})$. But then $(\operatorname{ad} \hat{\tau})^2 \hat{\tau} = -2\hat{\tau} \neq 0$, a contradiction. □

As a consequence we derive from (2.3.2) that $\partial_\alpha \hat{X} = 0$ for all $\hat{X} = (\alpha C(x), \hat{X}) \in \mathfrak{g}$. But this implies $b = 0$ in (2.2.4), whence $c = 0$ by (2.2.5b). This implies $C(x) = \text{const}$. Therefore

Corollary 1. *If* $\tilde{X} \in \mathfrak{g}$, *then* $\tilde{X}(\alpha, x) = (\alpha c_0, \hat{X}(x))$.

Corollary 2. $\mathfrak{g} = \mathbb{R}\tilde{\tau} + \breve{\mathfrak{g}}$, *where* $\breve{\mathfrak{g}} = \{\tilde{X} \in \mathfrak{g} : \tilde{X} = (0, \hat{X}(x))\}$.

3. In this section we draw geometric consequences from what we have shown so far.

3.1. By Theorem 2.4 the vector field τ is in the center of $\mathfrak{g}$. By definition, for every $X \in \mathfrak{g}$ we have $\left.\frac{\mathrm{d}}{\mathrm{d}t}\right|_{t=1} t^{-1}X(tx) = 0$, $x \in K$. This implies

(3.1.1) $$\mathrm{d}_x X(x) = X(x).$$

This is equivalent to X being homogeneous of degree 1

$$X(tx) = tX(x) \quad x \in K,\ t > 0. \tag{3.1.2}$$

Let φ_ε denote the 1-parameter group associated with $x \in \mathfrak{g}$; then $\frac{\mathrm{d}}{\mathrm{d}\varepsilon} t^1 \varphi_\varepsilon(tx) = X(t^{-1}\varphi_\varepsilon(tx))$. Since $\varphi_\varepsilon(x)$ solves the same differential equation with the same initial condition, we see that also φ_ε is homogeneous of degree 1. Since G is generated by transformations of type φ_ε we obtain

$$\varphi(tx) = t\varphi(x) \quad \text{for all } t > 0,\ x \in K \text{ and all } \varphi \in G. \tag{3.1.3}$$

As a consequence

$$\mathrm{d}_x\varphi(x) = \varphi(x) \quad \text{for all } x \in K. \tag{3.1.4}$$

Further differentiation yields

$$\mathrm{d}_x^2\varphi(x, u) = 0 \quad \text{for all } x \in K,\ u \in \mathbb{R}^n. \tag{3.1.5}$$

3.2. Next we consider $\check{X} \in \check{\mathfrak{g}}$. From Corollary 2.4.2 we see that $\varphi_\varepsilon(\mathcal{M}) = \mathcal{M}$, if φ_ε is the 1-parameter group associated with $\check{X}$. Let $\check{G}$ denote the connected Lie group with the Lie algebra $\check{\mathfrak{g}}$. Then $\varphi(\mathcal{M}) = \mathcal{M}$ for all $\varphi \in \check{G}$. More precisely

Theorem. a) *$\check{G}$ is a closed subgroup of G.*
b) $\check{G} = \{\varphi \in G : \varphi\mathcal{M} = \mathcal{M}\}$.

Corollary. *$\eta(\varphi(x)) = \eta(x)$ for all $x \in K$, $\varphi \in \check{G}$.*

Proof. From the theorem above we know $\varphi(\mathcal{M}) = \mathcal{M}$ for all $\varphi \in \check{G}$. This means $\eta(\varphi(x)) = \eta(x)$ for all $x \in \mathcal{M}$. Let $x \in K$ be arbitrary; then $x = \alpha x_0$, $x_0 \in \mathcal{M}$, $\alpha > 0$. Using (3.1.3) we obtain

$$\eta(\varphi(x)) = \eta(\varphi(\alpha x_0)) = \eta(\alpha\varphi(x_0)) = \alpha^{-n}.$$

On the other hand $\eta(x) = \eta(\alpha x_0) = \alpha^{-n}$. □

3.3. From Corollary 3.2 we know

$$\eta(\varphi(x)) = \eta(x) \quad \text{for all } x \in K,\ \varphi \in \check{G}. \tag{3.3.1}$$

A differentiation yields for $\varphi \in \check{G}$.

$$\mathrm{d}_{\varphi(x)}\eta\, \mathrm{d}_x\varphi(u) = \mathrm{d}_x\eta(u) \quad \text{for all } x \in K,\ u \in \mathbb{R}^n. \tag{3.3.2}$$

Another differentiation gives

(3.3.3) $$\mathrm{d}^2_{\varphi(x)}\eta(\mathrm{d}_x\varphi(v),\mathrm{d}_x\varphi(u))+\mathrm{d}_{\varphi(x)}\eta\mathrm{d}^2_x\varphi(v,u)=\mathrm{d}^2_x\eta(v,u)$$

for all $x \in K$, $u, v \in \mathbb{R}^n$. Since φ is an isometry relative to (K, g), we know

(3.3.4) $$g_{\varphi(x)}(\mathrm{d}_x\varphi(v),\mathrm{d}_x\varphi(u))=g_x(v,u) \quad \text{for all } x \in K,\ u,v\in\mathbb{R}^n.$$

Using (1.2.1) in (3.3.4) we get

$$\{\mathrm{d}^2_{\varphi(x)}\eta(\mathrm{d}_x\varphi(u),\mathrm{d}_x\varphi(v))\cdot\eta(\varphi(x))-$$

$$\mathrm{d}_{\varphi(x)}\eta(\mathrm{d}_x\varphi(u))\mathrm{d}_{\varphi(x)}\eta(\mathrm{d}_x\varphi(v))\}\cdot\eta(\varphi(x))^{-2}=$$

$$\{\mathrm{d}^2_x\eta(u,v)\cdot\eta(x)-\mathrm{d}_x\eta(u)\cdot\mathrm{d}_x\eta(v)\}\cdot\eta(x)^{-2}.$$

In view of (3.3.1) we can cancel the factors $\eta(x)^{-2}$ and $\eta(\varphi(x))^{-2}$. Because of (3.3.2) and (3.3.3) we see now that (3.3.4) is equivalent to

(3.3.5) $$\mathrm{d}_{\varphi(x)}\eta\mathrm{d}^2_x\varphi(u,v)=0 \quad \text{for all } x\in K,\ u,v\in\mathbb{R}^n.$$

This implies

(3.3.6) $$\mathrm{d}^2_x\varphi(u,v)\in \mathrm{T}_{\varphi(x)}\mathcal{M} \quad \text{for all } x\in\mathcal{M},\ u,v\in\mathbb{R}^n.$$

Remark. We feel that (3.3.6) expesses a remarkable property of an isometry relative to a hypersurface.

4. In this section we introduce a special commutative algebra on $V = \mathbb{R}^n$ and show how it can be used to describe $\mathfrak{g}$.

4.1. We fix some point $e \in \mathcal{M}$ and set

(4.1.1.) $$\sigma(u,v)=\mathrm{d}^2_e\eta(u,v), \quad u,v\in V.$$

Following [2, §3], we also set

(4.1.2) $$\sigma(h(x),u)=-\mathrm{d}_x\eta(u), \quad u\in V,$$

(4.1.3) $$\sigma(H(x)u,v)=\mathrm{d}^2_x\eta(u,v), \quad u,v\in V.$$

It is easy to see that h is homogeneous of degree -1 and H is homogeneous of degree -2. Moreover,

$$\mathrm{d}_x h(u) = -H(x)u, \quad u \in V, \tag{4.1.4}$$

$$H(x) = H(x)^\sigma, \tag{4.1.5}$$

where the superscript "σ" denotes the forming of the adjoint transformation.

For every $x \in K$ we set

$$\mathrm{d}_x^2\eta(A(x;u)v,w) = -\frac{1}{2}\mathrm{d}_x^3\eta(u,v,w). \tag{4.1.6}$$

Then from [2, §8] we obtain

$$A(x;u) = -\frac{1}{2}H(x)^{-1}\mathrm{d}_x H(u), \quad u \in V. \tag{4.1.7}$$

We will also need the derivative of $A(x;u)$. We obtain
(4.1.8)

$$\mathrm{d}_x(A(x;u)w)(v) = 2A(x;v)A(x;u)w + \frac{1}{2}H(x)^{-1}\mathrm{d}_x^3 h(u,v,w) \quad u,v,w \in V.$$

For $x = e$ we abbreviate $A(e;u) = A(u)$. We note

$$A(e) = I \quad \text{and} \quad A(u) = A(u)^\sigma \qquad \text{for all } u \in V. \tag{4.1.9}$$

We will write as usual

$$A(u)v = uv$$

and in particular $u^2 = uu$ etc. Then (4.1.9) implies

$$\sigma(uv,w) = \sigma(u,vw) \quad \text{for } u,v,w \in V. \tag{4.1.10}$$

Moreover, the algebra structure $\mathfrak{A}$ given by $A(u)$ on V is commutative.

4.2. Since $\mathcal{M}$ is a symmetric space, there exists a Cartan involution of $\breve{\mathfrak{g}}$ such that $\breve{\mathfrak{g}} = \mathfrak{k} + \breve{\mathfrak{p}}$ and $\breve{\mathfrak{p}}$ and $\mathrm{T}_e\mathcal{M}$ are naturally isomorphic. More precisely, $\exp : \breve{\mathfrak{p}} \to \mathcal{M}$ is a diffeomorphism.

It will be convenient to introduce the abbreviation

$$\varphi_t(x) = (\exp tU)(x),$$

where we will use a superscript U where necessary. We note that the vector field on K corresponding to U is given by

$$U(x) = \frac{\mathrm{d}}{\mathrm{d}t}\Big|_0 \varphi_t(x).$$

More generally, $\frac{\mathrm{d}}{\mathrm{d}t}\varphi_t(x) = U(\varphi_t(x))$. In particular, the isomorphism $\check{\mathfrak{p}} \to \mathrm{T}_e\mathcal{M}$ is given by $U \to U(e)$. It is easy to see that these relations can be extended naturally to $\mathfrak{g} = \mathfrak{k} + \mathfrak{p}$, where $\mathfrak{p} = \check{\mathfrak{p}} + \mathbb{R}E$ and E corresponds to e, i.e., $E(x) = x$, $x \in K$.

In the rest of the paper, by U we will always denote the element of $\mathfrak{p}$ associated with $u = U(e)$. Only "x" plays a different role. For the Riemann symmetric space (K, g) we have

(4.2.1) For $U \in \mathfrak{p}$, $Y = \mathrm{d}_e(\exp tU)(Y_0)$ is parallel along $\varphi_t(e) = \exp tU \cdot e$. In particular, $\varphi_t(e)$ is a geodesic of $\mathcal{M}$.

From [2, 8.7] we know that in K, the covariant derivative of the (symmetric) Riemann metric is given by

$$(\nabla_v U)_x = \mathrm{d}_x U(v) - A(x; v)U. \tag{4.2.2}$$

As a consequence, "Y is parallel" means

$$\dot{Y} = A(\varphi_t(e); U(\varphi_t(e))Y. \tag{4.2.3}$$

Computing $\dot{Y}$ directly from (4.2.1) thus yields

$$\mathrm{d}_{\varphi_t(e)}U\mathrm{d}_e\varphi_t(Y_0) = A(\varphi_t(e); U(\varphi_t(e)))\mathrm{d}_e\varphi_t(Y_0). \tag{4.2.4}$$

Evaluating this at $t = 0$ yields

$$\mathrm{d}_e U = A(u). \tag{4.2.5}$$

Next we consider an arbitrary $U \in \mathfrak{p}$. Then $\varphi_t^U(e) = (\exp tU)(e)$ is a geodesic of K. Therefore, if $Y \in \mathfrak{g}$ is arbitrary, the isometry φ_s^Y, $s \in \mathbb{R}$, maps geodesics to geodesics. In particular, for fixed $s \in \mathbb{R}$,

$$\varphi_s^Y \varphi_t^U(e) \quad \text{is a geodesic of } K. \tag{4.2.6}$$

In view of (4.2.2) this is equivalent with

(4.2.7) $$\frac{d^2}{dt^2}\varphi_s^Y\varphi_t^U(e) = A\left(\varphi_s^Y\varphi_t^U(e); \frac{d}{dt}\varphi_s^Y\varphi_t^U(e)\right)\frac{d}{dt}\varphi_s^Y\varphi_t^U(e).$$

We compute

$$\frac{d}{dt}\varphi_s^Y\varphi_t^U(e) = d_{\varphi_t^U(e)}\varphi_s^Y U(\varphi_t^U(e))$$

and

$$\begin{aligned}\frac{d^2}{dt^2}\varphi_s^Y\varphi_t^U(e) &= d^2_{\varphi_t^U(e)}\varphi_s^Y(U(\varphi_t^U(e)), U(\varphi_t^U(e))) \\ &\quad + d_{\varphi_t^U(e)}\varphi_s^Y d_{\varphi_t^U(e)}U(U(\varphi_t^U(e))).\end{aligned}$$

For $t = 0$ this yields

$$\left.\frac{d}{dt}\right|_0 \varphi_s^Y\varphi_t^U(e) = d_e\varphi_s^Y(u)$$

and

$$\frac{d^2}{dt^2}\varphi_s^Y\varphi_t^U(e) = d_e^2\varphi_s^Y(u,u) + d_e\varphi_s^Y d_eU(u).$$

Thus (4.2.2) means

(4.2.8) $$d_e^2\varphi_s^Y(u,u) + d_e\varphi_s^Y d_eU(u) = A(\varphi_s^Y(e); d_e\varphi_s^Y(u))d_e\varphi_s^Y(u).$$

We note that for $s = 0$ this yields in particular

(4.2.9) $$d_eU(u) = A(u)u.$$

Next we differentiate for s at $s = 0$. Then we obtain using (4.1.8)

$$\begin{aligned}d_e^2Y(u,u) + d_eY(u^2) &= 2A(Y(e))A(u)u \\ &\quad + \frac{1}{2}d_e^3h(u,u,Y(e)) + 2A(d_eYu)u.\end{aligned}$$

Differentiating this for u finally yields

(4.2.10) $$d_e^2Y(u,v) + d_eY(uv) = 2A(Y(e))A(uv)$$

$$+\frac{1}{2}d_e^3h(u,v,Y(e)) + A(d_eYu)v + A(d_eYv)u.$$

4.3. We evaluate (4.2.10). First we assume $Y \in \mathfrak{k}$. Then $Y(e) = 0$ and we obtain

$$d_e^2Y(v,\cdot) = A(d_eYv) - [d_eY, A(v)]. \tag{4.3.1}$$

If we choose $Y \in \mathfrak{p}$, then (4.2.10) yields

$$d_e^2Y(u,v) = \frac{1}{2}d_e^3h(u,v,y) + y(uv) + v(yu) + u(yv). \tag{4.3.2}$$

Obviously, the right side is totally symmetric in u, v, y, therefore

$$d_e^2Y(u,v) = d_e^2U(y,v). \tag{4.3.3}$$

Moreover, reading (4.3.2) as a linear map in v, we see

$$d_e^2Y(u,\cdot) = -\frac{1}{2}d_e^2H(u,y) + A(y)A(u) + A(yu) + A(u)A(y). \tag{4.3.4}$$

In particular, since $H(x)$ is self-adjoint relative to σ

$$d_e^2Y(u,\cdot) = d_e^2Y(u,\cdot)^\sigma. \tag{4.3.5}$$

We would also like to note that (4.3.4) yields with the notation of [2, §8]

$$Q_e(y,u) = d_e^2Y(u,\cdot).$$

Therefore, [2, 8.10] shows that (K, g) is symmetric if and only if for all $e \in \mathcal{M}$ and $u, w \in V$

$$[d_e^2Y(u,\cdot), A(w)] = [d_e^2Y(w,\cdot), A(u)]. \tag{4.3.6}$$

We have not been able to evaluate this identity effectively directly.

Let now $U, Y \in \mathfrak{p}$, then $[U, Y] \in \mathfrak{k}$. We compute

$$d_e[U,Y] = d_e(d_xU(Y) - d_xY(U)) \tag{4.3.7}$$

$$= d_e^2U(Y,\cdot) - d_e^2Y(U,\cdot) + [d_eU, d_eY].$$

Using (4.3.3) we thus obtain for $U, Y \in \mathfrak{p}$

$$d_e[U,Y] = [d_eU, d_eY] = [A(u), A(y)]. \tag{4.3.8}$$

Since $[\mathfrak{p}, \mathfrak{p}] = \mathfrak{k}$, we know that for every $Y \in \mathfrak{k}$ the differential $d_e Y$ is a sum of commutators of the form (4.3.8). In particular, $d_e\mathfrak{k}$ consists of skew adjoint endomorphisms of V.

Let now $U \in \mathfrak{p}$ and $Y \in \mathfrak{k}$. Then $[U, Y] \in \mathfrak{p}$. But then $d_e[U, Y] = A([U, Y](e))$ by (4.2.5). Now, $[U, Y](e) = -d_e Y(u)$ and we obtain for $U \in \mathfrak{p}$, $Y \in \mathfrak{k}$

$$d_e[Y, U] = A(d_e Y u). \tag{4.3.9}$$

Finally, let $U, Y \in \mathfrak{k}$. Then (4.3.7) shows

$$d_e[U, Y] = [d_e U, d_e Y]. \tag{4.3.10}$$

4.4. The results of the last section allow us now to describe the Lie algebra $\mathfrak{g}$. We identify $\mathfrak{k}$ with $\hat{\mathfrak{k}} = d_e\mathfrak{k}$ and $\mathfrak{p}$ with V. Then we set

$$\hat{\mathfrak{g}} = \hat{\mathfrak{k}} \oplus V. \tag{4.4.1}$$

Using this isomorphism of vector spaces and (4.3.8), (4.3.9) and (4.3.10) we obtain for the commutators in $\hat{\mathfrak{g}}$

$$[v, u] = [A(v), A(u)], \quad u, v \in V, \tag{4.4.2}$$

$$[\hat{k}, u] = \hat{k}u, \quad u \in V,\ \hat{k} \in \hat{\mathfrak{k}}, \tag{4.4.3}$$

$$[\hat{k}, k'] = \hat{k}k' - k'\hat{k}, \quad \hat{k}, k' \in \hat{\mathfrak{k}}. \tag{4.4.4}$$

These formulas thus define on $\hat{\mathfrak{g}}$ the structure of a Lie algebra such that $\mathfrak{g} \to \hat{\mathfrak{g}}$ is an isomorphism of Lie algebras.

To understand somewhat better what this means we evaluate the Jacobi identity in $\hat{\mathfrak{g}}$. Let $u, v, w \in V$. Then $[u, [v, w]] = [u, [A(v), A(w)]] = -[A(v), A(w)]u$ by (4.4.2) and (4.4.3). From this the Jacobi identity follows trivially.

Let now $u, v \in V$, $k \in \hat{\mathfrak{k}}$. Then $[k, [u, v]] = [k, [A(u), A(v)]]$ and $[u, [k, v]] = [u, kv] = [A(u), A(kv)]$. Hence, the Jacobi identity is equivalent with

$$[k, [A(u), A(v)]] = [A(ku), A(v)] + [A(u), A(kv)]. \tag{4.4.5}$$

In view of (4.4.2) we thus have in particular for all $a, b, u, v \in V$

$$\begin{aligned}(4.4.6)\qquad &[[A(a), A(b)], [A(u), A(v)]] \\ &= [A([A(a), A(b)]u), A(v)] + [A(u), A([A(a), A(b)]v)].\end{aligned}$$

Next we consider $u \in V$, $k, k' \in \hat{\mathfrak{k}}$. Then $[[k, k'], u] = [[k, u], k'] + [k, [k', u]]$ just rephrases the definition of $[k, k']$ in $\hat{\mathfrak{g}}$.

Finally, if $k, k', k'' \in \hat{\mathfrak{g}}$, then the Jacobi identity simply states that $\hat{\mathfrak{k}}$ is a Lie algebra of (skew adjoint) endomorphisms of V.

Note however, that (4.4.6) already implies (4.4.5), and this in turn implies that $\hat{\mathfrak{k}} = \text{span}\{[A(u), A(v)] : u, v \in V\}$ is a Lie algebra of skew adjoint endomorphisms of V.

5. In this section we study algebras $\mathfrak{A}$ satisfying the conditions listed in section 4.4.

5.1. Let us start now, conversely, with a vector space V with inner product σ and let $A(u)$, $u \in V$, define on V the structure of a commutative algebra $\mathfrak{A}$ with identity e. Assume, moreover, $A(u) = A(u)^\sigma$ for all $u \in V$ and that (4.4.6) holds. Then we set

$$(5.1.1)\qquad \hat{\mathfrak{k}} = \text{span}\{[A(u), A(v)] : u, v \in V\}$$

and

$$(5.1.2)\qquad \hat{\mathfrak{g}} = \hat{\mathfrak{k}} + V.$$

By what was said in section 4.4,

(5.1.3) $\hat{\mathfrak{k}}$ *is a Lie algebra of skew adjoint endomorphisms of* V.

Moreover, using (4.4.2), (4.4.3) and (4.4.4) as definitions, it follows from the arguments in section 4.4 that

(5.1.4) $\hat{\mathfrak{g}}$ *is a Lie algebra.*

We want to investigate the structure of $\hat{\mathfrak{g}}$. First we note

$$(5.1.5)\qquad [\hat{\mathfrak{k}}, V] \subset V \quad \text{and} \quad [V, V] \subset \hat{\mathfrak{k}},$$

whence

$$(5.1.6)\qquad \pi : \hat{\mathfrak{g}} \to \hat{\mathfrak{g}}, \quad k \oplus u \to k - u$$

is an automorphism of $\mathfrak{g}$ of order 2.

Now we consider a Levi decomposition of $\hat{\mathfrak{g}}$

$$\hat{\mathfrak{g}} = \hat{\mathfrak{s}} + \hat{\mathfrak{r}} \tag{5.1.7}$$

where $\hat{\mathfrak{r}}$ denotes the radical of $\hat{\mathfrak{g}}$ and $\hat{\mathfrak{s}}$ is a semi-simple subalgebra of $\hat{\mathfrak{g}}$. By a result of Mostow-Taft we can assume $\pi\hat{\mathfrak{s}} \subset \hat{\mathfrak{s}}$. Since we have anyway $\pi\hat{\mathfrak{r}} \subset \hat{\mathfrak{r}}$, we thus have

$$\hat{\mathfrak{s}} = \hat{\mathfrak{s}}_k + \hat{\mathfrak{s}}_V, \quad \hat{\mathfrak{r}} = \hat{\mathfrak{r}}_k + \hat{\mathfrak{r}}_V, \tag{5.1.8}$$

where $\hat{\mathfrak{s}}_k = \hat{\mathfrak{s}} \cap \hat{\mathfrak{k}}$ etc. From [1, Chapter 1, §5.3] we know that $\operatorname{ad}[\hat{\mathfrak{g}}, \hat{\mathfrak{r}}]$ consists of nilpotent endomorphisms of $\hat{\mathfrak{g}}$. In particular, $\operatorname{ad}[\hat{\mathfrak{s}}_V, \hat{\mathfrak{r}}_V] \subset \operatorname{ad}\hat{\mathfrak{k}}$ consists of nilpotent endomorphisms of $\hat{\mathfrak{g}}$. On the other hand, these endomorphisms are semisimple, whence $\operatorname{ad}[\hat{\mathfrak{s}}_V, \hat{\mathfrak{r}}_V] = 0$. Applying this to V yields

$$[\hat{\mathfrak{s}}_V, \hat{\mathfrak{r}}_V] = 0. \tag{5.1.9}$$

Now note that $\hat{\mathfrak{k}} = [V, V]$ by definition. Therefore, $\hat{\mathfrak{r}}_k = [\hat{\mathfrak{s}}_V, \hat{\mathfrak{r}}_V] + [\hat{\mathfrak{r}}_V, \hat{\mathfrak{r}}_V]$. Here the first summand vanishes by (5.1.9). To the second we can apply the same argument and obtain

$$[\hat{\mathfrak{r}}_V, \hat{\mathfrak{r}}_V] = 0, \tag{5.1.10}$$

$$\hat{\mathfrak{r}}_k = 0. \tag{5.1.11}$$

Finally, since $\hat{\mathfrak{s}}_k = [\hat{\mathfrak{s}}_V, \hat{\mathfrak{s}}_V]$, we derive from (5.1.9)

$$[\hat{\mathfrak{s}}_k, \hat{\mathfrak{r}}_V] = 0. \tag{5.1.12}$$

This shows

(5.1.13) $\hat{\mathfrak{g}} = \hat{\mathfrak{s}} + \hat{\mathfrak{r}}_V$ *is a reductive Lie algebra with center* $\hat{\mathfrak{r}}_V$.

A similar situation was considered in an algebraic context by K. Meyberg and U. Hirzebruch [5, §3].

5.2. We return to the discussion of symmetric cones (K, g). But we will only use the properties of the algebra $\mathfrak{A}$ as stated in 5.1. We investigate the algebra $\mathfrak{A}$ given by $A(u)v = uv$ more closely. First we show:

(5.2.1) *There exists some* $w \in V$, $\sigma(w, e) = 0$, $\sigma(w, w) = 1$, *such that* $w^2 + \beta w + \alpha e = 0$ *for some* $\beta, \alpha \in \mathbb{R}$.

To show this, we consider $V' = \{u \in V : \sigma(u,e) = 0\}$ and the map $f : V' \to \mathbb{R}$, $f(u) = \sigma(u^2,u)$. Clearly, on the unit sphere $\mathbb{S}' = \{u \in V' : \sigma(u,u) = 1\}$ the function f attains a minimum and a maximum. Therefore, there exists some $w \in \mathbb{S}'$, such that $3\sigma(w^2,v) - \lambda 2\sigma(w,v) = 0$ for all $v \in V'$, i.e., $w^2 + \beta w + \alpha e = 0$, for some $\beta, \alpha \in \mathbb{R}$. We look at this equation more closely. From the definition of σ we know $\sigma(e,e) = n$, therefore $0 = \sigma(u^2,e) + \beta\sigma(u,e) + \alpha\sigma(e,e) = 1 + n\alpha$, i.e., $\alpha = -1/n$. We set $d = rw + se$ and want to show

(5.2.2) *r and s can be determine so that $d^2 = d$ and $d \neq e$.*

Note $d^2 = d$ iff $r^2w^2 + 2rsw + s^2e = rw + se$. This is equivalent with

$$r^2w^2 + (2rs - r)w + (s^2 - s)e = 0.$$

It suffices now to determine $r \neq 0$ and s so that

$$\frac{2s-1}{r} = \beta,$$

$$\frac{s^2 - s}{r^2} = \alpha = -\frac{1}{n}.$$

The latter equation is $s^2 - s + r^2(1/n) = 0$, i.e., $(s-(1/2))^2 + r^2(1/n) = 1/4$. Obviously, this equation has many solutions. It suffices to show that also $(2s-1)/r = \beta$ can be satisfied. But this equation is equivalent with $s - (1/2) = (1/2)r\beta$, so that it suffices to solve $(1/4)r^2\beta^2 + r^2(1/n) = 1/4$, i.e., $r = (1/2)((1/4)\beta^2 + (1/n))^{-1/2}$. Note $r \neq 0$. This proves (5.2.2).

5.3. In the last section we have shown that there exists at least one idempotent $c \in V$, $c \neq e$, in $\mathfrak{A}$. We note

(5.3.1) $$c^2 = c \Leftrightarrow (e-c)^2 = e - c,$$

(5.3.2) $$c(e-c) = 0.$$

Since $A(c)$ is self-adjoint, we can decompose V into the eigenspaces V_λ of $A(c)$ relative to the eigenvalue λ. We claim

(5.3.3) *If λ is an eigenvalue of $A(c)$, then $0 \leq \lambda \leq 1$.*

To verify this we note that the sectional curvature in (K, g) is given (see e.g., [2, §8]) by

$$\begin{aligned} -K_e(c, u) &= \sigma([A(u), A(c)]c, u) \\ &= \sigma(uc^2 - c \cdot uc, u) \\ &= (\lambda - \lambda^2)\sigma(u, u), \end{aligned}$$

where $0 \neq u \in V_\lambda$. Since $K_e(c, u) \leq 0$ for a non compact symmetric space, $\lambda - \lambda^2 \geq 0$ follows, hence the claim.

In view of (4.4.6) it will be useful to consider the "Lie triple product" $\{uvw\} = [[u, v], w] = [A(u), A(v)]w$. If $u = w = c$, $v \in V_\lambda$, then

$$\{cv_\lambda c\} = (\lambda^2 - \lambda)v_\lambda = \lambda(\lambda - 1)V_\lambda. \tag{5.3.4}$$

We note that $\{c, \cdot, c\}$ does not distinguish all V_λ. But we have

$$\lambda(\lambda - 1) = \mu(\mu - 1) \Leftrightarrow \begin{cases} \lambda = 1 - \mu, & \text{or} \\ \lambda = \mu. \end{cases} \tag{5.3.5}$$

In particular

$$u \in V_1 + V_0 \Leftrightarrow \{cuc\} = 0. \tag{5.3.6}$$

Next we show

(5.3.7) *V_1 and V_0 are subalgebras of $\mathfrak{A}$. Moreover, $V_1 V_0 = 0$.*

To see this we apply (4.4.6) and get for $u, v \in V_1 + V_0$

$$[[A(c), A(u)], [A(c), A(u)]] = [A(\{cuc\}), A(v)] + [A(c), A(\{cuv\})].$$

Here the middle term vanishes by assumption. Applying both sides to c we obtain

$$0 = [A(c), A(\{cuv\})]c = c(c(c \cdot uv - u \cdot cv)) - c(c \cdot uv - u \cdot cv).$$

If $v \in V_\lambda$, $\lambda = 0$, 1, then we obtain from this

$$(c - e)(c((c - \lambda e)uv)) = 0.$$

But it is easy to see that $(c - e)(cw) = 0$ is equivalent with $w \in V_1 + V_0$. Hence

$$(c - \lambda e)uv \in V_1 + V_0.$$

Since $\lambda = 0, 1$, we have shown that $V_1 + V_0$ is a subalgebra of *goa*.

Next we consider $v_1 \in V_1$ and $w_0 \in V_0$. Then

$$[[A(c), A(v_1)], [A(v_1), A(w_0)]] = [A(c \cdot v_1^2 - v_1^2), A(w_0)] + [A(v_1), A(c \cdot v_1 w_0)].$$

Here the middle term vanishes. Applying the remaining terms to c, we see that the last term vanishes, since $c \cdot v_1 w_0 = c(c \cdot v_1 w_0)$, because of $V_1 V_0 \subset V_1 + V_0$. Therefore, from the first term we obtain $[A(c), A(v_1)](w_0 v_1) = 0$. Hence $c(v_1 \cdot w_0 v_1) = v_1(c \cdot w_0 v_1)$. As a consequence, we have

$$\sigma(w_0, c(v_1 \cdot w_0 v_1))\sigma(w_0, v_1(c \cdot w_0 v_1)).$$

Since $cw_0 = 0$, the left side vanishes. On the right side we obtain $\sigma(v_1 w_0, c \cdot w_0 v_1) = 0$, hence $(w_0 v_1)_1 = 0$. Therefore $w_0 v_1 \in V_0$. Interchanging c and $e - c$, and thus V_0 and V_1 we obtain $w_0 v_1 \in V_1$, whence $w_0 v_1 = 0$. This shows $V_0 V_1 = 0$. From this we obtain $\sigma(w_0^2, v_1) = \sigma(w_0, w_0 v_1) = 0$ for all $v_1 \in V_1$, whence $V_0 V_0 \subset V_0$. Similarly one shows $V_1 V_1 \subset V_1$. As a consequence of this we can prove

If $\mathfrak{A}$ is an algebra satisfying the assumptions of section 5.1 and $\sigma(u^2, v^2) - \sigma(uv, uv) \geq 0$ then there exist $c_1, \ldots, c_r \in V$ such that

(5.3.8) a) $c_i c_j = \delta_{ij} c_i$ *for all* i, j;

b) $\sum_{i=1}^{r} c_i = e$;

c) *If* $c_i u = \alpha u$, *then* $\alpha = 1$ *and* $u \in \mathbb{R} c_i$.

We prove this by induction on n. For $n = 1$ the claim is clear. Assume now $n > 1$. Let $c \neq e$ be an idempotent as in 5.2. Then we have seen above that V_1 and V_0 are subalgebras of $\mathfrak{A}$ satisfying the assumptions of the claim and $\dim V_j < n$, $j = 0, 1$. Therefore, in V_1 and V_0 we can find idempotents $c_i^{(j)}$ as in the claim. Since $V_1 V_0 = 0$, the union $\{c_i^{(1)}\}_i \cup \{c_i^{(0)}\}_i$ satisfies (5.3.8).

6. In this section we prove that symmetric cones have root systems of type A.

6.1. Let $c_1, \ldots c_r$ be as in 5.3. Then we claim

(6.1.1) $$[A(c_i), A(c_j)] = 0 \quad \text{for all } i, j.$$

This is equivalent with $[c_i, c_j] = 0$ in $\hat{\mathfrak{g}}$. But, since $[c_i, c_j] \in \mathfrak{k}$, it suffices to show $\beta([c_i, c_j], [c_i, c_j]) = 0$, where β denotes the Killing form of $\hat{\mathfrak{g}}$. But now

$$\beta([c_i, c_j], [c_i, c_j]) = \beta(c_i, [c_j, [c_i, c_j]])$$

$$= -\beta(c_i[A(c_i), A(c_j)]c_j) = 0,$$

since $c_i c_j = 0$, $c_j^2 = c_j$.

Next we claim

(6.1.2) $\mathfrak{c} = \bigoplus_{j=1}^{r} \mathbb{R}c_j$ *is a maximal abelian subspace of* V.

Let $x \in V$, $[A(x), A(c_i)] = 0$ for all i. Denote by $V = \bigoplus V_\lambda$ the root space decomposition relative to $A(\mathfrak{c})$. Then $x = \sum x_\lambda$. Applying the commutator above to c_j we obtain

$$c_i c_j \cdot x = c_i \cdot c_j x = \sum \lambda_i \lambda_j x_\lambda.$$

If $i = j$, then this implies $\lambda_i x_\lambda = \lambda_i \lambda_j x_\lambda$. For all λ, such that $\lambda_i \neq 0$ and $x_\lambda \neq 0$, we thus obtain $c_j x_\lambda = x_\lambda$, i.e., $x_\lambda = \alpha c_j$ by c) of (5.3.8). We can subtract such terms from x and assume $\lambda_i = 0$ for all λ such that $x_\lambda \neq 0$. Since i was arbitrary, we have shown that we can assume $c_i x = 0$ for all i. But then $x = ex = \sum c_i x = 0$.

6.2. We consider the root space decomposition

(6.2.1) $$V = \mathfrak{c} + \bigoplus_{\lambda \in R} V_\lambda$$

of V relative to $A(\mathfrak{c})$ in more detail. As before we have $\mathfrak{c} = \bigoplus_{j=1}^{r} \mathbb{R}c_i$ with $\{c_i\}$ satisfying (5.3.8). First we show

(6.2.2) *If* $\lambda \in R$, *then* $\lambda_i = \lambda(c_i) \neq 0$ *for at most two different s.* *Moreover, if* $\lambda_i, \lambda_j \neq 0$, $i \neq j$, *then* $\lambda_i + \lambda_j = 1$.

For this we choose $v_\lambda \in V_\lambda$ and consider

$$\begin{aligned}
0 &= [[A(v_\lambda), A(c_i)], [A(c_j), A(c_k)]] \\
&= [A(v_\lambda \cdot c_i c_j - c_i \cdot v_\lambda c_j), A(c_k)] + [A(c_j), A(v_\lambda \cdot c_i c_k - c_i \cdot v_\lambda c_k)] \\
&= [A((\delta_{ij}\lambda_i - \lambda_i\lambda_j)v_\lambda, A(c_k)] + [A(c_j), A((\ delta_{ik}\lambda_i - \lambda_i\lambda_k)v_\lambda)] \\
&= [A(v_\lambda), A(c)]
\end{aligned}$$

where

$$c = \lambda_i(\delta_{ij} - \lambda_j)c_k - \lambda_i(\delta_{ik} - \lambda_k)c_j.$$

Note that $c = 0$ if $j = k$. Therefore we will assume $j \neq k$ now. Similarly we will assume $\lambda_i \neq 0$. Hence

(6.2.3) *If, $j \neq k$, $\lambda_i \neq 0$ and $0 \neq v_\lambda \in V_\lambda$, then* $[A(v_\lambda), A((\delta_{ij} - \lambda_j)c_k - (\delta_{ik} - \lambda_k)c_j)] = 0$. *cr*

Assume that i, j, k are pairwise different and $\lambda_i, \lambda_j, \lambda_k \neq 0$. Then we apply (6.2.3) to c_j and obtain $\lambda_j\lambda_k v_\lambda = 0$, a contradiction.

Assume now $i = j \neq k$, $\lambda_i \neq 0$. Then $[A(v_\lambda), A(1 - \lambda_i)c_k + \lambda_k c_i] = 0$ and an application to c_k yields

$$\lambda_k(1 - \lambda_i) = \lambda_k((1 - \lambda_i)\lambda_k + \lambda_k\lambda_i) = \lambda_k^2.$$

If $\lambda_i, \lambda_k \neq 0$, this implies $\lambda_i + \lambda_k = 1$.

As a consequence, we will label V_λ by V_{ij}, if $\lambda_i, \lambda_j \neq 0$ and V_{ii} if only $\lambda_i \neq 0$. From (5.3.8) we know $V_{ii} = \mathbb{R}c_i$. We write

$$V = \bigoplus_{i \leq j} V_{ij}.$$

In this notation it is not yet clear that $V_{ij} \neq 0$ for any choice of $1 \leq i \leq j \leq r$. We also note that $\lambda_j = 1 - \lambda_i$.

But we can show

(6.2.4) *For any $c = \sum_{k=1}^{r} \alpha_k c_k$ and any $v_{ij} \in V_{ij}$ we have* $[c, [c, v_{ij}]] = \lambda_i(1 - \lambda_i)(\alpha_i - \alpha_j)^2 v_{ij}$.

A straightforward computation yields

$$\begin{aligned}[c, [c, v_{ij}]] &= -[A(c), A(v_{ij})]c = c^2 \cdot v_{ij} - c \cdot cv_{ij} \\ &= \{\alpha_i^2\lambda_i + \alpha_j^2\lambda_j - (\alpha_i\lambda_i + \alpha_j\lambda_j)^2\}v_{ij}.\end{aligned}$$

Using $\lambda_i = 1 - \lambda_j$, the coefficient can be rewritten as

$$\alpha_i^2\lambda_i + \alpha_j^2(1 - \lambda_i) - \alpha_i^2\lambda_i^2 - 2\alpha_i\alpha_j\lambda_i(1 - \lambda_i) - \alpha_j^2(1 - \lambda_i)^2$$

$$= \lambda_i(1 - \lambda_i)\{\alpha_i^2 - 2\alpha_i\alpha_j + \alpha_j\}^2,$$

whence the claim.

Comparing (6.2.4) with Tabels 4 and 8 of [4] we obtain

Theorem. *If (K, g) is a symmetric cone, then the root system of the symmetric space (K, g) is if type A.*

7. Consider now an open convex cone $\hat{K}$ in $V = \mathbb{R}^n$ that does not contain any full straight line. Then

$$\operatorname{Aut} \hat{K} = \{W \in \operatorname{Gl}(n, \mathbb{R}) : W(\hat{K}) = \hat{K}\}$$

is a closed Lie group. A cone is called homogeneous, if $\operatorname{Aut} \hat{K}$ acts transitively on $\hat{K}$, and $\hat{K}$ is called a "domain of positivity", if there exists a reductive subgroup $\hat{G} \subset \operatorname{Aut} \hat{K}$ that acts transitively on $\hat{K}$ (see [2, Theorem 7.10]).

In view of the theorem in the last section we know [4, Theorem 2.4] that (K, g) is isomorphic with a domain of positivity $(\hat{K}, \hat{g})$.

Altogether we have thus shown

Theorem. *Let K denote an open convex cone in $\mathbb{R}^n$, with vertex at the origin, that does not contain any straight line. Let $\eta : K \to \mathbb{R}_{\geq 0}$ denote the 'Koecher Function' of K and g the associated Riemannian metric of K. Then, if (K, g) is a symmetric space, then K is isomorphic with a domain of positivity.*

References

[1] Bourbaki, N., *Groupes et Algèbres de Lie*, Chapters 1, 7, 8, Hermann, 1972.

[2] Dorfmeister, J., Koecher, M., *Reguläre Kegel*, Uber d. D. Math.-Verein. **81** (1979) 109–151.

[3] Helgason, S., *Differential Geometry, Lie Groups, and Symmetric Spaces*, Pure and Applied Mathematics Vol 80, Academic Press, 1978.

[4] Loos, O., *Symmetric Spaces II: Compact Spaces and Classification*, Benjamin, 1969.

[5] Meyberg, K., *Jordan-Tripelsysteme und die Koecher-Konstruction von Lie Algebren*, Habilitationsschrift, München, 1969.

Mathematics Institute, TU-München, Arcisstr. 21, 80290 München, Germany

Received October 1995

Pseudo-Hermitian Symmetric Spaces of Tube Type

Jacques Faraut and Simon Gindikin

1. Introduction

Let Ω be an open connected cone in a real vector space $V \simeq \mathbb{R}^n$. One defines

$$G(\Omega) = \{g \in GL(n, \mathbb{R}) \mid g\Omega = \Omega\}.$$

The cone Ω is said to be *homogeneous* if the group $G(\Omega)$ acts transitively on Ω. For the beginning let us assume that Ω is convex and that $\bar{\Omega}$ is pointed (this means that $\bar{\Omega} \cap (-\bar{\Omega}) = \{0\}$). The convex cone Ω is said to be *selfdual* if there exists a positive inner product on V such that $\Omega^* = \Omega$, where the open dual cone Ω^* is defined by

$$\Omega^* = \{x \in V \mid \forall y \in \bar{\Omega} \setminus \{0\},\ (x|y) > 0\}.$$

The open convex cone Ω is said to be *symmetric* if it is homogeneous and selfdual. Let us recall the connection between symmetric convex cones and Jordan algebras. A Jordan algebra V is a vector space equipped with a product, i.e. a bilinear map $V \times V \to V$ such that

(J1) $xy = yx$,

(J2) $x(x^2y) = x^2(xy)$.

A finite dimensional Jordan algebra V is said to be *Euclidean* if there exists on V a positive definite inner product which is associative, i.e.

(J3) $(xy|z) = (x|yz)$.

Theorem 1.1. (Koecher, Vinberg). *The interior of the cone of squares in a Euclidean Jordan algebra is a symmetric convex cone, and every symmetric convex cone is obtained in that way.*

Let Ω be a symmetric convex cone and let us consider the tube $T_\Omega = V + i\Omega$ in the complexified vector space $V^{\mathbb{C}}$. The tube T_Ω is a symmetric domain: every point z in T_Ω is an isolated fixed point for an involutive holomorphic automorphism of T_Ω. In fact, using a lemma of

Siegel, one proves that every element z in T_Ω is invertible in the complex Jordan algebra $V^{\mathbf{C}}$, and then the map

$$z \mapsto -z^{-1}$$

is a holomorphic automorphism of T_Ω with $z = ie$ as a unique fixed point (e is the identity element of the Jordan algebra).

There is on T_Ω a canonical Hermitian metric, the Bergman metric, which is invariant under the group $G(T_\Omega)$ of all holomorphic automorphisms of T_Ω. It is defined as follows. Let $\mathcal{H}^2(T_\Omega)$ be the Bergman space, i.e. the space of square integrable holomorphic functions on T_Ω with respect to the Lebesgue measure, and let $K(z,w)$ be the Bergman kernel, i.e. the reproducing kernel of $\mathcal{H}^2(T_\Omega)$. The Bergman metric is defined by

$$h_z(w,w') = \partial_w \bar{\partial}_{w'} \log K(z,z).$$

When the symmetric convex cone is irreducible, there exists a simple formula for the Bergman kernel,

$$K(z,w) = c\Delta\Big(\frac{z-\bar{w}}{i}\Big)^{-2\frac{n}{r}},$$

where $n = \dim V$, $r = \operatorname{rank} V$, and Δ is the determinant polynomial of the Jordan algebra $V^{\mathbf{C}}$.

In this paper we drop the hypothesis of convexity. We consider non convex symmetric cones (or more precisely non necessarily convex). For such a cone Ω in a real vector space V we consider the tube $T_\Omega = V + i\Omega \subset V^{\mathbf{C}}$, and equipped it with a pseudo-Hermitian metric. In general it is no longer a pseudo-Hermitian symmetric space, but, by adding points *at infinity*, we construct a pseudo-Hermitian symmetric space X containing T_Ω as an open dense subset. This space X is an open set in the conformal compactification Y of $V^{\mathbf{C}}$.

When Ω is a *satellite* of a convex cone, the corresponding space X has been studied by Kaneyuki [1991]. (The definition of a *satellite* will be given in Section 3; for example the cone Ω of symmetric matrices with signature (p,q) is a satellite of the convex cone of positive definite symmetric matrices).

Later D'Atri and Gindikin [1993] pointed out that there are other pseudo-Hermitian symmetric spaces corresponding to non-convex cones and studied in details five families of such spaces.

We shall give a unified presentation of these spaces, which are associated with simple real Jordan algebras, including exceptional cases which have not been considered before.

This paper can be seen as a geometric preliminary for a study of Hardy spaces of $\bar{\partial}$-cohomology associated with these tube domains, as introduced in [Gindikin, 1992].

In Section 2 we establish the correspondence between (non necessarily convex) symmetric cones and real semi-simple Jordan algebras. The classification of the simple ones is given in Section 3. We will see in Section 4 that in such a cone there is a maximal convex symmetric cone associated with a maximal Euclidean Jordan subalgebra. The conformal compactification of a simple complex Jordan algebra is considered in Section 5, and we recall there some properties of the conformal transformations which will be used later. In Section 6 we construct explicitly a homogeneous embedding of this conformal compactification into a projective space. In the last section we describe the pseudo-Hermitian symmetric spaces of tube type associated with symmetric cones.

2. Symmetric cones

A connected open cone Ω in a real vector space $V \simeq \mathbb{R}^n$ is said to be *symmetric* if it satisfies the properties (S1), (S2), and (S3).

(S1) There exists a connected subgroup L of $GL(V)$ which acts transitively on Ω. This means that Ω is homogeneous.

Let us fix a point e in Ω and let

$$A = L_e = \{g \in L \mid ge = e\}.$$

(S2) The pair (L, A) is symmetric: there is an involutive automorphism σ of L such that

$$(L^\sigma)^0 \subset A \subset L^\sigma,$$

where

$$L^\sigma = \{g \in L \mid \sigma(g) = g\},$$

and $(L^\sigma)^0$ is the identity component in L^σ. This means that $\Omega = L/A$ is a symmetric space.

(S3) There exists on V a non-degenerate bilinear form b such that, for all g in L,

$$b(gx, y) = b(x, \sigma(g)^{-1}y).$$

Let Ω be a connected open cone with the properties (S1) and (S2), and let $\mathfrak{l}$ and $\mathfrak{a}$ be the Lie algebras of L and A. Then

$$\mathfrak{a} = \{X \in \mathfrak{l} \mid X \cdot e = 0\} = \{X \in \mathfrak{l} \mid \sigma(X) = X\}.$$

Define

$$\mathfrak{q}_0 = \{X \in \mathfrak{l} \mid \sigma(X) = -X\}.$$

Then, as a direct sum,

$$\mathfrak{l} = \mathfrak{a} + \mathfrak{q}_0,$$

and

$$[\mathfrak{a}, \mathfrak{q}_0] \subset \mathfrak{q}_0, \quad [\mathfrak{q}_0, \mathfrak{q}_0] \subset \mathfrak{a}.$$

The map

$$\mathfrak{q}_0 \to V, \quad X \mapsto X \cdot e,$$

is an linear isomorphism. If $X \cdot e = u$ we write $X = L(u)$, and we consider on V the product defined by

$$u \cdot v = L(u)v.$$

This product is commutative, in fact

$$[L(u); L(v)] \in \mathfrak{a},$$

therefore

$$[L(u), L(v)]e = 0,$$

or $u \cdot v = v \cdot u$. It follows that e is an identity element,

$$e \cdot u = u \cdot e = L(u)e = u,$$

or $L(e) = I$, the identity map.

One defines the associator $[u, v, w]$ of three elements in V by

$$\begin{aligned}[u, v, w] &= u \cdot (v \cdot w) - (u \cdot v) \cdot w \\ &= [L(u), L(w)]v,\end{aligned}$$

and the symmetric bilinear form τ by

$$\tau(u,v) = \operatorname{Tr} L(uv).$$

Proposition 2.1. *Under the assumptions* (S1) *and* (S2), *the algebra structure associated with the cone* Ω *has the following properties:* (i) $\forall X \in \mathfrak{a}$, $[X, L(u)] = L(Xu)$ $(u \in V)$. (ii) *Any* X *in* $\mathfrak{a}$ *is a derivation.* (iii) $L([u,v,w]) = \big[[L(u), L(w)], L(v)\big]$. (iv) *The bilinear form* τ *is associative, i.e.*

$$\tau(u\cdot v, w) = \tau(u, v\cdot w).$$

Proof.

(i) Since $X \in \mathfrak{a}$, $L(u) \in \mathfrak{q}_o$, $[X, L(u)]$ belongs to $\mathfrak{q}_o$ and

$$[X, L(u)]e = Xu,$$

therefore

$$[X, L(u)] = L(Xu).$$

(ii) For X in $\mathfrak{a}$,

$$\begin{aligned} X(u\cdot v) &= XL(u)v \\ &= [X, L(u)]v + L(u)Xv \\ &= L(Xu)v + L(u)Xv \\ &= (Xu)\cdot v + u\cdot (Xv). \end{aligned}$$

(iii) Let $X = [L(u), L(w)] \in \mathfrak{a}$, then

$$Xv = [L(u), L(w)]v = [u,v,w],$$

and, by (i),

$$\big[[L(u), L(w)], L(v)\big] = L([u,v,w]).$$

(iv) The trace of a commutator is zero; therefore

$$\tau(u, v\cdot w) - \tau(u\cdot v, w) = \operatorname{Tr} L([u,v,w]) = 0.$$

■

Now let us assume (S3), then all $X = L(u) \in \mathfrak{q}_\circ$ are selfadjoint with respect to the bilinear form b. This means that the bilinear form b is associative,

$$b(u \cdot v, w) = b(u, v \cdot w).$$

We can apply the following result of Vinberg [1960].

Theorem 2.2. *Let V be a finite dimensional commutative algebra with an identity element satisfying:* (i) *For all $u, v, w \in V$,*

$$L([u, v, w]) = \big[[L(u), L(w)], L(v)\big].$$

(ii) *There exists on V a non-degenerate symmetric bilinear form which is associative. Then V is a semi-simple Jordan algebra.*

Recall that a Jordan algebra V is said to be *semi-simple* if there exists a non-degenerate symmetric bilinear form which is associative. One shows that a semi-simple Jordan algebra is the direct sum of simple ideals.

Conversely let V be a semi-simple Jordan algebra with an identity element e, and let Ω be the connected component of e in the set $\mathcal{I}$ of invertible elements. Then Ω is a symmetric cone. In fact let L be the identity component of the structure group $Str(V)$ of V (see [Faraut-Koranyi, 1994], p.147, for the definition), then one proves that Ω satisfies the properties (S1), (S2), and (S3), with $b = \tau$.

Summarizing we can state.

Theorem 2.3. *Let V be a semi-simple real Jordan algebra with an identity element e. The connected component Ω of e in the set of invertible elements is a symmetric cone, and every symmetric cone is obtained in that way.*

The symmetric cone Ω is convex and proper if and only if the Jordan algebra V is Euclidean, and then the group A is compact.

The following result is due to Shima [1975].

Theorem 2.4. *Let Ω be a connected open cone in V with the properties* (S1) *and* (S2)*. Then* (S3) *is equivalent to* (S3′)*: the action of L in V is completely reducible.*

Proof.

(a) Il Ω is a symmetric cone in a vector space V, then, by Theorem 2.2, V is equipped with an algebra structure such that V is a semi-simple Jordan algebra. Since a semi-simple Jordan algebra is the direct sum of simple ideals, the action of L is completely reducible.

(b) Assume (S1), (S2), and (S3′). By Proposition 2.1 we know that the symmetric bilinear form τ is associative. We will show that it is non-degenerate. Let

$$V_0 = \{v \in V \mid \forall u \in V,\ \tau(u, v) = 0\}.$$

Since τ is associative, V_0 is an ideal, i.e. $\mathfrak{q} \cdot V_0 = V_0$. For $X \in \mathfrak{a}$, $u \in V$, $v \in V_0$,

$$\begin{aligned}\tau(Xv, u) &= \tau(Xv, u) + \tau(v, Xu)\\ &= \operatorname{Tr} L\big((Xv)\cdot u + v \cdot (Xu)\big).\end{aligned}$$

By using Proposition 2.1 (i) and (ii) we obtain

$$\tau(Xu, v) = \operatorname{Tr}[X, L(u \cdot v)] = 0.$$

Therefore $\mathfrak{a} \cdot V_0 \subset V_0$, and V_0 is an invariant subspace for the action of $\mathfrak{l}$. Since the action of $\mathfrak{l}$ in V is completely reducible, there exists an invariant complementary subspace V_1,

$$V = V_0 + V_1,\ V_0 \cap V_1 = \{0\},$$

and V_1 is an ideal. We have $V_0 \cdot V_1 = 0$. In fact, for $u \in V_0$, $v \in V_1$, $u \cdot v$ belongs to V_0 and to V_1, therefore $u \cdot v = 0$.

We can decompose the identity element e,

$$e = a + b,\ a \in V_0,\ b \in V_1.$$

Then

$$e = e^2 = a^2 + b^2,$$

therefore a and b are idempotents, and

$$\operatorname{Tr} L(a) = \operatorname{Tr} L(a^2) = \tau(a, a) = 0.$$

Since $L(a)$ is the projection onto V_0, it follows that $L(a) = 0$, and $V_0 = \{0\}$. ∎

3. Classification

A simple real Jordan algebra is either a real form of a simple complex Jordan algebra, or a simple complex Jordan algebra considered as a real algebra. A simple complex Jordan algebra $V^{\mathbb{C}}$ has a Euclidean real form, and all Euclidean real forms of $V^{\mathbb{C}}$ are conjugate under $Aut(V^{\mathbb{C}})$ (see [Faraut-Koranyi,1994], Theorem VIII.5.2). We give below the list of the simple complex Jordan algebras $V^{\mathbb{C}}$ and their Euclidean real forms V.

$V^{\mathbb{C}}$	$\mathbb{C}^n$	$Sym(m,\mathbb{C})$	$M(m,\mathbb{C})$	$Skew(2m,\mathbb{C})$	$Herm(3,\mathbb{O})\otimes\mathbb{C}$
V	$\mathbb{R}^n$	$Sym(m,\mathbb{R})$	$Herm(m,\mathbb{C})$	$Herm(m,\mathbb{H})$	$Herm(3,\mathbb{O})$

For $V^{\mathbb{C}} = \mathbb{C}^n$, or $V = \mathbb{R}^n$, the product is given by

$$z = x \cdot y \text{ if } z_1 = \sum_{j=1}^{n} x_j y_j,$$
$$z_j = x_1 y_j + x_j y_1, \; j = 2, \ldots, n.$$

In the other cases $V^{\mathbb{C}}$ and V are equipped with the Jordan product

$$x \cdot y = \tfrac{1}{2}(xy + yx),$$

with the exception of $V^{\mathbb{C}} = Skew(2m,\mathbb{C})$ for which

$$x \cdot y = \tfrac{1}{2}(xJy + yJx),$$

with

$$J = \begin{pmatrix} 0 & I_m \\ -I_m & 0 \end{pmatrix}.$$

The real forms of simple complex Jordan algebras have been classified by Hellwig [1967] (see also [Kayoya, 1994]). The principle of the classification is the following. Let $V^{\mathbb{C}}$ be a simple complex Jordan algebra and fix a Euclidean real form W of $V^{\mathbb{C}}$. Let α be an involutive automorphism of W and define

$$W_\pm = \{x \in W \mid \alpha(x) = \pm x\}.$$

Then $V = V_\alpha = W_+ + iW_-$ is a real form of $V^{\mathbb{C}}$ and every real form of $V^{\mathbb{C}}$ is obtained in that way.

For $V^{\mathbb{C}} = \mathbb{C}^n$, fix $W = \mathbb{R}^n$ with the product defined as above. Then $Aut(W) \simeq O(n-1)$. For

$$\alpha = I_{p,q} = \begin{pmatrix} I_p & 0 \\ 0 & -I_q \end{pmatrix}, \quad p+q = n-1,$$

the real form $V = V_\alpha$ is isomorphic to $\mathbb{R}^n$ equipped with the product

$$z = x \cdot y \text{ if } z_1 = \sum_{j=1}^{p+1} x_j y_j - \sum_{j=p+2}^{n} x_j y_j,$$
$$z_j = x_1 y_j + x_j y_1, \ 2 \le j \le n.$$

Then, if $q = 0$,

$$\Omega = \{x \in \mathbb{R}^n \mid x_1^2 - x_2^2 - \cdots - x_n^2 > 0, \ x_1 > 0\},$$

if $1 \le q \le n-2$,

$$\Omega = \{x \in \mathbb{R}^n \mid x_1^1 + \cdots + x_{p+1}^2 - x_{p+2}^2 - \ldots - x_n^2 > 0\},$$

and, if $q = n-1$, $\Omega = \mathbb{R}^n \setminus \{0\}$.

If $V^{\mathbb{C}}$ is the complexification of $Herm(m, \mathbb{F})$ ($\mathbb{F} = \mathbb{R}$, $\mathbb{C}$, $\mathbb{H}$, or $\mathbb{O}$), one classifies the real forms in three types (see [Kayoya, 1994]).

Type I

The involutive automorphism α has the form

$$\alpha(x) = I_{p,q} x I_{p,q} \ (p+q = m).$$

The real form $V = V_\alpha$ is isomorphic to $Herm(m, \mathbb{F})$ equipped with the product

$$x \cdot y = \tfrac{1}{2}\left(x I_{p,q} y + y I_{p,q} x\right),$$

and

$$\Omega = \Omega(p,q) = \{x \in Herm(m, \mathbb{F}) \mid \operatorname{sgn}(x) = (p,q)\}.$$

The cone $\Omega = \Omega(p,q)$ is said to be a *satellite* of the convex symmetric cone $\Omega(m, 0)$.

Type II

The involutive automorphism α has the form

$$\alpha\big((x_{ij})\big) = \big(\alpha_0(x_{ij})\big),$$

where α_0 is a non-trivial involutive automorphism of $\mathbb{F}$.

If $\mathbb{F} = \mathbb{C}$, $\alpha_0(u) = \bar{u}$, and V_α is isomorphic to $V = M(m, \mathbb{R})$ equipped with the Jordan product.

If $\mathbb{F} = \mathbb{H}$, $\alpha_0(u) = -juj$, and V_α is isomorphic to $V = Skew(2m, \mathbb{R})$ equipped the product

$$x \cdot y = \frac{1}{2}(xJy + yJx),$$

with

$$J = \begin{pmatrix} 0 & I_m \\ -I_m & 0 \end{pmatrix}.$$

In the case of a real form of type II

$$\Omega = \{x \in V \mid \Delta(x) > 0\}.$$

Type III

Assume m even, $m = 2\ell$. The involutive automorphism α has the form

$$\alpha(x) = -JxJ.$$

If $\mathbb{F} = \mathbb{R}$, then $V = V_\alpha \simeq Sym(2\ell, \mathbb{C}) \cap M(\ell, \mathbb{H})$, and, if $\mathbb{F} = \mathbb{C}$, then $V = V_\alpha \simeq M(\ell, \mathbb{H})$.

In the case of a real form of type III

$$\Omega = \{x \in V \mid \Delta(x) \neq 0\}.$$

4. Maximal convex cones in a symmetric cone

Let V be a simple real Jordan algebra and let $\Omega = L \cdot e \simeq L/A$ be the associated symmetric cone. The pair (L, A) is symmetric with respect to the involution σ. The group L is reductive and there exists a Cartan involution θ of L which commutes with σ, and $K_0 = L^\theta$ is a maximal compact subgroup of L. We consider the corresponding decompositions of $\mathfrak{l}$,

$$\mathfrak{l} = \mathfrak{k}_0 + \mathfrak{p}_0 = \mathfrak{a} + \mathfrak{q}_0.$$

Since $V \simeq \mathfrak{q}_0$, and since θ commutes with σ, the involution θ induces an involution ν of V,

$$L(\nu(x)) = -L(x)^{\theta},$$

and ν is an involutive automorphism of the Jordan algebra V. We define

$$V_+ = \{x \in V \mid \nu(x) = x\} \simeq \mathfrak{q}_0 \cap \mathfrak{p}_0,$$
$$V_- = \{x \in V \mid \nu(x) = -x\} \simeq \mathfrak{q}_0 \cap \mathfrak{k}_0.$$

Proposition 4.1. (i) *V_+ is a maximal Euclidean subalgebra of V.* (ii) *$W = V_+ + iV_-$ is a Euclidean real form of $V^{\mathbb{C}}$, and ν is the restriction to V of the conjugation of $V^{\mathbb{C}}$ with respect to W.*

Example

Take $V = M(m, \mathbb{R})$. Then $\nu(x) = x^T$, and

$$V_+ = Sym(m, \mathbb{R}), \quad V_- = Skew(m, \mathbb{R}).$$

The conjugation of $V^{\mathbb{C}} = M(m, \mathbb{C})$ is $x \mapsto x^*$.

The quadratic form $\mathrm{Tr}(x^2)$ is positive definite on V_+, negative definite on V_-. Let Ω_+ be the convex symmetric cone associated with V_+.

Proposition 4.2. (i) $\Omega_+ + V_- \subset \Omega$. (ii) *$\Omega_+ + V_-$ is a maximal convex subset of Ω.*

Proof.

(i) Let Ω_1 be the convex symmetric cone associated with the Euclidean Jordan algebra $W = V_+ + iV_-$. Then $\Omega_+ = \Omega_1 \cap V$, and

$$\Omega_+ + V_- = (\Omega_1 + iW) \cap V.$$

Every element $z \in \Omega_1 + iW$ is invertible (this fact is a generalization of a lemma of Siegel, see [Faraut-Koranyi,1994], Theorem X.1.1). Therefore every element x in $\Omega_+ + V_-$ is invertible. Since Ω is the connected component of e in the set of invertible elements of V,

$$\Omega_+ + V_- \subset \Omega.$$

(ii) Let U be a convex set such that

$$\Omega_+ + V_- \subset U \subset \Omega.$$

Assume that there exists $x \in U \setminus (\Omega_+ + V_-)$. We can assume that $x \in V_+$. For any $y \in \Omega_+$ there exists $z \in [x,y]$ (the segment joining x and y) such that $\Delta(z) = 0$, leading to a contradiction. ■

5. Conformal compactification of a simple complex Jordan algebra

Let $V^{\mathbf{C}}$ be a simple complex Jordan algebra, and $P^{\mathbf{C}} = L^{\mathbf{C}} \ltimes N^{\mathbf{C}}$ be the semi-direct product of the group $N^{\mathbf{C}}$ of complex translations and the structure group $L^{\mathbf{C}}$ of $V^{\mathbf{C}}$, and let $G^{\mathbf{C}}$ be the group generated by $P^{\mathbf{C}}$ and the symmetry s,

$$s(x) = -x^{-1}.$$

One shows that $G^{\mathbf{C}}$ is a Lie group whose Lie algebra is the space of vector fields on $V^{\mathbf{C}}$ of the form

$$\xi(x) = u + Tx - P(x)v,$$

with $u, v \in V^{\mathbf{C}}$, $T \in \mathfrak{l}^{\mathbf{C}} = Lie(L^{\mathbf{C}})$. Hence

$$\mathfrak{g} \simeq V^{\mathbf{C}} \times \mathfrak{l}^{\mathbf{C}} \times V^{\mathbf{C}},$$

and we will write $\xi = (u, T, v)$.

The group $P^{\mathbf{C}}$ is a maximal parabolic subgroup of $G^{\mathbf{C}}$. The manifold $Y = G^{\mathbf{C}}/P^{\mathbf{C}}$ is a compact complex manifold, and $Y \simeq U/K$ where U is a compact real form of $G^{\mathbf{C}}$, and $K = U \cap L^{\mathbf{C}}$ is a compact real form of $L^{\mathbf{C}}$. Hence Y is a compact Hermitian symmetric space.

Let y_0 be the base point in $Y = G^{\mathbf{C}}/P^{\mathbf{C}}$. The orbit $P^{\mathbf{C}} \cdot s \cdot y_0$ is dense in Y, and the map

$$x \mapsto \tau_x \cdot s \cdot m_0$$

(τ_x is the translation, $\tau_x : y \mapsto y + x$) is a diffeomorphism of $V^{\mathbf{C}}$ onto $P^{\mathbf{C}} \cdot s \cdot y_0$. One shows that $V^{\mathbf{C}}$ is Zariski open in Y. In fact $\Sigma = Y \setminus V^{\mathbf{C}}$ is a hypersurface (see Section 6). Hence Y is a compactification of $V^{\mathbf{C}}$. It is the *conformal compactification* of $V^{\mathbf{C}}$. Let us explain this fact. The cone

$$\Gamma = \{x \in V^{\mathbf{C}} \mid \Delta(x) = 0\}$$

defines a flat *conformal structure* on $V^{\mathbf{C}}$. A local diffeomorphism φ of $V^{\mathbf{C}}$ is said to be a *conformal transformation* if it preserves this conformal structure: for every x,

$$D\varphi_x(\Gamma) = \Gamma.$$

One shows that

$$\{g \in Gl(V^{\mathbf{C}}) \mid g(\Gamma) = \Gamma\} = L^{\mathbf{C}}.$$

Therefore a transformation φ is conformal if, for every x,

$$D\varphi_x \in L^{\mathbf{C}}.$$

The elements of $G^{\mathbf{C}}$ are conformal transformations, and the following converse, which has been proved by Bertram ([1996], Théorèmes 2.3.1, 2.4.1), is a generalisation of a theorem of Liouville.

Theorem 5.4. *Assume* $\dim V^{\mathbf{C}} \geq 2$. *Every conformal transformation of* $V^{\mathbf{C}}$ *extends to a diffeomorphism of* Y *which belongs to* $G^{\mathbf{C}}$.

Examples

1) Let $V^{\mathbf{C}}$ be with the product defined above. The determinant polynomial is the quadratic form

$$\Delta(z) = z_1^2 - z_2^2 - \cdots - z_n^2.$$

Then $G^{\mathbf{C}} = SO(n+2, \mathbb{C})$, and the complex manifold Y can be realized as the projective quadric in $P_{n+1}(\mathbb{C})$ with equation

$$\Delta(z) = uv.$$

The map

$$(z_1, \ldots, z_n) \mapsto (z_1, \ldots, z_n, \Delta(z), 1)$$

defines the embedding of $V^{\mathbf{C}}$ in Y. Furthermore

$$Y \simeq SO(n+2)/SO(n) \times SO(2).$$

2) Let $V^{\mathbf{C}} = M(n, \mathbb{C})$ with the Jordan product. Then $G^{\mathbf{C}} = SL(2m, \mathbb{C})$ (more precisely $SL(2m, \mathbb{C})/\{\pm I\}$) and Y can be realized as the Grassmannian manifold of m-planes in $\mathbb{C}^{2m}$,

$$Y \simeq Gr(m, 2m; \mathbb{C}).$$

It can also be seen as the quotient space

$$Y = \{Z \in M(2m, m; \mathbb{C}) \mid Z \text{ of rank } m\}/\sim,$$

where $\sim$ is the equivalence relation defined by

$$Z \sim Z' \text{ if } Z' = Zu \text{ with } u \in GL(m, \mathbb{C}).$$

The map

$$z \mapsto \begin{pmatrix} z \\ I_m \end{pmatrix} \in M(2m, m; \mathbb{C})$$

defines the embedding of $V^{\mathbb{C}}$ in Y. As a compact Hermitian symmetric space

$$Y \simeq SU(2m)/S\big(U(m) \times U(m)\big).$$

3) Let $V^{\mathbb{C}} = Sym(m, \mathbb{C})$ with the Jordan product. Then $G^{\mathbb{C}} = Sp(m, \mathbb{C})$ and Y can be realized as the Grassmannian manifold of Lagrangian m-planes in $\mathbb{C}^{2m}$. Let

$$J = \begin{pmatrix} 0 & I_m \\ -I_m & 0 \end{pmatrix}.$$

Then

$$Y \simeq \{Z \in M(2m, m; \mathbb{C}) \mid Z \text{ of rank } m, \;\; Z^T J Z = 0\}/\sim,$$

with the same equivalence as in Example 2. As a compact Hermitian symmetric space

$$Y \simeq Sp(m)/U(m).$$

We recall here some results of Loos which will be used in the next section ([1977], Proposition 8.13, see also [Bertram, 1996]). We saw that the Lie algebra $\mathfrak{g}^{\mathbb{C}}$ of $G^{\mathbb{C}}$ is the Lie algebra of the vector fields of the form

$$\xi(z) = u + Tz - P(z)v,$$

with $u, v \in V^{\mathbb{C}}$, $T \in \mathfrak{k}^{\mathbb{C}}$. The adjoint action of $G^{\mathbb{C}}$ on $\mathfrak{g}^{\mathbb{C}}$ can be written

$$\big(\mathrm{Ad}(g^{-1})\xi\big)(z) = \big(Dg\big)_z^{-1}\big(\xi(g \cdot z)\big).$$

Consider first the constant vector field $\xi_u(z) = u$ ($u \in V^{\mathbf{C}}$), and let

$$B(g,z)u = \big(\mathrm{Ad}(g^{-1})\xi_u\big)(z) = (Dg)_z^{-1}(u).$$

Then $B(g,z)$ is a polynomial in z of degree ≤ 2 with values in $End(V^{\mathbf{C}})$. Consider now the vector field η, $\eta(z) = z$, and let

$$A(g,z) = \big(\mathrm{Ad}(g^{-1})\eta\big)(z).$$

It is a polynomial in z of degree ≤ 2 with values in $V^{\mathbf{C}}$, and

$$g \cdot z = B(g,z)^{-1}A(g,z).$$

Proposition 5.5. (i) *For $g \in G^{\mathbf{C}}$, $z \in V^{\mathbf{C}}$, the element $g \cdot z$ belongs to $V^{\mathbf{C}}$ if and only if $B(g,z)$ is invertible.* (ii) *Recall that $N^{\mathbf{C}}$ is the group of complex translations.*

$$G^{\mathbf{C}} = N^{\mathbf{C}} \circ s \circ P^{\mathbf{C}} \circ s \circ N^{\mathbf{C}}.$$

(iii) *If $p, q \in Y$, there exists $g \in G^{\mathbf{C}}$ such that $g \cdot p$ and $g \cdot q \in V^{\mathbf{C}}$.*

Proof.

(i) follows from what we said above. Let us prove (ii) and let $g \in G^{\mathbf{C}}$. There is an $a \in V^{\mathbf{C}}$ such that $g \cdot a \in V^{\mathbf{C}}$,

$$g \circ \tau_a \cdot 0 \in V^{\mathbf{C}},$$

or

$$g \circ \tau_a \cdot 0 = \tau_b \cdot 0,$$

with some $b \in V^{\mathbf{C}}$. Therefore $\tau_{-b} \circ g \circ \tau_a$ belongs to the stabilizer of 0 in $G^{\mathbf{C}}$ which is $s \circ P^{\mathbf{C}} \circ s$.

(iii) amounts to saying that, for $p \in Y$, there exists $g \in s \circ P^{\mathbf{C}} \circ s$ such that $g \cdot p \in V^{\mathbf{C}}$, and this is equivalent to (ii). ∎

In the following table the groups are given up to local isomorphism.

$V^{\mathbb{C}}$	$G^{\mathbb{C}}$	U	K
$\mathbb{C}^n$	$SO(n+2,\mathbb{C})$	$SO(n+2)$	$SO(n)\times SO(2)$
$Sym(m,\mathbb{C})$	$Sp(m,\mathbb{C})$	$Sp(m)$	$U(m)$
$M(m,\mathbb{C})$	$SL(2m,\mathbb{C})$	$SU(2m)$	$S(U(m)\times U(m))$
$Skew(2m,\mathbb{C})$	$SO(4m,\mathbb{C})$	$SO(4m)$	$U(2m)$
$Herm(3,\mathbb{O})\otimes\mathbb{C}$	$E_7(\mathbb{C})$	E_7	$E_6\times SO(2)$

6. A realization of the conformal compactification of a simple complex Jordan algebra

In [1977], Section 7, Loos describes the conformal compactification Y of $V^{\mathbb{C}}$ in the more general setting of semi-simple Jordan pairs. Recently Roos gave a simplified version of Loos' construction [1994]. Following Roos' idea we present in this section a realization of the conformal compactification for a simple complex Jordan algebra. We will construct a homogeneous embedding of the conformal compactification Y in a projective space.

Denote by $\mathcal{P}(V^{\mathbb{C}})$ the space of holomorphic polynomial functions on $V^{\mathbb{C}}$, and let $\mathcal{W}\subset\mathcal{P}(V^{\mathbb{C}})$ be the finite dimensional subspace generated by the polynomials

$$p_a(z)=\Delta(z-a),\quad a\in V^{\mathbb{C}}.$$

We will define a projective representation of $G^{\mathbb{C}}$ on $\mathcal{W}$. We first look at the action of $L^{\mathbb{C}}$ on $\mathcal{W}$. Let us recall the expansion of the minimal polynomial of a regular element $x\in V^{\mathbb{C}}$,

$$\Delta(\lambda e-x)=\lambda^r-a_1(x)\lambda^{r-1}+\cdots+(-1)^r a_r(x),$$

and let $\mathcal{P}_{\langle j\rangle}$ be the $L^{\mathbb{C}}$-module generated by the polynomial a_j. (We have used the notation of [Faraut-Korányi,1994], p.28 and 236.) Notice that $a_r=\Delta$ and $\mathcal{P}_{\langle r\rangle}=\mathbb{C}\Delta$.

Proposition 6.1. *As a $L^{\mathbb{C}}$-module $\mathcal{W}$ decomposes into irreducible subspaces as*

$$\mathcal{W}=\mathbb{C}\oplus\mathcal{P}_{\langle 1\rangle}\oplus\cdots\oplus\mathcal{P}_{\langle r\rangle}.$$

Proof.

The space $\mathcal{W}$ is $L^{\mathbf{C}}$-invariant. Define

$$\mathcal{W}' = \mathbf{C} \oplus \mathcal{P}_{\langle 1\rangle} \oplus \cdots \oplus \mathcal{P}_{\langle r\rangle}.$$

Assume that $a \in V^{\mathbf{C}}$ is invertible.. We can write $a = b^{-2}$ (see Proposition VIII.3.4 in [Faraut-Korányi, 1994]), and then

$$\begin{aligned}\Delta(z-a) &= (-1)^r \Delta(a)\Delta\big(e - P(b)z\big)\\ &= \Delta(a)\sum_{j=1}^{r}(-1)^{r-j}a_j\big(P(b)z\big),\end{aligned}$$

and $a_j\big(P(b)z\big) \in \mathcal{P}_{\langle j\rangle}$. By continuity it follows that, for all $a \in V^{\mathbf{C}}$, $p_a \in \mathcal{W}'$. Therefore $\mathcal{W} \subset \mathcal{W}'$. Furthermore

$$a_j(z) = \frac{(-1)^j}{(r-j)!}\Big(\frac{d}{d\lambda}\Big)^{r-j}\Delta(\lambda e - z)\big|_{\lambda=0}.$$

Therefore $a_j \in \mathcal{W}$ and $\mathcal{P}_{\langle j\rangle} \subset \mathcal{W}$.

∎

(The space $\mathcal{W}$ is the subspace of $\mathcal{P}(V^{\mathbf{C}})$ denoted by M_0 for the value $\lambda = -1$ in [Faraut-Koranyi, 1990], p. 85.)

Let χ be the character of $L^{\mathbf{C}}$ corresponding to the semi-invariant Δ,

$$\Delta(g\cdot z) = \chi(g)\Delta(z) \quad (g \in L^{\mathbf{C}}).$$

(Recall that $\operatorname{Det} g = \chi(g)^{\frac{n}{r}}$.) If a conformal transformation $g \in G^{\mathbf{C}}$ is defined at z, and if its differential Dg is invertible at z, then $(Dg)_z \in L^{\mathbf{C}}$. We put

$$\alpha(g,z) = \chi\big((Dg)_z\big)^{-\frac{1}{2}}.$$

It is defined up to a sign. If g is a translation then $\alpha(g,z) = 1$, if g belongs to $L^{\mathbf{C}}$, then $\alpha(g,z) = \chi(g)^{-\frac{1}{2}}$, and if $g = s : z \mapsto -z^{-1}$, then $\alpha(g,z) = \Delta(z)$. Actually $\alpha(g,z)$ is a well defined cocycle on $\tilde{G}^{\mathbf{C}} \times V^{\mathbf{C}}$, for a double covering group $\tilde{G}^{\mathbf{C}}$ of $G^{\mathbf{C}}$.

For $p \in \mathcal{W}$, $g \in G^{\mathbf{C}}$, we put

$$\big(\pi(g)p\big)(z) = \alpha(g^{-1},z)p(g^{-1}\cdot z).$$

If g is a translation, $g = \tau_a : z \mapsto z + a$, then

$$\big(\pi(\tau_a)p\big)(z) = p(z - a),$$

if $g \in L^{\mathbf{C}}$, then

$$\big(\pi(g)p\big)(z) = \chi(g)^{\frac{1}{2}} p(g^{-1}z),$$

and if $g = s : z \mapsto -z^{-1}$, then

$$\big(\pi(s)p\big)(z) = \Delta(z)p(-z^{-1}).$$

Proposition 6.2. *π is an irreducible projective representation of G on $\mathcal{W}$.*

Proof.

Consider first the operators $\pi(g)$ as acting on the space of rational functions on $V^{\mathbf{C}}$. The representation property follows from the cocycle property of α,

$$\alpha(g_1 g_2, z) = \pm\alpha(g_1, g_2 \cdot z)\alpha(g_2, z).$$

We will show that $\mathcal{W}$ is stable under the representation π. It is clear that it is stable under the operators $\pi(\tau_a)$ $(a \in V^{\mathbf{C}})$, and $\pi(h)$ $(h \in L^{\mathbf{C}})$. The proof of the invariance of $\mathcal{W}$ under $\pi(s)$ has been communicated to the first author by A. Koranyi. It uses the following lemma, and the relation

$$s \circ h = \sigma(h) \circ s \quad (h \in L^{\mathbf{C}}).$$

Lemma 6.3

$$\Delta(z)a_j(z^{-1}) = a_{r-j}(z).$$

Proof.

One considers the following identity

$$\Delta(z)\Delta(\lambda e + z^{-1}) = \Delta(\lambda z + e) \quad (\lambda \in \mathbf{C}),$$

and expands both sides,

$$\sum_{j=0}^{r} \lambda^{r-j}\Delta(z)a_j(z^{-1}) = \sum_{j=0}^{r} \lambda^j a_j(z).$$

■

We show now that the representation π is irreducible. Let $\mathcal{W}_0 \neq \{0\}$ be an invariant subspace and $p_0 \not\equiv 0$ be a polynomial in $\mathcal{W}_0$. The derivatives of p_0 belong to $\mathcal{W}_0$, therefore the constants belong to $\mathcal{W}_0$, and so does the polynomial $\Delta = \pi(s)f_0$, with $f_0 \equiv 1$, and the polynomials $\big(\pi(\tau_a)\Delta\big)(z) = \Delta(z-a)$. Therefore $\mathcal{W}_0 = \mathcal{W}$.

■

As a consequence, for $g \in G^{\mathbb{C}}$,

$$\alpha(g,z) = \big(\pi(g^{-1})f_0\big)(z)$$

is, for g fixed, a polynomial in z which belongs to $\mathcal{W}$.

Proposition 6.4. *The stabilizer in $G^{\mathbb{C}}$ of the complex line $\mathbb{C}f_0$ is the maximal parabolic subgroup $P^{\mathbb{C}}$.*

Proof.

If $g^{-1} \in G^{\mathbb{C}}$ stabilizes $\mathbb{C}f_0$, then $\alpha(g,z)$ is a non-zero constant. Since

$$\det B(g,z) = \alpha(g,z)^{\frac{2n}{r}},$$

$B(g,z)$ is invertible for all $z \in V^{\mathbb{C}}$. By Proposition 5.5 it follows that

$$g(V^{\mathbb{C}}) \subset V^{\mathbb{C}},$$

and this implies that $g \in P^{\mathbb{C}}$.

■

It follows that the conformal compactification $Y = G^{\mathbb{C}}/P^{\mathbb{C}}$ can be identified with the orbit of $\mathbb{C}f_0$ in the projective space $\mathbb{P}(\mathcal{W})$. The embedding of $V^{\mathbb{C}}$ into Y is given by

$$V^{\mathbb{C}} \to \mathcal{W},$$
$$a \mapsto p_a = \pi(\tau_a \circ s)f_0,$$

with $p_a(z) = \Delta(z-a)$.

A polynomial p in W can be written

$$p(z) = c\Delta(z) + \text{ terms of degree } < r, \quad (c \in \mathbb{C}).$$

Let Λ be the linear form on $\mathcal{W}$ defined by

$$\Lambda(p) = c.$$

One shows easily that

$$\Lambda(p) = \big(\pi(s)p\big)(0).$$

Theorem 6.5. *In the realization of the manifold Y as the orbit of $\mathbb{C}f_0$ in $\mathbb{P}(\mathcal{W})$, the set $\Sigma = Y \setminus V^{\mathbb{C}}$ is defined by the equation*

$$\Lambda(p) = 0 \quad (p \in \mathcal{W}).$$

The part of Σ which is contained in the chart

$$V^{\mathbb{C}} \to Y,$$
$$w \mapsto \pi(s \circ \tau_w \circ s) f_0,$$

is defined by the equation

$$\Delta(w) = 0.$$

The set Σ is a hypersurface.

Proof.

A point a in $V^{\mathbb{C}}$ corresponds to the line $\mathbb{C}p_a \in \mathbb{P}(\mathcal{W})$, and

$$\Lambda(p_a) \neq 0.$$

Therefore we have to show that if a point $\mathbb{C}p$ lying on the orbit of $\mathbb{C}f_0$ is such that $\Lambda(p) \neq 0$, then there exists $a \in V^{\mathbb{C}}$ with $p \in \mathbb{C}p_a$. There exists $g \in G^{\mathbb{C}}$ such that $p(z) = c\alpha(g,z)$ $(c \neq 0)$ and the condition $\Lambda(p) \neq 0$ means that $\alpha(g \circ s, 0) \neq 0$. By Proposition 5.5, $a = (g \circ s) \cdot 0 \in V^{\mathbb{C}}$ because

$$\operatorname{Det} B(g,z) = \alpha(g,z)^{\frac{2n}{r}},$$

and, since the stabilizer of 0 in $G^{\mathbb{C}}$ is $s \circ P^{\mathbb{C}} \circ s$, it amounts to say that $g \in \tau_a \circ s \circ P^{\mathbb{C}}$.

Therefore

$$p \in \mathbb{C}\pi(\tau_a \circ s) f_0 = \mathbb{C}p_a.$$

■

Examples

1) Let $V^{\mathbb{C}} = \mathbb{C}^n$ equipped with a product which makes $V^{\mathbb{C}}$ a rank 2 simple complex Jordan algebra, and such that the determinant polynomial is the quadratic form

$$\Delta(z) = z_1^2 + \cdots + z_n^2.$$

The conformal group is $G^{\mathbb{C}} = SO(n+2, \mathbb{C})$. and $\mathcal{W} \simeq \mathbb{C}^{n+2}$ is the space of polynomials of the form

$$p(z) = \alpha\Delta(z) + 2\sum_{j=1}^{n} \beta_j z_j + \gamma.$$

We have, for $a \in V^{\mathbb{C}}$,

$$\Delta(z-a) = \Delta(z) - 2\sum_{j=1}^{n} a_j z_j + \Delta(a),$$

and the orbit of $\mathbb{C}f_0$ in $\mathbb{P}(\mathcal{W})$ is the projective quadric with equation

$$\alpha\gamma = \sum_{j=1}^{n} \beta_j^2.$$

2) Let $V^{\mathbb{C}} = M(m, \mathbb{C})$ with the Jordan product. The group $SL(2m, \mathbb{C})$ acts on $V^{\mathbb{C}}$ by the transformations

$$g \cdot z = (az+b)(cz+d)^{-1}, \quad g = \begin{pmatrix} a & b \\ c & d \end{pmatrix},$$

and the conformal group is $G^{\mathbb{C}} = SL(2m, \mathbb{C})/\{\pm I\}$. Then

$$\alpha(g, z) = \det(cz+d).$$

For $g_1, g_2 \in SL(2m, \mathbb{C})$, the polynomials

$$p_1(z) = \det(c_1 z + d_1), \quad p_2(z) = \det(c_2 z + d_2)$$

are proportionnal if and only if there exists $u \in GL(m, \mathbb{C})$ such that

$$c_2 = uc_1, \quad d_2 = ud_1,$$

and the orbit of $\mathbb{C}f_0$ in $\mathbb{P}(\mathcal{W})$ is isomorphic to the Grassmannian manifold $Gr(m, 2m; \mathbb{C})$.

The function $\Delta(z-w)$ defined on $V \times V$ cannot be extended to $Y \times Y$ as a function. We will see that it can be extended as a G-invariant section of a line bundle defined on $Y \times Y$. For this construction we need the following lemma.

Lemma 6.6. *Let $z, w \in V$, $g \in G^{\mathbf{C}}$. If $g \cdot z$ and $g \cdot w$ belong to V, then*

$$\alpha(g, z)\alpha(g, w)\Delta(g \cdot z - g \cdot w) = \Delta(z - w).$$

Proof.

If g is a translation or an element in $L^{\mathbf{C}}$ this relation is obvious. If $g = s$ it amounts to writing

$$\Delta(z)\Delta(w)\Delta(-z^{-1} + w^{-1}) = \Delta(z - w),$$

which is a part of Lemma X.4.4 in [Faraut-Korányi,1994]. As $G^{\mathbf{C}}$ is generated by $N^{\mathbf{C}}$, $L^{\mathbf{C}}$, and s, the results follows.

■

From Lemma 6.6 it follows that

$$\beta(g; z, w) = \alpha(g, z)\alpha(g, w)$$

is a well defined cocycle on $G \times V \times V$. We define the line bundle $\mathcal{L}$ on $Y \times Y$ as follows: a section f of $\mathcal{L}$ is a function $f(g; p, q)$ defined for $g \in G^{\mathbf{C}}$, $p, q \in g \cdot V^{\mathbf{C}}$, which satisfies

$$f(g; g \cdot z, g \cdot w) = \beta(g; z, w) f(e; g \cdot z, g \cdot w)$$

for $z, w, g \cdot z, g \cdot w \in V^{\mathbf{C}}$. Note that, by Proposition 5.5, for $p, q \in Y$, there exists $g \in G^{\mathbf{C}}$ such that $g \cdot p, g \cdot q \in V^{\mathbf{C}}$. The group $G^{\mathbf{C}}$ acts on the space of sections of $\mathcal{L}$ by

$$\big(\tau(g)f\big)(\gamma; p, q) = f(g^{-1}\gamma; g^{-1} \cdot p, g^{-1} \cdot q).$$

If f is an invariant section, then

$$\beta(g; z, w) f(e; g \cdot z, g \cdot w) = f(e; z, w).$$

Lemma 6.6 means that the section δ such that

$$\delta(e; z, w) = \Delta(z - w)$$

is invariant. From Proposition 6.5 it follows that, for $p, q \in Y$, the condition

$$\delta(g; p, q) \neq 0$$

does not depends on g, and, if this condition is satisfied, we will say that (p, q) is a *generic pair*. The equation of the hypersurface Σ can be written

$$\delta(p, p_\infty) = 0,$$

where p_∞ is the point of Y corresponding to the complexe line $\mathbb{C}f_0$.

7. Pseudo-Hermitian symmetric spaces of tube type

One considers on the tube $T_\Omega = V + i\Omega$ the pseudo-Hermitian metric defined by

$$h_z(w, w') = -\partial_w \bar{\partial}_{w'} \log \Delta\Big(\frac{z - \bar{z}}{2i}\Big).$$

This metric is invariant under the semi-direct product $P = V \rtimes L$ which acts on T_Ω transitively.

Proposition 7.1. *The geodesic symmetry with respect to ie is defined by*

$$s : z \mapsto -z^{-1},$$

for z close to ie.

Proof.

Let $w \in V^{\mathbb{C}}$ and let $\gamma_w(t)$ be the geodesic such that

$$\gamma_w(0) = ie, \quad \dot{\gamma}_w(0) = w.$$

Then, for small t,

$$\gamma_w(t) = i(e - i \operatorname{th} \frac{t}{2} w)(e + i \operatorname{th} \frac{t}{2} w)^{-1},$$

and the result follows.

■

Therefore T_Ω is a pseudo-Hermitian symmetric space if and only if the map $s : z \mapsto -z^{-1}$ is a holomorphic automorphism of T_Ω. If it is so, then every element $z \in T_\Omega$ is invertible. If V is Euclidean, or, equivalently, if Ω is convex and proper, then every $z \in T_\Omega$ is invertible.

Lemma 7.2. *If V is not Euclidean, there are elements $z = x + iy \in T_\Omega$ which are not invertible.*

Proof. An element $x \in V_-$ has purely imaginary eigenvalues. In fact ix belongs to the Euclidean real form $W = V_+ + iV_-$ of $V^{\mathbf{C}}$, hence has real eigenvalues. We can find elements $x \in V_-$ with the eigenvalue $-i$, and $\Delta(x + ie) = 0$, therefore $x + ie$ is not invertible.

∎

Therefore we can state the following theorem.

Theorem 7.3. *The tube T_Ω is a pseudo-Hermitian symmetric space if and only if V is Euclidean, or, equivalently, if Ω is a convex symmetric cone. In that case T_Ω is a Hermitian symmetric space.*

Let G be the subgroup of $G^{\mathbf{C}}$ generated by P and the symmetry s (G is the conformal group of V). One shows that G is connected and closed and is a real form of $G^{\mathbf{C}}$. The orbit $X = G \cdot (ie)$ is open in Y, and $T_\Omega = P \cdot (ie)$ is open in X. Since X is connected and $\Sigma = Y \setminus V^{\mathbf{C}}$ is a hypersurface (Theorem 6.6), $X \cap V^{\mathbf{C}} = X \setminus \Sigma$ is connected. Therefore $X \cap V^{\mathbf{C}} = X \setminus \Sigma = T_\Omega$, and T_Ω is dense in X. Let H be the stabilizer of ie in G,

$$H = \{g \in G \mid g \cdot (ie) = ie\}.$$

We consider the involution $\tilde{\sigma}$ of G defined by

$$\tilde{\sigma}(g) = s \circ g \circ s.$$

Then

$$H = \{g \in G \mid \tilde{\sigma}(g) = g\}.$$

The Lie algebra $\mathfrak{g}$ of G is the space of vector fields

$$\xi(x) = u + Tx - P(x)v,$$

with $u, v \in V$, $T \in \mathfrak{l} = Lie(L)$. At the Lie algebra level the involution $\tilde{\sigma}$ is given by

$$\tilde{\sigma}(\xi)(x) = -P(x)\xi(x^{-1}) = -v + \sigma(T)x + P(x)u,$$

or

$$\tilde{\sigma}(u,T,v) = (-v, \sigma(T), -u).$$

If $\mathfrak{h}$ denotes the Lie algebra of H, then

$$\mathfrak{h} = \{(u,T,-u) \mid u \in V,\ T \in \mathfrak{a} = Lie(A)\}.$$

The group H is a real form of $L^{\mathbb{C}}$.

The space $X \simeq G/H$ is a pseudo-Hermitian symmetric space. Since it contains the tube T_Ω as an open dense set, we will say that X is a *pseudo-Hermitian symmetric space of tube type.*

Examples

1) Let $V = \mathbb{R}^n$ equipped with the product

$$z = x \cdot y \text{ if } z_1 = \sum_{j=1}^{p+1} x_j y_j - \sum_{j=p+2}^{n} x_j y_j,$$
$$z_j = x_1 y_j + x_j y_1,\ 2 \le j \le n,$$

with $p+q = n-1$, $1 \le q \le n-2$. Then

$$\Delta(x) = x_1^2 - Q(x_2, \ldots, x_n),$$

where Q is a quadratic form of signature (p,q), and

$$\Omega = \{x \in V \mid \Delta(x) > 0\}.$$

We take $L = SO_0(q+1,p) \times \mathbb{R}_+$, then $A \simeq SO_0(q,p)$, and $G^{\mathbb{C}} = SO(n+2, \mathbb{C})$. As we already said, the compact complex manifold Y is realized as the projective quadric in $P_{n+1}(\mathbb{C})$ with equation

$$\Delta(z) = uv.$$

the map

$$(z_1, \ldots z_n) \mapsto (z_1, \ldots, z_n, \Delta(z), 1)$$

defines the embedding $V^{\mathbb{C}}$ in Y. Furthermore

$$G = SO_0(q+1, p+1),\ H = SO_0(q, p+1) \times SO(2),$$

and $X = G/H$ can be realized as the open set in Y defined by

$$\Delta(z, \bar{z}) - \Re(u\bar{v}) > 0.$$

In fact, in $V^{\mathbb{C}}$, it gives the following condition

$$\Delta(z, \bar{z}) - \Re\Delta(z) > 0,$$

or, if $z = x + iy$,

$$\Delta(x) + \Delta(y) - \big(\Delta(x) - \Delta(y)\big) = 2\Delta(y) > 0,$$

which means that $z = x + iy \in T_\Omega = V + i\Omega$.

2) Let $V = M(m, \mathbb{R})$ equipped with the Jordan product, then

$$\Omega = \{x \in M(m, \mathbb{R}) \mid \det(x) > 0\},$$

and $G^{\mathbb{C}} = SL(2m, \mathbb{C})$. Let us recall that the compact complex manifold is realized as the Grassmannian $Gr(m, 2m; \mathbb{C})$ or

$$Y = \{Z \in M(2m, m; \mathbb{C}) \mid Z \text{ of rank } m\}/\sim .$$

Furthermore $G = SL(2m, \mathbb{R})$,

$$H = \left\{ \begin{pmatrix} a & b \\ -b & a \end{pmatrix} \mid a + ib \in GL(m, \mathbb{C}),\ |\det(a + ib)| = 1 \right\},$$

and $X = G/H$ is realized as an open set in Y as follows: to a matrix $W = U + iV \in M(2m, m; \mathbb{C})$ one associates the matrix $(V\ U) \in M(2m, 2m; \mathbb{R})$. Then X is defined by the condition $\det(V\ U) > 0$. In fact this condition gives on $V^{\mathbb{C}}$

$$\det \begin{pmatrix} y & x \\ 0 & I \end{pmatrix} > 0 \text{ or } \det y > 0,$$

which means that $z = x + iy \in T_\Omega = V + i\Omega$.

Let $z \mapsto \bar{z}$ denote the conjugation of $V^{\mathbb{C}}$ with respect to V, and let $g \mapsto \bar{g}$ be the conjugation of $G^{\mathbb{C}}$ defined by

$$\overline{g \cdot z} = \bar{g} \cdot \bar{z}.$$

Since $\overline{P^{\mathbb{C}}} = P^{\mathbb{C}}$, the conjugation $z \mapsto z$ of $V^{\mathbb{C}}$ extends to Y. For $g \in G$,

$$\beta(g; z, \bar{z}) > 0,$$

therefore, for $p \in Y$, the condition $\delta(g, p, \bar{p}) > 0$ does not depend on g. In particular, for $p \in X$, $\delta(g, p, \bar{p}) > 0$. If V is a real form of $V^{\mathbf{C}}$ of type II, then

$$X = \{p \in Y \mid \delta(g, p, \bar{p}) > 0\}.$$

If V is a real form of $V^{\mathbf{C}}$ of type III, then, for all $p \in Y$, $\delta(g, p, \bar{p}) \geq 0$, and

$$X = \{p \in Y \mid \delta(g, p, \bar{p}) \neq 0\}.$$

Recall that a simple real Jordan algebra is either a real form of a simple complex Jordan algebra, or a simple complex Jordan algebra considered as a real algebra. Let us say a few words about the pseudo-Hermitian symmetric space $X = G/H$ in the second case. Let V be a simple complex Jordan algebra. We will use the same notation as when considering V as a real Jordan algebra. The cone Ω is then the set of invertible elements in V:

$$\Omega = \{x \in V \mid \Delta(x) \neq 0\}$$

and

$$\Omega = L/A$$

where L is a complex Lie group coinciding with the complex structure group of V. Let Y will be the conformal compactification of V and G be the conformal group ($G^{\mathbf{C}}$ in the notation of Section 5). We can identify the complexification of V (as a real algebra) with

$$V^{\mathbf{C}} = V \times \overline{V},$$

the conformal compactification of $V^{\mathbf{C}}$ with

$$Y^{\mathbf{C}} = Y \times \overline{Y}$$

and the conformal group of $V^{\mathbf{C}}$ with

$$G^{\mathbf{C}} = G \times \overline{G}.$$

In all cases bar means substitution of a complex structure by its conjugate.

Let us identify G with its diagonal imbedding in $G^{\mathbf{C}} = G \times \overline{G}$. Then G has a unique open orbit

$$X = G/H$$

in $Y^{\mathbf{C}}$ and

$$T_\Omega = \{(x_1, x_2) \in V^{\mathbf{C}} = V \times \bar{V} \mid \Delta(x_1 - x_2) \neq 0\}.$$

The stabilizer H of $(ie, -ie)$ is conjugate to L by the Cayley transform,

$$H = c \circ L \circ c^{-1},$$

with

$$c(z) = i(e + z)(e - z)^{-1}.$$

This orbit is a pseudo-Hermitian symmetric space which is also a symmetric Stein manifold (see [Matsushima, 1960]). It is possible to interpret X as a complexification of the compact Hermitian symmetric space Y (we compexify the compact group of automorphisms of Y and the isotropy subgroup) as well as a complexification of any pseudo-Hermitian symmetric space in Y.

Let V_0 be a Euclidean Jordan algebra and V be its complexification. Let Ω_0 be the convex symmetric cone in V_0, $\Omega_0 = L_0/A_0$, and let G_0 be the conformal group of V_0 generated by the translations in V_0, L_0 and the symmetry s. Then G_0/A_0 is a symmetric space which is called of Cayley type (see [Ólafsson, 1991], Definition 5.7). The space Y can be seen as a complexification of G_0/A_0.

Let us emphasize that the manifold X is irreducible although the dual compact symmetric manifold $Y^{\mathbf{C}}$ is reducible. The space X is of the tube type. Let us consider the open set

$$X_0 = \{(x_1, x_2) \in V = V \times \overline{V} \mid \Delta(x_1 - x_2) \neq 0\}.$$

Obviously X_0 is biholomorphically equivalent to the tube

$$T_\Omega = V + i\Omega$$

We can interpret X as the manifold of generic pairs of points of Y, i.e. pairs (p, q) for which

$$\delta(g; p, q) \neq 0,$$

where δ is the section of the line bundle $\mathcal{L}$ we introduced in Section 6.

Example

Let $V = M(m, \mathbb{C})$ and $Y = Gr(m, 2m; \mathbb{C})$ be the Grassmannian of m-subspaces in $\mathbb{C}^{2m}$. Then X consists of pairs of subspaces intersecting only at 0.

The space $X = G/H$ has two remarkable submanifolds: X_+ and X_-. The former is a Hermitian symmetric space of tube type, the latter is a compact Hermitian symmetric space.

The submanifold X_+ is the tube domain

$$X_+ = V_+ + i\Omega_+ \subset V_+^{\mathbb{C}}.$$

Let τ be the involution of G defined by

$$\tau(g) = \nu \circ g \circ \nu.$$

At the Lie algebra level it is given by

$$\tau(u, T, v) = \big(\nu(u), \theta(T), \nu(v)\big),$$

and $X_+ \simeq G^\tau / H^\tau$. Notice that

$$\mathfrak{g}^\tau = \{(u, T, v) \mid u, v \in V_+,\ T \in \mathfrak{k}_0\},$$
$$\mathfrak{h}^\tau = \{(u, T, -u) \mid u \in V^+,\ T \in \mathfrak{k}_0 \cap \mathfrak{a}\} \simeq \mathfrak{k}_\circ \cap \mathfrak{a} + i(\mathfrak{p}_0 \cap \mathfrak{q}_0),$$

and that H^τ is compact.

In order to define X_- one considers the Cartan involution $\tilde{\theta}$ of G defined by

$$\tilde{\theta}(g) = \nu \circ s \circ g \circ s \circ \nu = \tilde{\sigma} \circ \tau = \tau \circ \tilde{\sigma}.$$

At the Lie algebra level

$$\tilde{\theta}(u, T, v) = \big(-\nu(v), \theta(T), -\nu(u)\big),$$

and

$$\mathfrak{g}^\theta = \{(u, T, -\nu(u)) \mid u \in V,\ T \in \mathfrak{k}_0\}.$$

Then $X_- = G^\theta \cdot (ie) \simeq G^\theta / H^\theta$. Notice that $H^\theta = H^\tau$.

Example

If $V = M(m, \mathbb{R})$, then

$$G = SL(2m, \mathbb{R}),$$
$$H = \left\{ \begin{pmatrix} a & b \\ -b & a \end{pmatrix} \mid a + ib \in GL(m, \mathbb{C}),\ |\det(a + ib)| = 1 \right\}.$$

The manifold X_+ is the tube over the cone of positive definite symmetric matrices,

$$G^\tau = Sp(m, \mathbb{R}),\ H^\tau \simeq U(m).$$

The manifold X_- is a compactification of the space of complex skew-symmetric matrices,

$$G^\theta = SO(2m),\ H^\theta \simeq U(m).$$

We have considered the conformal compactification $Y = G^{\mathbb{C}}/P^{\mathbb{C}}$ of a complex simple Jordan algebra $V^{\mathbb{C}}$. More generally, for a real or complex simple Jordan algebra, G/P is its conformal complactification (see [Bertram, 1996]). In [Kantor-Sirota-Solodovnikov,1967], [Rivilis,1970], and [Makarevic,1973], one considers symmetric spaces which can be realized as open orbits in G/P of a subgroup of G. The pseudo-Hermitian symmetric spaces of tube type are special cases of such symmetric spaces.

$V^{\mathbb{C}}$		Type I	Type II	Type III ($m = 2\ell$)
$Sym(m,\mathbb{C})$	V	$Sym(m,\mathbb{R})$	×	$Sym(2\ell,\mathbb{C}) \cap M(\ell,\mathbb{H})$
	$\mathfrak{l}$	$\mathfrak{sl}(m,\mathbb{R}) \oplus \mathbb{R}$	×	$\mathfrak{sl}(\ell,\mathbb{H}) \oplus \mathbb{R}$
	$\mathfrak{a}$	$\mathfrak{so}(p,q)$	×	$\mathfrak{so}^*(2\ell)$
	$\mathfrak{g}$	$\mathfrak{sp}(m,\mathbb{R})$	×	$\mathfrak{sp}(\ell,\ell)$
	$\mathfrak{h}$	$\mathfrak{su}(p,q) \oplus i\mathbb{R}$	×	$\mathfrak{su}(\ell,\ell) \oplus i\mathbb{R}$
$M(m,\mathbb{C})$	V	$Herm(m,\mathbb{C})$	$M(m,\mathbb{R})$	$M(\ell,\mathbb{H})$
	$\mathfrak{l}$	$\mathfrak{sl}(m,\mathbb{C}) \oplus \mathbb{R}$	$\mathfrak{sl}(m,\mathbb{R}) \oplus \mathfrak{sl}(m,\mathbb{R}) \oplus i\mathbb{R}$	$\mathfrak{sl}(\ell,\mathbb{H}) \oplus \mathfrak{sl}(\ell,\mathbb{H}) \oplus \mathbb{R}$
	$\mathfrak{a}$	$\mathfrak{su}(p,q)$	$\mathfrak{sl}(m,\mathbb{R})$	$\mathfrak{sl}(l,\mathbb{H})$
	$\mathfrak{g}$	$\mathfrak{su}(m,m)$	$\mathfrak{sl}(2m,\mathbb{R})$	$\mathfrak{sl}(2\ell,\mathbb{H})$
	$\mathfrak{h}$	$\mathfrak{su}(p,q) \oplus \mathfrak{su}(p,q) \oplus i\mathbb{R}$	$\mathfrak{sl}(m,\mathbb{C}) \oplus i\mathbb{R}$	$\mathfrak{sl}(2\ell,\mathbb{C}) \oplus i\mathbb{R}$
$Skew(2m,\mathbb{C})$	V	$Herm(m,\mathbb{H})$	$Skew(2m,\mathbb{R})$	×
	$\mathfrak{l}$	$\mathfrak{sl}(m,\mathbb{H}) \oplus \mathbb{R}$	$\mathfrak{sl}(2m,\mathbb{R}) \oplus \mathbb{R}$	×
	$\mathfrak{a}$	$\mathfrak{su}(p,q,\mathbb{H})$	$\mathfrak{sp}(m,\mathbb{R})$	×
	$\mathfrak{g}$	$\mathfrak{so}^*(4m)$	$\mathfrak{so}(2m,2m)$	×
	$\mathfrak{h}$	$\mathfrak{su}(2p,2q) \oplus i\mathbb{R}$	$\mathfrak{su}(m,m) \oplus i\mathbb{R}$	×
$Herm(3,\mathbb{O}) \otimes \mathbb{C}$	V	$Herm(3,\mathbb{O})$	$Herm(3,\mathbb{O}_s)$	×
	$\mathfrak{l}$	$\mathfrak{e}_{6(-26)} \oplus \mathbb{R}$	$\mathfrak{e}_{6(6)} \oplus \mathbb{R}$	×
	$\mathfrak{a}$	$\mathfrak{f}_4, \mathfrak{f}_{4(-20)}$	$\mathfrak{f}_{4(4)}$	×
	$\mathfrak{g}$	$\mathfrak{e}_{7(-25)}$	$\mathfrak{e}_{7(7)}$	×
	$\mathfrak{h}$	$\mathfrak{e}_6 \oplus i\mathbb{R}, \mathfrak{e}_{6(-14)} \oplus i\mathbb{R}$	$\mathfrak{e}_{6(2)} \oplus i\mathbb{R}$	×

References

D'ATRI, J.E., and GINDIKIN, S. (1993). Siegel domains realization of pseudo-Hermitian symmetric manifolds, *Geometriae Dedicata*, **46**, 91–126.

BERTRAM, W. (1996). Un théorème de Liouville pour les algèbres de Jordan, *Bull. Soc. Math. France*, **124**, 101–129.

FARAUT, J. and KORÁNYI, A. (1990). Function spaces and reproducing kernels on bounded symmetric domains, *J. Funct. Anal.*, **88**, 64–89.

FARAUT, J., and KORÁNYI, A. (1994). *Analysis on symmetric cones.* Oxford University Press.

GINDIKIN, S. (1992). Fourier transform and Hardy spaces of $\bar{\partial}$-cohomology in tube domains, *C. R. Acad. Sci. Paris*, **315**, série I, 1139–1143.

KANTOR, I.L., SIROTA A.I., and SOLODOVNIKOV A.S. (1967). A class of symmetric spaces with extendable group of motions and a generalization of the Poincaré model, *Soviet. Math. Dokl.*, **8**, 423–426.

KANEYUKI, S. (1991). Pseudo-Hermitian symmetric spaces and Siegel domains over nondegenerate cones, *Hokkaido Math. J.*, **20**, 213–239.

KAYOYA, J.B. (1994). *Analyse sur les algèbres de Jordan simples réelles.* Thèse, Université de Paris VI.

LOOS, O. (1977). *Bounded symmetric domains and Jordan pairs.* University of California, Irvine.

MAKAREVIČ, B.O. (1973). Open symmetric orbits of reductive groups in symmetric spaces, *Math. Sbornik*, 20, 406–418.

MATSUSHIMA, Y. (1960). Espaces homogènes de Stein des groupes de Lie complexes, *Nagoya Math. J.*, **16**, 205–218.

ÓLAFSSON, G. (1991). Symmetric spaces of Hermitian type, *Differential Geometry and its Applications*, **1**, 195–233.

RIVILIS, A.A. (1970). Homogeneous locally symmetric regions in homogeneous spaces associated with semi-simple Jordan algebras, *Math. Sbornik*, **11**, 377-426.

ROOS, G. (1994). *Volume of bounded symmetric domains and compactification of Jordan triple systems.* Preprint.

SHIMA, H. (1975). On locally symmetric homogeneous domains of completely reducible linear Lie groups, *Math. Ann.*, **217**, 93–95.

VINBERG, E.B. (1960). Homogeneous cones, *Soviet Math. Dokl.*, **1**, 787–790.

Jacques Faraut
Institut de mathématiques,
Université Pierre et Marie Curie
Case 247, 4 place Jussieu, 75252 Paris cedex 05
email: faraut@mathp6.jussieu.fr

Simon Gindikin
Department of Mathematics, Rutgers University
New Brunswick, NJ 08903
e.mail gindikin@math.rutgers.edu

Received October 1995

Homogeneous Riemannian Manifolds Whose Geodesics Are Orbits

*Carolyn S. Gordon**

Let M be a homogeneous Riemannian manifold and G the isometry group of M. Then M can be viewed as a coset space G/H with a left-invariant Riemannian metric. M is said to be a *g. o. manifold* if every geodesic in M is an orbit of a one-parameter subgroup of G. In [KV1], O. Kowalski and L. Vanhecke showed that every g. o. manifold is a D'Atri space; i.e. the local geodesic symmetries are volume preserving up to sign. (See the survey article by Kowalski, Prufer and Vanhecke in this volume for a discussion of D'Atri spaces and additional comments about g. o. manifolds.) The simplest Riemannian homogeneous spaces are the naturally reductive manifolds; these form a subclass of the g. o. manifolds. The definition of naturally reductive (see §1) involves a purely algebraic condition on the isometry group.

In [DZ], J. D'Atri and W. Ziller constructed a class of left-invariant naturally reductive metrics on quotients of compact Lie groups G and showed that their construction included all left-invariant naturally reductive metrics on compact simple Lie groups. They also partially classified the Einstein metrics among these naturally reductive metrics. Motivated by their work, the author [G2] studied the structure of noncompact naturally reductive Riemannian manifolds. In contrast to the compact case, there are no nonsymmetric Einstein metrics among the noncompact naturally reductive metrics. Indeed, by combining results from [DZ] and [G2], Ziller and the author [GZ] showed that every naturally reductive metric of nonpositive Ricci curvature is symmetric.

In this article we study the structure of g. o. manifolds and compare our results with those of the more restrictive class of naturally reductive metrics. As we will see, the g. o. condition is considerably weaker than natural reductivity. However, we will still find that every g. o. manifold of nonpositive Ricci curvature is symmetric.

The paper is organized as follows: In §1, after reviewing the definitions and establishing some elementary properties of naturally reductive and g. o. manifolds, we obtain a number of structural conditions on the isometry groups of such manifolds. The main result of this section partially reduces the classification problem for g. o. manifolds to the three

---

* Partly supported by a grant from the National Science Foundation.

special cases in which (a) M is a nilmanifold (i.e. M is a nilpotent Lie group with a left-invariant Riemannian metric), (b) M is compact or (c) M admits a transitive group of isometries which is semisimple of non-compact type. The three special cases are studied in §§2, 3 and 4, respectively. In each case we also recall the known results for naturally reductive metrics. The most complete results are obtained for case (a) where we are able to completely characterize the g. o. nilmanifolds. Finally in §5, we prove that g. o. manifolds of non-positive Ricci curvature are symmetric.

Before proceeding, we remark:

You can't "hear" whether a Riemannian metric is a g. o. metric.

We explain the word "hear" in this context. In the case of closed Riemannian manifolds, the eigenvalue spectrum of the associated Laplace-Beltrami operator is discrete and is often referred to as the *spectrum* of the manifold. Following Mark Kac, we say a geometric property can be *heard* if it is determined by the spectrum of the manifold. It is not known whether one can hear the g. o. property for compact manifolds (at least not to the author's knowledge). We instead consider the noncompact case. In this case the spectrum need not be discrete. We will say, however, that two Riemannian manifolds M and M' are *isospectral* if there exists a unitary operator $T : L^2(M) \to L^2(M')$ which intertwines the Laplacians, i.e., such that $\Delta' = T \circ \Delta \circ T^{-1}$. In case M and M' are compact, this is the same as saying that they have the same spectrum. In [G3], an example is given of two simply-connected Riemannian nilmanifolds M and M', one of which is a g. o. manifold and the other not, which have isospectral compact quotients. The results of [GW2] show that the manifolds M and M' are themselves isospectral.

I would like to thank the editors for enabling me to contribute to this volume honoring Joseph D'Atri. My research was influenced not only by D'Atri's work but also by the encouragement he provided early in my career.

1. Structure of G. 0. Manifolds

Let M be a connected homogeneous Riemannan manifold. If G is any connected transitive group of isometries of M and L is the isotropy subgroup at a point $p \in M$, then M is naturally identified with the coset space G/L with a left-invariant Riemannian metric. We always assume that the action of G is effective, equivalently, L contains no nontrivial normal subgroups of G.

1.1 Let $I_0(M)$ be the identity component of the full isometry group of M.

If G is a transitive subgroup of $I_0(M)$, then the isotropy subgroup L of G is compact if and only if G is closed in $I_0(M)$. However, in any case, the Lie algebra $\mathfrak{l}$ of L is compactly embedded in $\mathfrak{g}$; i.e., $\mathfrak{g}$ admits an inner product relative to which ad(X) is skew-symmetric for all $X \in \mathfrak{l}$. (Here, ad(X) is the adjoint map $\mathrm{ad}_{\mathfrak{g}}(X)$ in $\mathfrak{g}$.) In particular, we can choose a complement $\mathfrak{q}$ of $\mathfrak{l}$ in $\mathfrak{g}$ with $\mathrm{Ad}(L)\mathfrak{q} \subseteq \mathfrak{q}$. The space $\mathfrak{q}$ is identified with the tangent space T_pM via the mapping $X \to \frac{d}{dt}|_{t=0} \exp tX \cdot p$, and the Riemannian structure defines an Ad(L)-invariant inner product $<,>$ on $\mathfrak{q}$.

1.2 Notational Conventions. We will always denote the Lie algebra of a Lie group by the corresponding fraktur letter. Given a decomposition $\mathfrak{g} = \mathfrak{l} + \mathfrak{q}$ as above, we will write $X_{\mathfrak{l}}$ and $X_{\mathfrak{q}}$ for the $\mathfrak{l}$ and $\mathfrak{q}$ components of X. If $\mathfrak{q} = \mathfrak{q}_1 + \mathfrak{q}_2$, direct sum of subspaces, we will also write $X_{\mathfrak{q}} = X_{\mathfrak{q}_1} + X_{\mathfrak{q}_2}$. We will let $\pi_{\mathfrak{q}} : \mathfrak{g} \to \mathfrak{q}$ be the projection with kernel $\mathfrak{l}$, i.e., $\pi_{\mathfrak{q}}(X) = X_{\mathfrak{q}}$, and similarly define $\pi_{\mathfrak{q}_i}, i = 1, 2$.

1.3 Definition. M is said to be *naturally reductive* if for some transitive connected group G of isometries and decomposition $\mathfrak{g} = \mathfrak{l} + \mathfrak{q}$ as above, $\pi_{\mathfrak{q}} \circ \mathrm{ad}(X)_{|\mathfrak{q}}$ is skew-symmetric for all $X \in \mathfrak{q}$. I.e., $< [X, Y]_{\mathfrak{q}}, Z >= - < Y, [X, Z]_{\mathfrak{q}} >$ for all $X, Y, Z \in \mathfrak{q}$.

We caution that we will say a metric on a homogeneous space G/L is naturally reductive even though the particular choice G of transitive isometry group may not satisfy the condition of Definition 1.2

1.4 Definition. M is said to be a *g. o. space* if every geodesic in M is an orbit of a one-parameter group of isometries. I.e., there exists a transitive group G of isometries such that every geodesic in M is of the form $\exp(tX)\cdot q$ with $X \in \mathfrak{g}$, $q \in M$. Note that we can always choose G to be $I_0(M)$ in this case. To check whether a homogeneous manifold is a g. o. space, it is enough to check whether all the geodesics through a single chosen point p are of the form $\sigma(t) = \exp(tX)\cdot p$. Indeed any other point in M is of the form $a\cdot p$, with $a \in G$, and the geodesics through $a\cdot p$ are then of the form $a\sigma(t) = \exp(t\mathrm{Ad(a)X})\cdot\mathrm{a}\cdot\mathrm{p}$.

1.5 Definition. Fix a choice of base point $p \in M$. For G a transitive group of isometries of M, we will say a nonzero element X of $\mathfrak{g}$ is a *geodesic vector* if $\exp(tX) \cdot p$ is a geodesic.

The following lemma is immediate from the expression for the Levi-Civita connection on a homogeneous Riemannian manifold.

1.6 Lemma [KV2]. *Let M be a connected homogeneous Riemannian manifold and G a transitive group of isometries. In the notation of 1.1, 1.2, and 1.5, a nonzero element X of $\mathfrak{g}$ is a geodesic vector if and only if $< [X, Y]_{\mathfrak{q}}, X_{\mathfrak{q}} >= 0$ for all $Y \in \mathfrak{q}$.*

The following three propositions are all elementary consequences of Lemma 1.6 and the definitions above.

1.7 Proposition. *We use the notation of 1.1-1.3.*

(a) *M is naturally reductive with respective to the transitive group G of isometries and the decomposition $\mathfrak{g} = \mathfrak{l} + \mathfrak{q}$ if and only if every nonzero element of $\mathfrak{q}$ is a geodesic vector.*

(b) *Every geodesic in M is an orbit of a one-parameter group of isometries of G if and only if for each $X \in \mathfrak{q}$, there exists $A \in \mathfrak{l}$ such that $< [X+A, Y]_{\mathfrak{q}}, X >= 0$ for all $Y \in \mathfrak{q}$. (I.e., $X + A$ is a geodesic vector.) In particular, M is a g. o. manifold if and only if this condition holds for $G = I_0(M)$.*

1.8 Proposition. *Suppose every geodesic in M is an orbit of a one-parameter subgroup of G. In the notation of 1.1-1.2, let $\mathfrak{q} = \mathfrak{q}_1 + \mathfrak{q}_2$ be an orthogonal decomposition of $\mathfrak{q}$ into $\mathrm{ad}(\mathfrak{l})$-invariant subspaces. Then for $X \in \mathfrak{q}_1$, the linear map $\pi_{\mathfrak{q}_2} \circ \mathrm{ad}(X)_{|\mathfrak{q}_2}$ is skew-symmetric. In particular, if an element X of $\mathfrak{q}$ commutes with $\mathfrak{l}$, then $\pi_{\mathfrak{q}} \circ \mathrm{ad}(X)_{|\mathfrak{q}}$ is skew-symmetric.*

Proof. For $X \in \mathfrak{q}_1$, choose $A \in \mathfrak{l}$ as in Proposition 1.7(b). For $U \in \mathfrak{q}_2$, we have by Lemma 1.6 and the hypothesis that

$$\begin{aligned} 0 &=< [A+X, U]_{\mathfrak{q}}, U > \\ &=< [A+X, U]_{\mathfrak{q}_2}, U > \\ &=< [A, U], U > + < \pi_{\mathfrak{q}_2} \circ \mathrm{ad}(X)(U), U > . \end{aligned}$$

Since $\mathrm{ad}(A)$ is skew-symmetric, the proposition follows.

1.9 Proposition. *Suppose every geodesic in $M = G/L$ is an orbit of a one-parameter subgroup of G. Let H be a subgroup of G normalized by L. Then the submanifold $H/(L \cap H)$ $(= HL/L)$ of M is totally geodesic and is itself a g. o. manifold.*

1.10 Notation. For any connected Lie group G, we will denote by $G = G_1 G_2$ a Levi decomposition of G. I.e., G_1 is a maximal connected semisimple subgroup of G, unique up to conjugacy, and G_2 is the solvable radical of G. The subgroup G_1 can be further decomposed $G_1 = G_{nc}G_c$ where G_{nc} and G_c are semisimple of noncompact type and compact type, respectively.

1.11 Definition. Let G be a connected transitive group of isometries of a Riemannian manifold M and let L be the isotropy subgroup at $p \in M$. A semisimple Levi factor G_1 is said to be *compatible* with L if $G_1 L$ is a reductive subgroup of G. (Recall that a Lie group H is said to be *reductive* if the radical H_2 is central in H. Equivalently, $[\mathfrak{h}, \mathfrak{h}]$ is semisimple and $\mathfrak{h} = [\mathfrak{h}, \mathfrak{h}] + z(\mathfrak{h})$ where $z(\mathfrak{h})$ is the center of $\mathfrak{h}$.) In particular if G_1 is compatible with L, then G_1, being a maximal semisimple subgroup of G and hence of G_1L, is the unique semisimple Levi factor of G_1L. We have $\mathfrak{g}_1 + \mathfrak{l} = \mathfrak{g}_1 \oplus \mathfrak{t}$ where $\mathfrak{t}$ is a compactly embedded abelian subalgebra of $\mathfrak{g}$ which commutes with $\mathfrak{g}_1$. (See [GW1].)

1.12 Lemma [GW1]. *Let G be a transitive connected group of isometries of a Riemannian manifold M. Given $p \in M$, there exists a semisimple Levi factor G_1 of G compatible with the isotropy subgroup of G at p.*

1.13 Remark. Suppose G_1 is compatible with L and define $\mathfrak{t}$ as in 1.11. Let $\mathfrak{s}$ be the orthogonal complement of $\mathfrak{g}_1 + \mathfrak{l}$ relative to the Killing form B of $\mathfrak{g}$. Then nilrad$(\mathfrak{g}) \subset \mathfrak{s} \subset \mathfrak{g}_2$. Since $\mathfrak{t}$ is compactly embedded in $\mathfrak{g}$, the Killing form of $\mathfrak{g}$ is negative semi-definite on $\mathfrak{t}$ and $\mathfrak{g}_2 = \mathfrak{t} + \mathfrak{s}$ with $\mathfrak{t} \cap \mathfrak{s}$ central in $\mathfrak{g}$.

1.14 Theorem. *Suppose every geodesic in $M = G/L$ is an orbit of a one-parameter group in G. Choose a semisimple Levi factor G_1 of G compatible with L and write $G_1 = G_{nc}G_c$ as in 1.10. Let $N =$* nilrad(G) . *Then $N \cap L = e$ and the submanifolds $G_{nc}/(G_{nc} \cap L)$, $G_c/(G_c \cap L)$, and N $(= N/(N \cap L)$ with the induced Riemannian metrics are g. o. manifolds and are totally geodesic in M.*

Proof. The statement that $N \cap L = \{e\}$ is true for any homogeneous Riemannian manifold $M = G/L$, since by 1.13, $N \cap L$ is central in G. (As always, we are assuming that G acts effectively on M.) The remaining statements follow from Proposition 1.9.

1.15 Theorem. *Suppose every geodesic in $M = G/L$ is an orbit of a one-parameter subgroup of G. Let $N =$* nilrad(G), *let G_1 be any semisimple Levi factor of G, and let G_{nc} be the noncompact part of G_1. Then:*

(i) G_1N acts transitively on M;

(ii) G_{nc} is normal in the full isometry group of M. In particular, G_{nc} commutes with N.

Theorem 1.15 is proven for naturally reductive manifolds in [G2] (Theorems 3.1 and 3.2). The proof of the general case requires only minor adjustments. The condition of Proposition 1.7(b) plays the role that the natural reductivity condition played in the earlier proof.

2. G. O. Nilmanifolds

A connected Riemannian manifold which admits a transitive nilpotent group N of isometries is called a *nilmanifold.* The action of N is necessarily simply transitive (see the proof of Theorem 1.14); thus the manifold may be identified with the group N endowed with a left-invariant metric. E. N. Wilson [W] proved that N is unique and is normal in the full connected isometry group G; in fact, $N = \text{nilrad}(G)$. (This fact is restated as Proposition 2.1 (a) below.) We say N is a *g. o. nilmanifold* if every geodesic is an orbit of a one-parameter subgroup of G.

2.1 Proposition [W].

(a) *Let N be a simply-connected Riemannian nilmanifold. The Lie algebra of the full isometry group is a semi-direct sum $\mathfrak{g} = \mathfrak{l} + \mathfrak{n}$ where the isotropy algebra $\mathfrak{l}$ consists of all derivations of $\mathfrak{n}$ which are skew-symmetric relative to the Riemannian inner product.*

(b) *Two Riemannian nilmanifolds N and N' are isometric if and only if there exists a Lie group isomorphism from N to N' which is also an isometry. Equivalently, there exists a linear map $\tau : \mathfrak{n} \to \mathfrak{n}'$ which is both a Lie algebra isomorphism and an inner product space isometry (relative to the Riemannian inner products on $\mathfrak{n}$ and $\mathfrak{n}'$).*

2.2. Theorem. *A g. o. nilmanifold N is at most 2-step nilpotent.*

This theorem was proven in [G2] in the special case that N is naturally reductive. However, the proof below is simpler even in this special case.

Proof. Let $\mathfrak{g}$ be the Lie algebra of the full isometry group. By Proposition 2.1, $\mathfrak{g} = \mathfrak{l} + \mathfrak{n}$, $\mathfrak{l} \cap \mathfrak{n} = (0)$ and $\mathfrak{n}$ is an ideal in $\mathfrak{g}$. Thus in the notation of 1.1, we can take $\mathfrak{q} = \mathfrak{n}$. Let $\mathfrak{a}$ be the orthogonal complement of $[\mathfrak{n}, \mathfrak{n}]$ in $\mathfrak{n}$. Since $\mathfrak{l}$ acts on $\mathfrak{n}$ by skew-symmetric derivations, $\mathfrak{l}$ leaves each of $[\mathfrak{n}, \mathfrak{n}]$ and $\mathfrak{a}$ invariant. Thus by Proposition 1.8, for $X \in \mathfrak{a}$, $\text{ad}(X)_{|[\mathfrak{n},\mathfrak{n}]}$ is skew-symmetric. Since it is also a nilpotent operator, $\text{ad}(X)_{|[\mathfrak{n},\mathfrak{n}]} = 0$. But $\mathfrak{a}$ generates $\mathfrak{n}$ as a Lie algebra. Hence $[\mathfrak{n}, \mathfrak{n}]$ is central, i.e., $\mathfrak{n}$ is at most 2-step nilpotent.

We will review the characterization of the naturally reductive nilmanifolds and then discuss the g. o. nilmanifolds. By Theorem 2.2, we may restrict our attention to two-step nilmanifolds N. For simplicity, we will assume throughout that N is simply-connected, though the general case requires only minor modification as in [G2].

2.3 Notation. Given a simply-connected Riemannian 2-step nilmanifold N, let $\mathfrak{n}$ be the Lie algebra of N and $<,>$ the Riemannian inner product on $\mathfrak{n}$. Set $\mathfrak{z} = [\mathfrak{n}, \mathfrak{n}]$ and $\mathfrak{a} = \mathfrak{z}^{\perp}$. Note that $\mathfrak{z}$ is central in $\mathfrak{n}$. We will denote by $so(\mathfrak{a})$ and $so(\mathfrak{z})$ the algebras of skew-symmetric transformations of $(\mathfrak{a}, <,>)$ and $(\mathfrak{z}, <,>)$. Given $Z \in \mathfrak{z}$ and $X \in \mathfrak{a}$, set $j(Z)X = (\text{ad}(X))^*(Z)$; i.e.,

$$< j(Z)X, Y > = < Z, [X, Y] > \qquad (*)$$

for all $Y \in \mathfrak{a}$. Then $j : \mathfrak{z} \to \mathrm{so}(\mathfrak{a})$ is an injective linear map. (We remark that our notation here differs from that of [G2] where $\mathfrak{z}$ was defined to be the center of $\mathfrak{n}$ and thus j was not necessarily injective. However, in the case of simply-connected N, the arguments of [G2] are valid with either interpretation of $\mathfrak{z}$.)

Conversely given inner product spaces $\mathfrak{a}$ and $\mathfrak{z}$ and an injective linear map $j : \mathfrak{z} \to \mathrm{so}(\mathfrak{a})$, we obtain a simply-connected Riemannian nilmanifold N by letting $(\mathfrak{n}, <,>)$ be the inner product space direct sum $\mathfrak{n} = \mathfrak{a} + \mathfrak{z}$, defining a Lie bracket on $\mathfrak{n}$ by $(*)$ together with the conditions $[\mathfrak{a}, \mathfrak{a}] \subset \mathfrak{z}$ and $[\mathfrak{n}, \mathfrak{z}] = 0$, and then letting N be the associated simply-connected Lie group with the left-invariant Riemannian metric defined by the inner product $<,>$. We will say $(\mathfrak{a}, \mathfrak{z}, j)$ is the *data triple* associated with the Riemannian nilmanifold N.

2.4 Proposition. *Let N and N' be simply-connected nilmanifolds and let $(\mathfrak{a}, \mathfrak{z}, j)$ and $(\mathfrak{a}', \mathfrak{z}', j')$ be the associated data triples as in 2.3. Then N is isometric to N' if and only if there exist orthogonal linear maps $\Phi : \mathfrak{a} \to \mathfrak{a}'$ and $\Psi : \mathfrak{z} \to \mathfrak{z}'$ such that*

$$j'(\Psi(Z)) = \Phi \circ j(Z) \circ \Phi^{-1}$$

for all $Z \in \mathfrak{z}$.

Proposition 2.4 is a corollary of Proposition 2.1(b).

2.5 Theorem. [G2]. *Let N be a two-step nilmanifold and $(\mathfrak{a}, \mathfrak{z}, j)$ the associated data triple. Then N is naturally reductive if and only if both the following conditions hold:*
(i) $j(\mathfrak{z})$ is a subalgebra of $\mathrm{so}(\mathfrak{a})$ *and*
(ii) $j^{-1} \circ \mathrm{ad}_{\mathrm{so}(\mathfrak{a})}(\mathrm{j}(\mathrm{Z})) \circ \mathrm{j} \in \mathrm{so}(\mathfrak{z})$ for all $Z \in \mathfrak{z}$.

Theorem 2.5 and Proposition 2.4 allow us to characterize the simply-connected naturally reductive nilmanifolds as follows:

2.6 Definition. Let $\mathfrak{w}$ be a subalgebra of so(n). We will say an inner product $<,>$ on $\mathfrak{w}$ is *invariant* if $\mathrm{ad}_{\mathfrak{w}}(\mathrm{X})$ is skew-symmetric relative to $<,>$ for all $X \in \mathfrak{w}$; equivalently, the group $\mathrm{Int}(\mathfrak{w})$ of inner automorphisms of $\mathfrak{w}$ leaves $<,>$ invariant.

Observe that the invariant inner products on $\mathfrak{w}$ are easily classified. Since $\mathfrak{w}$ lies in so(n), it is a reductive Lie algebra, i.e., $\mathfrak{w} = \mathfrak{w}_1 \oplus \ldots \oplus \mathfrak{w}_r \oplus z(\mathfrak{w})$ where $z(\mathfrak{w})$ is the center and the $\mathfrak{w}_i$'s are simple compact Lie algebras. This decomposition of $\mathfrak{w}$ is orthogonal relative to any admissible inner product $<,>$, and the restriction of $<,>$ to $\mathfrak{w}_i$ is unique up to scalar multiple. The restriction of $<,>$ to $z(\mathfrak{w})$ is arbitrary.

2.7 Definition. We will say a pair $(\mathfrak{w}, <,>)$ and $(\mathfrak{w}', <,>')$ of subalgebras of so(n) with invariant inner products are conjugate if $\mathfrak{w}$ and $\mathfrak{w}'$ are conjugate by an orthogonal transformation of $\mathbf{R}^n$ and the isomorphism from $\mathfrak{w}$ to $\mathfrak{w}'$ defined by the conjugation is an inner product space isometry. Let $C(n)$ denote the set of all conjugacy classes of pairs $(\mathfrak{w}, <,>)$ and let $C = \bigcup_{n=2}^{\infty} C(n)$.

2.8 Classification Theorem. *In the notation of 2.7, there is a one-to-one correspondence between C and the set of isometry classes of simply-connected naturally reductive two-step nilmanifolds given as follows: To $(\mathfrak{w}, <,>) \in C(n)$, we associate the simply-connected nilmanifold defined as in 2.3 by the data $\mathfrak{a} = \mathbf{R}^n$ with the standard inner product, $\mathfrak{z} = (\mathfrak{w}, <,>)$, and $j = Id$.*

We next consider the larger class of simply-connected two-step g. o. nilmanifolds N. To describe the full isometry group of N, we need only find the skew-symmetric derivations of $\mathfrak{n}$ and then apply Proposition 2.1.

2.9 Lemma. *Let N be a simply-connected two-step Riemannian nilmanifold and $(\mathfrak{a}, \mathfrak{z}, j)$ the associated data triple as in 2.3. Then a skew-symmetric linear map $D : \mathfrak{n} \to \mathfrak{n}$ is a derivation if and only if the following conditions hold:*
(i) D leaves each of $\mathfrak{a}$ and $\mathfrak{z}$ invariant and
(ii) $j(D(Z)) = [D_{|\mathfrak{a}}, j(Z)]$ for all $Z \in \mathfrak{z}$ where the bracket is that of $\mathrm{so}(\mathfrak{a})$.
Thus the isotropy algebra $\mathfrak{l}$ is isomorphic to the subalgebra of $\mathrm{so}(\mathfrak{a})$ *given by*

$$\{\delta \in \mathrm{so}(\mathfrak{a}) : [\delta, \mathrm{j}(\mathfrak{z})] \subset \mathrm{j}(\mathfrak{z}) \quad \text{and} \quad \mathrm{j}^{-1} \circ \mathrm{ad}(\delta) \circ \mathrm{j} \in \mathrm{so}(\mathfrak{z})\}.$$

2.10 Theorem. *Let N be a simply-connected 2-step nilmanifold. In the notation of 2.3, N is a g. o. manifold if and only if for each $X \in \mathfrak{a}$ and $Z \in \mathfrak{z}$, there exists a skew-symmetric derivation D of $\mathfrak{n}$ such that $D(Z) = 0$ and $D(X) = j(Z)X$.*

Proof. Let $D \in \mathfrak{l}$ and $U \in \mathfrak{n}$. Depending on whether we are viewing D as an element of the Lie algebra $\mathfrak{g} = \mathfrak{l} + \mathfrak{n}$ or as a derivation of $\mathfrak{n}$, we will write, respectively $[D, U]$ or $D(U)$ for the action of D on $U \in \mathfrak{n}$. We set $\mathfrak{q} = \mathfrak{n}$ in 1.1.

Given $X \in \mathfrak{a}$, $Z \in \mathfrak{z}$ and $D \in \mathfrak{l}$, then by Lemma 1.6, $X + Z + D$ is a geodesic vector if and only if both the following conditions hold:

(i) for all $U \in \mathfrak{a}$ we have

$$\begin{aligned} 0 &=< [X+Z+D,U], X+Z > \\ &=< [X,U], Z > + < D(U), X > \\ &=< J(Z)X - D(X), U > . \end{aligned}$$

(The second equality uses the fact that D normalizes $\mathfrak{a}$ and that $\mathfrak{a} \perp [\mathfrak{n},\mathfrak{n}] = \mathfrak{z}$.)

(ii) for all $W \in \mathfrak{z}$, we have

$$\begin{aligned} 0 &=< [X+Z+D,W], X+Z > \\ &=< [D,W], Z > \\ &= - < D(Z), W > . \end{aligned}$$

The theorem follows.

A. Kaplan [K] obtained Proposition 2.10 for nilmanifolds of Heiseberg type. The nilmanifolds of *Heisenberg type* are those for which $j(Z)^2 = -\|Z\|^2 Id$ for all $Z \in \mathfrak{z}$ in the notation of 2.3. C. Riehm [R] then classified all g. o. nilmanifolds of Heisenberg type.

We can now given an analog of 2.8 for g. o. nilmanifolds.

2.11 Definition.

(i) Let V be a subspace of so(n, **R**) and let $N(V)$ be the normalizer of V in so(n, **R**). We will say that V satisfies the *transitive normalizer condition* if for each $\alpha \in V$, $x \in \mathbf{R}^n$, there exists an element $\beta_{\alpha,x} \in N(V)$ such that $[\beta, \alpha] = 0$ and $\beta(x) = \alpha(x)$.

(ii) Suppose V satisfies the transitive normalizer condition. We will say an inner product $<,>$ on V is *admissible* if for each $\alpha \in V$, $x \in \mathbf{R}^n$, there eixsts a choice of $\beta_{\alpha,x}$ as in (i) such that $\mathrm{ad}(\beta_{\alpha,\mathrm{x}})_{|\mathrm{V}}$ is skew-symmetric with respect to $<,>$. (Here *ad* denotes the bracket in so(n, **R**).)

(iii) Let V and V' be subspaces of so(n, **R**) satisfying the transitive normalizer condition and let $<,>$ and $<,>'$ be admissible inner products on V and V', respectively. We will say $(V, <,>)$ is equivalent to $(V', <,>')$ if V is conjugate to V' by an element of O(n, **R**) and the isomorphism from V to V' defined by this conjugation is an inner product space isometry. Let $E(n)$ denote the set of all such equivalence classes and $E = \bigcup_{n=1}^{\infty} E(n)$.

2.12 Remark. If V is a subalgebra of so(n, **R**), then V trivially satisfies the transitive normalizer condition. In fact, we can choose $\beta_{\alpha,x} = \alpha$ for all $\alpha \in V$, $x \in \mathbf{R}^n$. An invariant inner product on V in the sense of Definition

2.6 is necessarily admissible. (However, the converse is not true as example 2.14 below shows.) In particular, the set C defined in 2.7 is contained in E.

2.13 Theorem. *There is a one-to-one correspondence between the set E, defined in 2.11, and the set of isometry classes of simply-connected g. o. two-step nilmanifolds given as follows: To the equivalence class of $(V, <,>)$ in E, we associate the nilmanifold defined as in 2.3 by the data triple $(\mathfrak{a}, \mathfrak{z}, j)$, where $\mathfrak{a} = \mathbf{R}^n$ with the standard inner product, $\mathfrak{z} = (V, <,>)$ and $j = Id$.*

Theorem 2.13 follows from Proposition 2.4, Lemma 2.9 and Theorem 2.10.

2.14 Example. We classify those g. o. simply-connected two-step nilmanifolds for which $\dim(\mathfrak{a}) \leq 4$ in the notation of 2.3. If $n \leq 3$, the only proper subspaces of so(3) satisfying the transitive normalizer condition are one-dimensional, and thus the associated nilmanifolds are naturally reductive. For $n = 4$, we have $so(4) \simeq so(3) \oplus so(3)$. The two ideals are conjugate under an orthogonal transformation T of $\mathbf{R}^4$. (In fact, $\mathbf{R}^4$ can be viewed as the quaternions and the two so(3) factors act as left, respectively right, multiplication by pure quaternions. The transformation T is the quaternionic conjugation.) As before, if V is one-dimensional then every inner product on V is invariant and the associated nilmanifolds are naturally reductive. We consider higher-dimensional subspaces V case-by-case:

Case 1. $\dim(V) = 2$.

There are two types of two-dimensional subspaces of so(4) satisfying the transitive normalizer condition. First we have abelian subalgebras of so(4) intersecting each of the two ideals in a one-dimensional subspace. All associated nilmanifolds are naturally reductive, so we don't consider these further.

Secondly, any two-dimensional subspace V lying in one of the ideals so(3) satisfies the transitive normalizer condition; indeed we can choose all the $\beta_{\alpha,x}$'s to lie in the centralizer of V (the other ideal). All such subspaces of so(4) are conjugate by elements of O(4), so we need only consider one such V. Every inner product on V is admissible since the $\beta_{\alpha,x}$'s commute with V. The associated nilmanifolds are g. o. spaces but are not naturally reductive since V is not a subalgebra of $\mathfrak{g}$. These are the six-dimensional examples found in [KV2].

Case 2. $\dim(V) = 3$.

The only 3-dimensional subspaces satisfying the transitive normalizer condition are subalgebras of so(4). Writing elements of so(4) as pairs (α_1, α_2) with $\alpha_i \in so(3)$, the 3-dimensional subalgebras of so(4) consist of the two ideals and the subalgebras of the form $\{(\alpha, T(\alpha)\}$ where

$T : \mathrm{so}(3) \to \mathrm{so}(3)$ is an isomorphism. The latter ones are self-normalizing and the only admissible inner products are invariant. Thus the associated nilmanifolds are naturally reductive. However, in case the subalgebra is one of the ideals, we can choose the $\beta_{\alpha,x}$'s to lie in the other ideal. Thus every inner product is admissible, whereas there is only a one-parameter family of invariant inner products. Thus we obtain 7-dimensional g. o. nilmanifolds which are not naturally reductive.

Case 3. $\dim(V) \geq 4$.

The only 4-dimensional subspaces satisfying the transitive normalizer condition are subalgebras $\mathrm{so}(3) \oplus \mathbf{R} \simeq \mathrm{u}(2)$, and these are self-normalizing. All admissible inner products are invariant, so we obtain no new examples of g. o. nilmanifolds which aren't naturally reductive.

There are no 5-dimensional subspaces satisfying the transitive normalizer condition.

We are left only to consider $V = \mathrm{so}(4)$. Again all admissible inner products are invariant, so no new examples are obtained.

3. Compact G. 0. Manifolds

D'Atri and Ziller constructed a large class of naturally reductive compact manifolds. They also showed that all naturally reductive manifolds which admit a simply transitive compact simple group of isometries are included in this class. However, the complete classification of naturally reductive compact manifolds is unknown. Likewise, we will not attempt to classify the compact g. o. manifolds but will only give some methods for obtaining examples.

We first review the construction of naturally reductive metrics.

3.1 Theorem [DZ]. *Let H be a compact Lie group and L a closed subgroup containing no normal subgroups of H. Let $\mathfrak{k}$ be a subalgebra of the centralizer of $\mathfrak{l}$ in $\mathfrak{h}$ such that $\mathfrak{l} \cap \mathfrak{k} = (0)$. Let g be a bi-invariant metric on H with $g(\mathfrak{l}, \mathfrak{k}) = 0$ and let $\mathfrak{a} = (\mathfrak{l} + \mathfrak{k})^{\perp}$. Write $\mathfrak{k} = \mathfrak{k}_0 \oplus \mathfrak{k}_1 \oplus \ldots \oplus \mathfrak{k}_r$ with $\mathfrak{k}_i$ simple, $1 \leq i \leq r$, and with $\mathfrak{k}_0$ the center of $\mathfrak{k}$. Let $\alpha_1, \ldots, \alpha_r, \alpha \in \mathbf{R}^+$ and let h be an arbitrary inner product on $\mathfrak{k}_0$. Then the left-invariant metric on H/L associated with the inner product*

$$<,>= h_{|\mathfrak{k}_0} + \alpha_1 g_{|\mathfrak{k}_1} + \ldots + \alpha_r g_{|\mathfrak{k}_r} + \alpha g_{|\mathfrak{a}}$$

is naturally reductive with respect to the group $H \times K$ of isometries, where K, the connected subgroup of H with Lie algebra $\mathfrak{k}$, acts by right translations on H/L.

We next consider g. o. metrics. The following result directly generalizes Theorem 3.1 but is too complicated to be practical. Thus we will omit the proof and instead give a simpler construction afterwards.

3.2 Proposition. *In the notation of Theorem 3.1, let* $\mathfrak{a} = \mathfrak{a}_1 \oplus \ldots \oplus \mathfrak{a}_s$ *be a decomposition of* $\mathfrak{a}$ *into* $(\mathfrak{l} + \mathfrak{k})$*-invariant subspaces. Suppose there exist positive constants* $\beta_1, \ldots, \beta_s$ *such that for each* $X = X_1 + \ldots + X_s \in \mathfrak{a}$*, there exists* $A \in \mathfrak{l}$ *with* $\sum_{j=1}^{s}(\beta_j \pi_{\mathfrak{a}}([A + X, X_j])) = 0$ *where* $\pi_{\mathfrak{a}}$ *is the orthogonal projection onto* $\mathfrak{a}$*. Then for* h, α_i *as in Theorem 3.1, the left-invariant metric on* H/L *defined by*

$$<,>= h_{|\mathfrak{k}_0} + \alpha_1 g_{|\mathfrak{k}_1} + \ldots + \alpha_r g_{|\mathfrak{k}_r} + \beta_1 g_{|\mathfrak{a}_1} + \ldots + \beta_s g_{|\mathfrak{a}_s}$$

is a g. o. metric.

Note that Theorem 3.1 is the special case $s = 1$ (so A can be taken to be zero and $\alpha = \beta_1$ is arbitrary).

We now give a more practical construction.

3.3 Theorem. *Let* H *be a compact Lie group with bi-invariant metric* g*, and let* L *be a compact subgroup containing no nontrivial normal subgroups of* H*. Let* $\mathfrak{k}$ *be a subalgebra of* $\mathfrak{h}$ *containing* $\mathfrak{l}$ *and let* $\mathfrak{u} = \mathfrak{k} \cap \mathfrak{l}^{\perp}$ *and* $\mathfrak{a} = \mathfrak{k}^{\perp}$ *relative to* g*. Assume that for each* $U \in \mathfrak{u}$ *and* $X \in \mathfrak{a}$*, there exists* $A \in \mathfrak{l}$ *such that* $[A, U] = 0$ *and* $[A, X] = [U, X]$*. Then for any* $\alpha, \beta \in \mathbf{R}^+$*, the left-invariant Riemannian metric on* H/L *defined by*

$$<,>= \alpha g_{|\mathfrak{u}} + \beta g_{|\mathfrak{a}}$$

is a g. o. metric.

Proof. Since $\mathfrak{k} = \mathfrak{u} + \mathfrak{l}$ is a subalgebra of $\mathfrak{h}$, the g-orthogonal complement $\mathfrak{a}$ is $\mathrm{ad}(\mathfrak{l} + \mathfrak{u})$-invariant. Extend $<,>$ to $\mathfrak{h}$ so that $\mathfrak{l} = \ker(<,>)$. Set $\mathfrak{q} = \mathfrak{u} + \mathfrak{a}$ and let $\pi_{\mathfrak{q}}$ be the projection of $\mathfrak{h}$ onto $\mathfrak{q}$ along $\mathfrak{l}$. For $U \in \mathfrak{u}$, $\pi_{\mathfrak{q}} \circ \mathrm{ad}(U)_{|\mathfrak{q}}$ normalizes both $\mathfrak{u}$ and $\mathfrak{a}$ and is easily seen to be skew-symmetric.

Let $U \in \mathfrak{u}$, $X \in \mathfrak{a}$ and choose $A \in \mathfrak{l}$ as in the theorem. Set $A' = (\frac{\alpha - \beta}{\beta})A$. We show that $A' + U + X$ is a geodesic vector.

For $V \in \mathfrak{u}$, $[A', V] \perp U$ since $[A', U] = 0$ so by the skew-symmetry of $\pi_{\mathfrak{q}} \circ \mathrm{ad}(V)_{|\mathfrak{q}}$, we have

$$< [V, A' + U + X], U + X >= 0.$$

Next for $Y \in \mathfrak{a}$, we have

$$\begin{aligned} &< [Y, A' + U + X], U + X > \\ &= < [Y, A' + U], X > + < [Y, X], U > + < [Y, X], X > \\ &= < [A' + U, X], Y > + \alpha g([Y, X], U) + \beta g([Y, X], X) \\ &= < [A' + U, X], Y > - \alpha g([U, X], Y) + 0 \end{aligned}$$

since ad(X) is skew-symmetric with respect to g. Thus

$$\begin{aligned} &< [Y, A' + U + X], U + X > \\ &= g([\beta A' + (\beta - \alpha)U, X], Y) \\ &= g([(\alpha - \beta)A + (\beta - \alpha)U, X], Y) = 0 \end{aligned}$$

by the hypothesis. Hence by Lemma 1.6, $A' + U + X$ is a geodesic vector.

3.4 Example. Let H = SO(5) and view SO(4) as a subgroup of H via the standard embedding. As in Example 2.14, view u(2) as a subalgebra of so(4) and set $\mathfrak{l}$ = u(2). Let L be the corresponding connected subgroup of SO(4) $\subset$ SO(5). In the notation of Theorem 3.3, we take $g = -B$, where B is the killing form of so(5), and we take $\mathfrak{k}$ = so(4) and $\mathfrak{a} = \mathfrak{k}^{\perp}$. Note that the bracket action of $\mathfrak{k}$ on $\mathfrak{a}$ is equivalent to the standard representation of so(4) on $\mathbf{R}^4$. Since the subspace $\mathfrak{u}$ = so(4) $\ominus$ u(2) satisfies the transitive normalizer condition (see Definition 2.11 and Example 2.14), the hypothesis of Theorem 3.3 holds. Taking $<,> = \alpha g_{|\mathfrak{u}} + \beta g_{|\mathfrak{q}}$, we obtain g. o. metrics on SO(5)/U(2). These are the six-dimensional compact g. o. manifolds found in [KV2].

4. Quotients of Noncompact Semisimple Lie Groups

Before discussing the naturally reductive and g. o. metrics on manifolds of the form H/L, where H is semisimple of noncompact type, we first describe the full isometry group of any left-invariant metric on H/L.

4.1 Notation and Remarks. Let $H = KP$ be a Cartan decomposition with $L \subset K$. Let $N = N_K(L)$ be the normalizer of L in K. For $a \in N$, the right action R_a of a on H induces a diffeomorphism $\overline{R_a}$ of $M := H/L$. Let $A = \{a \in N : \overline{R_a}$ is an isometry$\}$. At the Lie algebra level, we can choose an $ad(\mathfrak{l})$-invariant complement $\mathfrak{q}$ of $\mathfrak{l}$ by $\mathfrak{q} = \mathfrak{u} + \mathfrak{p}$ where $\mathfrak{u} = \mathfrak{k} \ominus \mathfrak{l}$ relative to the killing form of $\mathfrak{g}$. For $a \in N$, Ad(a) leaves $\mathfrak{q}$ invariant, and $a \in A$ if and only if Ad(a) preserves the Riemannian inner product on $\mathfrak{q}$. The Lie algebra $\mathfrak{a}$ of A is given by $\mathfrak{a} = \mathfrak{l} + \mathfrak{f}$ where $\mathfrak{f}$ consists of all elements X of the centralizer of $\mathfrak{l}$ in $\mathfrak{u}$ such that $\mathrm{ad}(X)_{|\mathfrak{q}}$ is skew-symmetric. The corresponding connected subgroup F of H is isomorphic to the identity componenet of A/L. Observe that $(H \times F)/D \subset I_0(M)$ where F acts by right translations and D is the discrete "diagonal" subgroup consisting of all central elements of H which lie in F.

4.2 Proposition [G1]. *Let $M = H/L$ be a Riemannian homogeneous space of a connected semisimple Lie group H of noncompact type. Then in the notation of 4.1, $I_0(M) = (H \times F)/D$.*

We now review the classification of naturally reductive metrics.

4.3 Theorem [G2]. *Let $M = H/L$ be a Riemannian homogeneous space, where H is a connected semisimple Lie group of noncompact type. Let $H = KP$ be a Cartan decomposition with $L \subset K$. Then M is naturally reductive if and only if L is normal in K and, in the notation of 4.1, $A = K$. In this case, the metric is naturally reductive with respect to $I_0(M)$.*

Theorem 4.3 can be rephrased as follows.

4.4 Corollary [G2]. *A left-invariant metric g on a homogeneous space $M = H/L$ of a connected semisimple Lie group H of noncompact type is naturally reductive if and only if the following conditions are satisfied:*

(i) Letting $\mathfrak{k}$ be a maximal compactly embedded subalgebra of $\mathfrak{h}$ containing $\mathfrak{l}$, then $\mathfrak{l}$ is a $\mathfrak{k}$-ideal.

(ii) Let $\mathfrak{h} = \mathfrak{k} + \mathfrak{p}$ be a Cartan decomposition and let $\mathfrak{f}$ be an ideal in $\mathfrak{k}$ complementary to $\mathfrak{l}$. Let $\mathfrak{f} = \mathfrak{f}_0 \oplus \mathfrak{f}_1 \oplus \ldots \oplus \mathfrak{f}_r$ be the decomposition of $\mathfrak{f}$ into simple ideals and center $\mathfrak{f}_0$, let $\mathfrak{h} = \mathfrak{h}_1 \oplus \ldots \oplus \mathfrak{h}_n$ be the decomposition of $\mathfrak{h}$ into simple ideals, and let $\mathfrak{p}_j = \mathfrak{p} \cap \mathfrak{h}_j$. Then

$$g = h_{|\mathfrak{f}_0} - \alpha_1 B_{|\mathfrak{f}_1} - \ldots - \alpha_r B_{|\mathfrak{f}_r} + \beta_1 B_{|\mathfrak{p}_1} + \ldots + \beta_n B_{|\mathfrak{p}_n}$$

where B is the killing form of $\mathfrak{g}$, $\alpha_i, \beta_j \in \mathbf{R}^+$ for all i, j and h is an arbitrary inner product.

We will not classify the g. o. metrics on homogeneous spaces of semisimple Lie groups. We instead show that the class of such metrics is quite restrictive (Theorem 4.5) and then give one method for constructing examples.

4.5 Theorem. *Let $M = H/L$ be a homogeneous Riemannian g. o. manifold where H is a connected semisimple Lie group of noncompact type. Let $H = KP$ be a Cartan decomposition with $L \subset K$. Then*

(i) In the notation 4.1, the Lie algebra $\mathfrak{a}$ of A is a self-normalizing subalgebra of $\mathfrak{k}$, i.e., $N_K(A)/A$ is discrete where $N_K(A)$ is the normalizer of A in K.

(ii) *Viewing* $\mathfrak{p}$ *as a subspace of the tangent space* $T_{eL}(M)$, *then the Riemannian inner product on* $\mathfrak{p}$ *is* Ad(K)*-invariant. Thus if* $\mathfrak{h} = \mathfrak{h}_1 \oplus \ldots \oplus \mathfrak{h}_n$ *is the decomposition of* $\mathfrak{h}$ *into simple ideals and* $\mathfrak{p} = \mathfrak{p}_1 \oplus \ldots \oplus \mathfrak{p}_n$ *the corresponding decomposition of* $\mathfrak{p}$, *then*

$$<,>_{\mathfrak{p}} = \alpha_1 B_{|\mathfrak{p}_1} + \ldots + \alpha_n B_{|\mathfrak{p}_n}$$

where B *is the killing form of* $\mathfrak{h}$ *and* $\alpha_1, \ldots, \alpha_r \in \mathbf{R}^+$.

(iii) K/L *with the induced Riemannian metric is a g. o. manifold.*

Proof. Write $\mathfrak{a} = \mathfrak{l} + \mathfrak{f}$. By Proposition 4.2, the Lie algebra $\mathfrak{g}$ of the full connected isometry group $I_0(M)$ is given by $\mathfrak{g} = \mathfrak{h} \oplus \mathfrak{f}$ with isotropy algebra

$$\tilde{\mathfrak{l}} = \mathfrak{l} + \{(X, X) : X \in \mathfrak{f}\}.$$

An ad($\tilde{\mathfrak{l}}$) -invariant complement $\tilde{\mathfrak{q}}$ of $\tilde{\mathfrak{l}}$ in $\mathfrak{g}$ is given by $\mathfrak{q} = \bar{\mathfrak{f}} + \mathfrak{b} + \mathfrak{p}$ where

$$\bar{\mathfrak{f}} = \{(X, -X) : X \in \mathfrak{f}\}$$

and $\mathfrak{b} = \mathfrak{k} \ominus \mathfrak{a}$ relative to the Killing form of $\mathfrak{h}$.

(i) Note that $[\mathfrak{a}, \mathfrak{b}] \subset \mathfrak{b}$, so the normalizer of $\mathfrak{a}$ in $\mathfrak{k}$ is given by $\mathfrak{a} \oplus \mathfrak{c}$ where

$$\mathfrak{c} = \{X \in \mathfrak{b} : [\mathfrak{a}, X] = 0\}.$$

For $X \in \mathfrak{c}$, we have $[\tilde{\mathfrak{l}}, X] = 0$, so by Proposition 1.8, $\mathrm{ad}_{\mathfrak{g}}(X)_{|\tilde{\mathfrak{q}}}$ is skew-symmetric. The smaller algebra $\mathfrak{h}$ decomposes as $\mathfrak{h} = \mathfrak{l} + \mathfrak{q}$ with $\mathfrak{q} = \mathfrak{f} + \mathfrak{b} + \mathfrak{p}$ and thus $\mathrm{ad}_{\mathfrak{h}}(X)_{|\mathfrak{q}}$ is skew-symmetric as well. But then by 4.1, $X \in \mathfrak{a}$. Thus $\mathfrak{c} \subset \mathfrak{a}$; i.e. $\mathfrak{c} = (0)$.

(ii) and (iii). Observe that K is normalized by the isotropy group $\tilde{L}$, so (ii) and (iii) follow from Propositions 1.8 and 1.9, respectively.

4.6 Theorem. *Let* H *be a connected semisimple Lie group of noncompact type and* L *a compact subgroup of* H. *Let* $\mathfrak{h} = \mathfrak{k} + \mathfrak{p}$ *be a Cartan decomposition, let* B *be the Killing form of* $\mathfrak{h}$, *and let* $\mathfrak{u}$ *be the* B*-orthogonal complement of* $\mathfrak{l}$ *in* $\mathfrak{k}$. *Assume that for each* $U \in \mathfrak{u}$ *and* $X \in \mathfrak{p}$, *there exists* $A \in \mathfrak{l}$ *such that* $[A, U] = 0$ *and* $[A, X] = [U, X]$. *For any* $\alpha, \beta \in \mathbf{R}^+$, *the left-invariant Riemannian metric on* H/L *defined by*

$$<,> = -\alpha B_{|\mathfrak{u}} + \beta B_{|\mathfrak{p}}$$

is a g. o. metric.

Note that the metric above is naturally reductive if and only if $\mathfrak{k} = \mathfrak{l} + \mathfrak{u}$, direct sum of ideals.

The Killing form B can be replaced by arbitrary positive multiples of the Killing form on each simple factor of $\mathfrak{h}$. The proof of Theorem 4.6 is identical to that of Theorem 3.3.

4.7 Example. The constructions in Theorems 3.3 and 4.6 can be viewed as dual to each other. Dualizing example 3.4, we take $H = \mathrm{SO}(4,1)$ and $L = \mathrm{U}(2) \subset \mathrm{SO}(4)$. Letting $\mathfrak{h} = \mathfrak{k} + \mathfrak{p}$ be a Cartan decomposition with $\mathfrak{k} = \mathrm{so}(4)$, we obtain g. o. metrics by setting

$$<,>= -\alpha B_{|\mathfrak{u}} + \beta B|\mathfrak{p}$$

with $\alpha, \beta \in \mathbf{R}^+$ and with $\mathfrak{u} = \mathfrak{k} \ominus \mathfrak{l}$ with respect to the Killing form B. These are the 6-dimensional examples found in [KV2].

5. G. O. Manifolds of Nonpositive Ricci Curvature

5.1 Theorem. *Every g. o. manifold M of nonpositive Ricci curvature is symmetric.*

Proof. Let $G = I_0(M)$ and let L be the isotropy subgroup at a point $p \in M$. As in 1.11-12, choose a semisimple Levi factor $G_1 = G_{nc}G_c$ compatible with L. Let $\mathfrak{g}_{nc} = \mathfrak{k} + \mathfrak{p}$ be a Cartan decomposition. Then a maximal compactly embedded subalgebra $\mathfrak{u}$ of $\mathfrak{g}$ containing $\mathfrak{l}$ is of the form

$$\mathfrak{u} = \mathfrak{k} + \mathfrak{g}_c + \mathfrak{a} \tag{1}$$

where $\mathfrak{a} \subset \mathrm{rad}(\mathfrak{g})$ is abelian.

Choose an $\mathrm{ad}(\mathfrak{l})$-invariant complement $\mathfrak{q}$ of $\mathfrak{l}$ in $\mathfrak{g}$ with $\mathfrak{p} + \mathfrak{n} \subset \mathfrak{q}$, where $\mathfrak{n} = \mathrm{nilrad}(\mathfrak{g})$. Write $\mathfrak{q} = \mathfrak{q}_1 + \mathfrak{q}_2$ where $\mathfrak{u} = \mathfrak{q}_1 + \mathfrak{l}$ and $\mathfrak{q}_1 \perp \mathfrak{q}_2$ with respect to the Riemannian inner product on $\mathfrak{q}$. The Ricci curvature of an arbitrary homogeneous Riemannian manifold is computed in [B]. The expression is simplified in case $G = I_0(M)$ is unimodular, i.e., in case $\mathrm{trace}(\mathrm{ad}(X)) = 0$ for all $X \in \mathfrak{g}$. To see that this condition always holds when M is a g. o. manifold, we recall Theorem 1.15: $\mathfrak{g} = \mathfrak{g}_1 + \mathfrak{l} + \mathfrak{n}$ (not a direct sum, of course) with $\mathfrak{g}_1 = \mathfrak{g}_{nc} + \mathfrak{g}_c$, and $\mathfrak{g}_{nc}$ is an ideal in $\mathfrak{g}$. Since every semisimple Lie algebra is unimodular, we have $\mathrm{trace}(\mathrm{ad}(X)) = \mathrm{trace}((\mathrm{ad}(X)_{|\mathfrak{g}_{nc}} = 0$ when $X \in \mathfrak{g}_{nc}$. Since $\mathfrak{g}_c$ and $\mathfrak{l}$ are compactly embedded in $\mathfrak{g}$ and since $\mathfrak{n} = \mathrm{nilrad}(\mathfrak{g})$, we also have $\mathrm{trace}(\mathrm{ad}(X)) = 0$ for $X \in \mathfrak{g}_c + \mathfrak{l} + \mathfrak{n}$. Thus G is unimodular.

We now recall the expression for the Ricci curvature (see [B], Lemma 7.32 and Corollary 7.38). Let $\{X_1, \ldots, X_n\}$ be an orthonormal basis of $\mathfrak{q}$, and let $B_{\mathfrak{g}}$ be the killing form of $\mathfrak{g}$. For $Y \in \mathfrak{q}$,

$$\begin{aligned}\mathrm{Ric}(Y,Y) = &-\frac{1}{2}B_{\mathfrak{g}}(Y,Y) \\ &-\frac{1}{2}\sum_{i=1}^{n}\|[Y,X_i]_{\mathfrak{q}}\|^2 \\ &+\frac{1}{2}\sum_{i<j} < [X_i,X_j]_{\mathfrak{q}}, Y >^2 .\end{aligned} \tag{2}$$

Since $\mathfrak{u}$ is a compact Lie algebra, $\mathfrak{u} = [\mathfrak{u},\mathfrak{u}] + z(\mathfrak{u})$ where $z(\mathfrak{u})$ is the center of $\mathfrak{u}$ and $[\mathfrak{u},\mathfrak{u}]$ is semisimple. We will show that $[\mathfrak{u},\mathfrak{u}] \subset \mathfrak{l}$.

There exists a compact Lie group U^* with Lie algebra $\mathfrak{u}$. Let L^* be the connected subgroup with Lie algebra $\mathfrak{l}$. Since $\mathfrak{u} = \mathfrak{l} + \mathfrak{q}_1$ with $[\mathfrak{l}, \mathfrak{q}_1] \subset \mathfrak{q}_1$, the inner product on $\mathfrak{q}_1$ defines a left-invariant Riemannian metric on U^*/L^*. Now U^* may not act effectively on U^*/L^*. However, we can mod out from U^* the largest normal subgroup contained in L^* in order to get an effective action. The expression for the Ricci curvature is not affected, however, if we work with $\mathfrak{u}$ itself rather than the corresponding quotient.

Since U^*/L^* is a compact homogeneous space, its Ricci curvature Ric^* is nonnegative. To compare $\mathrm{Ric}^*(Y,Y)$ and $\mathrm{Ric}(Y,Y)$ for $Y \in \mathfrak{q}_1$, let $\{U_1,\ldots,U_r\}$ be an orthonormal basis of $\mathfrak{q}_1$ and $\{W_1,\ldots,W_s\}$ an orthonormal basis of $\mathfrak{q}_2$. In (2), we can take $X_i = U_i, 1 \le i \le r$, and $X_{r+j} = W_j, 1 \le j \le s$.

Let $B_{\mathfrak{u}}$ be the killing form of $\mathfrak{u}$. Then

$$B_{\mathfrak{g}}(Y,Y) = B_{\mathfrak{u}}(Y,Y) + \sum_{j=1}^{s} < [Y,[Y,W_j]]_{\mathfrak{q}_2}, W_j > . \tag{3}$$

Since $Y \in \mathfrak{q}_1 \subset \mathfrak{u}$, we have $[Y,\mathfrak{u}]_{\mathfrak{q}} \subset \mathfrak{q}_1$, so

$$\begin{aligned}< [Y,[Y,W_j]]_{\mathfrak{q}_2}, W_j > &=< [Y,[Y,W_j]_{\mathfrak{q}_2}]_{\mathfrak{q}_2}, W_j > \\ &= -\|[Y,W_j]_{\mathfrak{q}_2}\|^2\end{aligned} \tag{4}$$

where the last equality uses Proposition 1.8. Thus by (3) and (4),

$$B_{\mathfrak{g}}(Y,Y) = B_{\mathfrak{u}}(Y,Y) - \sum_{j=1}^{s}\|[Y,W_j]_{\mathfrak{q}_2}\|^2. \tag{5}$$

By (2) and (5), we have

$$\begin{aligned}\mathrm{Ric}(Y,Y) =& \mathrm{Ric}^*(Y,Y) - \frac{1}{2}\sum_{j=1}^{s} \|[Y,W_j]_{\mathfrak{q}_1}\|^2 \\ &+ \frac{1}{2}\sum_{i=1}^{r}\sum_{j=1}^{s} < [U_i,W_j]_{\mathfrak{q}_1}, Y >^2 \\ &+ \frac{1}{2}\sum_{1\le i<j\le s} < [W_i,W_j]_{\mathfrak{q}_1}, Y >^2 .\end{aligned} \tag{6}$$

Using Proposition 1.8 again,

$$\begin{aligned}\|[Y,W_j]_{\mathfrak{q}_1}\|^2 &= \sum_{i=1}^{r} < [Y,W_j]_{\mathfrak{q}_1}, U_i >^2 \\ &= \sum_{i=1}^{r} < [U_i,W_j]_{\mathfrak{q}_1}, Y >^2 .\end{aligned} \tag{7}$$

Substituting (7) into (6), we obtain

$$\mathrm{Ric}^*(Y,Y) = \mathrm{Ric}(Y,Y) - \frac{1}{2}\sum_{1\le i<j\le s} < [W_i,W_j]_{\mathfrak{q}_1}, Y >^2 . \tag{8}$$

The conditions $\mathrm{Ric}(Y,Y) \le 0$ and $\mathrm{Ric}^*(Y,Y) \ge 0$ thus force $\mathrm{Ric}^*(Y,Y) = 0$. Moreover, we must also have

$$< [W_i,W_j]_{\mathfrak{q}_1}, Y >= 0$$

for all $Y \in \mathfrak{q}_1$, so

$$[\mathfrak{q}_2,\mathfrak{q}_2] \subset \mathfrak{l} + \mathfrak{q}_2. \tag{9}$$

Since $\mathrm{Ric}^* = 0$, U^*/L^* is Euclidean (see ([B]) and $\mathfrak{l}$ contains the maximal semisimple ideal $[\mathfrak{u},\mathfrak{u}]$ of $\mathfrak{u}$. The $\mathrm{ad}(\mathfrak{l})$-invariant complement $\mathfrak{q}_1$ of $\mathfrak{l}$ in $\mathfrak{u}$ thus lies in the center of $\mathfrak{u}$. In particular, $[\mathfrak{l},\mathfrak{q}_1] = 0$

We now show that $\mathfrak{k} \subset \mathfrak{l}$. Proposition 1.8 implies that $\pi_{\mathfrak{q}} \circ \mathrm{ad}(Y)_{|\mathfrak{q}}$ is skew-symmetric for all $Y \in \mathfrak{q}_1$. Noting that $\mathfrak{u} = \mathfrak{l} + \mathfrak{q}_1$, we conclude that $\pi_{\mathfrak{q}} \circ \mathrm{ad}(X)_{|\mathfrak{q}}$ is skew-symmetric for all $X \in \mathfrak{u}$, in particular, for all $X \in \mathfrak{k}$. Recall that we chose $\mathfrak{q}$ so that $\mathfrak{p} \subset \mathfrak{q}$. Since $[\mathfrak{k},\mathfrak{q}_1] = 0$ and $[\mathfrak{k},\mathfrak{p}] = \mathfrak{p}$, we must have $\mathfrak{q}_1 \perp \mathfrak{p}$; i.e., $\mathfrak{p} \subset \mathfrak{q}_2$. Finally by (9), $\mathfrak{k} = [\mathfrak{p},\mathfrak{p}] \subset \mathfrak{k}\cap(\mathfrak{l}+\mathfrak{q}_2) = \mathfrak{k}\cap\mathfrak{l}$, so $\mathfrak{k} \subset \mathfrak{l}$. Since also $[\mathfrak{u},\mathfrak{u}] \subset \mathfrak{l}$, we have

$$\mathfrak{l} = \mathfrak{k} + \mathfrak{g}_c + (\mathfrak{l}\cap\mathfrak{g}_2). \tag{10}$$

The remainder of the proof is identical to the last part of the proof given in [GZ] for the naturally reductive case. We use (10) and Theorem 1.15 to deduce that $\mathfrak{q} = \mathfrak{p} + \mathfrak{n}$. Also by Theorem 1.15, $[\mathfrak{g}_{nc}, \mathfrak{n}] = 0$. Thus we have $[\mathfrak{k}, \mathfrak{p}] = \mathfrak{p}$ and $[\mathfrak{k}, \mathfrak{n}] = 0$, so $\mathfrak{n} \perp \mathfrak{p}$. Hence $M \simeq (G_{nc}/K) \times N$, Riemannian direct product. The first factor is a symmetric space. The second factor is a nilmanifold of nonpositive Ricci curvature. Recall [M] that a nilmanifold has Ricci curvatures of both signs unless it is flat. Thus N is Euclidean and M is a symmetric space.

Bibliography

[B] A. L. Besse, *Einstein Manifolds*, Springer Verlag, Berlin, 1978.

[DZ] J. E. D'Atri and W. Ziller, Naturally reductive metrics and Einstein metrics on compact Lie groups, *Mem. Amer. Math. Soc.* **18**, No. 215 (1979).

[G1] C. S. Gordon, Riemannian isometry groups containing transitive reductive subgroups, *Math. Ann.* **248** (1980), 185-192.

[G2] _____, Naturally reductive homogeneous Riemannian manifolds, *Can. J. Math.* **37** (1985), 467-487.

[G3] _____, Isospectral closed Riemannian manifolds which are not locally isometric, *J. Diff. Geom.* **37** (1993), 639-649.

[GW1] C. S. Gordon and E. N. Wilson, The fine structure of transitive Riemannian isometry groups, *Trans. Amer. Math. Soc.* **289** (1985), 367-380.

[GW2] _____, Continuous families of isospectral Riemannian manifolds which are not locally isometric, preprint.

[GZ] C. S. Gordon and W. Ziller, Naturally reductive metrics of nonpositive Ricci curvature, *Proc. Amer. Math. Soc.* **91** (1984), 287-290.

[K] A. Kaplan, On the geometry of groups of Heisenberg type, *Bull. London Math. Soc.* **15** (1983), 35-42.

[KV1] O. Kowalski and L. Vanhecke, A generalization of a theorem on naturally reductive homeogeneous spaces, *Proc. Amer. Math. Soc.* **91**, (1984), 433-435.

[KV2] _____, Riemannian manifolds with homogeneous geodesics, *Boll. Un. Math. Ital. B (7)* **5** (1991), 189-246.

[M] J. Milnor, Curvatures of left-invariant metrics on Lie groups, *Adv. in Math.* **21** (1976), 293-329.

[R] C. Riehm, Explicit spin representations and Lie algebras of Heisenberg type, *J. London Math. Soc.* **29** (1984), 49-62.

[W] E. N. Wilson, Isometry groups on homogeneous nilmanifolds, *Geom. Dedicata* **12** (1982), 337-346.

Dartmouth College
Hanover, NH 03755
carolyn.s.gordon@dartmouth.edu

Received June 1995

On the $\mathcal{D}$-Module and Formal-Variable Approaches to Vertex Algebras

Yi-Zhi Huang and James Lepowsky

1. Introduction

In a program to formulate and develop two-dimensional conformal field theory in the framework of algebraic geometry, Beilinson and Drinfeld [BD] have recently given a notion of "chiral algebra" in terms of $\mathcal{D}$-modules on algebraic curves. This definition consists of a "skew-symmetry" relation and a "Jacobi identity" relation in a categorical setting, and it leads to the operator product expansion for holomorphic quantum fields in the spirit of two-dimensional conformal field theory, as expressed in [BPZ]. Because this operator product expansion, properly formulated, is known to be essentially a variant of the main axiom, the "Jacobi identity" [FLM], for vertex (operator) algebras ([Borc], [FLM]; see [FLM] for the proof), the chiral algebras of [BD] amount essentially to vertex algebras.

In this paper, we show directly that the chiral algebras of [BD] are essentially the same as vertex algebras without vacuum vector (and without grading), by establishing an equivalence between the skew-symmetry and Jacobi identity relations of [BD] and the (similarly-named, but different) skew-symmetry and Jacobi identity relations in the formal-variable approach to vertex operator algebra theory (see [FLM], [FHL]). In particular, among the equivalent formulations of the notion of vertex (operator) algebra, the $\mathcal{D}$-module notion of chiral algebra corresponds the most closely to the formal-variable notion, rather than to, say, the operator-product-expansion notion (based on the "commutativity" and "associativity" relations, as explained in [FLM], [FHL]) or to the geometric or operadic notion ([Hu1], [Hu2], [HL1]).

More precisely, we prove that for any nonempty open subset X of $\mathbb{C}$, the category of vertex algebras without vacuum over X (see Definitions 2.2 and 5.1) and the category of chiral algebras over X (see Definition 4.1) are equivalent (see Section 5). Beilinson-Drinfeld's notion of chiral algebra is in general formulated over higher-genus curves. But since chiral algebras in this sense are essentially local objects, the equivalence proved in this paper shows that the notion in [BD] is indeed essentially equivalent to the notion of vertex algebra without vacuum.

We hope that the present expository exercise helps to illuminate the relations between the theories and philosophies of $\mathcal{D}$-modules and of vertex operator algebras. For example, in the Jacobi identity for vertex algebras, the three formal variables are on equal footing because of an intrinsic S_3-symmetry (see [FHL], Section 2.7), while in the $\mathcal{D}$-module approach, there are only two (complex) variables, as in the operator-product-expansion approach. (See Remark 3.17.) To see the S_3-symmetry explicitly in the algebro-geometric framework, we would have to introduce an analogue of the notion of $\mathcal{D}$-module allowing global translations of a variable rather than just "infinitesimal translations." The Jacobi identity would then be interpreted as an identity in terms of such "modified $\mathcal{D}$-modules," so that the three variables involved would play symmetric roles. This will be discussed in future publications. The Jacobi identity for vertex operator algebras and its S_3-symmetry in fact play a central role in the theory of vertex operator algebras, in particular, in the construction of "vertex tensor categories" (see [HL2], [HL3], [HL4]).

Even without the introduction of such global translations of variables, all of the many calculations in this paper involving binomial expansions can be greatly simplified if we systematically introduce formal (not complex) variables playing the role of "formal global translations." For instance, the expression $(z_1 - z_2)^n(A_1 \otimes A_2)$, $n \in \mathbb{Z}$, occurring starting in Section 4 can be viewed as the coefficient of x^{-n-1} in $x^{-1}\delta(\frac{z_1-z_2}{x})(A_1 \otimes A_2)$, where x is a formal variable and $\delta(\frac{z_1-z_2}{x})$ is defined in Section 2.

In order to make this work reasonably self-contained, we include elementary definitions and notions needed in both theories. The reader can consult [FLM] and [FHL], for example, for the motivation and development of the theory of vertex operator algebras, and [Ha] for sheaves and [Borel] for algebraic $\mathcal{D}$-modules, whose theory was developed by Beilinson and Bernstein.

This paper is organized as follows: In Section 2, we recall some basic notations and elementary tools and give the definitions of vertex algebra and vertex algebra without vacuum. In Section 3, we recall some basic concepts in the theory of $\mathcal{D}$-modules and give examples which we shall need later. Beilinson-Drinfeld's notion of chiral algebra over X for a nonempty open subset $X \subset \mathbb{C}$ is given in Section 4. In Section 5, we define the notion of vertex algebra without vacuum over X and prove the equivalence theorem stated above.

We would like to thank P. Deligne and especially A. Beilinson for explaining the unpublished work [BD] to us. We are also grateful to F. Knop, whose Rutgers lecture notes on $\mathcal{D}$-modules and representation theory were very helpful to us. Y.-Z. H. is supported in part by NSF grant DMS-9596101 and by DIMACS, an NSF Science and Technology Center funded under contract STC-88-09648, and J. L by NSF grant DMS-9401851.

2. Vertex algebras and vertex algebras without vacuum

Following the treatment in [FLM] and [FHL], we describe the basic notations and elementary tools needed to formulate the notion of vertex algebra. We work over $\mathbb{C}$. In this paper, the symbols $x, x_0, x_1, \ldots$ are independent commuting formal variables, and all expressions involving these variables are to be understood as formal Laurent series. (Later we shall also use the symbols $z, z_1, \ldots$, which will denote complex numbers, not formal variables.) We use the "formal δ-function"

$$\delta(x) = \sum_{n\in\mathbb{Z}} x^n,$$

which has the following simple and fundamental property: For any Laurent polynomial $f(x) \in \mathbb{C}[x, x^{-1}]$,

$$f(x)\delta(x) = f(1)\delta(x).$$

This property has many important variants. For example, for any

$$X(x_1, x_2) \in (\text{End } W)[[x_1, x_1^{-1}, x_2, x_2^{-1}]]$$

(where W is a vector space) such that

$$\lim_{x_1\to x_2} X(x_1, x_2) = X(x_1, x_2)\Big|_{x_1=x_2}$$

exists, we have

$$X(x_1, x_2)\delta\left(\frac{x_1}{x_2}\right) = X(x_2, x_2)\delta\left(\frac{x_1}{x_2}\right). \tag{2.1}$$

The existence of this "algebraic limit" means that for an arbitrary vector $w \in W$, the coefficient of each power of x_2 in the formal expansion $X(x_1, x_2)w\Big|_{x_1=x_2}$ is a finite sum.

We use the convention that negative powers of a binomial are to be expanded in nonnegative powers of the second summand. For example,

$$x_0^{-1}\delta\left(\frac{x_1 - x_2}{x_0}\right) = \sum_{n\in\mathbb{Z}} \frac{(x_1 - x_2)^n}{x_0^{n+1}} = \sum_{m\in\mathbb{N},\, n\in\mathbb{Z}} (-1)^m \binom{n}{m} x_0^{-n-1} x_1^{n-m} x_2^m.$$

We have the following elementary identities:

$$x_1^{-1}\delta\left(\frac{x_2 + x_0}{x_1}\right) = x_2^{-1}\left(\frac{x_1 - x_0}{x_2}\right), \tag{2.2}$$

$$x_0^{-1}\delta\left(\frac{x_1 - x_2}{x_0}\right) - x_0^{-1}\delta\left(\frac{x_2 - x_1}{-x_0}\right) = x_2^{-1}\delta\left(\frac{x_1 - x_0}{x_2}\right). \tag{2.3}$$

Here and below, it is important to note that the relevant sums and products, etc., of formal series, are well defined. See [FLM] and [FHL] for extensive discussions of the calculus of formal δ-functions.

The following version of the definition of vertex algebra, using formal variables and the Jacobi identity of [FLM] is equivalent to Borcherds' original definition [Borc]:

Definition 2.1 A *vertex algebra* is a vector space V, equipped with a linear map $V \otimes V \to V[[x, x^{-1}]]$, or equivalently,

$$\begin{aligned} V &\to (\mathrm{End}\, V)[[x, x^{-1}]] \\ v &\mapsto Y(v, x) = \sum_{n \in \mathbf{Z}} v_n x^{-n-1} \quad (\text{where } v_n \in \mathrm{End}\, V), \end{aligned}$$

$Y(v, x)$ denoting the *vertex operator associated with* v, and equipped also with a distinguished homogeneous vector $\mathbf{1} \in V_{(0)}$ (the *vacuum*) and a linear map $D : V \to V$. The following conditions are assumed for $u, v \in V$: the *lower truncation condition* holds:

$$u_n v = 0 \quad \text{for } n \text{ sufficiently large}$$

(or equivalently, $Y(u, x)v \in V((x))$);

$$Y(\mathbf{1}, x) = 1 \quad (1 \text{ on the right being the identity operator});$$

the *creation property* holds:

$$Y(v, x)\mathbf{1} \in V[[x]] \quad \text{and} \quad \lim_{x \to 0} Y(v, x)\mathbf{1} = v$$

(that is, $Y(v, x)\mathbf{1}$ involves only nonnegative integral powers of x and the constant term is v); the *Jacobi identity* (the main axiom) holds:

$$\begin{aligned} x_0^{-1}\delta\left(\frac{x_1 - x_2}{x_0}\right) Y(u, x_1)Y(v, x_2) - x_0^{-1}\delta\left(\frac{x_2 - x_1}{-x_0}\right) Y(v, x_2)Y(u, x_1) \\ = x_2^{-1}\delta\left(\frac{x_1 - x_0}{x_2}\right) Y(Y(u, x_0)v, x_2) \end{aligned}$$

(note that when each expression in (2.8) is applied to any element of V, the coefficient of each monomial in the formal variables is a finite sum; on the right-hand side, the notation $Y(\cdot, x_2)$ is understood to be extended in the obvious way to $V[[x_0, x_0^{-1}]]$); and

$$\frac{d}{dx} Y(v, x) = Y(Dv, x)$$

(the *D-derivative property*).

The vertex algebra just defined is denoted by $(V, Y, \mathbf{1}, D)$ (or simply by V). Homomorphisms of vertex algebras are defined in the obvious way.

A consequence of the definition above is the *skew-symmetry* [Borc]:

$$Y(u,x)v = e^{xD}Y(v,-x)u \tag{2.4}$$

for $u, v \in V$. The proof uses the Jacobi identity, properties of the δ-function, the D-derivative property and the creation property (see [FHL]).

In the definition above, we have required that the vertex algebra has a vacuum, but in Section 5 we shall see that the notion of chiral algebra formulated using $\mathcal{D}$-modules does not give a vacuum. Thus we need the following notion:

Definition 2.2 A *vertex algebra without vacuum* is a vector space V, a vertex operator map $Y : V \otimes V \to V[[x, x^{-1}]]$ and an operator D on V satisfying all the axioms for a vertex algebra which do not involve the vacuum and in addition the skew-symmetry (2.4).

We denote the vertex algebra without vacuum by (V, Y, D) or simply by V.

3. $\mathcal{D}$-modules

We recall the elementary notions in the theories of sheaves (see [Ha]) and $\mathcal{D}$-modules (see [Borel]). We begin with the definition of presheaf. Following [Borel], we shall work over nonsingular quasi-projective varieties over $\mathbb{C}$ of pure dimension. Below, by a variety we shall mean a variety of this type.

Definition 3.1 Let X be a variety. A *presheaf $\mathcal{F}$ of abelian groups on X* consists of the data

1. for every open subset $U \subset X$, an abelian group $\mathcal{F}(U)$,
2. for every inclusion $V \subset U$ of open subsets of X, a morphism of abelian groups $\rho_{UV} : \mathcal{F}(U) \to \mathcal{F}(V)$,

satisfying the conditions

1. $\mathcal{F}(\emptyset) = 0$,
2. ρ_{UU} is the identity map from $\mathcal{F}(U)$ to itself,
3. if $W \subset V \subset U$ are open subsets of X, then $\rho_{UW} = \rho_{VW} \circ \rho_{UV}$.

We call the elements of $\mathcal{F}(U)$ *sections over* U and the maps ρ_{UV} the *restriction maps.* If $\mathcal{F}$ is a presheaf on X, and if P is a point of X, we define the *stalk* $\mathcal{F}_P$ of $\mathcal{F}$ at P to be the direct limit of the groups $\mathcal{F}(U)$ for all open sets U containing P via the restriction maps ρ. An element of $\mathcal{F}_P$ is called a *germ of sections of* $\mathcal{F}$ *at the point* P.

Presheaves with values in any fixed category can be defined by replacing the references to abelian groups in the definition by the analogous references for the given category. Most of the considerations below hold for *presheaves of (commutative or noncommutative) rings, presheaves of vector spaces* and *presheaves of (commutative or noncommutative) algebras*, for example; when we consider the zero object or subobjects or quotient objects, it will be understood that we are always working in a category where these exist. We sometimes write $s|_V$ instead of $\rho_{UV}(s)$ for $s \in \mathcal{F}(U)$. When it is necessary to distinguish the restriction maps of different sheaves, we shall write the restriction maps for a sheaf $\mathcal{F}$ as $\rho^{\mathcal{F}}_{UV}$.

Definition 3.2 A presheaf on a variety X is a *sheaf* if it satisfies the following additional conditions:

1. If U is an open subset of X, $\{V_i\}$ an open covering of U and $s \in \mathcal{F}(U)$ satisfying $s|_{V_i} = 0$ for all i, then $s = 0$.

2. If U is an open subset of X, $\{V_i\}$ an open covering of U and we have elements $s_i \in \mathcal{F}(V_i)$ for each i, with the property that for each i, j, $s_i|_{V_i \cap V_j} = s_j|_{V_i \cap V_j}$, then there is an element $s \in \mathcal{F}(U)$ such that $s_{V_i} = s_i$ for each i.

Example 3.3 Let X be a variety. We define the *sheaf* $\mathcal{O}_X$ *of regular functions on* X as follows: For any open set U of X, let $\mathcal{O}_X(U)$ be the ring of regular functions on U, which we shall also write as $\mathcal{O}(U)$, and for any open subset $V \subset U$ and $f \in \mathcal{F}(U)$, let $\rho_{UV}(f)$ be the restriction of f to V. It is clear that this gives a sheaf of rings (actually of commutative associative algebras).

Definition 3.4 *Morphisms* and *isomorphisms of presheaves* are defined in the obvious way. *Morphisms* and *isomorphisms of sheaves* are defined to be morphisms and isomorphisms of the underlying presheaves. *Presheaf kernels, presheaf cokernels* and *presheaf images of morphisms of presheaves* are defined in the obvious ways.

Note that presheaf kernels of morphisms of sheaves are sheaves but in general presheaf cokernels and presheaf images of morphisms of sheaves are not sheaves.

The following result associates a sheaf naturally to a presheaf in terms of a universal property:

Proposition 3.5 *Given a presheaf $\mathcal{F}$, there is a sheaf $\mathcal{F}^+$ and a morphism $\theta : \mathcal{F} \to \mathcal{F}^+$ with the property that for any sheaf $\mathcal{G}$ and any morphism $\varphi : \mathcal{F} \to \mathcal{G}$, there is a unique morphism $\psi : \mathcal{F}^+ \to \mathcal{G}$ such that $\varphi = \psi \circ \theta$. The pair $(\mathcal{F}^+, \theta)$ is unique up to unique isomorphism.*

See [Ha], for example, for a proof. The sheaf $\mathcal{F}^+$ is called the *sheaf associated with the presheaf $\mathcal{F}$*. It is determined completely by the germs of sections of $\mathcal{F}$. Thus we shall also call $\mathcal{F}^+$ the *sheaf of germs of sections of $\mathcal{F}$*.

Definition 3.6 A *subsheaf* of a sheaf $\mathcal{F}$ is a sheaf $\mathcal{F}'$ such that for every open set $U \subset X$, $\mathcal{F}'(U)$ is a subgroup (or subobject in the category we are discussing) of $\mathcal{F}(U)$, and the restriction maps of $\mathcal{F}'$ are induced from those of $\mathcal{F}$. Let $\varphi : \mathcal{F} \to \mathcal{G}$ be a morphism of sheaves. The *kernel* of φ is the presheaf kernel of φ, which is a sheaf, and in fact a subsheaf. The *image* of φ is the sheaf associated to the presheaf image of φ. The image of φ can be identified with a subsheaf of $\mathcal{G}$ by means of the universal property of the sheaf associated to a presheaf. Let $\mathcal{F}'$ be a subsheaf of a sheaf $\mathcal{F}$. The *quotient sheaf $\mathcal{F}/\mathcal{F}'$* is the sheaf associated to the presheaf given by $U \mapsto \mathcal{F}(U)/\mathcal{F}'(U)$. The *cokernel* of φ is the sheaf associated to the presheaf cokernel of φ.

Definition 3.7 Let X and Y be varieties and $f : X \to Y$ a morphism. For any sheaf $\mathcal{F}$ on X, The *direct image* $f_*(\mathcal{F})$ is the sheaf on Y given by $V \mapsto (f_*(\mathcal{F}))(V) = \mathcal{F}(f^{-1}(V))$ for any open set $V \subset Y$. For any sheaf $\mathcal{G}$ on Y, the *inverse image* $f^{-1}(\mathcal{G})$ is the sheaf associated to the presheaf given by $U \mapsto \lim_{V \supset f(U)} \mathcal{G}(V)$, where U is any open set of X and the limit is taken over all open sets V of Y containing $f(U)$.

For any open subset $V \subset Y$ and $s \in \mathcal{F}(f^{-1}(V))$, we denote s by $f_*(s)$ when it is viewed as an element of $(f_*(\mathcal{F}))(V)$, and we shall call s the *preimage of $f_*(s)$*. For any open subset $U \subset X$ and any open subset $V \subset Y$ such that $V \supset f(U)$, an element $s \in \mathcal{G}(V)$ determines an element of $(f^{-1}(\mathcal{G}))(U)$. We denote this element by $f^{-1}(s)$.

We now define the notions of left and right $\mathcal{O}$-module for any sheaf $\mathcal{O}$ of rings. Modules will always be left modules unless "right module" is specified. The concepts below have obvious analogues for right modules.

Definition 3.8 Let $\mathcal{O}$ be a sheaf of rings on X. A sheaf $\mathcal{F}$ on X is an $\mathcal{O}$-module if for any open set U, $\mathcal{F}(U)$ is an $\mathcal{O}(U)$-module, and for any open sets $V \subset U$ and $f \in \mathcal{O}(U)$, $s \in \mathcal{F}(U)$, we have $\rho^{\mathcal{F}}_{UV}(fs) = \rho^{\mathcal{O}}_{UV}(f)\rho^{\mathcal{F}}_{UV}(s)$.

Morphisms and *isomorphisms of $\mathcal{O}$-modules* are defined in the obvious ways. Note that kernels, images and cokernels of morphisms of $\mathcal{O}$-modules are again $\mathcal{O}$-modules. If $\mathcal{F}'$ is a subsheaf of $\mathcal{O}$-modules of an $\mathcal{O}$-module $\mathcal{F}$, then the quotient sheaf $\mathcal{F}/\mathcal{F}'$ is also an $\mathcal{O}$-module.

Now we restrict our attention to the sheaf $\mathcal{O}_X$ (recall Example 3.3), for which left and right modules are the same.

Definition 3.9 An $\mathcal{O}_X$-module $\mathcal{F}$ is said to be *quasi-coherent* if for any open affine subset U of X and $f \in \mathcal{O}_X(U)$, the following conditions hold:

1. Let V be the open set $\{x \in U \mid f(x) \neq 0\}$. For any $s \in \mathcal{F}(V)$, there exists $n \in \mathbb{N}$ and $\bar{s} \in \mathcal{F}(U)$ such that $\rho_{UV}(\bar{s}) = f^n s$.

2. For any $s \in \mathcal{F}(U)$ such that $\rho_{UV}(s) = 0$, there exists $n \in \mathbb{N}$ such that $f^n s = 0$.

Definition 3.10 Let $\mathcal{F}$ and $\mathcal{G}$ be $\mathcal{O}_X$-modules. The *tensor product* $\mathcal{F} \otimes_{\mathcal{O}_X} \mathcal{G}$ (or simply $\mathcal{F} \otimes \mathcal{G}$) is the sheaf associated to the presheaf given by $U \mapsto \mathcal{F}(U) \otimes_{\mathcal{O}_X(U)} \mathcal{G}(U)$.

Definition 3.11 Let X and Y be varieties, $f : X \to Y$ a morphism. This induces a natural morphism $f^{\#} : \mathcal{O}_Y \to f_*(\mathcal{O}_X)$ of sheaves. If $\mathcal{F}$ is an $\mathcal{O}_X$-module, then $f_*(\mathcal{F})$ is an $f_*(\mathcal{O}_X)$-module, and thus an $\mathcal{O}_Y$-module via $f^{\#}$. This $\mathcal{O}_Y$-module $f_*(\mathcal{F})$ is called the *direct image of $\mathcal{F}$ by f*. If $\mathcal{G}$ is a $\mathcal{O}_Y$-module, then $f^{-1}(\mathcal{G})$ is an $f^{-1}(\mathcal{O}_Y)$-module. Since $f^{\#}$ induces a natural morphism from $f^{-1}(\mathcal{O}_Y)$ to $\mathcal{O}_X$, we can form the tensor product $f^{-1}(\mathcal{G}) \otimes_{f^{-1}(\mathcal{O}_Y)} \mathcal{O}_X$ and it is an $\mathcal{O}_X$-module. This $\mathcal{O}_X$-module is called the *inverse image of $\mathcal{G}$ by f* and is denoted $f^*(\mathcal{G})$.

For any open subset $U \subset X$, any open subset $V \subset Y$ such that $V \supset f(U)$, and any element $s \in \mathcal{G}(V)$, we denote $f^{-1}(s) \otimes_{(f^{-1}(\mathcal{O}_Y))(U)} 1 \in (f^*(\mathcal{G}))(U)$ by $f^*(s)$.

Example 3.12 Let X be a variety. Here we define the sheaf of (noncommutative) algebras $\mathcal{D}_X$—the *sheaf of germs of algebraic differential operators on X* (see [Borel]). If X is an affine variety, we define $\mathcal{D}_X(X)$ to be the algebra $\mathcal{D}(X)$ of operators on $\mathcal{O}(X) = \mathcal{O}_X(X)$ generated by the elements of $\mathcal{O}(X)$ (acting by multiplication) and the derivations of $\mathcal{O}(X)$. This is the *ring of (algebraic) differential operators on X*. For any open affine subset U of X, we have $\mathcal{D}(U) = \mathcal{O}(U) \otimes_{\mathcal{O}(X)} \mathcal{D}(X)$, and there exists a unique quasi-coherent $\mathcal{O}_X$-module $\mathcal{D}_X$ on X such that $\mathcal{D}(U)$ is the $\mathcal{O}(U)$-module of sections over an open affine subset $U \subset X$. In general, for any variety X (of the type we are considering), there exists a unique sheaf $\mathcal{D}_X$ of algebras on X whose restriction to every open affine subset $U \subset X$ is $\mathcal{D}_U$. The algebra $\mathcal{D}_X(X)$ consists of the *algebraic differential operators on X*. The sheaf $\mathcal{D}_X$ is quasi-coherent as an $\mathcal{O}_X$-module and its restriction to an open subset $U \subset X$ is $\mathcal{D}_U$.

We shall use the following notion of (algebraic) $\mathcal{D}$-module:

Definition 3.13 Let X be a variety. A *$\mathcal{D}_X$-module* is a $\mathcal{D}_X$-module as defined above which is also quasi-coherent as an $\mathcal{O}_X$-module. Similarly for a *right $\mathcal{D}_X$-module.*

Note that a (left or right) $\mathcal{D}_X$-module is a $\mathcal{D}_X$-module in the sense of Definition 3.8 satisfying an extra condition. *Morphisms* and *isomorphisms of $\mathcal{D}_X$-modules* are defined in the obvious ways. It is clear that the set of all morphisms from a $\mathcal{D}_X$-module to another one has a natural abelian group structure. For any (left) $\mathcal{D}_X$-modules $\mathcal{F}_1, \mathcal{F}_2$ and right $\mathcal{D}_X$-modules $\mathcal{G}_1, \mathcal{G}_2$, $\mathcal{F}_1 \otimes_{\mathcal{O}_X} \mathcal{F}_2$ and $\mathrm{Hom}_{\mathcal{O}_X}(\mathcal{G}_1, \mathcal{G}_2)$ are naturally (left) $\mathcal{D}_X$-modules, and $\mathcal{F}_1 \otimes_{\mathcal{O}_X} \mathcal{G}_1$ and $\mathrm{Hom}_{\mathcal{O}_X}(\mathcal{F}_1, \mathcal{G}_2)$ are naturally right $\mathcal{D}_X$-modules. In particular, we have a tensor product operation in the category of (left) $\mathcal{D}_X$-modules.

Example 3.14 Let ω_X be the sheaf of germs of differential forms of top degree on X. Let U be an open affine subset of X. A derivation $\xi \in \mathcal{D}_X(U)$ acts on $\omega_X(U)$ from the left by the Lie derivative L_ξ, but this does not give a (left) $\mathcal{D}_X$-module structure to ω_X. Rather, the action $-L_\xi$ of ξ on $\omega_X(U)$ gives a right $\mathcal{D}_X$-module structure to ω_X. Thus for any (left) $\mathcal{D}_X$-module $\mathcal{F}$ and right $\mathcal{D}_X$-module $\mathcal{G}$, $\mathcal{F} \otimes_{\mathcal{O}_X} \omega_X$ is a right $\mathcal{D}_X$-module and $\mathrm{Hom}_{\mathcal{O}_X}(\omega_X, \mathcal{G})$ is a (left) $\mathcal{D}_X$-module. (See [Borel].)

We have the following notions of inverse image and direct image of a $\mathcal{D}_X$-module:

Definition 3.15 Let X and Y be varieties and $f : X \to Y$ a morphism. For any $\mathcal{D}_Y$-module $\mathcal{G}$, we have the inverse image $\mathcal{O}_X$-module $f^*(\mathcal{G})$, which is quasi-coherent. There is a natural $\mathcal{D}_X$-module structure on $f^*(\mathcal{G})$ (see [Borel]), which we call the *inverse image $\mathcal{D}_X$-module* and write as $f^\circ(\mathcal{G})$. Now, to define the direct image, we begin with the $\mathcal{O}_X$-module $f^*(\mathcal{D}_Y)$, which has a natural (left) $\mathcal{D}_X$-module structure, as above. On the other hand, the right multiplication on $\mathcal{D}_Y$ carries over to a right $f^{-1}(\mathcal{D}_Y)$-module structure on $f^*(\mathcal{D}_Y)$. We denote $f^*(\mathcal{D}_Y)$ by $\mathcal{D}_{X\to Y}$ when it is equipped with this $\mathcal{D}_X \times f^{-1}(\mathcal{D}_Y)$-module structure. Let $\mathcal{D}_{Y\leftarrow X} = \mathcal{D}_{X\to Y} \otimes_{\mathcal{O}_X} \mathrm{Hom}_{\mathcal{O}_X}(f^*(\omega_Y), \omega_X)$. Then $\mathcal{D}_{Y\leftarrow X}$ is a (left) $f^{-1}(\mathcal{D}_Y)$-module and a right $\mathcal{D}_X$-module. Let $\mathcal{F}$ be a (left) $\mathcal{D}_X$-module. Then $\mathcal{D}_{Y\leftarrow X} \otimes_{\mathcal{D}_X} \mathcal{F}$ (defined in the obvious way) is a (left) $f^{-1}(\mathcal{D}_Y)$-module. We also have a canonical morphism $\tilde{f} : \mathcal{D}_Y \to f_*(f^{-1}(\mathcal{D}_Y))$ of sheaves of rings (recall Definition 3.7). We define the *direct image $\mathcal{D}_Y$-module* $f_\circ(\mathcal{F})$ to be $f_*(\mathcal{D}_{Y\leftarrow X} \otimes_{\mathcal{D}_X} \mathcal{F})$, viewed as a $\mathcal{D}_Y$-module via $\tilde{f}$ (cf. the definition of direct image in Definition 3.11).

Example 3.16 Let X be a nonempty open subset of $\mathbb{C}$, $Y = X \times X$, $f = \Delta : X \to Y = X \times X$ the diagonal map, defined by $\Delta(z) = (z, z)$, and $\mathcal{F}$ a $\mathcal{D}_X$-module. Note that X is necessarily affine, and that for a local theory over curves, such varieties X are enough. In this case, $\omega_X(X)$ and $\omega_{X\times X}(X \times X)$ are generated by dz and $dz_1 \wedge dz_2$, respectively, and

$$(\mathrm{Hom}_{\mathcal{O}_X}(\Delta^*(\omega_{X\times X}), \omega_X))(X) = \mathrm{Hom}_{\mathcal{O}_X(X)}(\Delta^*(\omega_{X\times X}(X \times X)), \omega_X(X))$$

is generated by $\phi \in \mathrm{Hom}_{\mathcal{O}_X(X)}((\Delta^*(\omega_{X\times X}(X \times X)), \omega_X(X))$ defined by

$$\phi(\Delta^*(dz_1 \wedge dz_2)) = dz.$$

Thus $(\Delta_\circ(\mathcal{F}))(X \times X)$ is generated by elements of the form

$$\Delta_*((\Delta^*(1) \otimes_{\mathcal{O}_X(X)} \phi) \otimes_{\mathcal{D}_X(X)} A)$$

for $A \in \mathcal{F}(X)$. The element of $\frac{\partial}{\partial z_1} + \frac{\partial}{\partial z_2}$ of $\mathcal{D}_{X\times X}(X \times X)$ is in fact in $(\Delta_*(\Delta^{-1}(\mathcal{D}_{X\times X})))(X \times X)$ and from the definition, we have

$$\begin{aligned}
&\left(\frac{\partial}{\partial z_1} + \frac{\partial}{\partial z_2}\right) \Delta_*((\Delta^*(\xi) \otimes_{\mathcal{O}_X(X)} \phi) \otimes_{\mathcal{D}_X(X)} A) \\
&\quad = \Delta_* \left(\left(\Delta^* \left(\xi \left(\frac{\partial}{\partial z_1} + \frac{\partial}{\partial z_2} \right) \right) \otimes_{\mathcal{O}_X(X)} \phi \right) \otimes_{\mathcal{D}_X(X)} A \right) \\
&\quad = \Delta_* \left((\Delta^*(\xi) \otimes_{\mathcal{O}_X(X)} \phi) \otimes_{\mathcal{D}_X(X)} \frac{\partial}{\partial z} A \right)
\end{aligned}$$

for any $\xi \in \mathcal{D}_{X\times X}(X \times X)$ commuting with $\frac{\partial}{\partial z_1} + \frac{\partial}{\partial z_2}$ and $A \in \mathcal{F}(X)$, so that the elements

$$\frac{\partial^n}{\partial z_1^n} \Delta_* \left((\Delta^*(1) \otimes_{\mathcal{O}_X(X)} \phi) \otimes_{\mathcal{D}_X(X)} A \right), \quad n \in \mathbb{Z},$$

span $(\Delta_\circ(\mathcal{F}))(X \times X)$. Note that since X is an open subset of $\mathbb{C}$, $(\Delta_*(\mathcal{D}_X))(X \times X)$ can be embedded into $\mathcal{D}_{X\times X}(X \times X)$ such that $\Delta_*(\frac{\partial}{\partial z})$ is mapped to $\frac{\partial}{\partial z_1} + \frac{\partial}{\partial z_2}$. Thus we have a $\mathcal{D}_{X\times X}(X \times X)$-module

$$\mathcal{D}_{X\times X}(X \times X) \otimes_{(\Delta_*(\mathcal{D}_X))(X\times X)} (\Delta_*(\mathcal{F}))(X \times X),$$

and if $\{A_\alpha\}$ is a basis of $\mathcal{F}(X)$, this $\mathcal{D}_{X\times X}(X \times X)$-module has as a basis $\frac{\partial^n}{\partial z_1^n} \otimes_{(\Delta_*(\mathcal{D}_X))(X\times X)} A_\alpha$ for $n \in \mathbb{N}$. It is easy to see that

$$\frac{\partial^n}{\partial z_1^n} \Delta_* \left((\Delta^*(1) \otimes_{\mathcal{O}_X(X)} \phi) \otimes_{\mathcal{D}_X(X)} A \right) \mapsto \frac{\partial^n}{\partial z_1^n} \otimes_{(\Delta_*(\mathcal{D}_X))(X\times X)} A$$

gives an isomorphism from $(\Delta_\circ(\mathcal{F}))(X \times X)$ to

$$\mathcal{D}_{X\times X}(X \times X) \otimes_{(\Delta_*(\mathcal{D}_X))(X\times X)} (\Delta_*(\mathcal{F}))(X \times X).$$

Thus $(\Delta_\circ(\mathcal{F}))(X \times X)$ has as a basis the elements of the form

$$\frac{\partial^n}{\partial z_1^n}\Delta_*\left((\Delta^*(1) \otimes_{\mathcal{O}_X(X)} \phi) \otimes_{\mathcal{D}_X(X)} A_\alpha\right)$$

for $n \in \mathbb{N}$, and we can identify the $\mathcal{D}_{X\times X}$-module $\Delta_\circ(\mathcal{F})$ with $\mathcal{D}_{X\times X} \otimes_{\Delta_*(\mathcal{D}_X)} \Delta_*(\mathcal{F})$. Now let us take $\mathcal{F} = \mathcal{O}_X$. Then we have the element

$$\Delta_*((\Delta^*(1) \otimes_{\mathcal{O}_X(X)} \phi) \otimes_{\mathcal{D}_X(X)} 1). \tag{3.1}$$

For any $f(z_1, z_2) \in \mathcal{O}_{X\times X}(X \times X)$, we have

$$\begin{aligned} f(z_1, z_2)&\Delta_*((\Delta^*(1) \otimes_{\mathcal{O}_X(X)} \phi) \otimes_{\mathcal{D}_X(X)} 1) \\ &= \Delta_*(f(z, z)((\Delta^*(1) \otimes_{\mathcal{O}_X(X)} \phi) \otimes_{\mathcal{D}_X(X)} 1)) \\ &= f(z_1, z_1)\Delta_*((\Delta^*(1) \otimes_{\mathcal{O}_X(X)} \phi) \otimes_{\mathcal{D}_X(X)} 1). \end{aligned} \tag{3.2}$$

Similarly, the left-hand side of (3.2) is also equal to

$$f(z_2, z_2)\Delta_*((\Delta^*(1) \otimes_{\mathcal{O}_X(X)} \phi) \otimes_{\mathcal{D}_X(X)} 1).$$

So we see that the element (3.1) has the property similar to the property (2.1) of the δ-function $x_2^{-1}\delta\left(\frac{x_1}{x_2}\right)$. In fact it is easy to show that the derivatives

$$\frac{\partial^m}{\partial z_1^m}\Delta_*((\Delta^*(1) \otimes_{\mathcal{O}_X(X)} \phi) \otimes_{\mathcal{D}_X} 1),$$

$m \in \mathbb{N}$, and

$$\frac{\partial^n}{\partial z_2^n}\Delta_*((\Delta^*(1) \otimes_{\mathcal{O}_X(X)} \phi) \otimes_{\mathcal{D}_X} 1),$$

$n \in \mathbb{N}$, also have the corresponding properties of the derivatives of $x_2^{-1}\delta\left(\frac{x_1}{x_2}\right)$ (see [FLM], Proposition 8.2.2). Thus we can identify (3.1) with $x_2^{-1}\delta\left(\frac{x_1}{x_2}\right)$. In particular, we obtain a $\mathcal{D}_{X\times X}$-module generated by $x_2^{-1}\delta\left(\frac{x_1}{x_2}\right)$ which is isomorphic to $\Delta_\circ(\mathcal{O}_X)$.

Remark 3.17 In the example above, we have given a $\mathcal{D}$-module-theoretic interpretation of the δ-function $x_2^{-1}\delta\left(\frac{x_1}{x_2}\right)$ and its derivatives. But in the formal variable approach to vertex algebras, it is most natural to use $x_2^{-1}\delta\left(\frac{x_1-x_0}{x_2}\right)$ (see above), which is in fact a formal infinite linear combination of derivatives of $x_2^{-1}\delta\left(\frac{x_1}{x_2}\right)$, by a formal Taylor's theorem (see e.g. [FLM], Propositions 8.2.2 and 8.3.1). This δ-function and the other two in the Jacobi identity for a vertex algebra apparently do not have direct interpretations in terms of $\mathcal{D}$-modules. The S_3-symmetry of (2.3) and of the

Jacobi identity for vertex algebras (see [FLM], [FHL]) involves all the three formal variables x_0, x_1, x_2 on equal footing, and we note that the $\mathcal{D}$-module approach does not naturally have three variables on equal footing.

We also need the concept of exterior tensor product:

Definition 3.18 Let X and Y be varieties and $\mathcal{F}$ and $\mathcal{G}$ $\mathcal{D}_X$- and $\mathcal{D}_Y$-modules, respectively. Let $(\mathcal{F} \boxtimes \mathcal{G})(U \times V) = \mathcal{F}(U) \otimes \mathcal{G}(V)$ for all open affine subsets $U \subset X$ and $V \subset Y$. Then there is a unique $\mathcal{D}_{X\times Y}$-module $\mathcal{F}\boxtimes\mathcal{G}$ such that $(\mathcal{F}\boxtimes\mathcal{G})(U_1 \times U_2)$ is the $\mathcal{O}_{X\times Y}(U\times V)$-module of sections on $U \times V$ for any open affine subsets $U \subset X$ and $V \subset Y$. We call this $\mathcal{D}_{X\times Y}$-module the *exterior tensor product of $\mathcal{F}$ and $\mathcal{G}$*. From the definition, we see that $\mathcal{F} \boxtimes \mathcal{G}$ is determined by sections on open affine subsets of the form $U \times V$.

4. Beilinson-Drinfeld's chiral algebra

We now give Beilinson-Drinfeld's definition of chiral algebra. For simplicity and for the purpose of comparing it with the definition of vertex algebra without vacuum, we only give the definition of chiral algebra over a nonempty open subset $X \subset \mathbb{C}$. But since the definition is sheaf-theoretic and thus is local in nature, this definition carries over naturally to the general case. The axioms in the definition are natural in the language of quasi-tensor categories in Beilinson-Drinfeld's formulation.

Let $X \subset \mathbb{C}$ be a nonempty open subset. Let F^2 be the complement of the diagonal in $X \times X$. We denote the embedding map from F^2 to $X \times X$ by j and the diagonal map from X to $X \times X$ by Δ. Let

$$F^3 = \{(z_1, z_2, z_3) \in X \times X \times X \mid z_k \neq z_l \text{ for } k \neq l, k, l = 1, 2, 3\}.$$

We denote the embedding map from F^3 to $X \times X \times X$ by j_3 and the diagonal map from X to $X \times X \times X$ by Δ_3.

Let $\mathcal{A}$ be a $\mathcal{D}_X$-module.

The exterior tensor product $\mathcal{D}_{X\times X}$-module $\mathcal{A} \boxtimes \mathcal{A}$ is by definition determined by

$$\begin{aligned} &U_1 \times U_2 \subset X \times X \\ &\quad\mapsto \{\sum_{i=1}^{n} f_i(A_i \otimes B_i) \mid A_i \in \mathcal{A}(U_1), B_i \in \mathcal{A}(U_2), \\ &\qquad\qquad\qquad f_i \in \mathcal{O}(U_1 \times U_2) \text{ for } i = 1, \dots, n\} \end{aligned}$$

for all open subsets $U_1, U_2 \subset X$. The inverse image $\mathcal{O}_{F^2}$-module $j^*(\mathcal{A} \boxtimes \mathcal{A})$ is the $\mathcal{O}_{F^2}$-module obtained by restricting the sections of $\mathcal{A} \boxtimes \mathcal{A}$ over open subsets of $X \times X$ to open subsets of F^2, and the direct image $\mathcal{O}_{X\times X}$-module $j_*j^*(\mathcal{A} \boxtimes \mathcal{A})$ is the $\mathcal{O}_{X\times X}$-module determined by

$$\begin{aligned} &U_1 \times U_2 \subset X \times X \\ &\qquad \mapsto \{\sum_{i=1}^{n} f_i(A_i \otimes B_i) \mid A_i \in \mathcal{A}(U_1), B_i \in \mathcal{A}(U_2), \\ &\qquad\qquad\qquad\qquad f_i \in \mathcal{O}((U_1 \times U_2) \cap F^2) \text{ for } i = 1, \dots, n\} \end{aligned}$$

for all open subsets $U_1, U_2 \subset X$. It has a natural $\mathcal{D}_{X\times X}$-module structure. As an $\mathcal{O}_{X\times X}(U_1 \times U_2)$-module, $j_*j^*(\mathcal{A} \boxtimes \mathcal{A})(U_1 \times U_2)$ is generated by the elements of the form

$$(z_1 - z_2)^m (A_1 \otimes A_2)$$

for $m < 0$, $A_1 \in \mathcal{A}(U_1)$ and $A_2 \in \mathcal{A}(U_2)$. Similarly, if $U = U_1 \times U_2 \times U_3$ where U_1, U_2, U_3 are open subsets of X, then $((j_3)_* j_3^*(\mathcal{A} \boxtimes \mathcal{A} \boxtimes \mathcal{A}))(U)$ as an $\mathcal{O}_{X\times X\times X}(U)$-module is generated by the elements of the form

$$(z_1 - z_2)^{m_1} (z_2 - z_3)^{m_2} (z_1 - z_3)^{m_3} (A_1 \otimes A_2 \otimes A_3)$$

for $m_i < 0$, $A_i \in \mathcal{A}(U_i)$, $i = 1, 2, 3$. This has a natural $\mathcal{D}_{X\times X\times X}(U)$-module structure, and so we have the $\mathcal{D}_{X\times X\times X}$-module $(j_3)_* j_3^*(\mathcal{A} \boxtimes \mathcal{A} \boxtimes \mathcal{A})$.

In the definition of Lie algebra, we need compositions $[[\cdot,\cdot],\cdot]$ and $[\cdot,[\cdot,\cdot]]$ of the Lie bracket $[\cdot,\cdot]$ to formulate the Jacobi identity. Here we need analogues of these compositions.

For any morphism $\mu : j_*j^*(\mathcal{A} \boxtimes \mathcal{A}) \to \Delta_\circ \mathcal{A}$ of $\mathcal{D}_{X\times X}$-modules, we define a natural morphism

$$\mu(\mu(\cdot,\cdot),\cdot) : (j_3)_* j_3^*(\mathcal{A} \boxtimes \mathcal{A} \boxtimes \mathcal{A}) \to (\Delta_3)_\circ(\mathcal{A})$$

of $\mathcal{D}_{X\times X\times X}$-modules as follows: Let U be an open subset of $X \times X \times X$ of the form $U_1 \times U_2 \times U_3$ where U_1, U_2 and U_3 are open subsets of X. We identify $\Delta_\circ(\mathcal{A})$ with $\mathcal{D}_{X\times X} \otimes_{(\Delta_*(\mathcal{D}_X))} \Delta_*(\mathcal{A})$ (see Example 3.16). In particular, $(\Delta_\circ(\mathcal{A}))(U_1 \times U_2)$ is spanned by the elements of the form

$$\frac{\partial^n}{\partial z_1^n} \otimes_{(\Delta_*(\mathcal{D}_X))(U)} \Delta_*(A)$$

for $n \in \mathbb{N}$ and $A \in \mathcal{A}(\Delta^{-1}(U_1 \times U_2))$. Thus we see that for any $n_1 \in \mathbb{Z}$ there exist $p_{n_1} \in \mathbb{N}$ and

$$B_k^{n_1} \in \mathcal{A}(\Delta^{-1}(U_1 \times U_2)) = \mathcal{A}(U_1 \cap U_2), \quad k = 0, \dots, p_{n_1},$$

such that

$$\mu((z_1 - z_2)^{n_1}(A_1 \otimes A_2)) = \sum_{k=0}^{p_{n_1}} \frac{\partial^k}{\partial z_1^k} \otimes_{(\Delta_*(\mathcal{D}_X))(U_1\times U_2)} \Delta_*(B_k^{n_1}); \tag{4.1}$$

the nonzero elements $B_k^{n_1}$ are uniquely determined. Note that when $n_1 \geq 0$, $(z_1 - z_2)^{n_1}$ is regular, so that when $n_1 > p_0$, we have

$$\begin{aligned}
&\mu((z_1 - z_2)^{n_1}(A_1 \otimes A_2)) = \\
&\quad = (z_1 - z_2)^{n_1}\mu(A_1 \otimes A_2) \\
&\quad = (z_1 - z_2)^{n_1} \sum_{k=0}^{p_0} \frac{\partial^k}{\partial z_1^k} \otimes_{(\Delta_*(\mathcal{D}_X))(U_1 \times U_2)} \Delta_*(B_k^0) \\
&\quad = 0.
\end{aligned}$$

Similarly, for any $n_2 \in \mathbb{Z}$ we have, for the elements $B_k^{n_1}$,

$$\mu((z_2 - z_3)^{n_2}(B_k^{n_1} \otimes A_3)) = \sum_{l=0}^{q_{n_1,n_2}} \frac{\partial^l}{\partial z_2^l} \otimes_{(\Delta_*(\mathcal{D}_X))(U_2 \times U_3)} \Delta_*(C_{kl}^{n_1 n_2}) \quad (4.2)$$

where $q_{n_1,n_2} \in \mathbb{N}$ and $C_{kl}^{n_1 n_2} \in \mathcal{A}(\Delta^{-1}((U_1 \cap U_2) \times U_3) = \mathcal{A}(U_1 \cap U_2 \cap U_3)$, $k = 0, \ldots, p_{n_1}$, $l = 0, \ldots, q_{n_1,n_2}$. Since $\frac{\partial}{\partial z_2} + \frac{\partial}{\partial z_3} \in (\Delta_*(\mathcal{D}_X))(U_2 \times U_3)$ and its preimage in $\mathcal{D}_X(U_2 \cap U_3)$ is $\frac{\partial}{\partial z}$, the right-hand side of (4.2) is equal to

$$\begin{aligned}
&\sum_{l=0}^{q_{n_1,n_2}} \left(\left(\frac{\partial}{\partial z_2} + \frac{\partial}{\partial z_3}\right) - \frac{\partial}{\partial z_3}\right)^l \otimes_{(\Delta_*(\mathcal{D}_X))(U_2 \times U_3)} \Delta_*(C_{kl}^{n_1 n_2}) \\
&\quad = \sum_{l=0}^{q_{n_1,n_2}} \sum_{j=0}^{l} (-1)^j \binom{l}{j} \frac{\partial^j}{\partial z_3^j} \left(\frac{\partial}{\partial z_2} + \frac{\partial}{\partial z_3}\right)^{l-j} \\
&\qquad \otimes_{(\Delta_*(\mathcal{D}_X))(U_2 \times U_3)} \Delta_*(C_{kl}^{n_1 n_2}) \\
&\quad = \sum_{l=0}^{q_{n_1,n_2}} \sum_{j=0}^{l} (-1)^j \binom{l}{j} \frac{\partial^j}{\partial z_3^j} \otimes_{(\Delta_*(\mathcal{D}_X))(U_2 \times U_3)} \Delta_* \left(\frac{\partial^{l-j}}{\partial z^{l-j}} C_{kl}^{n_1 n_2}\right).
\end{aligned} \quad (4.3)$$

Let $C_{kl}^{n_1 n_2} = 0$ for $k > p_{n_1}$ or $l > q_{n_1,n_2}$. As in Example 3.16, $(\Delta_3)_\circ(\mathcal{A})$ is canonically isomorphic to $\mathcal{D}_{X \times X \times X} \otimes_{(\Delta_3)_*(\mathcal{D}_X)} \mathcal{A}$, and we shall identify $(\Delta_3)_\circ(\mathcal{A})$ with $\mathcal{D}_{X \times X \times X} \otimes_{(\Delta_3)_*(\mathcal{D}_X)} \mathcal{A}$. We define

$$\begin{aligned}
&(\mu(\mu(\cdot,\cdot),\cdot))((z_1 - z_2)^{m_1}(z_2 - z_3)^{m_2}(z_1 - z_3)^{m_3}(A_1 \otimes A_2 \otimes A_3)) \\
&\quad = \sum_{i \in \mathbb{N}} \sum_{k,l \in \mathbb{N}} \sum_{j=0}^{l} \binom{m_3}{i} (-1)^j \binom{l}{j} \frac{\partial^k}{\partial z_1^k} \frac{\partial^j}{\partial z_3^j} \\
&\qquad \otimes_{((\Delta_3)_*(\mathcal{D}_X))(U)} (\Delta_3)_* \left(\frac{\partial^{l-j}}{\partial z^{l-j}} C_{kl}^{m_1+i, m_2+m_3-i}\right),
\end{aligned} \quad (4.4)$$

using the expansion of

$$(z_1 - z_3)^{m_3} = ((z_2 - z_3) + (z_1 - z_2))^{m_3}$$

in nonnegative powers of $z_1 - z_2$. Note that the right-hand side of (4.4) is in fact a finite sum and is indeed an element of $((\Delta_3)_{\circ}(\mathcal{A}))(U)$. It is easy to verify directly that (4.4) indeed gives a morphism of $\mathcal{D}_{X\times X\times X}$-modules. Thus we obtain the morphism we want.

The morphism given by (4.4) is natural but we need another expression of the right-hand side of (4.4) in Section 5. Note that $\frac{\partial}{\partial z_1} + \frac{\partial}{\partial z_2} + \frac{\partial}{\partial z_3} \in ((\Delta_3)_*(\mathcal{D}_X))(U)$ and its preimage in $\mathcal{D}_X(U_1 \cap U_2 \cap U_3)$ is $\frac{\partial}{\partial z}$. So the right-hand side of (4.4) is equal to

$$\begin{aligned}
&\sum_{i\in\mathbb{N}}\sum_{k,l\in\mathbb{N}}\sum_{j=0}^{l}\binom{m_3}{i}(-1)^j\binom{l}{j}\frac{\partial^k}{\partial z_1^k}\frac{\partial^j}{\partial z_3^j}\left(\frac{\partial}{\partial z_1}+\frac{\partial}{\partial z_2}+\frac{\partial}{\partial z_3}\right)^{l-j}\\
&\qquad\otimes_{((\Delta_3)_*(\mathcal{D}_X))(U)}(\Delta_3)_*(C_{kl}^{m_1+i,m_2+m_3-i})\\
&\quad=\sum_{i\in\mathbb{N}}\sum_{k,l\in\mathbb{N}}\binom{m_3}{i}\frac{\partial^k}{\partial z_1^k}\left(\frac{\partial}{\partial z_1}+\frac{\partial}{\partial z_2}\right)^{l}\\
&\qquad\otimes_{((\Delta_3)_*(\mathcal{D}_X))(U)}(\Delta_3)_*(C_{kl}^{m_1+i,m_2+m_3-i})\\
&\quad=\sum_{i\in\mathbb{N}}\sum_{k,l\in\mathbb{N}}\sum_{j=0}^{l}\binom{m_3}{i}\binom{l}{j}\frac{\partial^{k+j}}{\partial z_1^{k+j}}\frac{\partial^{l-j}}{\partial z_2^{l-j}}\\
&\qquad\otimes_{((\Delta_3)_*(\mathcal{D}_X))(U)}(\Delta_3)_*(C_{kl}^{m_1+i,m_2+m_3-i}).
\end{aligned}\tag{4.5}$$

From

$$\mu((z_1-z_2)^{n+1}(A_1\otimes A_2)) = (z_1-z_2)\mu((z_1-z_2)^n(A_1\otimes A_2)) \tag{4.6}$$

and (4.1), we obtain

$$\begin{aligned}
&\sum_{k=0}^{p_{n_1+1}}\frac{\partial^k}{\partial z_1^k}\otimes_{(\Delta_*(\mathcal{D}_X))(U_1\times U_2)}\Delta_*(B_k^{n_1+1})\\
&\qquad=(z_1-z_2)\sum_{k=0}^{p_{n_1}}\frac{\partial^k}{\partial z_1^k}\otimes_{(\Delta_*(\mathcal{D}_X))(U_1\times U_2)}\Delta_*(B_k^{n_1}).
\end{aligned}\tag{4.7}$$

Since for any $k\in\mathbb{N}$,

$$\frac{\partial^k}{\partial z_1^k}(z_1-z_2)\otimes_{(\Delta_*(\mathcal{D}_X))(U_1\times U_2)}\Delta_*(B_m^{n_1})=0,$$

the right-hand side of (4.7) is equal to

$$-\sum_{k=0}^{p_{n_1}}k\frac{\partial^{k-1}}{\partial z_1^{k-1}}\otimes_{(\Delta_*(\mathcal{D}_X))(U_1\times U_2)}\Delta_*(B_k^{n_1}). \tag{4.8}$$

Note that for any $A_i \in \mathcal{A}(U_1 \cap U_2)$, $m_i \in \mathbb{N}$, $i = 1, \ldots, n$,

$$\frac{\partial^{m_i}}{\partial z_1^{m_i}} \otimes_{(\Delta_*(\mathcal{D}_X))(U_1 \times U_2)} \Delta_*(A_i),\ i = 1, \ldots, n$$

are linearly independent when $A_i \neq 0$ for $i = 1, \ldots, n$ and $m_i \neq m_j$ for $i \neq j$, $i, j = 1, \ldots, n$, and they are equal to 0 if and only if $A_i = 0$, $i = 1, \ldots, n$. Thus from (4.7) and (4.8), we obtain

$$B_k^{n_1+1} = -(k+1)B_{k+1}^{n_1} \tag{4.9}$$

for any $n_1 \in \mathbb{Z}$, $k = 0, \ldots, p_{n_1} - 1$. Similarly we have

$$\begin{aligned} C_{kl}^{n_1+1,n_2+1} &= -(k+1)C_{k+1,l}^{n_1,n_2+1} \\ &= -(l+1)C_{k,l+1}^{n_1+1,n_2}. \end{aligned} \tag{4.10}$$

Using (4.10) repeatedly, we can write the right-hand side of (4.5) as

$$\begin{aligned} &\sum_{i \in \mathbb{N}} \sum_{k,l \in \mathbb{N}} \sum_{j=0}^{l} \binom{m_3}{i} \binom{k+j}{j} \frac{\partial^{k+j}}{\partial z_1^{k+j}} \frac{\partial^{l-j}}{\partial z_2^{l-j}} \\ &\qquad \otimes_{((\Delta_3)_*(\mathcal{D}_X))(U)} (\Delta_3)_*(C_{k+j,l-j}^{m_1+i-j,m_2+m_3-i+j}) \\ &= \sum_{i' \in \mathbb{N}} \sum_{k',l' \in \mathbb{N}} \sum_{\substack{i+k=i' \\ 0 \le i,k \le i'}} \binom{m_3}{i} \binom{k'}{k} \frac{\partial^{k'}}{\partial z_1^{k'}} \frac{\partial^{l'}}{\partial z_2^{l'}} \\ &\qquad \otimes_{((\Delta_3)_*(\mathcal{D}_X))(U)} (\Delta_3)_*(C_{k',l'}^{m_1+i'-k',m_2+m_3-i'+k'}) \\ &= \sum_{i' \in \mathbb{N}} \sum_{k',l' \in \mathbb{N}} \binom{m_3+k'}{i'} \frac{\partial^{k'}}{\partial z_1^{k'}} \frac{\partial^{l'}}{\partial z_2^{l'}} \\ &\qquad \otimes_{((\Delta_3)_*(\mathcal{D}_X))(U)} (\Delta_3)_*(C_{k',l'}^{m_1+i'-k',m_2+m_3-i'+k'}), \end{aligned} \tag{4.11}$$

where in the last step, we have used the identity

$$\sum_{\substack{i+k=i' \\ 0 \le i,k \le i'}} \binom{m_3}{i} \binom{k'}{k} = \binom{m_3+k'}{i'}.$$

By (4.4), (4.5) and (4.11), we obtain

$$\begin{aligned} &(\mu(\mu(\cdot,\cdot),\cdot))((z_1-z_2)^{m_1}(z_2-z_3)^{m_2}(z_1-z_3)^{m_3}(A_1 \otimes A_2 \otimes A_3)) \\ &= \sum_{i \in \mathbb{N}} \sum_{k,l \in \mathbb{N}} \binom{m_3+k}{i} \frac{\partial^{k}}{\partial z_1^{k}} \frac{\partial^{l}}{\partial z_2^{l}} \\ &\qquad \otimes_{((\Delta_3)_*(\mathcal{D}_X))(U)} (\Delta_3)_*(C_{kl}^{m_1+i-k,m_2+m_3+k-i}). \end{aligned} \tag{4.12}$$

We also need another morphism (the other "composition")

$$\mu(\cdot,\mu(\cdot,\cdot)) : (j_3)_* j_3^*(\mathcal{A}\boxtimes\mathcal{A}\boxtimes\mathcal{A}) \to (\Delta_3)_\circ\mathcal{A}$$

of $\mathcal{D}_{X\times X\times X}$-modules defined naturally as follows: Let U be an open subset of $X \times X \times X$ of the form $U_1 \times U_2 \times U_3$ where U_1, U_2 and U_3 are open subsets of X. For any $n_2 \in \mathbb{Z}$ there exist $s_{n_2} \in \mathbb{N}$ and

$$D_l^{n_2} \in \mathcal{A}(\Delta^{-1}(U_2 \times U_3)) = \mathcal{A}(U_2 \cap U_3), \quad l = 0, \ldots, s_{n_2},$$

such that the element $\mu((z_2 - z_3)^{n_2}(A_1 \otimes A_2)) \in (\Delta_*(\mathcal{A}))(U_1 \times U_2)$ is equal to

$$\sum_{l=0}^{s_{n_2}} \frac{\partial^l}{\partial z_2^l} \otimes_{(\Delta_*(\mathcal{D}_X))(U_2\times U_3)} \Delta_*(D_l^{n_2}).$$

For any $n_1 \in \mathbb{Z}$, we have

$$\mu((z_1 - z_3)^{n_1}(A_1 \otimes D_l^{n_2})) = \sum_{k=0}^{r_{n_1,n_2}} \frac{\partial^k}{\partial z_1^k} \otimes_{(\Delta_*(\mathcal{D}_X))(U_1\times U_3)} \Delta_*(E_{kl}^{n_1 n_2})$$

where $r_{n_1,n_2} \in \mathbb{N}$ and $E_{kl}^{n_1 n_2} \in \mathcal{A}(\Delta^{-1}((U_1 \cap U_2) \times U_3) = \mathcal{A}(U_1 \cap U_2 \cap U_3)$, $k = 0, \ldots, r_{n_1,n_2}$, $l = 0, \ldots, s_{n_2}$. Let $E_{kl}^{n_1 n_2} = 0$ for $k > r_{n_1,n_2}$ or $l > s_{n_2}$. We define

$$\begin{aligned}(\mu(\cdot,\mu(\cdot,\cdot)))&((z_1 - z_2)^{m_1}(z_2 - z_3)^{m_2}(z_1 - z_3)^{m_3}(A_1 \otimes A_2 \otimes A_3)) \\ &= \sum_{i\in\mathbb{N}} (-1)^i \binom{m_1}{i} \sum_{k,l\in\mathbb{N}} \frac{\partial^k}{\partial z_1^k}\frac{\partial^l}{\partial z_2^l} \\ &\qquad \otimes_{((\Delta_3)_*(\mathcal{D}_X))(U)} (\Delta_3)_*(E_{kl}^{m_1+m_3-i,m_2+i}),\end{aligned} \tag{4.13}$$

using the expansion of

$$(z_1 - z_2)^{m_1} = ((z_1 - z_3) - (z_2 - z_3))^{m_1}$$

in nonnegative powers of $z_2 - z_3$.

We also have an action of the symmetric group S_2 on $j_*j^*(\mathcal{A}\boxtimes\mathcal{A})$ and an action of S_3 on $(j_3)_*j_3^*(\mathcal{A}\boxtimes\mathcal{A}\boxtimes\mathcal{A})$: In fact, the elements of S_2 act on $X\times X$, and we have direct image sheaves of $j_*j^*(\mathcal{A}\boxtimes\mathcal{A})$ by these actions. It is easy to see that these direct image sheaves are the same as $j_*j^*(\mathcal{A}\boxtimes\mathcal{A})$, so that we obtain an action of S_2 on $j_*j^*(\mathcal{A}\boxtimes\mathcal{A})$. Explicitly, this action can be described as follows: For any element $f(z_1, z_2)(z_1 - z_2)^n(A_1 \otimes A_2)$ of $(j_*j^*(\mathcal{A}\boxtimes\mathcal{A}))(U_1 \times U_2)$, we define $\sigma(f(z_1, z_2)(z_1 - z_2)^n(A_1 \otimes A_2))$ to be the element $f(z_1, z_2)(z_1 - z_2)^n(A_2 \otimes A_1)$ of $(j_*j^*(\mathcal{A}\boxtimes\mathcal{A}))(U_2 \times U_1)$. The action of S_3 on $(j_3)_*j_3^*(\mathcal{A}\boxtimes\mathcal{A}\boxtimes\mathcal{A})$ is defined similarly.

Definition 4.1 (Beilinson-Drinfeld [BD]) Let X be a nonempty open subset of $\mathbb{C}$. A *chiral algebra over* X is a $\mathcal{D}_X$-module $\mathcal{A}$ equipped with a morphism $\mu : j_*j^*(\mathcal{A} \boxtimes \mathcal{A}) \to \Delta_\circ\mathcal{A}$ of $\mathcal{D}_{X\times X}$-modules satisfying the following axioms:

1. (Skew-symmetry) Let σ_{12} be the nontrivial element of the symmetric group S_2. Then
$$\mu \circ \sigma_{12} = -\mu.$$

2. (Jacobi identity) Let σ_{12} be the element of S_3 permuting the first two letters. Then
$$\mu(\mu(\cdot,\cdot),\cdot) = \mu(\cdot,\mu(\cdot,\cdot)) - \mu(\cdot,\mu(\cdot,\cdot)) \circ \sigma_{12}.$$

5. Vertex algebras without vacuum and chiral algebras in the sense of Beilinson-Drinfeld

In this section, X is a nonempty open subset of $\mathbb{C}$. We need the following notion:

Definition 5.1 A *vertex algebra without vacuum over* X is a vertex algebra without vacuum (V, Y, D) equipped with an $\mathcal{O}(X)$-module structure on V such that for any $f, g \in \mathcal{O}(X)$ and $u, v \in V$, $Y(fu, x)gv = fgY(u, x)v$ and $D(fu) = (\frac{\partial}{\partial z}f)u + fDu$.

Any vertex algebra without vacuum tensored with the commutative associative algebra $\mathcal{O}(X)$, viewed as a vertex algebra (see [Borc]), is a vertex algebra without vacuum over X (see [Borc], [FHL]).

Let (V, Y, D) be a vertex algebra without vacuum over X. For any open (necessarily affine) subset $U \subset X$, let $\mathcal{A}(U) = \mathcal{O}(U) \otimes_{\mathcal{O}(X)} V$. We define $A(u, U) = 1 \otimes_{\mathcal{O}(X)} u \in \mathcal{A}(U)$. Note that $\mathcal{A}(X) = \mathcal{O}(X) \otimes_{\mathcal{O}(X)} V = V$ and $A(u, X) = u$. Since $\mathcal{O}_X$ is quasi-coherent, we see that the $\mathcal{O}_X$-module defined by $U \mapsto \mathcal{A}(U)$ for all open subsets $U \subset X$ is also quasi-coherent. We define $\frac{\partial}{\partial z}A(u, U) = A(Du, U)$, so that $\frac{\partial}{\partial z}$ acts on $\mathcal{A}(U)$. Thus the $\mathcal{O}_X$-module defined by $U \mapsto \mathcal{A}(U)$ for all open subsets $U \subset X$ is a $\mathcal{D}_X$-module. We denote this $\mathcal{D}_X$-module by $\mathcal{A}$.

For the vertex algebra V without vacuum, we write the Jacobi identity in the following form:

$$(x_1 - x_2)^n Y(u, x_1)Y(v, x_2) - (-x_2 + x_1)^n Y(u, x_2)Y(v, x_1)$$
$$= \mathrm{Res}_{x_0} x_0^n x_2^{-1}\delta\left(\frac{x_1 - x_0}{x_2}\right) Y(Y(u, x_0)v, x_2)$$

for $u, v \in V$, $n \in \mathbb{Z}$. We rewrite the right-hand side as

$$\begin{aligned}&\mathrm{Res}_{x_0} x_0^n e^{-x_0 \frac{\partial}{\partial x_1}} x_2^{-1}\delta\left(\frac{x_1}{x_2}\right) Y(Y(u, x_0)v, x_2)\\&= \sum_{m\in\mathbb{N}} \frac{(-1)^m}{m!} \frac{\partial^m}{\partial x_1^m} x_2^{-1}\delta\left(\frac{x_1}{x_2}\right) Y(u_{m+n}v, x_2).\end{aligned}$$

We define $\mu : j_*j^*(\mathcal{A}\boxtimes\mathcal{A}) \to \Delta_\circ(\mathcal{A})$ as follows: For any open subset U of the form $U_1 \times U_2 \subset X \times X$, $j_*j^*(\mathcal{A}\boxtimes\mathcal{A})(U)$ is spanned by elements of the form

$$(z_1 - z_2)^n (A(u, U_1) \otimes A(v, U_2)),$$

$u, v \in V$, $n \in \mathbb{Z}$ (or $n < 0$). We define

$$\begin{aligned}&\mu((z_1 - z_2)^n (A(u, U_1) \otimes A(v, U_2)))\\&= \sum_{m\in\mathbb{N}} \frac{(-1)^m}{m!} \frac{\partial^m}{\partial z_1^m} \otimes_{(\Delta_*(\mathcal{D}_X))(U)} \Delta_*(A(u_{m+n}v, \Delta^{-1}(U))). \qquad (5.1)\end{aligned}$$

It is easy to verify that μ is well-defined and is indeed a morphism of $\mathcal{D}_{X\times X}$-modules.

We have:

Proposition 5.2 *The pair $(\mathcal{A}, \mu)$ is a chiral algebra over X in the sense of Definition 4.1.*

Proof We first verify the skew-symmetry using the skew-symmetry for the vertex algebra without vacuum V. By definition, for $U = U_1 \times U_2$ and $u, v \in V$,

$$\begin{aligned}&(\mu \circ \sigma_{12})((z_1 - z_2)^n (A(u, U_1) \otimes A(v, U_2)))\\&= \mu((z_1 - z_2)^n (A(v, U_2) \otimes A(u, U_1)))\\&= (-1)^n \mu((z_2 - z_1)^n (A(v, U_2) \otimes A(u, U_1)))\\&= (-1)^n \sum_{m\in\mathbb{N}} \frac{(-1)^m}{m!} \frac{\partial^m}{\partial z_2^m} \otimes_{(\Delta_*(\mathcal{D}_X))(U)} \Delta_*(A(v_{m+n}u, \Delta^{-1}(\sigma_{12}(U)))\\&= \sum_{m\in\mathbb{N}} \frac{(-1)^{m+n}}{m!} \frac{\partial^m}{\partial z_2^m} \otimes_{(\Delta_*(\mathcal{D}_X))(U)} \Delta_*(A(v_{m+n}u, \Delta^{-1}(U)))\\&= \sum_{m\in\mathbb{N}} \frac{(-1)^{m+n}}{m!} \left(-\frac{\partial}{\partial z_1} + \left(\frac{\partial}{\partial z_1} + \frac{\partial}{\partial z_2}\right)\right)^m\\&\qquad \otimes_{(\Delta_*(\mathcal{D}_X))(U)} \Delta_*(A(v_{m+n}u, \Delta^{-1}(U)))\\&= \sum_{m\in\mathbb{N}} \sum_{k\in\mathbb{N}} \frac{(-1)^{m+n}}{m!} \binom{m}{k} \left(-\frac{\partial}{\partial z_1}\right)^{m-k} \left(\frac{\partial}{\partial z_1} + \frac{\partial}{\partial z_2}\right)^k\end{aligned}$$

$$\begin{aligned}
&\otimes_{(\Delta_*(\mathcal{D}_X))(U)}\Delta_*(A(v_{m+n}u,\Delta^{-1}(U)))\\
&=\sum_{l\in\mathbf{N}}\sum_{k\in\mathbf{N}}\frac{(-1)^{k+n}}{l!k!}\left(\frac{\partial}{\partial z_1}\right)^l\left(\frac{\partial}{\partial z_1}+\frac{\partial}{\partial z_2}\right)^k\\
&\quad\otimes_{(\Delta_*(\mathcal{D}_X))(U)}\Delta_*(A(v_{k+l+n}u,\Delta^{-1}(U))).
\end{aligned}\tag{5.2}$$

Since $\frac{\partial}{\partial z_1}+\frac{\partial}{\partial z_2}\in(\Delta_*(\mathcal{D}_X))(U)$ and its preimage in

$$\mathcal{D}_X(\Delta^{-1}(U))=\mathcal{D}_X(U_1\cap U_2)$$

is $\frac{\partial}{\partial z}$, the right-hand side of (5.2) is equal to

$$\begin{aligned}
&\sum_{l\in\mathbf{N}}\sum_{k\in\mathbf{N}}\frac{(-1)^{k+n}}{l!k!}\left(\frac{\partial}{\partial z_1}\right)^l\otimes_{(\Delta_*(\mathcal{D}_X))(U)}\Delta_*\left(\left(\frac{\partial}{\partial z}\right)^k A(v_{k+l+n}u,\Delta^{-1}(U))\right)\\
&\quad=\sum_{l\in\mathbf{N}}\sum_{k\in\mathbf{N}}\frac{(-1)^{k+n}}{l!k!}\left(\frac{\partial}{\partial z_1}\right)^l\\
&\quad\quad\otimes_{(\Delta_*(\mathcal{D}_X))(U)}\Delta_*\left(A(D^kv_{k+l+n}u,\Delta^{-1}(U))\right)\\
&\quad=-\sum_{l\in\mathbf{N}}\frac{(-1)^l}{l!}\left(\frac{\partial}{\partial z_1}\right)^l\\
&\quad\quad\otimes_{(\Delta_*(\mathcal{D}_X))(U)}\Delta_*\left(A\left(\sum_{k\in\mathbf{N}}\frac{(-1)^{k+l+n+1}}{k!}D^kv_{k+l+n}u,\Delta^{-1}(U)\right)\right).
\end{aligned}\tag{5.3}$$

But from the skew-symmetry (2.4), we obtain its component form

$$\sum_{k\in\mathbf{N}}\frac{(-1)^{k+m+1}}{k!}D^kv_{k+m}u=u_mv$$

for all $m\in\mathbb{Z}$. Thus the right-hand side of (5.3) is equal to

$$-\sum_{l\in\mathbf{N}}\frac{(-1)^l}{l!}\left(\frac{\partial}{\partial z_1}\right)^l\otimes_{(\Delta_*(\mathcal{D}_X))(U)}\Delta_*\left(A\left(u_{l+n}v,\Delta^{-1}(U)\right)\right).\tag{5.4}$$

On the other hand,

$$\begin{aligned}
&(-\mu)((z_1-z_2)^nA(u,U_1)\otimes A(v,U_2))\\
&\quad=-\mu((z_1-z_2)^nA(u,U_1)\otimes A(v,U_2))\\
&\quad=-\sum_{m\in\mathbf{N}}\frac{(-1)^m}{m!}\frac{\partial^m}{\partial z_1^m}\otimes_{(\Delta_*(\mathcal{D}_X))(U)}\Delta_*(A(u_{m+n}v,\Delta^{-1}(U))).
\end{aligned}\tag{5.5}$$

We see that the right-hand side of (5.5) is equal to (5.4). By the calculation from (5.2) to (5.5), we see that the left-hand side of (5.1) and the left-hand side of (5.4) are equal, proving the skew-symmetry.

Next we prove the Jacobi identity using the Jacobi identity for vertex algebra without vacuum. By (5.1) and (4.13), for any open subset U of the form $U_1 \times U_2 \times U_3 \subset X \times X \times X$,

$$\begin{aligned}
&(\mu(\cdot, \mu(\cdot, \cdot)))((z_1 - z_2)^{m_1}(z_2 - z_3)^{m_2}(z_1 - z_3)^{m_3} \cdot \\
&\qquad \cdot (A(u, U_1) \otimes A(v, U_2) \otimes A(w, U_3))) \\
&= \sum_{i \in \mathbb{N}} (-1)^i \binom{m_1}{i} \sum_{k,l \in \mathbb{N}} \frac{(-1)^{k+l}}{k!l!} \frac{\partial^k}{\partial z_1^k} \frac{\partial^l}{\partial z_2^l} \\
&\qquad \otimes_{((\Delta_3)_*(\mathcal{D}_X))(U)} (\Delta_3)_*(A(u_{m_1+m_3-i+k} v_{m_2+i+l} w, \Delta_3^{-1}(U))) \\
&= \sum_{k,l \in \mathbb{N}} \frac{(-1)^{k+l}}{k!l!} \frac{\partial^k}{\partial z_1^k} \frac{\partial^l}{\partial z_2^l} \\
&\qquad \otimes_{((\Delta_3)_*(\mathcal{D}_X))(U)} (\Delta_3)_* \Bigg(A\Bigg(\sum_{i \in \mathbb{N}} (-1)^i \binom{m_1}{i} \cdot \\
&\qquad\qquad \cdot u_{m_1+m_3-i+k} v_{m_2+i+l} w, \Delta_3^{-1}(U) \Bigg)\Bigg).
\end{aligned} \tag{5.6}$$

Similarly, we have

$$\begin{aligned}
&(\mu(\cdot, \mu(\cdot, \cdot)) \circ \sigma_{12})((z_1 - z_2)^{m_1}(z_2 - z_3)^{m_2}(z_1 - z_3)^{m_3} \cdot \\
&\qquad \cdot (A(u, U_1) \otimes A(v, U_2) \otimes A(w, U_3))) \\
&= \sum_{k,l \in \mathbb{N}} \frac{(-1)^{k+l}}{k!l!} \frac{\partial^k}{\partial z_1^k} \frac{\partial^l}{\partial z_2^l} \\
&\qquad \otimes_{((\Delta_3)_*(\mathcal{D}_X))(U)} (\Delta_3)_* \Bigg(A\Bigg((-1)^{m_1} \sum_{i \in \mathbb{N}} (-1)^i \binom{m_1}{i} \cdot \\
&\qquad\qquad \cdot v_{m_1+m_2-i+l} u_{m_3+i+k} w, \Delta_3^{-1}(U) \Bigg)\Bigg).
\end{aligned} \tag{5.7}$$

On the other hand, by (5.1) and (4.12), we have

$$\begin{aligned}
&(\mu(\mu(\cdot, \cdot), \cdot)))((z_1 - z_2)^{m_1}(z_2 - z_3)^{m_2}(z_1 - z_3)^{m_3} \cdot \\
&\qquad \cdot (A(u, U_1) \otimes A(v, U_2) \otimes A(w, U_3))) \\
&= \sum_{k,l \in \mathbb{N}} \frac{(-1)^{k+l}}{k!l!} \frac{\partial^k}{\partial z_1^k} \frac{\partial^l}{\partial z_2^l}
\end{aligned}$$

$$\otimes_{((\Delta_3)_*(\mathcal{D}_X))(U)}(\Delta_3)_*\Bigg(A\Bigg(\sum_{i\in\mathbb{N}}\binom{m_3+k}{i}\cdot$$
$$\cdot(u_{m_1+i}v)_{m_3+m_2-i+k+l}w, \Delta_3^{-1}(U)\Bigg)\Bigg). \quad (5.8)$$

From the Jacobi identity for vertex algebras without vacuum, we can obtain its component form

$$\sum_{i\in\mathbb{N}}\binom{m}{i}(u_{l+i}v)_{m+n-i}w =$$
$$= \sum_{i\in\mathbb{N}}(-1)^i\binom{l}{i}u_{m+l-i}v_{n+i}w$$
$$-(-1)^l\sum_{i\in\mathbb{N}}(-1)^i\binom{l}{i}v_{n+l-i}u_{m+i}w \quad (5.9)$$

for all $l, m, n \in \mathbb{Z}$ and $u, v, w \in V$ (see [FLM]). From (5.9), we obtain

$$\sum_{i\in\mathbb{N}}\binom{m_3+k}{i}(u_{m_1+i}v)_{m_3+m_2-i+k+l}w =$$
$$= \sum_{i\in\mathbb{N}}(-1)^i\binom{m_1}{i}u_{m_1+m_3-i+k}v_{m_2+i+l}w$$
$$-(-1)^l\sum_{i\in\mathbb{N}}(-1)^i\binom{m_1}{i}v_{m_1+m_2-i+l}u_{m_3+i+k}w \quad (5.10)$$

for any $m_1, m_2, m_3 \in \mathbb{Z}$, $k, l \in \mathbb{N}$ and $u, v, w \in V$. By (5.6)–(5.10), we obtain

$$(\mu(\mu(\cdot,\cdot),\cdot)))((z_1-z_2)^{m_1}(z_2-z_3)^{m_2}(z_1-z_3)^{m_3}\cdot$$
$$\cdot(A(u,U_1)\otimes A(v,U_2)\otimes A(w,U_3)))$$
$$= (\mu(\cdot,\mu(\cdot,\cdot)))((z_1-z_2)^{m_1}(z_2-z_3)^{m_2}(z_1-z_3)^{m_3}\cdot$$
$$\cdot(A(u,U_1)\otimes A(v,U_2)\otimes A(w,U_3)))$$
$$-(\mu(\cdot,\mu(\cdot,\cdot))\circ\sigma_{12})((z_1-z_2)^{m_1}(z_2-z_3)^{m_2}(z_1-z_3)^{m_3}\cdot$$
$$\cdot(A(u,U_1)\otimes A(v,U_2)\otimes A(w,U_3))),$$

proving the Jacobi identity for $\mathcal{A}$. □

Conversely, suppose that we have a chiral algebra $(\mathcal{A},\mu)$ over a nonempty open subset X of $\mathbb{C}$ in the sense of Definition 4.1, and let $V=\mathcal{A}(X)$ be the space of all global sections over X. Then V is an $\mathcal{O}(X)$-module and $\frac{\partial}{\partial z}$ acts on V. For any element $v\in V$, we define $Dv=\frac{\partial}{\partial z}v$. For any $f \in \mathcal{O}(X)$ and $u \in V$, we have $D(fu) = (\frac{\partial}{\partial z} f)u + fDu$. For $u, v \in V$, there exist $B_m^n(u, v) \in V$, $n \in \mathbb{Z}$, $m = 1, \ldots, p_n$, such that

$$\mu((z_1 - z_2)^n (u \otimes v)) = \sum_{m=0}^{p_n} \frac{\partial^m}{\partial z_1^m} \otimes_{(\Delta_*(\mathcal{D}_X))(X \times X)} \Delta_*(B_m^n(u, v)), \quad (5.11)$$

and the nonzero $B_m^n(u, v)$ are uniquely determined. The equation (4.9) in this case becomes

$$B_m^{n+1}(u, v) = -(m+1) B_{m+1}^n(u, v) \quad (5.12)$$

for any $u, v \in V$, $n \in \mathbb{Z}$ and $m = 0, \ldots, p_n - 1$.

We define $u_n v = B_0^n(u, v)$ and

$$Y(u, x)v = \sum_{n \in \mathbb{Z}} B_0^n(u, v) x^{-n-1}.$$

From the definition, we have $Y(fu, x)gv = fgY(u, x)v$ for any $f, g \in \mathcal{O}(X)$ and $u, v \in V$.

Proposition 5.3 *The triple (V, Y, D) is a vertex algebra without vacuum over X.*

Proof When $n > p_0$, we have

$$\begin{aligned}
&\mu((z_1 - z_2)^n (u \otimes v)) = \\
&\quad = (z_1 - z_2)^n \mu(u \otimes v) \\
&\quad = (z_1 - z_2)^n \sum_{m=0}^{p_0} \frac{\partial^m}{\partial z_1^m} \otimes_{(\Delta_*(\mathcal{D}_X))(X \times X)} \Delta_*(B_m^0(u, v)) \\
&\quad = 0.
\end{aligned}$$

Thus for $n > p_0$, $u_n v = B_0^n(u, v) = 0$.

To show the D-derivative property, we need only show its component form:

$$-(n+1) B_0^n(u, v) = B_0^{n+1}(Du, v) \quad (5.13)$$

for all $n \in \mathbb{Z}$. In fact, for any $n \in \mathbb{Z}$,

$$\begin{aligned}
&\frac{\partial}{\partial z_1} \mu((z_1 - z_2)^{n+1} (u \otimes v)) = \\
&\quad = (n+1)\mu((z_1 - z_2)^n (u \otimes v)) + \mu((z_1 - z_2)^{n+1} \frac{\partial}{\partial z_1} (u \otimes v)) \\
&\quad = (n+1)\mu((z_1 - z_2)^n (u \otimes v)) + \mu\left((z_1 - z_2)^{n+1} \left(\left(\frac{\partial}{\partial z} u\right) \otimes v\right)\right) \\
&\quad = (n+1)\mu((z_1 - z_2)^n (u \otimes v)) + \mu((z_1 - z_2)^{n+1}((Du) \otimes v)). \quad (5.14)
\end{aligned}$$

On the other hand,

$$\frac{\partial}{\partial z_1}\mu((z_1-z_2)^{n+1}(u\otimes v)) =$$

$$= \frac{\partial}{\partial z_1}\left(\sum_{m=0}^{p_n}\frac{\partial^m}{\partial z_1^m}\otimes_{(\Delta_*(\mathcal{D}_X))(X\times X)}\Delta_*(B_m^n(u,v))\right)$$

$$= \sum_{m=0}^{p_n}\frac{\partial^{m+1}}{\partial z_1^{m+1}}\otimes_{(\Delta_*(\mathcal{D}_X))(X\times X)}\Delta_*(B_m^n(u,v)). \quad (5.15)$$

From (5.14), (5.15) and the expansions of $\mu((z_1-z_2)^n(u\otimes v))$ and $\mu((z_1-z_2)^{n+1}((Du)\otimes v))$, we obtain

$$(n+1)\sum_{m=0}^{p_n}\frac{\partial^m}{\partial z_1^m}\otimes_{(\Delta_*(\mathcal{D}_X))(X\times X)}\Delta_*(B_m^n(u,v))$$
$$+\sum_{m=0}^{p_{n+1}}\frac{\partial^m}{\partial z_1^m}\otimes_{(\Delta_*(\mathcal{D}_X))(X\times X)}\Delta_*(B_m^{n+1}(u,v))$$
$$= (n+1)\mu((z_1-z_2)^n(u\otimes v)) + \mu((z_1-z_2)^{n+1}((Du)\otimes v))$$
$$= \sum_{m=0}^{p_n}\frac{\partial^{m+1}}{\partial z_1^{m+1}}\otimes_{(\Delta_*(\mathcal{D}_X))(X\times X)}\Delta_*(B_m^n(u,v)). \quad (5.16)$$

The equality (5.16) implies

$$(n+1)B_0^n(u,v) + B_0^{n+1}(u,v) = 0$$

which is equivalent to (5.13).

We now prove that the skew-symmetry for vertex algebras without vacuum is satisfied by V, using the skew-symmetry for $(\mathcal{A},\mu)$. By (5.11) and the skew-symmetry for $(\mathcal{A},\mu)$, we obtain

$$\sum_{m=0}^{p_n}\frac{\partial^m}{\partial z_1^m}\otimes_{(\Delta_*(\mathcal{D}_X))(X\times X)}\Delta_*(B_m^n(u,v))$$
$$= -(-1)^n\sum_{m=0}^{q_n}\frac{\partial^m}{\partial z_2^m}\otimes_{(\Delta_*(\mathcal{D}_X))(X\times X)}\Delta_*(B_m^n(v,u)). \quad (5.17)$$

The right-hand side of (5.17) is equal to

$$-(-1)^n\sum_{m=0}^{q_n}\left(-\frac{\partial}{\partial z_1}+\left(\frac{\partial}{\partial z_1}+\frac{\partial}{\partial z_2}\right)\right)^m\otimes_{(\Delta_*(\mathcal{D}_X))(X\times X)}\Delta_*(B_m^n(v,u))$$
$$= (-1)^{n+1}\sum_{m=0}^{q_n}\sum_{k=0}^{m}(-1)^{m-k}\binom{m}{k}\cdot$$

$$\cdot\frac{\partial^{m-k}}{\partial z_1^{m-k}}\left(\frac{\partial}{\partial z_1}+\frac{\partial}{\partial z_2}\right)^k \otimes_{(\Delta_*(\mathcal{D}_X))(X\times X)} \Delta_*(B_m^n(v,u))$$
$$= (-1)^{n+1}\sum_{m=0}^{q_n}\sum_{k=0}^{m}(-1)^{m-k}\binom{m}{k}\frac{\partial^{m-k}}{\partial z_1^{m-k}}$$
$$\otimes_{(\Delta_*(\mathcal{D}_X))(X\times X)}\Delta_*\left(\frac{d^k}{dz^k}B_m^n(v,u)\right)$$
$$= (-1)^{n+1}\sum_{m=0}^{q_n}\sum_{k=0}^{m}(-1)^{m-k}\binom{m}{k}\frac{\partial^{m-k}}{\partial z_1^{m-k}}$$
$$\otimes_{(\Delta_*(\mathcal{D}_X))(X\times X)}\Delta_*(D^k B_m^n(v,u)). \tag{5.18}$$

Comparing the left-hand side of (5.17) and the right-hand side of (5.18), the terms involving $\frac{\partial^0}{\partial z_1^0}$ give

$$B_0^n(u,v) = \sum_{k=0}^{q_n}(-1)^{n+1}D^k B_k^n(v,u). \tag{5.19}$$

Using (5.12) repeatedly, we obtain

$$B_k^n(v,u) = \frac{(-1)^k}{k!}B_0^{k+n}(v,u). \tag{5.20}$$

Combining (5.19) and (5.20), we obtain

$$B_0^n(u,v) = \sum_{k=0}^{q_n}\frac{(-1)^{k+n+1}}{k!}D^k B_0^{k+n}(v,u). \tag{5.21}$$

Since when $k > q_n$, we have

$$\mu((z_1-z_2)^{k+n}(v\otimes u)) =$$
$$= (z_1-z_2)^k\mu((z_1-z_2)^n(v\otimes u))$$
$$= (z_1-z_2)^k\sum_{m=0}^{q_n}\frac{\partial^m}{\partial z_2^m}\otimes_{(\Delta_*(\mathcal{D}_X))(X\times X)}\Delta_*(D^k B_m^n(v,u))$$
$$= 0.$$

Thus $B_0^{k+n}(v,u) = 0$ when $k > q_n$. So (5.21) becomes

$$B_0^n(u,v) = \sum_{k\in\mathbb{N}}\frac{(-1)^{k+n+1}}{k!}D^k B_0^{k+n}(v,u),$$

or equivalently

$$u_n v = \sum_{k\in\mathbb{N}}\frac{(-1)^{k+n+1}}{k!}D^k v_{k+n}u$$
.

which is the component form of the skew-symmetry.

Finally, we prove that the Jacobi identity for vertex algebra without vacuum is satisfied by V, using the Jacobi identity for $(\mathcal{A}, \mu)$. By (4.12), (4.13) and the Jacobi identity for $(\mathcal{A}, \mu)$, we obtain the following identity:

$$\begin{aligned}
&\sum_{i\in\mathbf{N}}\sum_{k,l\in\mathbf{N}}\binom{m_3+k}{i}\frac{\partial^k}{\partial z_1^k}\frac{\partial^l}{\partial z_2^l}\\
&\qquad\otimes_{((\Delta_3)_*(\mathcal{D}_X))(U)}(\Delta_3)_*(B_l^{m_2+m_3-k-i}(B_k^{m_1+i+k}(u,v),w))\\
&\quad=\sum_{i\in\mathbf{N}}(-1)^i\binom{m_1}{i}\sum_{k,l\in\mathbf{N}}\frac{\partial^k}{\partial z_1^k}\frac{\partial^l}{\partial z_2^l}\otimes_{((\Delta_3)_*(\mathcal{D}_X))(U)}\\
&\qquad\quad\otimes_{((\Delta_3)_*(\mathcal{D}_X))(U)}(\Delta_3)_*(B_k^{m_1+m_3-i}(u,B_l^{m_2+i}(v,w)))\\
&\qquad-(-1)^{m_1}\sum_{i\in\mathbf{N}}(-1)^i\binom{m_1}{i}\sum_{k,l\in\mathbf{N}}\frac{\partial^k}{\partial z_1^k}\frac{\partial^l}{\partial z_2^l}\otimes_{((\Delta_3)_*(\mathcal{D}_X))(U)}\\
&\qquad\quad\otimes_{((\Delta_3)_*(\mathcal{D}_X))(U)}(\Delta_3)_*(B_l^{m_1+m_2-i}(v,B_k^{m_3+i}(u,w))).
\end{aligned}\tag{5.22}$$

This equality implies

$$\begin{aligned}
&\sum_{i\in\mathbf{N}}\binom{m_3}{i}B_0^{m_2+m_3-i}(B_0^{m_1+i}(u,v),w)=\\
&\quad=\sum_{i\in\mathbf{N}}(-1)^i\binom{m_1}{i}B_0^{m_1+m_3-i}(u,B_0^{m_2+i}(v,w))\\
&\qquad-(-1)^{m_1}\sum_{i\in\mathbf{N}}(-1)^i\binom{m_1}{i}B_0^{m_1+m_2-i}(v,B_0^{m_3+i}(u,w)))
\end{aligned}$$

or equivalently, the component form of the Jacobi identity for vertex algebra without vacuum:

$$\begin{aligned}
&\sum_{i\in\mathbf{N}}\binom{m_3}{i}(u_{m_1+i}v)_{m_2+m_3-i}w=\\
&\quad=\sum_{i\in\mathbf{N}}(-1)^i\binom{m_1}{i}u_{m_1+m_3-i}v_{m_2+i}w\\
&\qquad-(-1)^{m_1}\sum_{i\in\mathbf{N}}(-1)^i\binom{m_1}{i}v_{m_1+m_2-i}u_{m_3+i}w.
\end{aligned}$$

Since (V, Y, D) satisfies the lower-truncation condition for vertex operators, the D-derivative property, the skew-symmetry and the Jacobi identity for vertex algebras without vacuum, it is a vertex algebra without vacuum. We already know that V is an $\mathcal{O}(X)$-module and for any $f, g \in \mathcal{O}(X)$,

$u, v \in V$, $Y(fu, x)gv = fgY(u, x)v$ and $D(fu) = (\frac{\partial}{\partial z} f)u + fDu$. So (V, Y, D) is a vertex algebra without vacuum over X. □

From the two propositions above, we obtain the following main result of this paper:

Theorem 5.4 *Let X be a nonempty open subset of $\mathbb{C}$. The category of vertex algebras without vacuum over X and the category of chiral algebras over X are equivalent.* □

References

[BPZ] A. A. Belavin, A. M. Polyakov and A. B. Zamolodchikov, Infinite conformal symmetries in two-dimensional quantum field theory, *Nucl. Phys.* **B241** (1984), 333–380.

[BD] A. Beilinson and V. Drinfeld, unpublished manuscript.

[Borc] R. E. Borcherds, Vertex algebras, Kac-Moody algebras, and the Monster, *Proc. Natl. Acad. Sci. USA* **83** (1986), 3068–3071.

[Borel] A. Borel et al., *Algebraic $\mathcal{D}$-modules*, Perspectives in Mathematics, Vol. 2, Academic Press, Boston, 1987.

[FHL] I. B. Frenkel, Y.-Z. Huang and J. Lepowsky, On axiomatic approaches to vertex operator algebras and modules, preprint, 1989; *Memoirs Amer. Math. Soc.* **104**, 1993.

[FLM] I. B. Frenkel, J. Lepowsky and A. Meurman, *Vertex Operator Algebras and the Monster*, Pure and Appl. Math., Vol. 134, Academic Press, Boston, 1988.

[Ha] R. Hartshorne, *Algebraic geometry*, Graduate Texts in Mathematics, Vol. 52, Springer-Verlag, New York, 1977.

[Hu1] Y.-Z. Huang, *On the geometric interpretation of vertex operator algebras*, Ph.D. thesis, Rutgers University, 1990.

[Hu2] Y.-Z. Huang, *Two-dimensional conformal geometry and vertex operator algebras*, Birkhäuser, to appear.

[HL1] Y.-Z. Huang and J. Lepowsky, Operadic formulation of the notion of vertex operator algebra, in: *Mathematical Aspects of Conformal and Topological Field Theories and Quantum Groups, Proc. Joint Summer Research Conference, Mount Holyoke, 1992*, ed. P. Sally,

M. Flato, J. Lepowsky, N. Reshetikhin and G. Zuckerman, Contemporary Math., Vol. 175, Amer. Math. Soc., Providence, 1994, 131-148.

[HL2] Y.-Z. Huang and J. Lepowsky, Tensor products of modules for a vertex operator algebras and vertex tensor categories, in: *Lie Theory and Geometry, in honor of Bertram Kostant,* ed. R. Brylinski, J.-L. Brylinski, V. Guillemin, V. Kac, Birkhäuser, Boston, 1994, 349–383.

[HL3] Y.-Z. Huang and J. Lepowsky, A theory of tensor products for module categories for a vertex operator algebra, I, *Selecta Mathematica* (New Series) **1** (1995).

[HL4] Y.-Z. Huang and J. Lepowsky, A theory of tensor products for module categories for a vertex operator algebra, II, *Selecta Mathematica* (New Series) **1** (1995).

Department of Mathematics, Rutgers University, New Brunswick, NJ 08903
E-mail address: yzhuang@math.rutgers.edu

Department of Mathematics, Rutgers University, New Brunswick, NJ 08903
E-mail address: lepowsky@math.rutgers.edu

Received December 1995; corrected February 1996

The Lowest Eigenvalue for Congruence Groups

Henryk Iwaniec

1. Introduction

Let $L^2(\Gamma\backslash\mathbb{H})$ be the space of square-integrable automorphic functions with respect to a group $\Gamma \subset SL_2(\mathbb{R})$ acting discontinuously on the hyperbolic plane $\mathbb{H}$ such that the quotient space $\Gamma\backslash\mathbb{H}$ has finite volume. The Laplace-Beltrami operator on $L^2(\Gamma\backslash\mathbb{H})$ has a discrete spectrum $\lambda_0 = 0 < \lambda_1 \leq \lambda_2 \leq \dots$ and a continuous spectrum $[\frac{1}{4}, \infty)$. The eigenpacket of continuous spectrum consists of Eisenstein series $E_{\mathfrak{a}}(z,s)$ on the line Re $s = \frac{1}{2}$ and the eigenfunctions of the discrete spectrum are Maass cusp forms together with a finite number of residues of $E_{\mathfrak{a}}(z,s)$ at poles in the segment $\frac{1}{2} < s \leq 1$. If Γ is a congruence group the only pole of Eisenstein series is at $s = 1$ which yields a constant eigenfunction for eigenvalue $\lambda_0 = 0$, the remaining subspace of discrete spectrum is cuspidal and infinite dimensional.

There are groups Γ for which the lowest positive eigenvalue λ_1 is arbitrarily small, however for congruence groups A. Selberg [Se] has shown that λ_1 is bounded below by an absolute constant, namely

$$\lambda_1 \geq \frac{3}{16}. \tag{1}$$

The original argument of Selberg uses estimates for Kloosterman sums, thus it appeals indirectly to the Riemann hypothesis for curves established by A. Weil, but this can be avoided as in [I1]. By a very different method S. Gelbart and H. Jacquet [GJ] showed that $\frac{3}{16}$ is not attained.

The eigenvalues in $0 < \lambda_j < \frac{1}{4}$ are called exceptional to emphasize they are not welcomed since they distort rather than simplify results in the same fashion as do the real zeros of Dirichlet L-functions. The fundamental conjecture of Selberg [Se] asserts there is no exceptional eigenvalue for congruence groups, in other words the cuspidal spectrum lies on the continuous one. Selberg's conjecture seems to be out of reach at present, whereas numerous substitutes of a kind of density theorems are established (see [DI] and [I2]) which suffice for applications to analytic number theory.

There is an analogy between Selberg's eigenvalue conjecture and the Ramanujan-Petersson conjecture about Fourier coefficients of Maass cusp forms pointed out two decades ago, but only recently this analogy was confirmed by matching progress in both problems. The best bound in the eigenvalue problem is due to W. Luo, Z. Rudnick, P. Sarnak [LRS], namely

$$\lambda_1 \geq \frac{171}{784} \tag{2}$$

which corresponds to the exponent $\frac{5}{28}$ in the bound for the Fourier coefficients of cusp forms due to D. Bump, W. Duke, J. Hoffstein, H. Iwaniec [BDHI] (actually $\frac{5}{28}$ was obtained earlier in somewhat abstract setting [DI1]). In retrospect it turns out that both works are based on similar ideas, i.e., both exploit analytic properties of the same L-functions (the Rankin-Selberg convolution on GL_3) twisted by Dirichlet characters, and in either work one appeals to Deligne's estimate for the same hyper-Kloosterman sums. Thus the arguments are quite advanced.

The purpose of this paper is to establish a slightly weaker result, namely

$$\lambda_1 \geq \frac{10}{49} \tag{3}$$

but using only the GL_2 theory. Therefore we shall break the Selberg bound (1) within the hyperbolic plane. Our arguments immitate [DI3] besides a few modifications which make the present work still more elementary; only simple estimates for Gauss sums will be needed in place of Kloosterman sums.

Acknowledgment. I would like to thank Wenzhi Luo, Zeév Rudnick and Peter Sarnak for numerous discussions on the subject of this paper.

2. The Symmetric Square L-functions

Suppose $\lambda = \frac{1}{4} - \nu^2$ with $0 < \nu \leq \frac{1}{4}$ is an exceptional eigenvalue for a cusp form f on a congruence group Γ of level N, we can assume that f is a new form. Here we collect basic facts about the symmetric square L-function attached to f, say

$$L(s) = \sum_1^\infty a(n)n^{-s}, \tag{4}$$

which are needed in the forthcoming sections.

First of all the series (4) converges absolutely in Re $s > 1$. It is also known that the series

$$L^{(2)}(s) = \sum_1^\infty a(n)^2 n^{-s} \tag{5}$$

converges absolutely in Re $s > 1$. This is not an obvious result, yet it has been proved by using only the GL_2 theory (cf. [DI2]). It is essential in our approach (as in the work by Luo-Rudnick-Sarnak) to include to $L(s)$ the product of gamma functions (the infinite place factor)

$$L_{\infty}(s) = \pi^{-\frac{3}{2}s}\Gamma(\frac{s}{2} - \nu)\Gamma(\frac{s}{2})\Gamma(\frac{s}{2} + \nu). \tag{6}$$

The key point is that the complete symmetric square, say $\Lambda(s) = L_{\infty}(s)L(s)$, is entire function and it satisfies the functional equation

$$\Lambda(s) = w(s)\Lambda(1-s) \tag{7}$$

where $w(s) = wD^{-s}$, D is a positive integer and w is a complex number with $|w| = D^{\frac{1}{2}}$. Moreover $\Lambda(s)$ behaves nicely under twisting by characters. Put

$$\Lambda(s, \chi) = L_{\infty}(s)\sum_{1}^{\infty} a(n)\chi(n)n^{-s}. \tag{8}$$

Note that the local factor at the infinite place is not altered by the character. For almost all primitive, even characters $\chi(\bmod\, q)$ with $(q, 2D) = 1$ (save for a finite number) the twisted function $\Lambda(s, \chi)$ is entire and it satisfies the functional equation

$$\Lambda(s, \chi) = w(s, \chi)\Lambda(1 - s, \bar{\chi}) \tag{9}$$

where $w(s, \chi) = w(s)\chi(D)\bar{\tau}(\chi)^3 q^{-3s}$ and $\tau(\chi)$ is the Gauss sum

$$\tau(\chi) = \sum_{a(\bmod\, q)} \chi(a)e(\frac{a}{q}). \tag{10}$$

All the above assorted properties can be established within GL_2 theory by Shimura's method of convolutions with theta function [Sh]. If one wished, the symmetric square L-function could be wiped out from the surface of this paper, however such an approach would require proving from scratch many formulas, consequently it would make the complete argument a lot longer and it would not reveal critical ideas.

From now on we assume the assorted properties and never again appeal to the automorphic nature of $\Lambda(s, \chi)$. It suffices to show that

$$\nu \leq \frac{3}{14}. \tag{11}$$

3. Summation Formulas

It is more effective to work with finite sums, or rapidly convergent series rather than with L-functions, thus we begin by taking Mellin's transform of the functional equation (9) to obtain the following summation formula

$$\sum_{1}^{\infty} a(m)\bar{\chi}(m)F\left(\frac{my}{Dq^3}\right) = wy^{-1}\tau(\chi)^3\sum_{1}^{\infty} a(n)\chi(n\bar{D})F\left(\frac{n}{y}\right).$$

with

$$F(y) = \frac{1}{2\pi i} \int_{(\sigma)} L_\infty(s) y^{-s} ds$$

for $y > 0$. Note that

$$F(y) \sim 2\pi^{-3\nu} \Gamma(\nu)\Gamma(2\nu) y^{-2\nu} \tag{12}$$

as $y \to 0$ and $F(y)$ has exponential decay at ∞, therefore for all $y > 0$ we have

$$F(y) \ll y^{-2\nu} G(y) \quad \text{with } G(y) = \frac{2}{\pi}(1+y^2)^{-1}. \tag{13}$$

The underlying strategy is to kill in the summation formula as many terms as possible by playing with characters, while saving the first term (which is the largest one) to produce a bound for ν.

Summing over all the primitive, even characters of conductor q we get

$$\sum_1^\infty a(m) V_q(m) F\left(\frac{my}{Dq^3}\right) = wy^{-1} \sum_1^\infty a(n) W_q(n\bar{D}) F\left(\frac{n}{y}\right)$$

where

$$V_q(m) = \sum_{\chi (\mathrm{mod}\ q)}^{*} (\chi(m) + \chi(-m)),$$

$$W_q(n) = \sum_{\chi (\mathrm{mod}\ q)}^{*} \tau(\chi)^3 (\chi(n) + \chi(-n)).$$

The first sum can be computed explicitly by way of Ramanujan sums, we obtain

$$V_q(m) = \sum_{d \mid (q, m \pm 1)} \varphi(d) \mu(q/d) \tag{14}$$

if $(m, q) = 1$ and $V_q(m) = 0$ otherwise. In particular

$$V_q(1) = q \prod_{p \| q} \left(1 - \frac{2}{p}\right) \prod_{p^2 \mid q} \left(1 - \frac{1}{p}\right)^2. \tag{15}$$

The second sum can be written in terms of hyper-Kloosterman sums (in three variables) but we do not need such expression.

Next we average over moduli q of special type to be chosen later subject to $Q < q \leq 4Q$ where Q is another parameter to be made optimal together with y. We obtain

$$\sum_1^\infty a(m) V(m) = wy^{-1} \sum_1^\infty a(n) W(n\bar{D}) F(\frac{n}{y}) \tag{16}$$

where

$$V(m) = \sum_q V_q(m) F\left(\frac{my}{Dq^3}\right),$$
$$W(n) = \sum_q W_q(n).$$

Let $\mathcal{L}$ and $\mathcal{R}$ denote the left and the right sides of (16) respectively.

Remarks. Contemplate how $V_q(m)$ loses its order of magnitude as $m > 1$. The point is that every q divides zero but only few q divide a given positive integer. This situation is orchestrated by playing with characters and their moduli so a lot of functional equations are employed, far more than these in the s-aspect alone. An immediate advantage is that we don't require the coefficients of L-functions to be positive. Another gain is a flexibility in sums of hyper-Kloosterman sums which allows us to improve on Deligne's bound (see Section 5).

4. Estimation of $\mathcal{L}$

The main contribution comes from the first term $m = 1$ which by (12) and (15) satisfies

$$V(1) \gg x^{2\nu} \sum_q \varphi(q)^2 q^{-1} \tag{17}$$

where $xy = Q^3$. Here and thereafter we assume that x, y are large, so is Q, while ν, D are fixed. For $m > 1$ we have

$$\sum_q |V_q(m)| \le Q \sum_{d|(m\pm 1)} \varphi(d) d^{-1} \ll Qm^\varepsilon,$$

whence by (13)

$$V(m) \ll QG\left(\frac{m}{x}\right)\left(\frac{x}{m}\right)^{2\nu} m^\varepsilon. \tag{18}$$

Therefore these terms contribute

$$\sum_{m>1} a(m)V(m) \ll Qx^{1+\varepsilon} \tag{19}$$

because the series (4) converges absolutely if $s > 1$. Hence

$$\mathcal{L} = V(1) + O(QX^{1+\varepsilon}). \tag{20}$$

The leading term will be estimated in (28) after having specialized the set of moduli.

5. Estimation of $\mathcal{R}$

On the right side of (16) we estimate $F(\frac{n}{y})$ by (13) and apply Cauchy's inequality getting

$$\mathcal{R} \ll y^{2\nu-1}\mathcal{A}^{\frac{1}{2}}\mathcal{B}^{\frac{1}{2}}$$

where

$$\mathcal{A} = \sum_1^\infty |a(n)|^2 n^{-4\nu} G(\frac{n}{y}) \ll y^{1-4\nu+\varepsilon}$$

because the series (5) converges absolutely if $s > 1$, and

$$\mathcal{B} = \sum_1^\infty G(\frac{n}{y})|W(n\bar{D})|^2. \tag{21}$$

Hence

$$\mathcal{R} \ll y^{\varepsilon-\frac{1}{2}}\mathcal{B}^{\frac{1}{2}}. \tag{22}$$

It remains to estimate $\mathcal{B}$. One could apply Deligne's bound for hyper-Kloosterman sums getting

$$W_q(n) \ll q^{2+\varepsilon}. \tag{23}$$

whence quickly

$$W(n) \ll Q^{3+\varepsilon}. \tag{24}$$

However, this estimate is not sufficient to break Selberg's bound (1), it just gives (1) by appealing to the more sophisticated Riemann hypothesis for varieties.

6. Estimation of $\mathcal{B}$

Needless to say that (23) is best possible, nevertheless we shall improve on (24) by observing cancellation in the sum $W(n)$ over special moduli q due to a variation in the argument of $W_q(n)$. We take moduli of type $q = pr$ with p, r ranging over primes, $P < p \leq 2P$, $R < r \leq 2R$. We shall choose P, R later with $PR = Q$ to optimize the results. Since P will be much larger than R we have p, r co prime, therefore every primitive character $\chi(\bmod q)$ has unique factorization $\chi = \psi\eta$ into primitive characters $\psi(\bmod p)$ and $\eta(\bmod r)$. Accordingly the Gauss sum factors as follows

$$\tau(\chi) = \psi(r)\eta(p)\tau(\psi)\tau(\eta).$$

Hence

$$|W(n)| \leq \sum_r 3r^{\frac{3}{2}} \sum_{\eta(\bmod r)} |\sum_p \eta(p) \sum_{\psi(\bmod p)}^{*} \tau(\psi)^3\psi(\pm nr^3)|.$$

Notice we have taken all characters $\eta(\text{mod } r)$, not only the cubes of primitive ones. By Cauchy's inequality and the orthogonality of the characters $\eta(\text{mod } r)$ we obtain

$$|W(n)|^2 \leq 9R^6 \sum_r C_r(n)$$

where

$$C_r(n) = \sum_{p_1 \equiv p_2 (\text{mod } r)} \sum \; \sum_{\substack{\psi_1 (\text{mod } p_1) \\ \psi_2 (\text{mod } p_2)}}^{*} \sum^{*} \tau(\psi_1)^3 \bar{\tau}(\psi_2)^3 \psi_1 \bar{\psi}_2(\pm n r^3).$$

Hence

$$\mathcal{B} \leq 9R^6 \sum_r S_r, \tag{25}$$

where

$$\begin{aligned} S_r &= \sum_1^{\infty} C_r(n\bar{D}) G(\frac{n}{y}) \\ &= \sum_{p_1 \equiv p_2 (\text{mod } r)} \sum \; \sum_{\substack{\psi_1 (\text{mod } p_1) \\ \psi_2 (\text{mod } p_2)}}^{*} \sum^{*} \tau(\psi_1)^3 \bar{\tau}(\psi_2)^3 \bar{\psi}_1 \psi_2(D) T(\psi_1 \bar{\psi}_2) \end{aligned}$$

and

$$T(\psi_1 \bar{\psi}_2) = \sum_{-\infty}^{\infty} \psi_1 \bar{\psi}_2(n) G\left(\frac{n}{y}\right).$$

If $\psi_1 = \psi_2$ we obtain the trivial character $\psi_1 \bar{\psi}_2 = \psi_0$ and the trivial estimate $T(\psi_0) \ll y$, therefore the diagonal terms contribute to S_r at most $O(yP^5)$. If $\psi_1 \neq \psi_2$ but $p_1 = p_2 = p$ we have a primitive character $\psi_1 \bar{\psi}_2 = \psi$ to modulus p. By Poisson's summation

$$T(\psi) = y \frac{\tau(\bar{\psi})}{p} \sum_h \bar{\psi}(h) H\left(\frac{hy}{p}\right)$$

where $H(v) = e^{-2\pi|v|}$ is the Fourier transform of $G(u)$. Hence $T(\psi) \ll p^{\frac{1}{2}}$, therefore these terms contribut to S_r at most $O(P^{13/2})$. If $p_1 \neq p_2$ we have a primitive character $\psi_1 \psi_2$ to modulus $p_1 p_2$. By Poisson's summation

$$T(\psi_1 \bar{\psi}_2) = y \frac{\tau(\bar{\psi}_1 \psi_2)}{p_1 p_2} \sum_h \bar{\psi}_1 \psi_2(h) H\left(\frac{hy}{p_1 p_2}\right).$$

Here $\tau(\bar{\psi} \psi_2) = \bar{\psi}_1(-p_2) \psi_2(p_1) \bar{\tau}(\psi_1) \tau(\psi_2)$ therefore we have

$$\tau(\psi_1)^3 \bar{\tau}(\psi_2)^3 \tau(\bar{\psi}_1 \psi_2) = \bar{\psi}_1(-p_1) \psi_2(p_1) \tau(\psi_1)^2 \bar{\tau}(\psi_2)^2 p_1 p_2.$$

Inserting the above formulas to S_r we obtain

$$S_r = \sum_{\substack{p_1 \equiv p_2 (\mathrm{mod}\ r) \\ p_1 \neq p_2}} \sum S_r(p_1, p_2) + O(yP^5 + P^{13/2})$$

where

$$S_r(p_1, p_2) = y \sum_{h \neq 0} K_{p_1}(-Dhp_2\bar{r}^3)\overline{K}_{p_2}(Dhp_1\bar{r}^3) H\left(\frac{hy}{p_1p_2}\right)$$

and

$$K_p(a) = \sum_{\psi (\mathrm{mod}\ p)}^{*} \tau(\psi)^2 \bar{\psi}(a).$$

Notice (this is crucial!) that the cubes of Gauss sums are reduced to squares. The resulting sum is essentially a Kloosterman sum in two variables instead of three, more precisely

$$K_p(a) = (p-1) \sum_{uv \equiv a (\mathrm{mod}\ p)} e_p(u+v) - 1$$

if $(a,p) = 1$. By Weil's estimate one obtains $|K_p(a)| < 2p^{\frac{3}{2}}$, whence

$$S_r(p_1, p_2) \ll y(p_1p_2)^{\frac{3}{2}} \sum_{h \neq 0} H\left(\frac{hy}{p_1p_2}\right) \ll P^5.$$

This bound would suffice to establish (11) but we wish to avoid any application of advanced results. In the next section we shall establish essentially the same bound for $S_r(p_1, p_2)$ on average over $p_1 \neq p_2$ by elementary arguments.

7. Estimation of S_r

In place of the estimate $|K_p(a)| < 2p^{\frac{3}{2}}$ we use the following elementary formula

$$\sum_{a (\mathrm{mod}\ p)}^{*} |K_p(a)|^2 = (p-1) \sum_{\psi (\mathrm{mod}\ p)}^{*} |\tau(\psi)|^4 = (p-2)(p-1)p^2.$$

We proceed as follows (employ $2|K_{p_1}K_{p_2}| \leq |K_{p_1}|^2 + |K_{p_2}|^2$),

$$\sum_{\substack{p_1 \equiv p_2 (\mathrm{mod}\ r) \\ p_1 \neq p_2}} \sum |S_r(p_1, p_2)| \leq y \sum_{\substack{p_1 \equiv p_2 (\mathrm{mod}\ r) \\ p_1 \neq p_2}} \sum \sum_{h \neq 0} |K_{p_2}(Dhp_1\bar{r}^3)|^2 H\left(\frac{hy}{4P^2}\right)$$

$$= y \sum_{p} \sum_{a (\mathrm{mod}\ p)}^{*} |K_p(a)|^2 J_p(a),$$

where

$$J_p(a) = \sum_{\substack{h\neq 0, p_1\equiv p (\mathrm{mod}\ r)\\ hp_1\equiv ar^3\bar{D}(\mathrm{mod}\ p)}}\sum H(hy/4P^2).$$

Here write $p_1 = p + kr$ with $0 < |k| \leq Pr^{-1}$ by virtue of the first congruence, interpret the second congruence as $hk \equiv \alpha (\mathrm{mod}\ p)$ where $\alpha = ar^2\overline{D}$ and consider $\ell = hk$ as one variable with multiplicity bounded by $\tau(|\ell|)H\left(\frac{\ell ry}{4P^3}\right)$. Hence

$$J_p(a) < \sum_{\ell\equiv\alpha(\mathrm{mod}\ p)} \tau(|\ell|)H\left(\frac{\ell ry}{4P^3}\right) \ll \left(1+\frac{P^2}{ry}\right)P^{\varepsilon}.$$

Combining the above results we obtain

$$S_r \ll (y + r^{-1}P^2)P^{5+\varepsilon} + P^{\frac{13}{2}}. \tag{26}$$

8. Conclusion

By (25) and (26) we get

$$\mathcal{B} \ll R^6(yR + P^2)P^{5+\varepsilon} + R^7P^{\frac{13}{2}}.$$

We choose $P = y^{\frac{1}{3}}Q^{\frac{1}{3}}$ and $R = y^{-\frac{1}{3}}Q^{\frac{2}{3}}$ subject to $Q < y < Q^2$ getting

$$\mathcal{B} \ll y^{\frac{1}{3}}Q^{\frac{19}{3}+\varepsilon}.$$

Hence by (22)

$$\mathcal{R} \ll y^{-\frac{1}{3}}Q^{\frac{19}{6}+\varepsilon} = x^{\frac{1}{3}}Q^{\frac{13}{6}+\varepsilon}. \tag{27}$$

Comparing (20) with (27) we get

$$V(1) \ll (xQ + x^{\frac{1}{3}}Q^{\frac{13}{6}})Q^{\varepsilon}.$$

In the other direction by (17) for our choice of moduli we have

$$V(1) \gg x^{2\nu}Q^2(\log Q)^{-2}. \tag{28}$$

Therefore

$$x^{2\nu} \ll \left(xQ^{-1} + x^{\frac{1}{3}}Q^{\frac{1}{6}}\right)Q^{\varepsilon}(\log Q)^2 \ll x^{\frac{3}{7}+\varepsilon}$$

by choosing $Q = x^{\frac{4}{7}}$. Since x can be arbitrarily large this implies (11).

Remark. Our choice of the involved parameters is admissible, expressing in terms of Q they are: $R = Q^{\frac{1}{4}}$, $P = Q^{\frac{3}{4}}$, $y = Q^{\frac{5}{4}}$, $x = Q^{\frac{7}{4}}$.

REFERENCES

[BDHI] D. Bump, W. Duke, J. Hoffstein and H. Iwaniec, *An estimate for the Hecke eigenvalues of Maass forms*, Inter.Math.Res.Notices **4** (1992), 75–81.

[DI] J. -M. Deshouillers and H. Iwaniec, *Kloosterman sums and Fourier coefficients of cusp forms*, Invent.Math. **70** (1982), 219–288.

[DI 1-3] W. Duke and H. Iwaniec, *Estimates for coefficients of L-functions*, I (Montreal 1989), II (Amalfi 1992), III (Paris 1990).

[GJ] S. Gelbart and H. Jacquet, *A relation between automorphic representations of $GL(2)$ and $GL(3)$*, Ann.Sci.École Norm.Sup. 4e série **11** (1978), 471–552.

[I 1] H. Iwaniec, *Selberg's lower bound of the first eigenvalue for congruence groups,* in Number Theory, Trace Formulas and Discrete Groups, Academic Press (1989), San Diego, 371–375.

[I 2] H. Iwaniec, *Small eigenvalues of Laplacian for* $\Gamma_0(N)$, Acta Arith. **56** (1990), 65–82.

[LRS] W. Luo, Z. Rudnick and P. Sarnak, *On Selberg's eigenvalue conjecture,*, Geom. Funct. Anal. **5** (1995), 387–401.

[Se] A. Selberg, *On the estimation of Fourier coefficients of modular forms*, AMS. Proc.Symp. Pure Math. **VII** (1965), 1–15.

[Sh] G. Shimura, *On the holomorphy of certain Dirichlet series*, Proc. London Math.Soc.(3) **31** (1975), 79–98.

Department of Mathematics
Rutgers University
New Brunswick, NJ 08903

Received August 1995; revised January 1996

Signatures of Roots and a New Characterization of Causal Symmetric Spaces

Soji Kaneyuki

Introduction

Let $\mathfrak{g}$ be a real semisimple Lie algebra, let τ be a Cartan involution of $\mathfrak{g}$. Choose a maximal abelian subspace $\mathfrak{a}$ in the (-1)-eigenspace of τ, and let Δ be the root system for $(\mathfrak{g}, \mathfrak{a})$. Oshima-Sekiguchi [11] introduced the notion of a signature function ϵ on Δ, which is a map of $\Delta \cup (0)$ to $\{\pm 1\}$ satisfying a multiplicative property. Using ϵ and τ, they define a new involution τ_ϵ of $\mathfrak{g}$ by putting $\tau_\epsilon(X) = \epsilon(\alpha)\tau(X)$, where X is a root vector for a root $\alpha \in \Delta \cup (0)$. The resulting symmetric pair $(\mathfrak{g}, \tau_\epsilon)$ is said to be a *symmetric pair of type K_ϵ*. A $(\mathbb{Z})$-graded Lie algebra (or shortly GLA) $\mathfrak{g} = \sum_{k=-\nu}^{\nu} \mathfrak{g}_k$ is said to be *of the ν-th kind*, if $\mathfrak{g}_{-1} \neq (0)$ and $\mathfrak{g}_{\pm\nu} \neq (0)$. In [7, 6], we have worked out the classification and construction of real semisimple GLA's.

Let G be a connected Lie group and H a closed subgroup of G. Suppose that G/H is a symmetric space. If G/H admits a G-invariant causal structure (for the definition, see §4), then G/H is said to be a *causal symmetric space*, which is an interesting and growing object (see for instance [2, 8, 4]). The infinitesimal classification of simple causal symmetric spaces was given by Ol'shansky [10], and the rigorous proof for it was given by 'Olafsson [8].

In this paper we first show that every signature function on Δ arises from a gradation (Lemma 2.1). From this point of view, we can give a root-theory-free characterization of symmetric pairs of type K_ϵ (Proposition 2.1). Using this and a result of Oshima-Sekiguchi [11], we conclude (§2) that simple symmetric pairs of type K_ϵ break up into two classes: types $K_\epsilon I$ and $K_\epsilon II$, whose lists are give in §2. Our second aim is to prove the coincidence of simple symmetric pairs of type $K_\epsilon I$ with symmetric pairs of Ol'shansky (Theorem 3.1), which leads to a new characterization of simple causal symmetric spaces in terms of symmetric pairs of type $K_\epsilon I$ (Theorem 4.1). This also provides a geometric meaning of simple symmetric spaces of type $K_\epsilon I$. It seems to be an interesting problem to find the geometric meaning of symmetric pairs of type $K_\epsilon II$. For this direction, we only know that all simple irreducible pseudo-hermitian symmetric spaces of type K_ϵ are of type $K_\epsilon II$ (For the proof, see Lemmas 3.3 and 3.4 in [5]).

The author is grateful to the Department of Mathematics, University of Massachusetts at Amherst for their hospitality during the final stage of preparation of this paper. The author also expresses his hearty thanks to Professor S. G. Gindikin for inviting him to contribute this paper to this volume.

Note. $\mathfrak{g}^{\mathbf{C}}$ denotes the complexification of a Lie algebra $\mathfrak{g}$. $\mathfrak{c}_{\mathfrak{g}}(X)$ denotes the centralizer of an element X in a Lie algebra $\mathfrak{g}$. Sometimes we simply write $\mathfrak{c}(X)$ for $\mathfrak{c}_{\mathfrak{g}}(X)$.

1. Signatures of roots and graded Lie algebras

1.1. Let $\mathfrak{g}$ be a real semisimple Lie algebra and τ a Cartan involution of $\mathfrak{g}$, and let

$$\mathfrak{g} = \mathfrak{k} + \mathfrak{p} \tag{1.1}$$

be the corresponding Cartan decomposition, where $\tau|_{\mathfrak{k}} = 1$ and $\tau|_{\mathfrak{p}} = -1$. Let $\mathfrak{a}$ be a maximal abelian subspace of $\mathfrak{p}$ and Δ be the (restricted) root system of $(\mathfrak{g}, \mathfrak{a})$. Then we have the root space decomposition of $\mathfrak{g}$:

$$\mathfrak{g} = \sum_{\alpha \in \Delta \cup (0)} \mathfrak{g}^{\alpha}, \tag{1.2}$$

where $\mathfrak{g}^0 = \mathfrak{c}(\mathfrak{a})$, the centralizer of $\mathfrak{a}$ in $\mathfrak{g}$, and $\mathfrak{g}^\alpha$ is the root space for $\alpha \in \Delta$.

Definition 1.1. (Oshima-Sekiguchi [11]) A map $\epsilon \colon \Delta \cup (0) \to \{\pm 1\}$ is called a *signature function* on Δ, if the condition

$$\epsilon(\alpha + \beta) = \epsilon(\alpha)\epsilon(\beta) \quad , \qquad \alpha, \beta \in \Delta \cup (0) \tag{1.3}$$

is satisfied, whenever $\alpha + \beta \in \Delta \cup (0)$.

This definition is slightly different from the Oshima-Sekiguchi's original one. But we have easily from (1.3) that

$$\begin{aligned} &\epsilon(0) = 1\,, \\ &\epsilon(-\alpha) = \epsilon(\alpha)\,, \qquad \alpha \in \Delta. \end{aligned} \tag{1.4}$$

Let $\Pi = \{\alpha_1, \ldots, \alpha_\ell\}$ be a fundamental system for Δ. Then one can express a root $\alpha \in \Delta$ as

$$\alpha = \sum_{i=1}^{\ell} m_i(\alpha)\alpha_i. \tag{1.5}$$

Lemma 1.1. *Let ϵ be a signature function on Δ, and let $\alpha \in \Delta$. Then we have*

$$\epsilon(\alpha) = \prod_{i=1}^{\ell} \epsilon(\alpha_i)^{m_i(\alpha)} \qquad \alpha \in \Delta \cup (0). \tag{1.6}$$

In particular, ϵ extends to a $\{\pm 1\}$-valued multiplicative character of the root lattice Q generated by Δ.

Proof. Suppose that $\alpha \in \Delta^+$ (= the positive roots). Then the assertion can be proved by the induction on the height of the root α in the expression (1.5). Note that a positive root $\alpha \in \Delta^+ - \Pi$ can be written as $\alpha = \beta + \alpha_i$, where $\beta \in \Delta^+$ and $\alpha_i \in \Pi$. If α is negative, use the property (1.4). □

Thus a signature function on Δ is nothing but a $\{\pm 1\}$-valued character of the root lattice Q.

1.2. Let us recall some basic facts on graded Lie algebras from our previous paper [7]. Let

$$\mathfrak{g} = \sum_{k=-\nu}^{\nu} \mathfrak{g}_k \tag{1.7}$$

be a real semisimple GLA of the ν-th kind. We always assume that $\mathfrak{g}_{-1} \neq (0)$ and $\mathfrak{g}_{\pm\nu} \neq (0)$. The family of subspaces $(\mathfrak{g}_k)_{-\nu \leq k \leq \nu}$ is called a *gradation* in $\mathfrak{g}$.

Definition 1.2. Let $\mathfrak{g}$ be a semisimple Lie algebra and let $Z \in \mathfrak{g}$. A triple $(\mathfrak{g}, Z, \tau)$ is called a *graded triple*, if the following conditions are satisfied:

(i) Z is the *characteristic element* of a gradation in $\mathfrak{g}$ of the form (1.7), that is, each subspace $\mathfrak{g}_k$ is given by

$$\mathfrak{g}_k = \{X \in \mathfrak{g}: (adZ)X = kX\}. \tag{1.8}$$

(ii) τ is a Cartan involution of $\mathfrak{g}$ satisfying $\tau(Z) = -Z$, or equivalently, τ is *grade-reversing*, i.e. $\tau(\mathfrak{g}_k) = \mathfrak{g}_{-k}$ for each k.

It is well-known (for example, Tanaka [13]) that to a semisimple GLA one can always associate a graded triple. In order to express a gradation in $\mathfrak{g}$, we will use the notation $(\mathfrak{g}, Z)$ rather than (1.7). We say that a semisimple GLA (1.7) is *of type* α_0, if the negative part $\sum_{k<0} \mathfrak{g}_k$ is generated by $\mathfrak{g}_{-1}$.

Now let $(\mathfrak{g}, Z, \tau)$ be a graded triple associated with (1.7). Let $\mathfrak{g} = \mathfrak{k} + \mathfrak{p}$ be the Cartan decomposition (1.1) by τ. Then $Z \in \mathfrak{p}$. Let $\mathfrak{a}$ be a maximal abelian subspace of $\mathfrak{p}$ containing Z, and let Δ be the root system for $(\mathfrak{g}, \mathfrak{a})$. We denote by $(\,,)$ the Killing form of $\mathfrak{g}$. Then the subspaces $\mathfrak{g}_k$ can be written as ([7])

$$\begin{aligned} \mathfrak{g}_0 &= \sum_{\substack{\alpha \in \Delta \cup (0) \\ (\alpha, Z) = 0}} \mathfrak{g}^\alpha, \\ \mathfrak{g}_k &= \sum_{\substack{\alpha \in \Delta \\ (\alpha, Z) = k}} \mathfrak{g}^\alpha, \quad k \neq 0. \end{aligned} \tag{1.9}$$

Therefore we have a partition of Δ:

$$\begin{aligned} \Delta &= \coprod_{k=-\nu}^{\nu} \Delta_k, \\ \Delta_k &= \{\alpha \in \Delta: (\alpha, Z) = k\}. \end{aligned} \tag{1.10}$$

Choose a linear order in Δ in such a way that

$$\coprod_{k=1}^{\nu} \Delta_k \subset \Delta^+ \subset \coprod_{k=0}^{\nu} \Delta_k, \tag{1.11}$$

where Δ^+ denotes the positive roots in Δ with respect to that order. Now choose a fundamental system $\Pi = \{\alpha_1, \ldots, \alpha_\ell\}$ of Δ with respect to this order. Then we have a partition of Π:

$$\Pi = \coprod_{k=0}^{s} \Pi_k , \qquad \Pi_k : = \Delta_k \cap \Pi . \tag{1.12}$$

Note that $\Pi_1 \neq \emptyset$, since $\mathfrak{g}_1 \neq (0)$.

Theorem 1.1 ([7]). *The gradation (1.7) is of type α_0, if and only if $\Pi_k = \emptyset$ for $k \geq 2$, that is,*

$$\Pi = \Pi_0 \coprod \Pi_1 . \tag{1.13}$$

Later on, we always assume that gradations are of type α_0. We define the height function $h_{\Pi_1} : \Delta \to \mathbb{Z}$ relative to the subset $\Pi_1 \subset \Pi$ by putting (cf. (1.5))

$$h_{\Pi_1}(\alpha) = \sum_{\alpha_i \in \Pi_1} m_i(\alpha) , \qquad \alpha \in \Delta . \tag{1.14}$$

Then we have

$$h_{\Pi_1}(\alpha) = (\alpha, Z) , \qquad \alpha \in \Delta . \tag{1.15}$$

Two subsets Π_1 and Π_1' of Π are said to be *equivalent*, if there exists a diagram automorphism of Π sending Π_1 to Π_1'. Then we have the following classification theorem for gradations (see [7] for more general setting).

Theorem 1.2 ([7]). *Let $\mathfrak{g}$ be a real semisimple Lie algebra and $\Pi = \{\alpha_1, \ldots, \alpha_\ell\}$ be a fundamental system for a restricted root system Δ of $\mathfrak{g}$ as in 1.1. Then there exists a bijection between the set of isomorphism classes of gradations of type α_0 in $\mathfrak{g}$ and the set of equivalence classes of subsets Π_1 in Π.*

Let Π_1 be a subset of Π, and put

$$\begin{aligned} \mathfrak{g}_0 &= \mathfrak{c}(\mathfrak{a}) + \sum_{h_{\Pi_1}(\alpha)=0} \mathfrak{g}^\alpha , \\ \mathfrak{g}_k &= \sum_{h_{\Pi_1}(\alpha)=k} \mathfrak{g}^\alpha , \qquad (k \neq 0) \\ \nu &= h_{\Pi_1}(\vartheta) , \end{aligned} \tag{1.16}$$

where ϑ denotes the highest root of Δ with respect to Π. Then we have a gradation $\mathfrak{g} = \sum_{k=-\nu}^{\nu} \mathfrak{g}_k$ of the ν-th kind of type α_0 corresponding to the subset Π_1.

2. Symmetric pairs of type K_ϵ

2.1.

Lemma 2.1. *Let* $(\mathfrak{g}, Z)$ *be a semisimple GLA of type* α_0, *and* $\tau, \mathfrak{a}$ *and* Δ *the same as in 1.2. Let* Q *be the root lattice for* Δ *in* $\mathfrak{a}$. *If we define* ϵ_Z *as*

$$\epsilon_Z(\alpha) := e^{\pi i(\alpha, Z)} \qquad \alpha \in Q, \tag{2.1}$$

then ϵ_Z *is a* $\{\pm 1\}$*-valued character of* Q. *Conversely, for any* $\{\pm 1\}$*-valued non-trivial character* ϵ *of* Q, *there exists a gradation* $(\mathfrak{g}, Z)$ *of* $\mathfrak{g}$ *of type* α_0 *such that* $\epsilon = \epsilon_Z$.

Proof. Let $\alpha \in \Delta$. Then, since (α, Z) is an integer (cf. (1.9)), the first assertion is clear. To show the second assertion, choose a fundamental system Π for Δ. Put

$$\begin{aligned} \Pi_0 &:= \{\alpha_i \in \Pi \colon \epsilon(\alpha_i) = 1\}, \\ \Pi_1 &:= \{\alpha_i \in \Pi \colon \epsilon(\alpha_i) = -1\}. \end{aligned} \tag{2.2}$$

Then we have a partition $\Pi = \Pi_0 \amalg \Pi_1$. Since ϵ is non-trivial, Π_1 is not empty. Therefore it follows from Theorem 1.2 and (1.16) that to the subset Π_1 there corresponds a gradation of $\mathfrak{g}$, whose characteristic element $Z \in \mathfrak{a}$ is determined by the equations

$$\begin{cases} (\alpha_i, Z) = 0, & \alpha_i \in \Pi_0, \\ (\alpha_j, Z) = 1, & \alpha_j \in \Pi_1. \end{cases} \tag{2.3}$$

The equality $\epsilon = \epsilon_Z$ follows from (2.1)–(2.3). □

We denote by $\mathrm{Aut}(\mathfrak{a}, \Delta)$ the group of orthogonal transformations of $\mathfrak{a}$ (with respect to $(\,,)$) which leave Δ stable. Let ϵ and ϵ' be two signature functions on Δ. We say that ϵ is equivalent to ϵ', if there exists an element $\phi \in \mathrm{Aut}(\mathfrak{a}, \Delta)$ such that $\epsilon' = \epsilon \circ \phi$. The classification of signature functions up to equivalence for real simple Lie algebras was given by Oshima-Sekiguchi [11]. Their results together with Lemma 2.1 and $(1.16)_3$ imply that for each real simple Lie algebra $\mathfrak{g}$, a signature function arises either from a gradation of the first kind or from that of the second kind.

2.2. Let $\mathfrak{g}$ be a real semisimple Lie algebra, and let $\tau, \mathfrak{a}$ and Δ be as in 1.1. Let ϵ be a signature function on Δ. In view of the root space decomposition (1.2), we define a map $\tilde{\epsilon}\colon \mathfrak{g} \to \mathfrak{g}$ by putting

$$\tilde{\epsilon}(X) = \epsilon(\alpha)X\,, \quad X \in \mathfrak{g}^{\alpha}, \quad \alpha \in \Delta \cup (0)\,. \tag{2.4}$$

Then it follows from (1.3) that $\tilde{\epsilon}$ is an involutive automorphism of $\mathfrak{g}$. $\tilde{\epsilon}$ is called a *signature involution* (associated with $\tau, \mathfrak{a}$ and ϵ). (1.4) implies that $\tilde{\epsilon}$ commutes with τ, and hence we have a new symmetric pair $(\mathfrak{g}, \tilde{\epsilon}\tau)$ which is called a *symmetric pair of type K_ϵ* (cf. Introduction).

Proposition 2.1. *A semisimple symmetric pair $(\mathfrak{g}, \sigma)$ is of type K_ϵ, if and only if there exists a graded triple $(\mathfrak{g}, Z, \tau)$ such that*

$$\sigma = (Ad \exp \pi i Z)\tau\,. \tag{2.5}$$

Proof. Suppose that $(\mathfrak{g}, \sigma)$ is of type K_ϵ. Then σ can be written as $\sigma = \tilde{\epsilon}\tau$. By Lemma 2.1, there exists a gradation $(\mathfrak{g}, Z)$ such that $\tilde{\epsilon} = \tilde{\epsilon}_Z$. Hence it follows from (1.9) and (2.4) that $\tilde{\epsilon}_Z$ is expressed as $\mathrm{Ad} \exp \pi i Z$ (cf. Lemma 2.1 [5]). Since Z lies in $\mathfrak{a} \subset \mathfrak{p}, \tau$ is grade-reversing. □

Definition 2.1. Let $(\mathfrak{g}, \sigma)$ be a semisimple symmetric pair of type K_ϵ, where $\sigma = (\mathrm{Ad} \exp \pi i Z)\tau$. We say that $(\mathfrak{g}, \sigma)$ is *of type $K_\epsilon I$* (resp. *of type $K_\epsilon II$*), if the gradation $(\mathfrak{g}, Z)$ can be chosen to be of the first (resp. second) kind.

From what is mentioned at the end of 2.1, it follows that simple symmetric pairs of type K_ϵ break up into two classes: type $K_\epsilon I$ and type $K_\epsilon II$. The following theorem follows from Oshima-Sekiguchi [11] together with (1.14) and $(1.16)_3$.

Theorem 2.1. *Simple symmetric pairs $(\mathfrak{g}, \mathfrak{h})$ of type $K_\epsilon I$ and those of type $K_\epsilon II$ are given by the following table. The corresponding gradation of $\mathfrak{g}$ is given by the subset Π_1 of the fundamental system Π of $\mathfrak{g}$.*

TYPE $K_\epsilon I$

	$(\mathfrak{g}, \mathfrak{h})$	Π	Π_1
I1	$(\mathfrak{sl}(p+q,\mathbb{C}), \mathfrak{su}(p,q)), 1 \le p \le q$	A_{p+q-1}	$\{\alpha_p\}$
I2	$(\mathfrak{so}(2n,\mathbb{C}), \mathfrak{so}^*(2n)), n \ge 4$	D_n	$\{\alpha_n\}$
I3	$(\mathfrak{so}(n+2,\mathbb{C}), \mathfrak{so}(2,n))$	$B_{(n+1)/2}(n$ odd $\ge 3)$	$\{\alpha_1\}$
		$D_{n/2+1}(n$ even $\ge 4)$	$\{\alpha_1\}$
I4	$(\mathfrak{sp}(n,\mathbb{C}), \mathfrak{sp}(n,\mathbb{R})), n \ge 2$	C_n	$\{\alpha_1\}$
I5	$(E_6^{\mathbb{C}}, E_{6(-14)})$	E_6	$\{\alpha_1\}$
I6	$(E_7^{\mathbb{C}}, E_{7(-25)})$	E_7	$\{\alpha_7\}$
I7	$(\mathfrak{sl}(p+q,\mathbb{R}), \mathfrak{so}(p,q)), 1 \le p \le q$	A_{p+q-1}	$\{\alpha_p\}$
I8	$(\mathfrak{su}(n,n), \mathfrak{sl}(n,\mathbb{C})+\mathbb{R}), n \ge 3$	C_n	$\{\alpha_n\}$
I9	$(\mathfrak{sl}(p+q,\mathbb{H}), \mathfrak{sp}(p,q)), 1 \le p \le q$	A_{p+q-1}	$\{\alpha_p\}$
I10	$(\mathfrak{so}(n,n), \mathfrak{so}(n,\mathbb{C})), n \ge 4$	D_n	$\{\alpha_n\}$
I11	$(\mathfrak{so}^*(4n), \mathfrak{gl}(n,\mathbb{H})), n \ge 3$	C_n	$\{\alpha_n\}$
I12	$(\mathfrak{so}(p+1,q+1), \mathfrak{so}(1,p)+\mathfrak{so}(1,q)), 1 \le p \le q,$	$B_p(p+q$ odd$)$	$\{\alpha_1\}$
		$D_{p+1}(p=q)$	$\{\alpha_1\}$
I13	$(\mathfrak{sp}(n,\mathbb{R}), \mathfrak{gl}(n,\mathbb{R})), n \ge 3$	C_n	$\{\alpha_n\}$
I14	$(\mathfrak{sp}(n,n), \mathfrak{sp}(n,\mathbb{C})), n \ge 2$	C_n	$\{\alpha_n\}$
I15	$(E_{6(6)}, \mathfrak{sp}(2,2))$	E_6	$\{\alpha_1\}$
I16	$(E_{6(-26)}, F_{4(-20)})$	A_2	$\{\alpha_1\}$
I17	$(E_{7(7)}, \mathfrak{sl}(4,\mathbb{H}))$	E_7	$\{\alpha_7\}$
I18	$(E_{7(-25)}, E_{6(-26)}+\mathbb{R})$	C_3	$\{\alpha_3\}$

TYPE $K_\epsilon II$

	$(\mathfrak{g}, \mathfrak{h})$	Π	Π_1
II1	$(\mathfrak{so}(n,\mathbb{C}), \mathfrak{so}(n-2j,2j))$,	$BD_{[n/2]}$	$\{\alpha_j\},\ 2 \le j \le [n/2]$
II2	$(\mathfrak{sp}(n,\mathbb{C}), \mathfrak{sp}(n-j,j))$	C_n	$\{\alpha_j\},\ 2 \le j \le [n/2]$
II3	$(E_6^{\mathbb{C}}, E_{6(2)})$	E_6	$\{\alpha_2\}$
II4	$(E_7^{\mathbb{C}}, E_{7(-5)})$	E_7	$\{\alpha_1\}$
II5	$(E_7^{\mathbb{C}}, E_{7(7)})$	E_7	$\{\alpha_2\}$
II6	$(E_8^{\mathbb{C}}, E_{8(8)})$	E_8	$\{\alpha_1\}$
II7	$(E_8^{\mathbb{C}}, E_{8(-24)})$	E_8	$\{\alpha_8\}$
II8	$(F_4^{\mathbb{C}}, F_{4(4)})$	F_4	$\{\alpha_1\}$
II9	$(F_4^{\mathbb{C}}, F_{4(-20)})$	F_4	$\{\alpha_4\}$
II10	$(G_2^{\mathbb{C}}, G_{2(2)})$	G_2	$\{\alpha_2\}$
II11	$(\mathfrak{su}(p,q), \mathfrak{su}(p-j,j)$	$BC_p(p<q)$	$\{\alpha_j\},\ 1 \le j \le p$
	$+\mathfrak{su}(j,q-j)+i\mathbb{R}),\ 1 \le p \le q$	$C_p(p=q)$	$\{\alpha_j\},\ 1 \le j \le [p/2]$
II12	$(\mathfrak{so}(p,q), \mathfrak{so}(p-j,j)+\mathfrak{so}(j,q-j))$,	$B_p(p<q)$	$\{\alpha_j\},\ 2 \le j \le p$
	$1 \le p \le q$	$D_p(p=q)$	$\{\alpha_j\},\ 2 \le j \le [p/2]$
II13	$(\mathfrak{sp}(n,\mathbb{R}), \mathfrak{u}(n-j,j))$	C_n	$\{\alpha_j\},\ 1 \le j \le [n/2]$
II14	$(\mathfrak{sp}(p,q), \mathfrak{sp}(p-j,j)+\mathfrak{sp}(j,q-j))$,	$BC_p(p<q)$	$\{\alpha_j\},\ 1 \le j \le p$
	$1 \le p \le q$	$C_p(p=q)$	$\{\alpha_j\},\ 1 \le j \le [p/2]$
II15	$(\mathfrak{so}^*(2n), \mathfrak{u}(2n-2j,2j))$	$BC_\ell(n=2\ell+1)$	$\{\alpha_j\},\ 1 \le j \le \ell$
		$C_\ell(n=2\ell)$	$\{\alpha_j\},\ 1 \le j \le [\ell/2]$
II16	$(E_{6(6)}, \mathfrak{sp}(4,\mathbb{R}))$	E_6	$\{\alpha_2\}$
II17	$(E_{6(2)}, \mathfrak{su}(3,3)+\mathfrak{sl}(2,\mathbb{R}))$	F_4	$\{\alpha_1\}$
II18	$(E_{6(2)}, \mathfrak{su}(2,4)+\mathfrak{su}(2))$	F_4	$\{\alpha_4\}$
II19	$(E_{6(-14)}, \mathfrak{so}^*(10)+i\mathbb{R})$	BC_2	$\{\alpha_1\}$
II20	$(E_{6(-14)}, \mathfrak{so}(2,8)+i\mathbb{R})$	BC_2	$\{\alpha_2\}$
II21	$(E_{7(7)}, \mathfrak{su}(4,4))$	E_7	$\{\alpha_1\}$
II22	$(E_{7(7)}, \mathfrak{sl}(8,\mathbb{R}))$	E_7	$\{\alpha_2\}$
II23	$(E_{7(-5)}, \mathfrak{so}^*(12)+\mathfrak{sl}(2,\mathbb{R}))$	F_4	$\{\alpha_1\}$
II24	$(E_{7(-5)}, \mathfrak{so}(4,8)+\mathfrak{su}(2))$	F_4	$\{\alpha_4\}$
II25	$(E_{7(-25)}, E_{6(-14)}+i\mathbb{R})$	C_3	$\{\alpha_1\}$
II26	$(E_{8(8)}, \mathfrak{so}(8,8))$	E_8	$\{\alpha_1\}$
II27	$(E_{8(8)}, \mathfrak{so}^*(16))$	E_8	$\{\alpha_8\}$
II28	$(E_{8(-24)}, E_{7(-25)}+\mathfrak{sl}(2,\mathbb{R}))$	F_4	$\{\alpha_1\}$
II29	$(E_{8(-24)}, E_{7(-5)}+\mathfrak{su}(2))$	F_4	$\{\alpha_4\}$
II30	$(F_{4(4)}, \mathfrak{sp}(3,\mathbb{R})+\mathfrak{sl}(2,\mathbb{R}))$	F_4	$\{\alpha_1\}$
II31	$(F_{4(4)}, \mathfrak{sp}(1.2)+\mathfrak{su}(2))$	F_4	$\{\alpha_4\}$
II32	$(F_{4(-20)}, \mathfrak{so}(1.8))$	BC_1	$\{\alpha_1\}$
II33	$(G_{2(2)}, \mathfrak{sl}(2,\mathbb{R})+\mathfrak{sl}(2,\mathbb{R}))$	G_2	$\{\alpha_2\}$

In the above table, we employ the numbering of simple roots given in Bourbaki [1]. (II11), (II12) with $p = j = 2$, (II13), (II15), (II19), (II20) and (II25) exhaust all simple irreducible pseudo-hermitian symmetric spaces of type K_ϵ (cf. Introduction).

3. Characterization of symmetric triples of type $K_\epsilon I$

3.1. Let $\mathfrak{g}$ be a simple Lie algebra of hermitian type. Let τ be a Cartan involution of $\mathfrak{g}$ and $\mathfrak{g} = \mathfrak{k} + \mathfrak{p}$ the corresponding Cartan decomposition given in (1.1). Then $\mathfrak{g}_u = \mathfrak{k} + i\mathfrak{p}$ is a compact real form of the complexification $\mathfrak{g}^{\mathbf{C}}$. τ extend to $\mathfrak{g}^{\mathbf{C}}$ as the conjugation (denoted again by τ) with respect to $\mathfrak{g}_u$. Let σ be the conjugation of $\mathfrak{g}^{\mathbf{C}}$ with respect to $\mathfrak{g}$. Since $[\sigma, \tau] = 0$ on $\mathfrak{g}^{\mathbf{C}}$, we have the (σ, τ)-decomposition ([3]) of $\mathfrak{g}^{\mathbf{C}}$:

$$\mathfrak{g}^{\mathbf{C}} = \mathfrak{k} + i\mathfrak{p} + \mathfrak{p} + i\mathfrak{k}. \tag{3.1}$$

Also we have the decomposition by the involution $\sigma\tau$:

$$\mathfrak{g}^{\mathbf{C}} = \mathfrak{k}^{\mathbf{C}} + \mathfrak{p}^{\mathbf{C}}. \tag{3.2}$$

Note that $\sigma\tau = 1$ on $\mathfrak{k}^{\mathbf{C}}$ and -1 on $\mathfrak{p}^{\mathbf{C}}$.

Definition 3.1 (Ol'shansky [9]). Let $\mathfrak{g}, \sigma$ and τ be as above, and let $\mathfrak{z}$ be the center of $\mathfrak{k}$. Let $\mathfrak{g}^*$ be a real form of $\mathfrak{g}^{\mathbf{C}}$ and θ be the conjugation of $\mathfrak{g}^{\mathbf{C}}$ with respect to $\mathfrak{g}^*$. $\mathfrak{g}^*$ is called the *regular real form associated with* $\mathfrak{g}$, if the following conditions are satisfied:

$$\text{(R1)} \qquad [\theta, \sigma] = [\theta, \tau] = 0 \qquad \text{on } \mathfrak{g}^{\mathbf{C}},$$

$$\text{(R2)} \qquad i\mathfrak{z} \subset \mathfrak{g}^*.$$

Note that a regular real form is defined in association with a real form of hermitian type. Now let $\mathfrak{g}^*$ be a regular real form of $\mathfrak{g}^{\mathbf{C}}$. Then it follows from (R1) that $\mathfrak{g}^*$ is stable under σ and τ. $\tau|_{\mathfrak{g}^*}$ is a Cartan involution of $\mathfrak{g}^*$. We have the three decompositions of $\mathfrak{g}^*$ by τ, σ and $\sigma\tau$, respectively:

$$\mathfrak{g}^* = \mathfrak{g}^* \cap \mathfrak{g}_u + \mathfrak{g}^* \cap i\mathfrak{g}_u, \tag{3.3}$$

$$\mathfrak{g}^* = \mathfrak{g}^* \cap \mathfrak{g} + \mathfrak{g}^* \cap i\mathfrak{g}, \tag{3.4}$$

$$\mathfrak{g}^* = \mathfrak{g}^* \cap \mathfrak{k}^{\mathbf{C}} + \mathfrak{g}^* \cap \mathfrak{p}^{\mathbf{C}}. \tag{3.5}$$

Let Z_0 be the element of $\mathfrak{z}$ such that $\mathrm{ad}_\mathfrak{p} Z_0$ gives a complex structure of $\mathfrak{p}$. We have the decomposition $\mathfrak{p}^{\mathbf{C}} = \bar{\mathfrak{g}}_{-1} + \bar{\mathfrak{g}}_1$, where $\bar{\mathfrak{g}}_{\pm 1}$ are the $\pm i$-eigenspaces under the operator $\mathrm{ad} Z_0$ on $\mathfrak{p}^{\mathbf{C}}$. $\mathfrak{g}^{\mathbf{C}}$ can be written as a GLA

$$\mathfrak{g}^{\mathbf{C}} = \bar{\mathfrak{g}}_{-1} + \bar{\mathfrak{g}}_0 + \bar{\mathfrak{g}}_1 , \tag{3.6}$$

where $\bar{\mathfrak{g}}_0 = \mathfrak{k}^{\mathbf{C}}$. The element $Z = -iZ_0 \in \bar{\mathfrak{g}}_0$ is the characteristic element of the GLA. Since Z lies in $\mathfrak{g}^*$ by (R2), $\mathfrak{g}^*$ can be expressed as a GLA with characteristic element Z:

$$\mathfrak{g}^* = \mathfrak{g}^*_{-1} + \mathfrak{g}^*_0 + \mathfrak{g}^*_1 , \tag{3.7}$$

where $\mathfrak{g}^*_k = \bar{\mathfrak{g}}_k \cap \mathfrak{g}^*, k = 0, \pm 1$.

Lemma 3.1. *The simple symmetric triple* $(\mathfrak{g}^{\mathbf{C}}, \mathfrak{g}, \sigma)$ *is of type* $K_\epsilon I$.

Proof. Since $Z \in i\mathfrak{g}_u$, $(\mathfrak{g}^{\mathbf{C}}, Z, \tau)$ is a graded triple associated with (3.6). In view of the decomposition (3.2) by $\sigma\tau$ and the relations $\bar{\mathfrak{g}}_0 = \mathfrak{k}^{\mathbf{C}}$ and $\bar{\mathfrak{g}}_{-1} + \bar{\mathfrak{g}}_1 = \mathfrak{p}^{\mathbf{C}}$, we have that the equality $\sigma\tau = \mathrm{Ad} \exp \pi i Z$ is valid on $\mathfrak{g}^{\mathbf{C}}$. Therefore the lemma follows from Proposition 2.1. □

Lemma 3.2. *Let* $\mathfrak{h} = \mathfrak{g}^* \cap \mathfrak{g}$. *Then* $(\mathfrak{g}^*, \mathfrak{h}, \sigma)$ *is a symmetric triple of type* $K_\epsilon I$.

Proof. Since $Z \in \mathfrak{g}^* \cap i\mathfrak{g}_u$, $(\mathfrak{g}^*, Z, \tau)$ is a graded triple associated with (3.7). From (3.5) and the inclusions $\mathfrak{g}^*_0 \subset \mathfrak{k}^{\mathbf{C}}$ and $\mathfrak{g}^*_{-1} + \mathfrak{g}^*_1 \subset \mathfrak{p}^{\mathbf{C}}$, it follows that $\sigma\tau = \mathrm{Ad} \exp \pi i Z$ on $\mathfrak{g}^*$. Therefore the lemma follows from the Proposition 2.1. □

3.2. We wish to give a (classification-free) characterization of simple symmetric triple of type $K_\epsilon I$.

Theorem 3.1. *Let* $(\mathfrak{l}, \mathfrak{l}_\sigma, \sigma)$ *be a simple symmetric triple. Then* $(\mathfrak{l}, \mathfrak{l}_\sigma, \sigma)$ *is of type* $K_\epsilon I$, *if and only if it is either one of the following two cases:*
Case 1: $\mathfrak{l}$ *is a complex Lie algebra, and* $(\mathfrak{l}, \mathfrak{l}_\sigma, \sigma) = (\bar{\mathfrak{g}}, \mathfrak{g}, \sigma)$, *where* $\mathfrak{g}$ *is simple of hermitian type,* $\bar{\mathfrak{g}}$ *is the complexification of* $\mathfrak{g}$, *and* σ *is the conjugation of* $\bar{\mathfrak{g}}$ *with respect to* $\mathfrak{g}$.
Case 2: $\mathfrak{l}$ *is not complex, and* $(\mathfrak{l}, \mathfrak{l}_\sigma, \sigma) = (\mathfrak{g}^*, \mathfrak{g}^* \cap \mathfrak{g}, \sigma)$, *where* $\mathfrak{g}$ *is a real form of hermitian type of the complexification* $\mathfrak{g}^{*\mathbf{C}}, \mathfrak{g}^*$ *is a regular*

real form of $\mathfrak{g}^{*\mathbf{C}}$ *associated with* $\mathfrak{g}$, *and* σ *is the restriction to* $\mathfrak{g}^*$ *of the conjugation of* $\mathfrak{g}^{*\mathbf{C}}$ *with respect to* $\mathfrak{g}$.

Proof. It was already proved in Lemma 3.1 and Lemma 3.2 that the symmetric triples in Cases 1 and 2 were of type $K_\epsilon I$. Let us consider the converse assertion. Consider Case 1 first. Suppose that $\mathfrak{l} = \bar{\mathfrak{g}}$ is complex simple. Since $(\mathfrak{l}, \mathfrak{l}_\sigma \sigma)$ is of type $K_\epsilon I, \bar{\mathfrak{g}}$ is a complex GLA of the first kind:

$$\bar{\mathfrak{g}} = \bar{\mathfrak{g}}_{-1} + \bar{\mathfrak{g}}_0 + \bar{\mathfrak{g}}_1 , \tag{3.8}$$

and σ can be expressed as

$$\sigma = (\operatorname{Ad}\exp \pi i Z)\tau , \tag{3.9}$$

where $(\bar{\mathfrak{g}}, Z, \tau)$ is a graded triple associated with (3.8). Note that τ is the conjugation with respect to a compact real form of $\bar{\mathfrak{g}}$. By (3.9), σ is also a grade-reversing conjugation of $\bar{\mathfrak{g}}$. Let

$$\mathfrak{g} = \{X \in \bar{\mathfrak{g}} \colon \sigma(X) = X\} . \tag{3.10}$$

Then $\mathfrak{g}$ is a real form of $\bar{\mathfrak{g}}$. Put $Z_0 = iZ \in \bar{\mathfrak{g}}$. Then we have $\tau(Z_0) = Z_0$, and hence, by using (3.9), we have $\sigma(Z_0) = Z_0$, that is, $Z_0 \in \mathfrak{g}$. Now let $\mathfrak{g}_0 := \bar{\mathfrak{g}}_0 \cap \mathfrak{g}$. Since $\bar{\mathfrak{g}}_0 = \mathfrak{c}_{\bar{\mathfrak{g}}}(Z) = \mathfrak{c}_{\bar{\mathfrak{g}}}(Z_0)$, we have $\mathfrak{c}_{\mathfrak{g}}(Z_0) = \mathfrak{c}_{\bar{\mathfrak{g}}}(Z_0) \cap \mathfrak{g} = \bar{\mathfrak{g}}_0 \cap \mathfrak{g} = \mathfrak{g}_0$. Since τ commutes with σ, τ leaves $\mathfrak{g}$ stable, and the restriction $\tau|_{\mathfrak{g}}$ is a Cartan involution of $\mathfrak{g}$. Let $\mathfrak{k} = \{X \in \mathfrak{g} \colon \tau(X) = X\}$. We want to prove $\mathfrak{g}_0 = \mathfrak{k}$. Let $X \in \mathfrak{g}_0$. Then $X = \sigma(X) = (\operatorname{Ad}\exp \pi i Z)\tau(X) = \tau(\operatorname{Ad}\exp \pi i Z)X = \tau(X)$, since Z commutes with X. This implies that $X \in \mathfrak{k}$, or equivalently $\mathfrak{g}_0 \subset \mathfrak{k}$. The converse inclusion follows easily from the fact that $\bar{\mathfrak{g}}_0$ is the 1-eigenspace of the operator $\operatorname{Ad}\exp \pi i Z$. We have thus proved that $Z_0 \in \mathfrak{g}_0$ is the central element of the maximal compact subalgebra $\mathfrak{k}$ of $\mathfrak{g}$.

Next suppose that $\mathfrak{l} =: \mathfrak{g}^*$ is not complex. Since $(\mathfrak{l}, \sigma)$ is of type $K_\epsilon I, \mathfrak{g}^*$ is expressed as a GLA:

$$\mathfrak{g}^* = \mathfrak{g}^*_{-1} + \mathfrak{g}^*_0 + \mathfrak{g}^*_1 , \tag{3.11}$$

and $\sigma = (\operatorname{Ad}\exp \pi i Z)\tau$, where $(\mathfrak{g}^*, Z, \tau)$ is a graded triple associated with (3.11). Let us consider the complexification of the GLA (3.11):

$$\bar{\mathfrak{g}}^* := \mathfrak{g}^{*\mathbf{C}} = \bar{\mathfrak{g}}^*_{-1} + \bar{\mathfrak{g}}^*_0 + \bar{\mathfrak{g}}^*_1 , \tag{3.12}$$

which is simple and has the same characteristic element Z. We extend τ to $\bar{\mathfrak{g}}^*$ so as to be conjugate-linear with respect to $\mathfrak{g}^*$. Then σ extends to $\bar{\mathfrak{g}}^*$. The extensions of τ and σ are denoted by the same letters. Then $(\bar{\mathfrak{g}}^*, Z, \tau)$ becomes a graded triple associated with (3.12). σ is a grade-reversing conjugation of $\bar{\mathfrak{g}}^*$. Let

$$\mathfrak{g} = \{X \in \bar{\mathfrak{g}}^* : \sigma(X) = X\}. \tag{3.13}$$

Then $\mathfrak{g}$ is a real form of $\bar{\mathfrak{g}}^*$. By the same arguments as in Case 1, we conclude that $\mathfrak{g}$ is simple of hermitian type. To be more precise, since τ leaves $\mathfrak{g}$ stable, $\tau|_{\mathfrak{g}}$ is a Cartan involution of $\mathfrak{g}$. Let $\mathfrak{g} = \mathfrak{k} + \mathfrak{p}$ be the corresponding Cartan decomposition as in (1.1). As was proved in Case 1, we have $\mathfrak{k} = \mathfrak{c}_{\mathfrak{g}}(Z_0)$, where $Z_0 = iZ$. Since $\mathfrak{g}$ is simple, the center $\mathfrak{z}$ of $\mathfrak{k}$ is one-dimensional, and is spanned by the element Z_0. Since $Z \in \mathfrak{g}^*$, we have $i\mathfrak{z} \subset \mathfrak{g}^*$ and hence $\mathfrak{g}^*$ satisfies (R2). The (σ, τ)-decomposition (cf. (3.1)) of $\bar{\mathfrak{g}}^* = \mathfrak{g}^{\mathbf{C}}$ induces the (σ, τ)-decomposition of $\mathfrak{g}^*$:

$$\mathfrak{g}^* = \mathfrak{g}^* \cap \mathfrak{k} + \mathfrak{g}^* \cap i\mathfrak{p} + \mathfrak{g}^* \cap \mathfrak{p} + \mathfrak{g}^* \cap i\mathfrak{k}. \tag{3.14}$$

Let θ be the conjugation of $\bar{\mathfrak{g}}^*$ with respect to $\mathfrak{g}^*$. Then it follows that σ, τ and θ are 1 or -1 on each subspace in the right-hand side of (3.14), as well as on $i(\mathfrak{g}^* \cap \mathfrak{k}), i(\mathfrak{g}^* \cap i\mathfrak{p}), i(\mathfrak{g}^* \cap \mathfrak{p})$ and $i(\mathfrak{g}^* \cap i\mathfrak{k})$ of $\bar{\mathfrak{g}}^*$. From this it follows that $\mathfrak{g}^*$ satisfies (R1). We have thus proved that $\mathfrak{g}^*$ is a regular real form of $\bar{\mathfrak{g}}^*$ associated with $\mathfrak{g}$. □

Corollary 3.1. *A real simple Lie algebra $\mathfrak{g}^*$ is a regular real form of a complex simple Lie algebra if and only if $\mathfrak{g}^*$ admits a gradation of the first kind.*

Proof. This follows from 3.1 and the proof of Theorem 3.1. □

Remark. In the table of simple symmetric pairs of type $K_\epsilon I$ in §1, (I1)–(I6) correspond to Case 1 in Theorem 3.1 and (I7)–(I18) correspond to Case 2. (I8), (I11), (I12) for $p = 1$, (I13) and (I18) exhaust all simple parahermitian symmetric pairs of Silov type ([3]). They are isomorphic to their c-duals (cf. §4).

4. Characterization of causal symmetric spaces

4.1. Let $(\mathfrak{l}, \mathfrak{m}, \omega)$ be a semisimple symmetric triple. We have then the decomposition $\mathfrak{l} = \mathfrak{m} + \mathfrak{n}$ by the involution ω, where $\omega|_{\mathfrak{m}} = 1$ and $\omega|_{\mathfrak{n}} = -1$. Consider the subalgebra of the complexification $\mathfrak{l}^{\mathbb{C}}$:

$$\mathfrak{l}^{\#} = \mathfrak{m} + i\mathfrak{n}, \tag{4.1}$$

which is a real form of $\mathfrak{l}^{\mathbb{C}}$. Let $\omega^{\#}$ be the $\mathbb{C}$-linear extension of ω onto $\mathfrak{l}^{\mathbb{C}}$. Then (4.1) is the decomposition into eigenspaces of $\omega^{\#}$. The symmetric triple $(\mathfrak{l}^{\#}, \mathfrak{m}, \omega^{\#})$ is called the *c-dual* of $(\mathfrak{l}, \mathfrak{m}, \omega)$ (cf. 'Olafsson [8]).

Lemma 4.1. *Let $\mathfrak{g}$ be a simple Lie algebra of hermitian type, $\mathfrak{g}^*$ be the regular real form of $\mathfrak{g}^{\mathbb{C}}$ associated with $\mathfrak{g}$. Let σ and θ be the conjugations of $\mathfrak{g}^{\mathbb{C}}$ given in 3.1, and let $\mathfrak{h} = \mathfrak{g} \cap \mathfrak{g}^*$. Then the symmetric triple $(\mathfrak{g}, \mathfrak{h}, \theta)$ is the c-dual of $(\mathfrak{g}^*, \mathfrak{h}, \sigma)$.*

Proof. For the symmetric triple $(\mathfrak{g}^*, \mathfrak{h}, \sigma)$, we have the decomposition (3.4). Consequently we see that $(\mathfrak{g}^*)^{\#} = \mathfrak{h} + \mathfrak{g} \cap i\mathfrak{g}^*$. The right-hand side is identical with $\mathfrak{g}$, since $\mathfrak{g}$ is stable under θ by the property (R1). It follows easily that $\sigma^{\#} = \theta$ is valid on $\mathfrak{g}$. □

The list on the following page was taken from 'Olafsson [8] and Oshima-Sekiguchi [12]. $Ik^*(k = 1, \ldots, 18)$ denotes the c-dual of Ik in the table of type $K_{\epsilon}I$ in §2.

C-DUALS OF SIMPLE SYMMETRIC PAIRS OF TYPE $K_\epsilon I$

I1*	$(\mathfrak{su}(p,q) \times \mathfrak{su}(p,q), \mathfrak{su}(p,q))$
I2*	$(\mathfrak{so}^*(2n) \times \mathfrak{so}^*(2n), \mathfrak{so}^*(2n))$
I3*	$(\mathfrak{so}(2,n) \times \mathfrak{so}(2,n), \mathfrak{so}(2,n))$
I4*	$(\mathfrak{sp}(n,\mathbb{R}) \times \mathfrak{sp}(n,\mathbb{R}), \mathfrak{sp}(n,\mathbb{R}))$
I5*	$(E_{6(-14)} \times E_{6(-14)}, E_{6(-14)})$
I6*	$(E_{7(-25)} \times E_{7(-25)}, E_{7(-25)})$
I7*	$(\mathfrak{su}(p,q), \mathfrak{so}(p,q))$
I8*	I8
I9*	$(\mathfrak{su}(2p,2q), \mathfrak{sp}(p,q))$
I10*	$(\mathfrak{so}^*(2n), \mathfrak{so}(n,\mathbb{C}))$
I11*	I11
I12*	$(\mathfrak{so}(2,p+q), \mathfrak{so}(1,p) \times \mathfrak{so}(1,q))$
I13*	I13
I14*	$(\mathfrak{sp}(2n,\mathbb{R}), \mathfrak{sp}(n,\mathbb{C}))$
I15*	$(E_{6(-14)}, \mathfrak{sp}(2,2))$
I16*	$(E_{6(-14)}, F_{4(-20)})$
I17*	$(E_{7(-25)}, \mathfrak{sl}(4,\mathbb{H}))$
I18*	I18

4.2. Let V be a (finite dimensional) real vector space. A subset $C \subset V$ is called a *causal cone*, if it satisfies the three conditions: 1) C is a closed convex cone in V, 2) the interior of C is non-empty and 3) $C \cap (-C) = (0)$. Let M be a smooth manifold. By a *causal structure* $\mathcal{C}$ on M we mean a smooth assignment $\mathcal{C}$ of a causal cone C_p in the tangent space T_pM at p to each point $p \in M$ (cf. Faraut [2]). The pair $(M, \mathcal{C})$ is called a *causal manifold*. We denote the causal structure by $\mathcal{C} = \{C_p\}_{p \in M}$. Let A be a Lie group acting on a causal manifold $(M, \mathcal{C})$. The causal structure $\mathcal{C}$ is called A-*invariant*, if the following condition is satisfied.

$$C_{g(p)} = g_{*p}C_p \qquad p \in M, \quad g \in A \tag{4.2}$$

Let G/H be a symmetric (coset) space, where G is a connected Lie group. G/H is called a *causal symmetric space*, if it admits a G-invariant causal structure. Let $\mathfrak{g}$ = Lie G and $\mathfrak{h}$ = Lie H, and let $(\mathfrak{g}, \mathfrak{h}, \sigma)$ be the symmetric triple corresponding to G/H, where σ is the involution of $\mathfrak{g}$ which characterizes $\mathfrak{h}$.

We can give a new characterization of a simple causal symmetric space.

Theorem 4.1. *Let G/H be an effective symmetric space of a connected simple Lie group G. Suppose that H is connected. Then G/H is a causal symmetric space, if and only if the symmetric triple $(\mathfrak{g}, \mathfrak{h}, \sigma)$ is either of type $K_\epsilon I$ or the c-dual of a symmetric triple of type $K_\epsilon I$.*

Proof. Let $\mathfrak{g} = \mathfrak{h} + \mathfrak{m}$ be the decomposition by σ, where $\sigma|_{\mathfrak{m}} = -1$, and let $Con_H(\mathfrak{m})$ denote the set of $Ad_{\mathfrak{m}}H$-invariant causal cones in $\mathfrak{m}$. G/H is a causal symmetric if and only if $Con_H(\mathfrak{m}) \neq \emptyset$. Since H is connected, it follows from Ol'shansky [10] and 'Olafsson [8] that $Con_H(\mathfrak{m}) \neq \emptyset$, if and only if $(\mathfrak{g}, \mathfrak{h}, \sigma)$ is either a symmetric triple given in Case 1 or Case 2 in Theorem 3.1, or the c-dual of a triple in those two cases. Therefore, by Theorem 3.1, G/H is causal symmetric if and only if $(\mathfrak{g}, \mathfrak{h}, \sigma)$ is either of type $K_\epsilon I$ or the c-dual of a triple of type $K_\epsilon I$. □

References

1. N. Bourbaki, Groupes et Algèbres de Lie, Chapitre 4, 5 et 6, Masson, Paris, 1981.
2. J. Faraut, Espaces symétriques ordonnés et algèbres de Volterra, *J. Math. Soc. Japan*, **43** (1991), 133–147.
3. S. Kaneyuki, On orbit structure of compactifications of parahermitian symmetric spaces, *Japan. J. Math.*, **13** (1987), 333-370.
4. S. Kaneyuki, On the causal structures of the Silov boundaries of symmetric bounded domains; Prospects in Complex Geometry, Lect. Notes in Math., 1468, pp. 127–159, Springer-Verlag, 1991.

5. S. Kaneyuki, Pseudo-hermitian symmetric spaces and Siegel domains over nondegenerate cones, *Hokkaido Math. J.*, **20** (1991), 213–239.
6. S. Kaneyuki, On the subalgebras $\mathfrak{g}_0$ and $\mathfrak{g}_{ev}$ of semisimple graded Lie algebras, *J. Math. Soc. Japan*, **45** (1993), 1–19.
7. S. Kaneyuki and H. Asano, Graded Lie algebras and generalized Jordan triple systems, *Nagoya Math. J.*, **112** (1988), 81–115.
8. G. 'Olafsson, Symmetric spaces of Hermitian type, *Differential Geometry and its Applications*, **1** (1991), 195–233.
9. G. I. Ol'shansky, Invariant cones in Lie algebras, Lie semigroups, and the holomorphic discrete series, *Functional Anal. Appl.*, **15** (1982), 275–285.
10. G. I. Ol'shansky, Convex cones in symmetric Lie algebras, Lie semigroups and invariant causal (order) structures on pseudo-Riemannian symmetric spaces, *Soviet Math. Dokl.*, **26** (1982), 97–101.
11. T. Oshima and J. Sekiguchi, Eigenvalues of invariant differential operators on an affine symmetric spaces, *Invent. Math.*, **57** (1980), 1–81.
12. T. Oshima and J. Sekiguchi, The restricted root system of a semisimple symmetric pair: Group Represnetations and Systems of Differential Equations, *Adv. Studeis in Pure Math.*, **4**, pp. 433–497, Kinokuniya, Tokyo and North-Holland, Amsterdam, 1984.
13. N. Tanaka, On non-degenerate real hypersurfaces, graded Lie algebras and Cartan connections, *Japan. J. Math.*, **2** (1976), 131–190.

Department of Mathematics
Sophia University
Tokyo 102, Japan
University of Massachusetts
Amherst, MA 01003
kaneyuki@mm.sophia.ac.jp

Received August 1995; corrected December 1995

Admissible Limit Sets of Discrete Groups On Symmetric Spaces of Rank One

*Adam Korányi**

1. Introduction

In recent years it has become increasingly clear that a number of important results about limit sets of discrete groups acting on real hyperbolic space $H^n_{\mathbf{R}}$ remain true also for the complex, quaternionic, and two-dimensional octonionic spaces, i.e. for all non-compact Riemannian symmetric spaces of rank one [Ka], [C], [CI], [BJ]. There are two essential elements in these generalizations. One is the description of the "non-tangential" or "conical" boundary approach domains of the Poincaré model of $H^n_{\mathbf{R}}$ as tubes of constant diameter around geodesic rays: We call these the *admissible domains.* The other is the intrinsic, in general non-isotropic, metric of the boundary and the way it is related to the admissible domains.

The purpose of the present paper is to describe these notions and the connections between them, with complete proofs, as simply as possible. Our goal is not to prove new results only to clarify things and to give a short self-contained exposition of the basic geometric facts.

We will describe the admissible domains in several equivalent ways not only in terms of the intrinsic geometry of the space, but also in terms of the ball model that all hyperbolic spaces have, so as to get something like a dictionary for generalizing the classical arguments about the Poincaré model of $H^n_{\mathbf{R}}$. We will show that with some stretching of the imagination, and of the metric, one can always think of the admissible domains as cones (even though one could also argue that they are more similar to paraboloids). This is a useful point of view, and it might justify talking about "conical limits" even in the general case. Still, in order to avoid possible misunderstandings we prefer to keep the old name "admissible limits" introduced in [K1],[K2]. (Rudin [R] and Kamiya [Ka] later used the name "K-limits", Corlette and Iozzi [C],[CI] the name "radial limits".)

In Section 1, we summarize some rather simple material about symmetric spaces of rank one from [H2]; this is all we assume in the sequel, we don't need the more sophisticated geometric results quoted in [C]. We note that there exists a still more elementary approach to our spaces in [CDKR], but here we will use only the standard approach. Section 2 contains our main

---

* Partially supported by the National Science Foundation and by a PSC-CUNY grant.

statements. Here we give complete proofs, mostly by repeating or specializing arguments from [K1],[K2],[KV],[K3],[K4]. In Section 3 we give an example of how the proof of a known result presents itself in our setup, and as a final remark we show how the admissible domains can be characterized in terms of the group-invariant Poisson kernel.

1. Notation and Preliminaries

We follow the notation, by now quite standard, of [H2]. G will denote a simple real Lie group with Lie algebra $\mathfrak{g}, \mathfrak{g} = \mathfrak{k} + \mathfrak{p}$ will be a Cartan decomposition corresponding to the involution θ, $\mathfrak{a}$ will be a maximal Abelian subspace of $\mathfrak{p}$ and it will be assumed that $\dim \mathfrak{a} = 1$. The $\mathfrak{a}$ - roots will be denoted $\pm\lambda$ and $\pm 2\lambda$, the corresponding root spaces $\mathfrak{g}^{\pm\lambda}$ and $\mathfrak{g}^{\pm 2\lambda}$, their dimensions p and q (p or q may be zero; this is the case of real hyperbolic space). $H \in \mathfrak{a}$ will be the element such that $\lambda(H) = 1$; $\mathfrak{a}^+$ will denote the set of non-negative multiples of H. We write $\mathfrak{n} = \mathfrak{g}^{\lambda} + \mathfrak{g}^{2\lambda}, \bar{\mathfrak{n}} = \mathfrak{g}^{-\lambda} + \mathfrak{g}^{-2\lambda}$ and define $\mathfrak{m}$ to be the centralizer of $\mathfrak{a}$ in $\mathfrak{k}$. A $\mathfrak{k}$ - invariant positive definite inner product on $\mathfrak{g}$ is given by $-B(X, \theta Y)$ we denote the corresponding norm by $| X |$.

The analytic subgroups of G corresponding to $\mathfrak{k}, \mathfrak{a}, \mathfrak{n}, \bar{\mathfrak{n}}$ are denoted by $K, A, N, \bar{N}$. The centralizer of $\mathfrak{a}$ in K is denoted M. We write A^+ for $exp\mathfrak{a}^+$. Clearly, A and M normalize N and $\bar{N}$ and the A-action (g^a stands for aga^{-1}) is given by

$$(1.1) \qquad (exp(X+Y))^{exptH} = exp(e^t X + e^{2t} Y), \quad (X \in \mathfrak{g}^{\lambda}, Y \in \mathfrak{g}^{2\lambda})$$

$$(1.2) \quad (exp(X+Y))^{exptH} = exp(e^{-t} X + e^{-2t} Y), \quad (X \in \mathfrak{g}^{-\lambda}, Y \in \mathfrak{g}^{-2\lambda})$$

(Note that $N, \bar{N}$ are nilpotent simply connected, hence exp is a diffeomorphism in them).

The rank one symmetric space X is uniquely determined by $\mathfrak{g}$ and can be identified with G/K. We write $o \in X$ for the point corresponding to the identity coset. X has a G-invariant Riemannian metric unique up to normalization. At o it is given, identifying the tangent space with $\mathfrak{p}$, by an arbitrarily fixed positive scalar multiple of $-B(X, \theta Y)$.

We have the Iwasawa decomposition $G = KAN$, meaning that every $g \in G$ can be uniquely written as a product

$$g = k(g)a(g)n(g)$$

with the factors depending continuously on g. One of the consequences is that $\bar{N} \times A \ni (\bar{n}, exptH) \mapsto \bar{n}exptH \cdot o$ is an analytic parametrization of X.

The Cartan decomposition $G = KA^+K$ amounts to saying that every $x \in X$ can be written as $x = kexptH \cdot o$, with $t \geq 0$ unique and $k \in K$ determined up to its right M-coset.

The Satake-Furstenberg boundary of X is the space $S = G/MAN \simeq K/M$; we denote its base point by u_0. By the Cartan decomposition S is a sphere. The topology of $X_\cup S$ is determined by the condition $k_\nu expt_\nu H \cdot o \rightarrow$ iff $k_\nu \cdot u_0 \rightarrow k \cdot u_0$ and $t_\nu \rightarrow \infty$ (so S is the "sphere at infinity").

There is a standard way to regard the unit ball $\mathcal{B}$ in $\mathbf{R}^n \simeq \mathfrak{p}$ as a model of X. The adjoint action of K on $\mathfrak{p}$ is transferred to $\mathbf{R}^n$ and the point $(\tanh t)k \cdot e_1$ (where $e_1 = (1,0,\dots,0) \in \mathbf{R}^n$) is made to correspond to $kexptH \cdot o$. The geometry of X and the G-action can be described entirely in terms of the model [T],[M],[CDKR], but here we don't need any of that. We don't really need the model at all, we will use it only to get a more intuitive picture of certain things and to make clear the connection with the classical results on real hyperbolic space [A],[B]. We note here that in the latter case the above construction gives two models, those of Cayley-Klein and of Poincaré, depending on whether we regard it as the case $\mathfrak{g}^{\pm 2\lambda} = 0$ or $\mathfrak{g}^{\pm\lambda} = 0$.

2. Admissible Limits

We consider the symmetric space X of rank one with an invariant Riemannian metric fixed on it in any way. For $\alpha > 0$ we denote by $B(o,\alpha)$ the open ball of center o and radius α in X. For any $g \in G$ and $\alpha > 0$ we define the admissible domain at $g \cdot u_0$, specializing the definition in [K2], p. 398, to the case of rank one, by

$$\mathcal{R}_\alpha(g \cdot u_o) = \{k(g)exptH \cdot x \mid x \in B(o,\alpha), t > 0\}.$$

This is clearly well defined and is the same thing as the set of all points in X whose distance from the geodesic ray $\{k(g)exptH \cdot o\}_{t \geq 0}$ is less than α. Given a set $E \subset X$, we say that $u \in S$ is an *admissible limit point* of E if there exists $\alpha > 0$ such that u is a limit point of $E \cap \mathcal{R}_\alpha(u)$.

The following crucial fact is a special case of Proposition 2.1 in [K2].

Proposition 2.1. *For all $g \in G$ and $\alpha > 0$ there exists $\alpha' > 0$ such that*

$$g \cdot \mathcal{R}_\alpha(u) \subset \mathcal{R}_{\alpha'}(g \cdot u)$$

for all $u \in S$.

For completeness we repeat the proof. We write u as $u = k \cdot u_0$, $(k \in K)$. By the Iwasawa decomposition, $gk = k(gk)an$ with a and n staying in compact sets $A_c \subset A$, $N_c \subset N$ as k varies. We have

$$g \cdot \mathcal{R}_\alpha(u) = gk \cdot \mathcal{R}_\alpha(u_0) = k(gk)an \cdot \mathcal{R}_\alpha(u_0),$$

and it suffices to show that $an \cdot \mathcal{R}_\alpha(u_0) \subset \mathcal{R}_{\alpha'}(u_0)$ for some $\alpha' > 0$. Now

$$\begin{aligned} an\mathcal{R}_\alpha(u_0) &= \{an \exp tH \cdot x \mid x \in B(o,\alpha),\, t > 0\} \\ &= \{\exp tH an^{\exp -tH} \cdot x \mid x \in B(o,\alpha),\, t > 0\}. \end{aligned}$$

As (1.1) shows, conjugation by $exp{-}tH$, $(t > 0)$ contracts n to the identity element. It follows that $an^{exp-tH} \cdot x$ stays in a compact set, hence in some ball $B(0, \alpha')$, finishing the proof.

This Proposition shows that if u is an admissible limit point of a set E, then $g \cdot u$ is an admissible limit point of $g \cdot E$. It also shows that if we consider the Riemannian space X as the primary object, and o (hence K) a matter of choice, the notion of admissible limit is independent of that choice.

We proceed to give reformulations of the definition of an admissible limit point replacing the sets $\mathcal{R}_\alpha(u)$ by other, equivalent families of sets. To be precise, for fixed $u \in S$, a family of subsets $\{\Gamma_\alpha(u)\}_{\alpha>0}$ of X will be said to be *equivalent* to the family $\{\mathcal{R}_\alpha(u)\}_{\alpha>0}$ if for every $\alpha > 0$ there exists $\alpha' > 0$ and $M > 0$ such that

$$\Gamma_\alpha(u) \cap CB(o, M) \subset \mathcal{R}_{\alpha'}(u) \tag{2.1}$$

and

$$\mathcal{R}_\alpha(u) \cap CB(o, M) \subset \Gamma_{\alpha'}(u) \tag{2.2}$$

where C denotes complement. The following statement is obvious:

Proposition 2.2. *In the definition of an admissible limit point the family* $\{\mathcal{R}_\alpha(u)\}$ *may be replaced by any equivalent family.*

It is an easy special case of the Bruhat decomposition ([H2], p. 403) that the orbit $\bar{N} \cdot u_0$ is all of S except one point and that $\bar{N}$ is simply transitive on this orbit. We can think of $\bar{N}$ as identified with $\bar{N} \cdot u_0$.

It is clear from (1.2) that it is possible to define (in may ways) a *homogeneous gauge* (cf. [KV]) on $\bar{N}$, i.e. a continuous non-negative function $\bar{n} \mapsto \mid \bar{n} \mid$ such that $\mid \bar{n}^{-1} \mid = \mid \bar{n} \mid$, $\mid \bar{n} \mid = 0$ iff $\bar{n} = e$, $\mid \bar{n}^{exptH} \mid = e^{-t} \mid \bar{n} \mid$, and $\mid \bar{n}\bar{n}' \mid \leq C(\mid \bar{n} \mid + \mid \bar{n}' \mid)$ with an absolute constant C (the last property is a consequence of the preceding ones). In fact, any symmetric compact neighborhood of e that is starlike with respect to the dilations can serve as the "unit ball" for such a gauge. Via the relation $\delta(\bar{n}, \bar{n}') = \mid \bar{n}'^{-1}\bar{n} \mid$, such a gauge is clearly the same thing as a semi-metric on $\bar{N}$ that is left-invariant under $\bar{N}$ and homogeneous in the sense that $\delta(\bar{n}^{\exp tH}, \bar{n}'^{exptH}) = e^{-t}\delta(\bar{n}, \bar{n}')$.

As in [K2], we now fix any gauge of the type just described and define for $\bar{n} \in \bar{N}$, $\alpha > 0$,

$$\Gamma_\alpha(\bar{n} \cdot u_0) = \{\bar{n}' \exp tH \cdot o \mid \delta(\bar{n}', \bar{n}) < \alpha e^{-t}, t > 0\}$$

Proposition 2.3.. *For every* $\bar{n} \in \bar{N}$, *the family* $\{\Gamma_\alpha(\bar{n} \cdot u_0)\}_{\alpha>0}$ *is equivalent to the family* $\{\mathcal{R}_\alpha(\bar{n} \cdot u_0)\}_{\alpha>0}$.

Proof. Clearly $\Gamma_\alpha(\bar{n} \cdot u_0) = \bar{n} \cdot \Gamma_\alpha(u_0)$. Therefore, in view of Proposition 2.1, we only have to prove (2.1) and (2.2) for $u = u_0$. We have

$$\begin{aligned}\Gamma_\alpha(u_0) &= \{exptH \cdot \bar{n}^{\exp -tH} \cdot o \mid |\bar{n}| < \alpha e^{-t}, t > 0\} \\ &= \{exptH \cdot \bar{n} \cdot o \mid \ \mid \bar{n} \mid < \alpha, t > 0\}.\end{aligned}$$

Now $\bar{n} \cdot o$, with $| \bar{n} |< \alpha$ is contained in some $B(o, \alpha')$, which proves (2.1). Conversely, given $\mathcal{R}_\alpha(u_0)$, every $x \in B(o, \alpha)$ can be written as $x = expt'H \cdot \bar{n}$ with $| t' |< T$ and $| \bar{n} |< \alpha'$ (for some $T, \alpha' > 0$). Hence every y in $\mathcal{R}_\alpha(u_0)$ can be written as $y = \exp tH \cdot \bar{n} \cdot o$ with $t > -T$, $| \bar{n} |< \alpha'$. When such a point is outside a sufficiently large compact set, i.e. a $B(o, M)$, we have $| t |> T$ and (2.2) follows, finishing the proof.

If we think of X as a half-space above $\bar{N}$, in which the height of the element $x = \bar{n} \exp tH \cdot o$ is measured by e^{-t}, then the sets $\Gamma_\alpha(u)$ begin to look very much like cones. This will be even more so if we use a semi-metric δ which is actually a metric.

Such metrics always exist, there are at least two natural ones: One is given by the gauge

$$(2.3) \qquad | \exp(X + Y) |= (c^2|X|^4 + 4c|Y|^2)^{1/4}$$

($X \in \mathfrak{g}^{-\lambda}$, $Y \in \mathfrak{g}^{-2\lambda}$; $c^{-1} = 4(p + 4q)$), as proved in [K4]. The other, more intrinsic one is a Carnot-Carathéodory (abbreviated CC) metric (also known as a "sub-Riemannian" or "contravariant Riemannian" metric), discussed in [K3] and [K4], which we now proceed to describe.

The left-invariant vector fields on N belonging to $\mathfrak{g}^{-\lambda}$ span a subspace of the tangent space at every point. (We temporarily exclude from consideration the real hyperbolic space interpreted via $\mathfrak{g}^{-\lambda} = 0$.) We call a C^1 curve $\gamma(s)$ in $\bar{N}$ *horizontal* if its tangent $\dot{\gamma}(s)$ belongs to this subspace for every s. Its arc length is defined by $\int | \dot{\gamma}(s) | ds$. The CC distance of two points in $\bar{N}$ is defined as the greatest lower bound of the lengths of all horizontal piecewise C^1 curves joining them. It is obvious from the definition that this distance is invariant under left translations by $\bar{N}$ and homogeneous under A; the only thing requiring proof is that it is always finite. For this what is crucial is the simple algebraic fact ([H2], p. 408) that $[\mathfrak{g}^{-\lambda}, \mathfrak{g}^{-\lambda}] = \mathfrak{g}^{-2\lambda}$. Knowing this, we could refer to the known general result that a CC distance is always finite when the brackets of the horizontal vector fields span the tangent space at all points, but in the present simple case we can also do everything explicitly:

By translation invariance and homogeneity it is enough to show that the horizontal curves of length one (or of some other fixed length) fill out a neighborhood of e. The curves $\gamma(s) = exp(sX + Y)$ for $X \in \mathfrak{g}^{-\lambda}$, $Y \in \mathfrak{g}^{-2\lambda}$ are clearly horizontal; they allow us, so to speak, to move in the $\mathfrak{g}^{-\lambda}$ direction. To see that we can also move in the $\mathfrak{g}^{-2\lambda}$ direction, let $Y \in \mathfrak{g}^{-2\lambda}$, and let $X_1, X_2 \in \mathfrak{g}^{-\lambda}$ be such that $[X_1, X_2] = Y$. (The lengths $| X_1 |$, $| X_2 |$ are controlled by $| Y |$, cf. [H2], p. 410.) Now the curve $\gamma(s) = \exp(X_1 \cos s + X_2 \sin s + \frac{1}{2}sY)$, $(0 \leq s \leq 2\pi)$ joins $\exp X_1$ to $\exp(X_1 + Y)$, and is horizontal by a standard formula ([H2], p. 105) giving the differential of the exponential map in any Lie group. It is clear that curves pieced together from curves of the two types just described fill out a neighborhood of e. The CC metric of $\bar{N}$ is studied in more precise detail in [K3] and [K4], but for our present purpose we already know enough.

In the case of real hyperbolic space, when it is interpreted as $\mathfrak{g}^{\pm 2\lambda} = 0$, both (2.3) and the CC construction give (up to constant) the ordinary Euclidean metric of $\mathfrak{g}^{-\lambda}$ transferred to $\bar{N}$ by the exponential map. When it is interpreted as $\mathfrak{g}^{\pm\lambda} = 0$, the CC metric is not defined, as a reasonable substitute for it we use, and still denote by δ, the square root of the Euclidean metric, which is also what (2.3) gives in this case.

We proceed to give another reformulation of admissible convergence, this time in K-invariant terms, adjusted to the ball model.

The tangent space of $S = G/MAN$ at u_0 is naturally identified with $\mathfrak{g}/\mathfrak{m}+\mathfrak{a}+\mathfrak{n}$. The cosets of $\mathfrak{g}^{-\lambda}$ determine a subspace of this, clearly invariant under the adjoint action of G; so we get a G-invariant subbundle of the tangent bundle, we still call the corresponding directions *horizontal*. Of course, after identifying $\bar{N} \cdot u_0$ with $\bar{N}$, these are still the same horizontal directions as before at the points of $\bar{N} \cdot u_0$. But now we define an inner product on the horizontal subspaces by propagating the metric at u_0 in a K-invariant way, i.e. we define, for any horizontal v at $k \cdot u_0$, $\|v\| = | X |$ where $X \in \mathfrak{g}^{-\lambda}$ corresponds to $k_*^{-1} \cdot v$ in the usual identification.

This construction gives rise to a CC metric on S which we denote by d. Again, we have to define d separately for the case $\mathfrak{g}^{\pm\lambda} = 0$: We define it as the square root of the standard Riemannian distance on S.

Now for any $x = k \exp tH \cdot o \in X (t \geq 0)$, we set $| x | = \tanh t$ and $x' = k \cdot u_0$. (These are the distance from o resp. the radial projection onto the boundary in the ball model.) We define, for $u \in S, \alpha > 0$

$$\mathcal{A}_\alpha(u) = \{x \in X \mid d(x', u) < \alpha(1 - | x |)^{1/2}\}$$

(this is a variant of the definition of $\Gamma'_\alpha(u)$ in [K1], p. 510) and we have:

Proposition 2.4. *For all $u \in S$, the families $\{\mathcal{A}_\alpha(u)\}_{\alpha>0}$ and $\{\mathcal{R}_\alpha(u)\}_{\alpha>0}$ are equivalent.*

Proof. By Proposition 2.1, and by the fact that K-invariance is built into the definition of $\mathcal{A}_\alpha(u)$, it suffices to consider $u = u_0$; it also suffices to show the equivalence of $\{\mathcal{A}_\alpha(u_0)\}$ with $\{\Gamma_\alpha(u_0)\}$.

For this we note that $(1 - |x|) \sim 2e^{-2t}$ for large t, and we show that on any compact neighborhood of u_0 contained in $\bar{N} \cdot u_0$ the distances δ and d are comparable. In fact, the horizontal curves for the two CC structures are the same, and, given the horizontal vector v at $\bar{n} \cdot u_0 = k(\bar{n}) \cdot u_0, v = \bar{n}_* X = k(\bar{n})_* a(\bar{n})_* n(\bar{n})_* X$ with some $X \in \mathfrak{g}^{-\lambda}$ the length of v in the first metric is $| X |$, in the second metric $| a(\bar{n})_* n(\bar{n})_* X | = | Ad(a(\bar{n})n(\bar{n}))X |$. These quantities are comparable on any compact subset of $\bar{N}$, implying our statement and finishing the proof.

Pursuing our arguments only a little further we could see that G acts conformally with respect to the CC metric (while, in general, it is not conformal for any Riemannian metric). Hence, in particular, the metrics δ and d on $\bar{N} \cdot u_0$ are conformally equivalent.

It is easy to see what the admissible domains look like in the ball model. Since $1-|x|$ is just the distance of x from the boundary, and since d (just like δ) is comparable with Euclidean distance d_E in the horizontal directions and with d_E^2 in the transversal directions, $\mathcal{A}_\alpha(u)$ touches the boundary as a paraboloid in the horizontal directions and as a cone in the transversal directions. The Cayley-Klein and Poincaré models of real hyperbolic space are the extreme cases where all directions are parabolic resp. conical.

More importantly, we can also think of $\mathcal{A}_\alpha(u)$ in all cases as a cone, if the metric on the boundary is taken to be d and the "height" of a point x over the boundary is measured by $(1-\mid x\mid)^{1/2}$ instead of the Euclidean height $1-\mid x\mid$. With this point of view, it becomes immediately clear (and easy to verify formally) that the "tent with tip at x",

$$T_\alpha(x) = \mathcal{C}\left(\cup_{x\notin\mathcal{A}_\alpha(u)}\mathcal{A}_\alpha(u)\right)$$

is just an inverted cone. In particular, the "base" of the tent,

$$TB_\alpha(x) = \{u \in S \mid x \in \mathcal{A}_\alpha(u)\}$$

is the ball $B_d(x', \alpha(1-\mid x\mid)^{1/2})$ with center x' and radius $\alpha(1-\mid x\mid)^{1/2}$ measured in the metric d.

From this the following statement is obvious:

Proposition 2.5. *A point $u \in S$ is an admissible limit point of a set $E \subset X$ iff, for some $\alpha > 0, u$ is contained in infinitely many balls $B_d(x'$, $\alpha(1-\mid x\mid)^{1/2}), x \in E$.*

3. An Example and Further Remarks.

As an example, we want to reformulate in our language the proof of a known result about the admissible limit set.

We consider a discrete subgroup Γ of G and list its elements as $\gamma_1, \gamma_2, \ldots$ The *critical exponent* [C] of Γ is the number

$$\delta_\Gamma = \inf\{\delta > 0 \mid \sum_\nu (1-\mid x_\nu\mid)^{\delta/2} < \infty\}$$

where $x_\nu = \gamma_\nu \cdot o$. The *admissible limit set* $L_\mathcal{A}(\Gamma)$ of Γ is the set of all admissible limit points of the sequence $\{x_\nu\}$.

From Proposition 2.5, the following statement is clear.

Proposition 3.1. *If $\{\alpha_\nu\}$ is a sequence of numbers such that $\alpha_\nu \to \infty$, then $L_\mathcal{A}(\Gamma)$ is covered infinitely many times by the family of balls $\{B_d(x'_\nu, r_\nu)\}$, where $r_\nu = \alpha_\nu(1-\mid x_\nu\mid)^{1/2}$.*

What we want to re-prove is the easy half of Theorem 5.2 from [C]. (Actually, the notion of "radial" limit used in [C] is more restrictive, but, as mentioned in [CI], the proof in [C] is also valid for a more general notion of "radial" limit which is equivalent to our admissible limit.)

Proposition 3.2. *(Corlette.) The Hausdorff dimension of $L_A(\Gamma)$ with respect to d does not exceed δ_Γ.*

Proof. By the definition of Hausdorff dimension it suffices to show the following: For every $\delta > \delta_\Gamma, \eta > 0$, $\epsilon > 0$, there exists a covering of L_A by a family of balls of radius $r_\nu < \epsilon$ such that $\sum r_\nu^\delta < \eta$.

As above, we denote $x_\nu = \gamma_\nu \cdot o$, and we set $\alpha_\nu = -\log(1- \mid x_\nu \mid)$. We write $r_\nu = \alpha_\nu(1- \mid x_\nu \mid)^{1/2}$ and consider the balls $B_\nu = B_d(x'_\nu, r_\nu)$. By definition of δ_Γ we have $\sum r_\nu^\delta < \infty$. Hence there exists a number N such that $r_\nu < \epsilon$ for $\nu > N$ and

$$\sum_{\nu > N} r_\nu^\delta < \eta.$$

By Proposition 3.1 the balls B_ν with $\nu > N$ form a covering of L_A, finishing the proof.

The converse inequality has been proved for increasingly general classes of subgroups Γ in [C],[CI] and [BJ]. We are not going to deal with this question here, since we only wanted to give one simple example.

We shall finish with one more characterization of admissible domains, this time in terms of the Poisson kernel of X. It was in connection with Poisson integrals that admissible limits were originally introduced in [K1] and [K2].

There are two forms in which the Poisson kernel appears, as a function P on $X \times S$ or as a function $\mathcal{P}$ on $X \times \bar{N}$; it is characterized by the property that for any continuous function f on S, the function F defined on X by

$$F(x) = \int_S P(x,u) f(u) d\mu(u) = \int_{\bar{N}} \mathcal{P}(x, \bar{n}) f(\bar{n} \cdot u_0) d\bar{n}$$

solves the Dirichlet problem for the G-invariant Laplacian; here μ is normalized K-invariant measure on S and $d\bar{n}$ is Haar measure on $\bar{N}$. It is clear that the two kernels differ only by a continuous positive numerical factor (namely $d\mu(\bar{n} \cdot u_0)/d\bar{n}$).

By Helgason's explicit formula [H1], p. 65, we have

$$\mathcal{P}(\exp tH \cdot o, \bar{n}) = \left(\frac{e^{2t}}{(1 + c \mid e^t X \mid^2)^2 + 4c \mid e^{2t} Y \mid^2} \right)^{p/2+q}$$

for $\bar{n} = \exp(X + Y)$, $X \in \mathfrak{g}^{-\lambda}$, $Y \in \mathfrak{g}^{-2\lambda}$. A very simple computation based on $\bar{N}$ - invariance and on (2.3) now shows that the family

$$\Gamma'_\alpha(\bar{n} \cdot u_0) = \left\{ \bar{n}' \exp tH \cdot o \mid e^{-(p+2q)t} \mathcal{P}(\bar{n}' \exp tH \cdot o, \bar{n} \cdot u_0) > \alpha \right\}$$

is equivalent to the family $\{\Gamma_\alpha(\bar{n} \cdot u_0)\}_{\alpha > 0}$ considered in Section 2, and therefore can be used to describe admissible convergence. Since P is just a positive multiple of $\mathcal{P}$, one sees at once that, for all $u \in S$,

$$\Gamma''_\alpha(u) = \left\{ x \in X \mid (1- \mid x \mid)^{-(\frac{p}{2}+q)} P(x,u) > \alpha \right\}$$

as $\alpha > 0$, is yet another equivalent family characterizing admissible convergence.

References

[A] L. Ahlfors, *Möbius transformations in several dimensions,* School of Mathematics University of Minnesota, 1981.

[B] A. F. Beardon, *The geometry of discrete groups,* Springer, New York, 1983.

[BJ] C. Bishop and P. Jones, Hausdorff dimension and Kleinian groups, to appear.

[C] K. Corlette, Hausdorff dimension of limit sets I, *Invent. Math.* **102** (1990), 521-542.

[CI] K. Corlette and A. Iozzi, Limit sets of isometry groups of exotic hyperbolic spaces, to appear.

[CDKR] M. Cowling, A. H. Dooley, A. Korányi and F. Ricci, An approach to symmetric spaces of rank one via groups of Heisenberg type, to appear.

[H1] S. Helgason, A duality for symmetric spaces with applications to group representations, *Adv. Math.* **5** (1970), 1-154.

[H2] S. Helgason, *Differential geometry, Lie groups, and symmetric spaces,* Academic Press, New York, 1978.

[Ka] S. Kamiya, Discrete groups of convergence type of $U(1, n; c)$, *Hiroshima Math. J.,* **21** (1991), 1-21.

[K1] A. Korányi, Harmonic functions on Hermitian hyperbolic space, *Trans. Amer. Math. Soc.* **140** (1969), 509-516.

[K2] A. Korányi, Boundary behaviour of Poisson integrals on symmetric spaces, *Trans. Amer. Math. Soc.* **140** (1969), 393-409.

[K3] A. Korányi, Geometric aspects of analysis on the Heisenberg group, in *Topics in Modern Harmonic Analysis,* pp. 209-258, Ist. Naz. Alta Matematica, Roma, 1983.

[K4] A. Korányi, Geometric properties of Heisenberg-type groups, *Adv. Math.* **56** (1985), 28-38.

[KV] A. Korányi and S. Vági, Singular integrals on homogeneous spaces and some problems of classical analysis, *Ann. Scuola Norm. Sup. Pisa, Classe di Scienze* **25** (1971), 575-648.

[M1] G.D. Mostow, *Strong rigidity of locally symmetric spaces,* Princeton Univ. Press, 1973.

[M2] G. D. Mostow, A remark on quasiconformal mappings on Carnot groups, *Michigan Math. J.* **41** (1994), 31-37.

[R] W. Rudin, *Function theory in the unit ball in C^n*, Springer-Verlag, New York, 1980.

[T] M. Takeuchi, Cell decompositions and Morse inequalities on certain symmetric spaces, *J. Fac. Sci. Univ. Tokyo, Sec. I,* **12** (1965), 81-192.

Adam Korányi
Dept. of Mathematics
H.H. Lehman College
Bronx, NY 10468

Received October 1995

D'Atri Spaces

Oldřich Kowalski, Friedbert Prüfer and Lieven Vanhecke*

1. Introduction

Riemannian symmetric spaces play an important role in many fields of mathematics and physics. They have been studied by several generations of mathematicians and from different viewpoints. Their classification is well-known. Usually they are characterized as Riemannian manifolds whose geodesic symmetries (that is, geodesic reflections with respect to all points) are globally defined isometries. Hence, these geodesic symmetries are also volume-preserving. This observation led D'Atri and Nickerson to the study of Riemannian and pseudo-Riemannian manifolds all of whose (local) geodesic symmetries are volume-preserving (up to sign) or equivalently, which are divergence-preserving. They are introduced in [DN1], [DN2], [D] where the first non-symmetric examples, namely all naturally reductive homogeneous spaces, are found and where various other characterizations are given. Following [VW1] such spaces are called *D'Atri spaces.*

Since then, many geometrical properties of such spaces have been discovered and a broad spectrum of examples has been found. This shows that the theory of D'Atri spaces is connected to the study of several classes of interesting Riemannian manifolds and a vast variety of concepts. We mention some of the most important examples : naturally reductive homogeneous spaces, Riemannian manifolds all of whose geodesics are orbits of one-parameter subgroups of isometries (called g.o. spaces), commutative and probabilistic commutative spaces, generalized Heisenberg groups, harmonic spaces and weakly symmetric spaces (called also ray symmetric spaces).

The main purpose of this paper is to give a survey of the results on D'Atri spaces together with a list of references for further details and information but we do not claim that both are exhaustive.

We start with some preliminaries in Section 2 and we present a list of various characterizations of D'Atri spaces in Section 3. Section 4 is entirely devoted to a brief information about special classes of D'Atri spaces. In

* Supported by grant GA ČR 201/93/0469.

Section 5 we then consider some inclusion relations between these classes. Their classification and some special aspects for low-dimensional D'Atri spaces are discussed in Section 6. The problem of homogeneity and the consequences of some curvature restrictions for D'Atri spaces are the main content of Section 7. Then, in Section 8, we introduce the notion of a D'Atri space of type k as a natural analog of a concept from the theory of harmonic spaces. Section 9 is devoted to a short discussion about D'Atri spaces from the point of view of symplectic, complex, contact and quaternionic geometry. The volumes of tubes about curves in D'Atri spaces are studied in Section 10. The main result here is related to a well-known result of H. Weyl for tubes in spaces of constant curvature. Finally, in Section 11 we treat some generalizations of the notion of a D'Atri space, in particular to pseudo-Riemannian and affine manifolds.

In the text we included several open problems and we hope that these problems as well as the whole survey will stimulate some fruitful further research in this direction.

2. Preliminaries

2.1. Let (M, g) be an n-dimensional connected Riemannian manifold. In what follows we shall suppose (M, g) to be of class C^ω although in several considerations class C^∞ will be sufficient. Furthermore, we shall mostly work in a normal neighborhood of a point $m \in M$.

Let T_mM denote the tangent space of M at m and let $v \in T_mM$. Further, let γ_v denote the maximal geodesic with initial conditions $\gamma_v(0) = m, \gamma_v'(0) = v$. Then the exponential map $\exp_m$ is determined by $v \mapsto \gamma_v(1)$ in a normal neighborhood U_0 of the origin in T_mM. Furthermore, let $\{e_i, i = 1, ..., n\}$ be an orthonormal basis of T_mM. Then the normal coordinates $(x_1, ..., x_n)$ are defined in $U_m = \exp_m(U_0)$ by

$$x_i(\exp_m \sum_{j=1}^{n} a_i e_i) = a_i$$

and the distance $r = d(m, p)$ from a point $p(x_1, ..., x_n)$ to m is given by

$$r^2 = \sum_{i=1}^{n} x_i^2.$$

In what follows we put $\Omega(m, p) = \frac{1}{2}d^2(m, p)$.

2.2. Since *Jacobi vector fields* play an important role in the work on D'Atri spaces we include here some details and refer to [V5] for further information and references. Let γ be a geodesic $r \mapsto \exp_m(r\xi)$ through m tangent to a fixed unit vector $\xi \in T_mM$. A vector field Y along γ is said to be a Jacobi vector field if and only if

$$Y'' + R_{\gamma' Y}\gamma' = 0. \tag{2.1}$$

Here R_{XY} denotes the curvature operator of (M, g) defined by

$$R_{XY} = \nabla_{[X,Y]} - [\nabla_X, \nabla_Y]$$

for all tangent vector fields X, Y and ∇ is the Levi Civita connection. The Jacobi equation (2.1) is a second order differential equation.

So, let $\{e_i, i = 1, ..., n\}$ be an orthonormal basis of T_mM such that $\xi = e_1$ and consider the $n-1$ Jacobi vector fields $Y_a, a = 2, ..., n$, along γ determined by the initial conditions

$$Y_a(0) = 0, \quad Y_a'(0) = e_a, \quad a = 2, ..., n. \tag{2.2}$$

For sufficiently small r, these vectors $Y_a(r)$ determine a basis for the space $\{\gamma'(r)\}^\perp \subset T_{\gamma(r)}M$. Furthermore, let $\{E_i, i = 1, ..., n\}$ be the orthonormal frame field along γ obtained by parallel translation of the basis $\{e_i, i = 1, ..., n\}$ along γ and put

$$Y_a(r) = (AE_a)(r), \qquad a = 2, ..., n.$$

Then $r \mapsto A(r)$ is an endomorphism-valued function. Each $A(r)$ is an endomorphism of the corresponding space $\{\gamma'(r)\}^\perp$ and, as we will often do, all these spaces may be identified via parallel translation along γ. We easily get

$$A'' + R \circ A = 0$$

with initial conditions

$$A(0) = 0 \quad , \qquad A'(0) = I.$$

Here, R denotes the *Jacobi operator* defined by $X \mapsto R_{\gamma' X}\gamma'$. Moreover, by using the frame field $\{\frac{\partial}{\partial x_i}, i = 1, ..., n\}$ in the given normal coordinate neighborhood, we have

$$Y_a(r) = r\frac{\partial}{\partial x_a}(\gamma(r)) \quad , \quad a = 2, ..., n,$$

for sufficiently small $r > 0$.

Now we consider the *metric in normal coordinates*, that is,

$$g_{ij} = g(\frac{\partial}{\partial x_i}, \frac{\partial}{\partial x_j}).$$

Then we have, for $p = \exp_m(r\xi)$, a matrix equality

$$g(p) = (g_{ij})(p) = \begin{pmatrix} 1 & 0 \\ 0 & \frac{1}{r^2}({}^tAA)(r) \end{pmatrix}$$

where the matrix of ${}^tAA$ is taken with respect to the parallel frame $\{E_i, i = 2, ..., n\}$. Moreover, for the *normal volume density function*

$$\omega(m,p) = \omega_m(p) = (\det(g_{ij}))^{1/2}(p)$$

we get

$$\omega_m(p) = r^{1-n}(\det A)(p).$$

Next, let $S^{n-1}(r)$ denote the $(n-1)$-dimensional sphere with sufficiently small radius $r > 0$ and center o in T_mM and consider the *geodesic sphere* $G_m(r) = \exp_m S^{n-1}(r)$ with center m and radius r. The extrinsic geometry of the hypersurface $G_m(r)$ is determined by the shape operator T_m defined at $p = \exp_m(r\xi) \in G_m(r)$ by

$$T_m(p)X = (\nabla_X \frac{\partial}{\partial r})(p)$$

for $X \in T_pG_m(r)$. Using the endormorphisms $A(r)$ we obtain

$$T_m(p) = (A'A^{-1})(r).$$

Here $r > 0$, that is, $p \neq m$. The *mean curvature* h_m of the geodesic sphere $G_m(r)$ at p is given by

$$h(m,p) = h_m(p) = \operatorname{tr} T_m(p).$$

In what follows, we put

$$C(m,p) = rT_m(p).$$

Here $C(m,p)$ can be defined also for $p = m$ by continuity.

2.3. Many of the results contained in the next sections may be formulated in terms of two-point functions. By a *two-point function* on M we mean a smooth function $F(x, y)$ defined on an open neighborhood $U \subset M \times M$ of the diagonal $\Delta(M \times M)$ such that, whenever $(x, y) \in U$, there is a unique shortest geodesic joining x to y. Many "geometric" two-point functions occur in a natural way in Riemannian geometry : the distance function Ω, the normal volume density function ω and the multiple of the mean curvature function $rh = \operatorname{tr} C$ considered above are examples. On the other hand, $h(m, p)$ itself can be considered as a two-point function *with singularities* on the diagonal $\Delta(M \times M)$. See [KV7] for more examples and further considerations. Power series expansions are a useful tool in this context. (See also [V1], [V5] for examples.) So, let $F(m, p)$ be a two-point function on (M, g). Then F determines an infinite sequence $\{F^{(k)}\}$ of polynomial functions on TM (or, equivalently, an infinite sequence of symmetric tensor fields on M) of degree $k = 0, 1, 2, \ldots$ respectively given by

$$F^{(k)}(\xi, \ldots, \xi) = \frac{d^k}{dt^k} F(m, \exp_m(t\xi))_{|t=0}.$$

For an analytic two-point function $F(m, p)$, F is uniquely determined by

$$F(m, \exp_m(t\xi)) = \sum_{k=o}^{\infty} \frac{1}{k!} F^{(k)}(\xi, \ldots, \xi) t^k.$$

Here we put $F^{(0)}(m) = F(m, m)$. We remark that our notation coincides with that from [KV7] up to a normalization.

If an analytic two-point function $F(m, p)$ is *symmetric*, that is, $F(m, p) = F(p, m)$, then the associated polynomial functions $F^{(k)}(\xi, \ldots, \xi)$ satisfy the recursion formula

$$2F^{(k)}(\xi, \ldots, \xi) = \sum_{\ell=1}^{k} (-1)^{k-\ell} \binom{k}{\ell} \nabla^{\ell}_{\xi \ldots \xi} F^{(k-\ell)}(\xi, \ldots, \xi)$$

for all *odd* k. Moreover, $F(m, p)$ is said to be *left centrally symmetric* (respectively *right centrally symmetric*) on M if $F(m, \exp_m(t\xi)) = F(m, \exp_m(-t\xi))$ (respectively $F(\exp_p(t\xi), p) = F(\exp_p(-t\xi), p)$ for any point m (respectively p) of M, any unit vector ξ of $T_m M$ (respectively $T_p M$) and any sufficiently small t. In the analytic case we have the following criterion [KV7] : *An analytic two-point function is symmetric and centrally symmetric if and only if it is a function depending only on the distance along any geodesic.* Note that this does not hold in the C^{∞} case (see [KV7] for a counterexample). In terms of the associated polynomial functions we have : *An analytic function $F(m, p)$ is symmetric and centrally* *symmetric if and only if we have, for all* $\xi \in TM$,

$$F^{(k)}(\xi, ..., \xi) = 0 \text{ for all odd } k,$$

and

$$\nabla_\xi F^{(k)}(\xi, ..., \xi) = 0 \text{ for all even } k.$$

Note that the second series of conditions means that all the corresponding symmetric tensor fields $F^{(2k)}$ are *Killing tensor fields.*

We mention, as is well-known, that the normal volume density function $\omega(m,p)$ is symmetric (see for example [BE], [V5]).

2.4. Now, we consider the *endomorphism-valued* two-point function $C(m,p)$ defined above. The power series expansion

$$C(m, \exp_m(t\xi)) = \sum_{k=0}^{\infty} \frac{1}{k!} C^{(k)}(\xi, ..., \xi) t^k$$

plays an important role in the theory of harmonic spaces and D'Atri spaces. Here, the $C^k(\xi, ..., \xi)$ are endomorphisms of $\{\xi\}^\perp \subset T_m M$. Ledger determined a recursion formula for the coefficients $C^k(\xi, ..., \xi)$ (see [BE], [CV], [RWW], [V5]) which makes it possible to express the $C^{(k)}$ explicitly in function of the curvature operator R and its covariant derivatives. This *recursion formula of Ledger* is given by

$$(k-1)C^{(k)}(0) = -k(k-1)R^{(k-2)}(0) - \sum_{\ell=0}^{k} \binom{k}{\ell} C^{(\ell)}(0) C^{(k-\ell)}(0)$$

with $C(0) = I, C'(0) = 0$ and where $C^{(k)}(0) = C^{(k)}(\xi, ..., \xi), R^{(k)}(0) = \nabla^k_{\xi...\xi} R_\xi.\xi$. In this context we note that the conditions

$$\text{tr}C^{(2k+1)}(\xi, ..., \xi) = 0 \qquad , k = 0, 1, 2, ...$$

respectively

$$\text{tr}C^{(2k)}(\xi, ..., \xi) = c_{2k} = \text{const.} \qquad , k = 0, 1, 2...$$

are called the *Ledger conditions of odd order*, respectively *even order*, at the point m.

Both series of Ledger conditions are obviously satisfied if and only if each small geodesic sphere has constant mean curvature.

2.5. Next, we turn to another important tool, namely the *mean-value operators*. Let $m \in M$ and let U be a normal neighborhood of m. For sufficiently small t the *first mean-value operator* M is defined by

$$M_t[\varphi](m) = \frac{1}{m_t} \int_{G_m(t)} \varphi do$$

where $m_t = \int_{G_m(t)} do$ is the volume of the geodesic sphere $G_m(t)$, do is the volume element induced by the Riemannian metric of $G_m(t)$ and φ is an L^2-function defined on U. The corresponding *unnormed* operator is given by

$$\hat{M}_t[\varphi](m) = \int_{G_m(t)} \varphi do.$$

The *second mean-value operator* L, which will play a fundamental role in the definition of probabilistic commutative spaces, is defined by

$$L_t[\varphi](m) = \frac{1}{\omega_{n-1}} \int_{S_m^{n-1}(1)} \varphi(\exp_m(t\xi))d\xi$$

where

$$\omega_{n-1} = \frac{n\pi^{n/2}}{\Gamma(\frac{n}{2}+1)}$$

is the volume of $S^{n-1}(1)$ in $\mathbb{R}^n = T_m M$ and $d\xi$ denotes the volume element of $S^{n-1}(1)$. Note that $M_t, \hat{M}_t, L_t$ are defined only locally, in general, and for small $t \geq 0$.

These mean-value operators play an interesting role via the related *generalized Pizetti formula*. Let φ be an analytic function defined in a neighborhood of m. Then we have

$$L_t[\varphi](m) = \varphi(m) + \sum_{k=1}^{\infty} \frac{\tilde{\Delta}^{(k)}[\varphi](m)}{2^k k! n(n+2)...(n+2k-2)} t^{2k}$$

and similarly

$$\hat{M}_t[\varphi](m) = 2\pi^{n/2} t^{n-1} \sum_{k=0}^{\infty} (\frac{t}{2})^{2k} \frac{1}{k!\Gamma(\frac{n}{2}+k)} \tilde{\Delta}^{(k)}[\varphi\omega_m](m).$$

Moreover, we have (see [KO2])

$$L_t[\varphi\omega_m](m) = \frac{1}{\omega_{n-1}t^{n-1}} \hat{M}_t[\varphi](m).$$

Here the operators $\tilde{\Delta}^{(k)}$, called *Euclidean Laplacians of higher order*, are globally defined differential operators of order $2k$ given by (see [GW] and [KOZ] for further details and also for further considerations)

$$\tilde{\Delta}^{(k)}[\varphi](m) = \frac{1}{(2k)!} \sum_{i,\dots i_{2k}=1}^{n} \sum_{\sigma} \{\delta_{i_{\sigma(1)} i_{\sigma(2)}} \cdots \delta_{i_{\sigma(2k-1)} i_{\sigma(2k)}} \nabla^{2k}_{i_1 \dots i_{2k}} \varphi\}(m)$$

where the summation is made over all permutations σ of the set $\{1, \dots, 2k\}$, δ_{ij} denotes the Kronecker symbol and where we have put $\nabla^k_{E_{i_1} \dots E_{i_k}} = \nabla^k_{i_1 \dots i_k}$ for any local orthonormal frame field $(E_1, ..., E_n)$. The right-hand side of this formula is indeed independent of the choice of such a frame field. For small k, it is not difficult to express $\tilde{\Delta}^{(k)}$ in terms of Δ^k and terms of lower order, where Δ is the Laplacian of (M, g). The explicit expressions for $k = 1, 2, 3$ have been computed in [GW] and for $k = 4$ in [KOZ]. For example, we have

$$\tilde{\Delta}^{(1)}[\varphi](m) = (\Delta\varphi)(m),$$
$$\tilde{\Delta}^{2}[\varphi](m) = (\Delta^2\varphi)(m) + \frac{1}{3} g^{ij}(m)(\nabla_i \tau)(m)(\nabla_i \varphi)(m) + \frac{2}{3} \rho^{ij}(m)(\nabla^2_{ij}\varphi)(m)$$

where ρ denotes the Ricci tensor of type (0,2) and τ the scalar curvature of (M, g).

We note that these Euclidean Laplacians may also be defined by

$$\tilde{\Delta}^{(k)}[\varphi](m) = [\sum_{i=1}^{n} (\frac{\partial}{\partial x_i})^2]^k [\varphi](m) \qquad , k = 1, 2, ...,$$

with respect to any *normal coordinates* $x_1, ..., x_n$ in U centered at m. Here the exponent k on the right-hand side indicates an iteration of a local differential operator, namely the Euclidean Laplacian in $T_m M = \mathbb{R}^n$.

3. Characterizations of D'Atri spaces

In this section we start our considerations about D'Atri spaces. We give the definition and several other characterizations. Let (M, g) be a Riemannian manifold and $m \in M$. Further, let s_m denote the local *geodesic symmetry* centered at m, that is

$$s_m : p = \exp_m(t\xi) \mapsto s_m(p) = \exp_m(-t\xi)$$

where $\xi \in T_m M$ is a unit vector. Such a map is always defined in a *symmetric* normal neighborhood of m. In [DN1], [DN2] the authors started to consider (M, g) such that each s_m is *volume-preserving* (up to sign), or equivalently, each s_m is *divergence-preserving.* This led in [VW1], [VW2] to the following

3.1. Definition. (M, g) is called a *D'Atri space* if each local geodesic symmetry preserves the Riemannian volume element (up to sign).

Note that we consider here only positive definite metrics. An extension to the pseudo-Riemannian case will be considered in Section 11.

3.2. In [DN1] several other characterizations are derived which can be expressed by using various two-point functions. In particular we have : *Let (M, g) be an analytic Riemannian manifold. Then the following statements are equivalent :*

(i) *(M, g) is a D'Atri space;*
(ii) *the normal volume density function $\omega(m,p)$ is left centrally symmetric for all $m \in M$;*
(iii) *$\Delta\Omega$ is left centrally symmetric;*
(iv) tr C *is left centrally symmetric.*

Note that $\Delta\Omega$ is indeed a two-point function (see [KV7] for the precise definition).

In [KV7] it is proved that (iii) and (iv) may be replaced by
(iii)' $\Delta\Omega$ *is symmetric or right centrally symmetric;*
(iv)' tr C *is symmetric or right centrally symmetric.*
See also [VW1].

3.3. From 3.2(iv) we reprove another result from [DN1] : *An analytic (M, g) is a D'Atri space if and only if the Ledger conditions of odd order are satisfied at all points $m \in M$.*

3.4. Here it is worthwhile to mention that the first Ledger condition of odd order is equivalent to

$$\nabla_\xi \rho_{\xi\xi} = 0,$$

that is, the Ricci tensor ρ is *cyclic-parallel* or equivalently, ρ is a *Killing tensor.* A smooth Riemannian manifold satisfying this condition is analytic with respect to any atlas of normal coordinates [KA], [SZ4]. (See [DK] for the corresponding result on Einstein spaces.) Since this condition follows from all the characterizations given above, and also from the ones we shall give later in this section, we can always replace the analyticity assumption by the smoothness of class C^∞. Note that the above condition also implies that *the scalar curvature is constant.*

3.5. Consider the two-point function $\mathrm{tr}(g^{ij})(m,p)$ where $(g^{ij})(m,p)$ is the matrix of the contravariant components of g at the point p expressed with respect to any normal coordinates centered at m. It is known that $\mathrm{tr}(g^{ij})$ is symmetric. The following characterization is proved in [KV7] : *(M,g) is a D'Atri space if and only if the symmetric two-point function $\mathrm{tr}(g^{ij})$ is centrally symmetric.*

3.6. Moreover, (iv)' means that *(M,g) is a D'Atri space if and only if $h(m,p) = h(p,m)$ for $p \neq m$*, that is, the mean curvature of a geodesic sphere h is a symmetric two-point function (singular on the diagonal $\Delta(M \times M)$).

3.7. This characterization using an extrinsic property for the geodesic spheres may be, for $n > 2$, replaced by the following intrinsic one. Let $\tau(m,p) = \tau_m(p)$ be the scalar curvature of the geodesic sphere $G_m(r)$ at $p = \exp_m(r\xi)$. Then, *(M,g) is a D'Atri space if and only if $\tau(m,p)$ is a symmetric two-point function, or equivalently, if it is centrally symmetric outside the singular set $\Delta(M \times M)$.* This result may be proved in a similar way as in [CV, Theorem 5.4].

3.8. More recently, some characterizations of different kind were given by using commutation properties of special operators. The following result has been proved in [P1] for manifolds with positive injectivity radius, but it is not difficult to generalize it to any (M,g) [P2] : *A Riemannian manifold is a D'Atri space if and only if the Laplacian for smooth functions commutes with each operator M_t or $\hat{M}_t$ (locally and for small $t \geq 0$).*

3.9. Next, let $\tilde{\Delta}_\omega^{(k)}$ be the differential operator $\square_\omega^k$ defined in [SZ4] (see also 8.6). Then we have at once ([SZ2], [SZ4]) : *(M,g) is a D'Atri space if and only if the Laplacian commutes with the density operators $\tilde{\Delta}_\omega^{(k)}$, $k \in \mathbb{N}$.*

3.10. Finally, consider the volume density function $\omega(m,p)$ and put

$$\omega(m,p) = \sum_{\ell=0}^{\infty} \frac{1}{\ell!} A_\ell(\xi, \dots, \xi) r^\ell, \quad p = \exp_m(r\xi).$$

It is well-known that $A_\ell = A_\ell(\xi, \dots, \xi)$ are certain polynomials of the curvature tensor and its covariant derivatives at each point m. (See, for example, [GRV1] and the references therein.) From 3.2(ii) we see at once that (M,g)

is a D'Atri space if and only if $A_{2k+1} = 0$ *for all* k *on* M. (These conditions are similar to the Ledger conditions of odd order and they are in fact equivalent to them for any finite sequence of indices $k = 0, 1, ..., k_0$.)

Next, let $\tilde{A}_\ell$ denote the symmetric tensor fields corresponding to the polynomial functions A_ℓ. Using again 3.2(ii), the symmetry of $\omega(m,p)$ and the recursion formula for symmetric two-point functions from Section 2, we obtain : (M, g) *is a D'Atri space if and only if the tensor fields* $\tilde{A}_{2k}$ *are Killing tensor fields.* (See also [H2] for another proof.)

4. Special classes of D'Atri spaces

Locally symmetric Riemannian spaces may be characterized by the property that all geodesic symmetries are isometries. Hence, such spaces are trivially D'Atri spaces. The purpose of the work of D'Atri and Nickerson was to provide some non-trivial classes of examples, and this is also the aim of this section.

4.1. The first class is formed by the *naturally reductive Riemannian manifolds.* Let G be a connected Lie group with Lie algebra $\mathfrak{g}$ and K a closed Lie subgroup of G with Lie algebra $\mathfrak{k}$. Further, let M be the space of left cosets G/K and assume that G acts effectively on M on the left. The canonical projection $\pi : G \to M$ maps $e \in G$ to $o = \pi(e) \in M$ and induces the map $\pi_* : T_eG \to T_oM$. Assume that the homogeneous space M is reductive, that is, there exists a vector space decomposition

$$\mathfrak{g} = \mathfrak{k} \oplus \mathfrak{p}, \tag{4.1}$$

such that

$$Ad(K)\mathfrak{p} \subseteq \mathfrak{p}. \tag{4.2}$$

Then, the canonical connection $\tilde{\nabla}$ on M is defined by $(\tilde{\nabla}_{X^*}Y)_o = [X^*, Y]_o$ for any vector field Y on M and any $X \in \mathfrak{p}$, where the global vector field X^* on M is generated by the one-parameter group $\{\mathrm{Exp}\ tX\}$. (We identify $\mathfrak{p}$ and T_oM by π_*.)

G/K, equipped with a G-invariant metric $<,>$ is called a *naturally reductive homogeneous space* if there exists a decomposition (4.1) satisfying (4.2) and

$$< [X,Y]_\mathfrak{p}, Z > + < [X,Z]_\mathfrak{p}, Y > = 0 \quad , \quad X, Y, Z \in \mathfrak{p} \tag{4.3}$$

where $<,>$ denotes here the inner product induced on $\mathfrak{p}$ (see [KN, Chapter 10]).

(4.3) is equivalent to the following more geometrical property : For every vector $X \in \mathfrak{p} \simeq T_oM$, the orbit $(\text{Exp}\, tX)(o)$ is a geodesic. From this we see that every geodesic on $(G/K, <,>)$ is an orbit of a one-parameter subgroup of isometries.

Further, we say that *a Riemannian manifold* (M,g) *is naturally reductive* if it is isometric to some naturally reductive homogeneous space $(G/K, <,>)$.

Obviously, every naturally reductive Riemannian manifold is homogeneous, that is, there exists a transitive group of isometries of it. But it can be difficult to decide whether a homogeneous Riemannian manifold is naturally reductive or not. One has to consider all transitive groups of isometries, all reductive decompositions (4.1), (4.2) and then the condition (4.3).

In [TV1], [TV2] some conditions are derived which make it sometimes much easier to decide about this problem. There the following criterion is derived : Let (M,g) be a connected, simply connected and complete Riemannian manifold with Levi Civita connection ∇ and curvature tensor R. Then (M,g) is a naturally reductive Riemannian manifold if and only if there exists a tensor field T of type $(1,2)$ such that

$$\begin{aligned}
&g(T_XY,Z) + g(Y,T_XZ) = 0,\\
&(\nabla_X R)_{YZ} = [T_X, R_{YZ}] - R_{T_XYZ} - R_{YT_XZ},\\
&(\nabla_X T)_Y = [T_X, T_Y] - T_{T_XY},\\
&T_XY + T_YX = 0
\end{aligned}$$

for all vector fields X, Y, Z on M. The first three conditions are the *Ambrose-Singer conditions* and the existence of a T satisfying these conditions is equivalent to the homogeneity of (M,g). (Note that putting $\tilde{\nabla} = \nabla - T$, these conditions are equivalent to $\tilde{\nabla}g = \tilde{\nabla}R = \tilde{\nabla}T = 0$ and this means that (M,g) is *locally* homogeneous.) The fourth condition is equivalent to the naturally reductivity. We refer to [DZ] for a classification of naturally reductive metrics on compact simple Lie groups and to [GO] for the non-compact semisimple case. There one will find many non-symmetric examples (see also [BTV2], [GR1], [KR], [N], [MA], [TV3], [WO], [Z]).

In [DN1], [DN2] it was proved that *all naturally reductive spaces are D'Atri spaces* and another more simple proof was provided in [D]. A very simple proof is given in [KV5] by using the above mentioned geometric characterization. In fact, the last proof extends the original result of [DN1], [DN2] to a broader class of Riemannian manifolds described in the next subsection.

[DN2] contains in fact a stronger result. To state it, we mention that a tensor field T of type $(1,2)$ given on a *general* (M,g) and satisfying the four conditions above is called a *naturally reductive homogeneous structure* on (M,g). It then follows at once from [DN2] that any Riemannian manifold which is equipped with a naturally reductive structure is a D'Atri space. This follows also from Proposition 10 in [BPV] (see also [BTV2]).

We note that there exist Riemannian manifolds with a naturally reductive homogeneous structure which are *not* locally isometric to any naturally reductive space [KO5].

4.2. This above mentioned broader class is formed by the *Riemannian g.o. spaces*, that is, Riemannian manifolds (M,g) all of whose geodesics are orbits of one-parameter subgroups of isometries [KV11]. This means that $M = G/H$ and for a geodesic $c = c(t)$ of M, we can always find an element A in the Lie algebra $\mathfrak{g}$ of G such that $c(t) = (\operatorname{Exp} tA) \cdot c(0)$ for $t \in \mathbb{R}$, where $\operatorname{Exp} : \mathfrak{g} \to G$ is the Lie exponential map. A classification of all g.o. spaces in dimension not greater than six and which are in no way naturally reductive is given in [KV11] and it turns out that the naturally reductive Riemannian manifolds form a proper subset of the set of all Riemannian g.o. spaces when $\dim M = 6$. For $\dim M \leq 5$ both classes coincide (see 6.5).

The first proof of the fact that g.o. spaces are D'Atri spaces was given in [KV5]. See also [KV7].

4.3. Next, let $I_0(M)$ denote the largest connected group of isometries of a Riemannian manifold (M,g). Then a homogeneous Riemannian manifold (M,g) is called a *commutative space* if all $I_0(M)$-invariant differential operators commute. As proved in [KV2], any commutative space is a D'Atri space. (See also [KV7].) For compact manifolds this result is implicitly included in [ROU]. A classification of the commutative spaces of dimension three, four and five is given in [KO3], [KV6] and [BI], respectively. For dimensions ≤ 5 it was shown that a homogeneous Riemannian manifold is commutative if and only if it is naturally reductive.

We note that the commutativity is sometimes defined with respect to the full isometry group $I(M)$. Of course, commutativity with respect to $I_0(M)$ implies commutativity with respect to $I(M)$ and thus the last class of spaces is (at least formally) broader. Yet, the same argument as in [KV2] shows that all these spaces are still D'Atri spaces.

4.4. The *weakly symmetric spaces* provide an interesting class of commutative spaces in this broader sense. They have been introduced by Selberg in

1956 [SE] in relation with his work on harmonic analysis. Here, an (M, g) is said to be weakly symmetric if there exists a subgroup G of the isometry group $I(M)$ acting transitively on M and an $f \in I(M)$ with $f^2 \in G$ and $fGf^{-1} = G$ such that for any two points p, q on M there exists an $h \in G$ satisfying $h(p) = f(q)$ and $h(q) = f(p)$. In [BV4] it is proved that this is equivalent to the following : for any two points $p, q \in M$ there exists an isometry of M mapping p to q and q to p. Furthermore, Szabó gives in [SZ4] still another characterization (for a proof see [BPV]) which motivated him to call these spaces *ray symmetric.* This means that on such a space (M, g) there exists, for every $m \in M$ and every maximal geodesic γ through m, an isometry of M such that its restriction to γ is an involution with m as fixed point.

Now we give a list of examples from [BV4] formed by weakly symmetric hypersurfaces in some rank one symmetric spaces.

ambient space	hypersurface
$\mathbb{C}P^n$	tube around $\{p\}, \mathbb{C}P^1, ..., \mathbb{C}P^{n-1}$
$\mathbb{H}P^n$	tube around $\{p\}, \mathbb{H}P^1, ..., \mathbb{H}P^{n-1}$
Cay P^2	tube around $\{p\}$, Cay P^1
$\mathbb{C}H^n$	horosphere; tube around $\{p\}, \mathbb{C}H^1, ..., \mathbb{C}H^{n-1}$
$\mathbb{H}H^n$	horosphere; tube around $\{p\}, \mathbb{H}^1, ..., \mathbb{H}^{n-1}$
Cay H^2	horosphere; tube around $\{p\}$, Cay H^1

In the compact case we consider only tubes with radius $r \in (0, \pi/\sqrt{c})$ where c is the maximal value of the sectional curvature of the ambient space. In the non-compact case r may be an arbitrary positive real number.

Note that it is still an open problem, stated by Selberg, whether any commutative space (in the broader sense) is weakly symmetric or not.

We refer to [BRV], [BTOV], [GV3] for further examples.

4.5. A. Kaplan introduced a new and very interesting class of examples which we will consider now (see [K2]). Let $\mathfrak{v}$ and $\mathfrak{z}$ be real vector spaces of dimensions $n, m \in \mathbb{N}$, respectively, and $\beta : \mathfrak{v} \times \mathfrak{v} \to \mathfrak{z}$ a skew-symmetric bilinear map. We endow the direct sum $\mathfrak{n} = \mathfrak{v} \oplus \mathfrak{z}$ with an inner product $<, >$ such that $\mathfrak{v}$ and $\mathfrak{z}$ are perpendicular and define an $\mathbb{R}$-algebra homomorphism

$$J : \mathfrak{z} \to \mathrm{End}(\mathfrak{v}), \quad Z \mapsto J_Z$$

by

$$< J_Z U, V >=< \beta(U, V), Z >$$

for all $U, V \in \mathfrak{v}, Z \in \mathfrak{z}$. Next, we define a Lie algebra structure on $\mathfrak{n}$ by

$$[U + X, V + Y] = \beta(U, V)$$

for all $U, V \in \mathfrak{v}; X, Y \in \mathfrak{z}$. Then the two-step nilpotent Lie algbra $\mathfrak{n}$ is said to be a *generalized Heisenberg algebra* (or *H-type algebra*) if

$$J_Z^2 = -\|Z\|^2 id_{\mathfrak{v}}$$

for all $Z \in \mathfrak{z}$. The associated simply connected Lie group N, equipped with the induced left-invariant Riemannian metric g, is called a *generalized Heisenberg (or H-type) group.*

Note that the simplest example is given by the group of all matrices of the form

$$\begin{pmatrix} 1 & x & y \\ 0 & 1 & z \\ 0 & 0 & 1 \end{pmatrix}, \qquad x, y, z \in \mathbb{R},$$

equipped with any left-invariant metric. But there are an infinite number of other examples. We refer to [K1], [K3] and [TV1], [TV2], [BTV1], [BTV2] for more information and references. A. Kaplan proved in [K2] that any generalized Heisenberg group is a D'Atri space. See [BTV2] for alternative proofs.

4.6. Now let $\mathfrak{n}$ be a generalized Heisenberg algebra, $\mathfrak{a}$ a one-dimensional real vector space and A a non-zero vector in $\mathfrak{a}$. We denote the inner product and the Lie bracket on $\mathfrak{n}$ by $<, >_{\mathfrak{n}}$ and $[\cdot, \cdot]_{\mathfrak{n}}$, respectively, and define a new vector space

$$\mathfrak{s} = \mathfrak{n} \oplus \mathfrak{a}$$

as the direct sum of $\mathfrak{n}$ and $\mathfrak{a}$. Each vector in $\mathfrak{s}$ can be written in a unique way in the form $V + Y + sA$ with some $V \in \mathfrak{v}, Y \in \mathfrak{z}$ and $s \in \mathbb{R}$ where $\mathfrak{v}$ and $\mathfrak{z}$ are as in 4.5.

We now define an inner product $<, >$ and a Lie bracket $[,]$ on $\mathfrak{s}$ by

$$< U + X + rA, V + Y + sA >=< U + X, V + Y >_{\mathfrak{n}} + rs$$

and

$$[U + X + rA, V + Y + sA] = [U, V]_{\mathfrak{n}} + \frac{1}{2}rV - \frac{1}{2}sU + rY - sX,$$

for all $U, V \in \mathfrak{v}, X, Y \in \mathfrak{z}$ and $r, s \in \mathbb{R}$. Then $\mathfrak{s}$ becomes a Lie algebra with an inner product. The corresponding simply connected Lie group, equipped with the induced left-invariant metric is denoted by S and is called a *Damek-Ricci space* (briefly, a *DR-space*). Any DR-space is a D'Atri space, again.

Note that S is a solvable Lie group. It turns out that with the given left-invariant metric S has non-positive curvature. Moreover, the generalized Heisenberg group, corresponding to $\mathfrak{n}$, is a horosphere in the attached S. In general S is not a symmetric space. In fact it can be proved that S is symmetric if and only if the J^2-condition is satisfied, that is, if for all $X, Y \in \mathfrak{z}$ with $< X, Y >= 0$ and all non-zero $U \in \mathfrak{v}$, there exists a $Z \in \mathfrak{z}$ so that $J_X J_Y U = J_Z U$. In this case the corresponding generalized Heisenberg algebras are isomorphic to the nilpotent part in the Iwasawa decomposition of the Lie algebra of the isometry group of the complex or quaternionic hyperbolic space or the Cayley hyperbolic plane, and the corresponding DR-spaces are isometric to these hyperbolic spaces.

We refer to [DR], [BTV1], [BTV2], [SZ4], [TV4], [TV5] for more details about the geometry of DR-spaces.

4.7. A further non-trivial class of D'Atri spaces has been introduced by Roberts and Ursell [ROU] from a probabilistic point of view. They call these "commutative spaces", but we prefer to call them *probabilistic commutative spaces.* Such a space has been defined originally as a compact Riemannian manifold on which any two random steps commute. But the non-invariant treatment in [ROU] makes it difficult to understand the general definition. A simple way to define such manifolds is as follows. A Riemannian manifold (M, g) is said to be a probabilistic commutative space if the second mean value operators L_t (defined only locally, in general) commute, that is, $L_r L_s = L_s L_r$ on $C^\infty(M)$ for sufficiently small radii r and s. An equivalent definition says : (M, g) is probabilistic commutative if and only if all Euclidean Laplacians $\tilde{\Delta}^{(k)}, k \in \mathbb{N}$, commute on $C^\infty(M)$ (see 2.5). See also [KO1], [KO3], [KP].

We refer to [KO1] for more explicit calculations concerning the commutativity of the Euclidean Laplacians $\tilde{\Delta}^{(k)}$ when $k = 1, 2, 3$. See also [SZ4] and [GI] for other considerations.

4.8. *Harmonic manifolds* are important examples of probabilistic commutative spaces. A harmonic manifold can be characterized as an (M, g) with the following mean-value property :

$$M_r[f](m) = f(m)$$

holds for each $m \in M$ and all sufficiently small $r > 0$, where f is harmonic in a neighborhood of m containing the geodesic ball of radius r and center m. An equivalent definition says that the normal volume density function $\omega(m,p)$ depends only on the distance $d(m,p)$. Further, (M,g) is harmonic if and only if it satisfies the Ledger conditions of odd and even order (each small geodesic sphere has constant mean curvature).

There are several other characterizations. We refer to [BE], [CV], [KO2], [RWW], [SZ3], [V2], [V5] for more details. We mention that all harmonic manifolds are Einstein spaces.

A long-standing conjecture was that any harmonic manifold is locally symmetric and locally isometric to a two-point homogeneous space. This is the so-called Lichnerowicz conjecture. In [SZ3] the author proved this conjecture for compact manifolds with finite fundamental group but two years later Damek and Ricci published their surprising result on the existence of non-compact, non-symmetric harmonic Riemannian spaces. These are the DR-spaces described in 4.6.

4.9. More recently, new classes of D'Atri spaces have been defined using properties of the shape operator of small geodesic spheres [BPV], [BTV2] [BV3].

Let $m, p, q \in M$ such that $q = s_m(p), d(m,p) = r$ where r is sufficiently small. Let $T_p(m)$, respectively $T_q(m)$, be the shape operator of the geodesic sphere $G_p(r)$, respectively $G_q(r)$, at m and with respect to the inward unit normal vector. Then, (M,g) is said to be a *$\mathcal{TC}$-space* if $T_p(m)$ and $T_q(m)$ have the same eigenvalues, that is, $G_p(r)$ and $G_q(r)$ have the same principal curvatures, for any such configuration. In [BV3] the authors proved that any $\mathcal{TC}$-space is a D'Atri space.

The same result holds for any *$\mathcal{SC}$-space.* Taking the notation as above, (M,g) is said to be an $\mathcal{SC}$-space if and only if $T_m(p)$ and $s_{m*}^{-1} \circ T_m(q) \circ s_{m*}$ have the same eigenvalues (counted with multiplicities), that is, all the local geodesic symmetries s_m preserve the principal curvatures of small geodesic spheres centered at m.

5. Relations between the special classes

In this section we shall consider several relations between the classes of D'Atri spaces given in Section 4.

5.1. We have

(1) { g.o. spaces } ∪ { commutative spaces } ∪ { $\mathcal{SC}$-spaces } ⊂ { D'Atri spaces };

(2) { commutative spaces } ⊂ { probabilistic commutative spaces };
{ harmonic spaces } ⊂ { probabilistic commutative spaces };

(3) { naturally reductive spaces } ⊂ { g.o. spaces };

(4) { weakly symmetric spaces } ⊆ { $\mathcal{TC}$-spaces };
{ weakly symmetric spaces } ⊆ { commutative spaces in the broader sense };
{ weakly symmetric spaces } ⊆ { g.o. spaces };

(5) { symmetric spaces } ⊂ { naturally reductive spaces } ∩ { weakly symmetric spaces } ∩ { commutative spaces }.

All the inclusions marked by ⊂ are strict, that is, if $A \subset B$, then an explicit example in $B \setminus A$ is known.

5.2. First, we give the references and examples for the inclusions of 5.1.

The non-symmetric DR-spaces are D'Atri spaces which are neither g.o. spaces, nor commutative spaces, nor $\mathcal{SC}$-spaces [BTV2], [TV4]. Hence the inclusion (1) is strict.

In a harmonic space the higher order Euclidean Laplacians $\tilde{\Delta}^{(k)}$ are polynomials of the Laplacian [SZ1], [SZ4] and thus they all commute. Therefore, any harmonic space is probabilistic commutative (see 4.6). Moreover, because the Euclidean Laplacians are invariant under isometries, it is obvious that each commutative space is probabilistic commutative. The strictness of the second inclusion in (2) follows from the fact that there exist commutative spaces which are not harmonic (for example, all non-symmetric naturally reductive spaces of dimension 3, cf. 6.2 below). The strictness of the first inclusion follows from the fact that any non-symmetric DR-space is probabilistic commutative but not commutative [BTV1], [BTV2].

As we already remarked, in 4.1 and 4.2 any naturally reductive space is a g.o. space and the inclusion in (3) is strict [K2], [KV11].

The second inclusion in (4) is given in [SE] for commutativity in the broader sense (see 4.4). Moreover, since weakly symmetric spaces are $\mathcal{TC}$-spaces [BPV] we get the first inclusion of (4). The last inclusion of (4) has been proved recently in [BKV].

Finally, the inclusions (5) are either well-known or immediate from the definitions. The strictness is also obvious from the given examples in Section 4.

5.3. These considerations give rise to some open problems :

- Is { probabilistic commutative spaces } $\subseteq$ { D'Atri } a strict inclusion?
- Is the class of commutative spaces or that of the g.o. spaces included in the class of $\mathcal{TC}$-spaces?
- Do there exist commutative spaces which are not g.o. spaces?
- Is the inclusion { weakly symmetric spaces } $\subseteq$ { commutative spaces in the broader sense } a strict inclusion?

5.4. We finish this section by giving some examples in order to show that some inclusions cannot hold.

Examples of naturally reductive spaces which are not commutative are given in [JI1] - [JI4]. For example, let $M = G/T$ where G is one of the following : $SU(n)(n > 2), SO(n)(n > 6), Sp(n)(n > 1)$, T a maximal torus of G. Endow M with the Riemannian metric induced by the negative of the Killing form of G. Then M is a non-commutative naturally reductive Riemannian manifold. This results also holds when G is a compact semi-simple Lie group (and $M = G/T$ not symmetric).

The small geodesic spheres in Cay P^2 are weakly symmetric spaces which are not naturally reductive [BV4].

The list of generalized Heisenberg groups which are g.o. spaces is determined in [RI2] and those which are commutative in [R] (see also [BTV2]). These lists show that each commutative generalized Heisenberg group is also a g.o. space. The list of g.o. spaces contains exactly one (of dimension 31) which is neither commutative nor naturally reductive.

6. D'Atri spaces in low dimensions

Now we shall mention several classification results relating to D'Atri spaces of low dimension.

6.1. As we already mentioned in 3.4 a D'Atri space has constant scalar curvature. Hence a *two-dimensional D'Atri manifold is a space of constant curvature* [DN1].

6.2. The case of three-dimensional manifolds is more complicated. A complete solution is given in [KO3]. More precisely, we have : *A three-dimensional, connected, complete and simply connected C^∞ Riemannian manifold (M, g) is a D'Atri space if and only if it is one of the following spaces :*

(i) *a Riemannian symmetric space;*

(ii) SU(2) *with a left-invariant Riemannian metric g such that the isotropy subgroup of $I_0(M,g)$ at any point is SO(2). All these metrics depend on two arbitrary parameters a, b (the extremal Ricci curvatures) where the only restrictions are $b > 0, a+b > 0, a \neq b$. Here, a is the double root and b the single root of the Ricci tensor;*

(iii) *the universal covering of* SL(2, $\mathbb{R}$) *with a left-invariant Riemannian metric such that the isotropy subgroup of $I_0(M,g)$ at any point is* SO(2). *The explicit form of the metric for $M = \mathbb{R}^3[t,x,y]$ is*

$$ds^2 = \frac{1}{|a+b|}dt^2 + |a+b|e^{-2t}dx^2 + (dy + \sqrt{2b}e^{-t}dx)^2$$

where $a = \lambda_1 = \lambda_2, b = \lambda_3$ are the extremal Ricci curvatures. The only restrictions are $b > 0, a+b < 0$.

(iv) *the three-dimensional Heisenberg group with any left-invariant metric. The explicit form for the metric on $M = \mathbb{R}^3[x,y,z]$ is*

$$ds^2 = \frac{1}{2b}\{dx^2 + dz^2 + (dy - xdz)^2\}$$

where $b > 0$ is arbitrary and $\lambda_1 = \lambda_2 = -b, \lambda_3 = b$ are the extremal Ricci curvatures.

If (M,g) is not complete or not simply connected, then it is locally isometric to one of these spaces.

It follows that all the connected, complete, simply connected D'Atri spaces of dimension three are *homogeneous* spaces. Moreover, they are *naturally reductive* and *commutative.*

6.3. In dimension four we do not have a complete classification result but we have some results for related classes.

First, a harmonic manifold of dimension four is locally isometric to a two-point homogeneous space [BE], [RWW], [W]. A generalization of this result is contained in [SV2]. There the authors have proved that a connected, four-dimensional 2-stein space with volume-preserving geodesic symmetries (up to sign) is locally flat or locally isometric to a rank one symmetric space.

Similarly, in [SV1] the authors showed that a four-dimensional Kählerian D'Atri space is locally symmetric.

Furthermore, in [KV3] the authors classified all simply connected naturally reductive Riemannian manifolds of dimension four. In this case (M,g) is either symmetric or a Riemannian product of the form $M = M_3 \times \mathbb{R}$ where M_3 is a D'Atri space of dimension three (see 6.2).

This result is completed in [KV6] where it is proved that a four-dimensional, simply connected, homogeneous Riemannian manifold is a commutative space if and only if it is naturally reductive. The analogous result for g.o. spaces is proved in [KV11].

It is not known whether the above spaces exhaust the class of homogeneous D'Atri spaces (in the simply connected case). Moreover, *we do not know if all D'Atri spaces, in particular in dimension four, are locally homogeneous.* A positive answer and the well-known result of Jensen about four-dimensional homogeneous Einstein spaces [JE] would show that the following conjecture of Sekigawa and the third author holds : Any four-dimensional Einsteinian D'Atri space is locally symmetric.

6.4. In dimension five the situation is rather similar to dimension four. First, we know the classification of the five-dimensional naturally reductive spaces. We use this opportunity to present a corrected version of the main theorem 2.1 from [KV8].

Every naturally reductive space of dimension five is either locally symmetric, or locally decomposable, or locally isometric to some member of the following families :

(i) the underlying homogeneous manifold is either

$$\frac{SO(3)\times SO(3)}{SO(2)_r}, \text{or } \frac{SO(3)\times SL(2,\mathbb{R})}{SO(2)_r}, \text{or } \frac{SL(2,\mathbb{R})\times SL(2,\mathbb{R})}{SO(2)_r}$$

where $r \in \mathbb{Q}$ and $SO(2)_r$ denotes the subgroup of all matrices of the form

$$\begin{pmatrix} \cos t & -\sin t & 0 \\ \sin t & \cos t & 0 \\ 0 & 0 & 1 \end{pmatrix} \times \begin{pmatrix} \cos rt & -\sin rt & 0 \\ \sin rt & \cos rt & 0 \\ 0 & 0 & 1 \end{pmatrix}, t \in \mathbb{R}.$$

On each underlying space there is a family of naturally reductive invariant metrics depending on two real parameters;

(ii) the underlying homogeneous manifold is either

$$\frac{H_3 \times SO(3)}{SO(2)^r} \quad \text{or} \quad \frac{H_3 \times SL(2,\mathbb{R})}{SO(2)^r}, r \in \mathbb{R},$$

where H_3 is the three-dimensional Heisenberg group and $SO(2)^r$ denotes the subgroup of all matrices of the form

$$\begin{pmatrix} 1 & 0 & t \\ 0 & 1 & 0 \\ 0 & 0 & 1 \end{pmatrix} \times \begin{pmatrix} \cos rt & -\sin rt & 0 \\ \sin rt & \cos rt & 0 \\ 0 & 0 & 1 \end{pmatrix}, t \in \mathbb{R}.$$

On each underlying space there is a family of naturally reductive invariant metrics depending on two real parameters;

(iii) the underlying homogeneous manifold is the five-dimensional generalized Heisenberg group (with one-dimensional center). The family of naturally reductive left-invariant metrics on it coincides with the family of all invariant metrics and it depends on two real parameters. The whole family of these Riemannian manifolds can be described as the Cartesian space $\mathbb{R}^5[x, y, z, u, v]$ with the family of metrics given by

$$g = \frac{1}{\rho}(du^2 + dx^2) + \frac{1}{\lambda}(dv^2 + dy^2) + (udx + vdy - dz)^2$$

where $\lambda, \rho > 0$ are real parameters;

(iv) the underlying homogeneous manifold is either

$$SU(3)/SU(2) \quad \text{or} \quad SU(1,2)/SU(2)$$

and on each space there is a family of naturally reductive invariant metrics depending on two real parameters.

In the same paper the authors prove that all simply connected five-dimensional naturally reductive homogeneous spaces are also commutative spaces. The converse has been proved in [BI]. Hence, in the simply connected case, the class of naturally reductive homogeneous spaces and the class of commutative spaces coincide in dimension five.

Finally, in [KV11], the authors proved the following results :

(i) Each five-dimensional Riemannian g.o. space $(M, g) = G/H$ is either naturally reductive or of isotropy type $SU(2)$.

(ii) In the second case one can always express (M, g) as a homogeneous space G'/H' of isotropy type $U(2)$, which is already naturally reductive.

Summarizing : all five-dimensional g.o. spaces are naturally reductive Riemannian manifolds.

Very recently [KOM] it has been proved that all 5-dimensional simply connected naturally reductive spaces are weakly symmetric. Because all weakly symmetric spaces are g.o. spaces (see 5.1) we see, in the simply connected case, that the classes of weakly symmetric spaces, g.o. spaces, naturally reductive spaces and commutative spaces coincide in dimension five (in fact, for all dimensions ≤ 5, see [BV4] and [KV11]). Yet, we do not know the classification of commutative spaces *in the broader sense* for dimensions bigger than three.

6.5. The situation is quite different in dimension six. The six-dimensional generalized Heisenberg group with two-dimensional center is a commutative space but it is not naturally reductive although it is a g.o. space (see [K2], [R], [RI2]).

The homogeneous space $SU(3)/T$, where T is a maximal torus, endowed with an arbitrary invariant metric is naturally reductive but not commutative [JI1].

In [KV11] one finds the explicit list of all simply connected g.o. spaces in dimension six which are in no way naturally reductive. One class of such spaces consists of two-step nilpotent Lie groups with two-dimensional center equipped with a three-parameter family of left invariant metrics. The other class is formed by the universal coverings of homogeneous Riemannian spaces of the form $SO(5)/U(2)$ or $SO(4,1)/U(2)$, respectively. Each of these spaces is equipped with a two-parameter family of Riemannian metrics. To our knowledge, the type $SO(5)/U(2)$ provides the first examples of *compact* simply connected Riemannian g.o. spaces which are in no way naturally reductive.

7. D'Atri spaces, homogeneity and non-positive curvature

7.1. All the examples we gave in Section 4 are (locally) homogeneous spaces. In fact, up to now, we do not know any example of a D'Atri space which is not locally homogeneous. This leads to the following problem : *Is any D'Atri space locally homogeneous or not ?*

7.2. This seems to be a rather difficult problem for which we only know some partial positive answers. As we already mentioned in 6.1, any two-dimensional D'Atri space has constant curvature and hence is locally homogeneous. Furthermore, the result given in 6.2 shows that three-dimensional D'Atri spaces have the same property. Finally, in 6.3 we gave some partial positive answers in the four-dimensional case but the general problem remains open.

7.3. Also for higher dimensions we have only partial results which we shall mention now.

An unpublished result of Tricerri, the second and third author gives a sufficient condition for a D'Atri space to be locally homogeneous : If the Laplacian of a Riemannian manifold commutes with a suitable subset all isometry-invariant differential operators of second order, then the manifold must be locally homogeneous.

Using 3.9 we get : If the Laplacian of (M, g) commutes with all isometry-invariant differential operators, then it is a locally homogeneous D'Atri space.

Note that the non-symmetric DR-spaces are Lie groups for which the Laplacian does not commute with all isometry-invariant differential operators.

Finally, we mention the following more special problem which is also still open : *Is any harmonic manifold* (M, g) *of dim*$M \geq 5$ *locally homogeneous or not ?*

7.4. A Riemannian manifold (M, g) is said to be *ball-homogeneous* if the volume $V_m(r)$ of any geodesic sphere $G_m(r)$ with sufficiently small radius r does not depend on the center m [KV1]. Of course, any locally homogeneous manifold is ball-homogeneous but what about the converse ? In [P1] the author proved that all *compact* D'Atri spaces are ball-homogeneous and using a similar idea as in [P1] this result has been extended to all D'Atri spaces in [GP]. This result implies that some curvature invariants of D'Atri spaces must be global constants. For example $3\|R\|^2 - 8\|\rho\|^2$ is constant but up to now it is not known if $\|R\|$ and $\|\rho\|$ are both constant.

7.5. The situation is more special for *D'Atri spaces of non-positive curvature.* In this case the D'Atri property seems to be rather restrictive. We mention now some results.

A *Hadamard manifold H* is a complete, simply connected Riemannian manifold of non-positive curvature. Note that all non-symmetric DR-spaces are homogeneous Hadamard manifolds [DR], [BTV2].

According to [AL], [AW1], [AW2], [HE], a homogeneous Hadamard manifold is isometric to a solvable Lie group S (with Lie algebra $\mathfrak{s}$) endowed with a left-invariant metric $<,>$. If $\mathfrak{n} = [\mathfrak{s}, \mathfrak{s}]$, then $\mathfrak{a} = \mathfrak{n}^\perp$ is Abelian and there is a semi-direct splitting of S into the corresponding Lie groups A and N. The *DR*-examples belong to the special class where dim $A = 1$ and N is a two-step nilpotent group of Heisenberg type.

Next, let $k = \dim \mathfrak{a}$. Then A is a k-flat, that is, a totally geodesic copy of $\mathbb{R}^k$ in H. If we assume that the flat de Rham factor of H is trivial, then k is uniquely determined by H and it is called the *algebraic rank* of H (see [AL], [H1]).

Now, let $\pi : U_1H \to H$ denote the unit tangent bundle of H. Then $v \in U_1H$ is called $I(H)$*-recurrent* if for suitable sequences $t_n \to +\infty, \phi_n \in I(H)$ we have $\lim_{n\to\infty} \phi_{n*}(g^{t_n}(v))$
$= v$ where g^t is the geodesic flow. $I(H)$ is said to satisfy the *duality condition* if the set of $I(H)$-recurrent vectors lies dense in U_1H. For example, if H admits a quotient of finite volume, then $I(H)$ satisfies the duality condition. See [H1], [H2] for more details and references.

In [H2] the author studies Hadamard manifolds satisfying one of the following properties :

(i) H is a homogeneous space;

(ii) the isometry group $I(H)$ satisfies the duality condition.

These both classes of Hadamard manifolds are essentially disjoint because in the case that H satisfies (i) and (ii), it is a symmetric space.

7.6. In [H2] several results are derived concerning Hadamard manifolds which satisfy the D'Atri condition. First we note that the *rank* of H is given by $rk(H) = \min\{rk(v) | v \in U_1 H\}$ where the rank of $v \in U_1 H$ is the dimension of the space of parallel Jacobi fields along the geodesic c_v. This notion coincides with the usual one when H is a symmetric space. We have [H2] :

- Let $I(H)$ satisfy the duality condition. Then H is a D'Atri space if and only if every non-flat de Rham factor H_j of H is either symmetric of non-compact type and higher rank or harmonic of rank equal to one.
- Let M be a D'Atri space of non-positive curvature and finite volume. If the universal covering of M is irreducible, then M is either locally symmetric of higher rank or harmonic of rank equal to one.
- A homogeneous Hadamard manifold H is a D'Atri space if and only if every non-flat de Rham factor of H is either symmetric of non-compact type and higher rank or harmonic of algebraic rank one.
- A homogeneous Riemannian manifold with $-4\delta \leq K \leq -\delta, \delta > 0$, is a symmetric space of rank one if and only if $\nabla_\xi \rho_{\xi\xi} = 0$ for all tangent ξ. From this and 3.4 we see that each homogeneous D'Atri space satisfying $-4\delta \leq K \leq -\delta$ is a symmetric space of rank one.

7.7. We note that in case the answer to the problem in 7.1 is positive, then any D'Atri-Hadamard manifold of finite volume would be locally symmetric.

In this context the following problem seems to be a reasonable one : Is any D'Atri space of non-positive curvature and finite volume locally symmetric ?

7.8. We mention that the considerations in [H2] are motivated by the fact that a Riemannian manifold with ergodic geodesic flow admits only trivial Killing tensors. As we have seen, there are a lot of Killing tensors on a D'Atri space. In case of symmetric tensors of type $(0, 2)$ this idea has been used in [MO] to show that a compact manifold of negative curvature and satisfying the condition $\nabla_\xi \rho_{\xi\xi} = 0$ for all ξ, is Einsteinian. (See also [H2] for related results and further references.)

7.9. Under stronger assumptions it is possible to get more. In [CA] it is proved that a compact harmonic Kähler manifold of negative curvature is isometric to a compact quotient of complex hyperbolic space up to a constant scalar factor. Using the ideas of Heber it is possible to substitute here the word "harmonic" by "D'Atri". For a further extension we refer to [BCG]. Finally, a combination of [BCG] and [H2] yields that any compact D'Atri space of negative curvature is locally symmetric.

8. Some modifications of the D'Atri property

8.1. In this section we will treat some analogs of D'Atri spaces. Their introduction is motivated by the following consideration. As follows from Section 2, the volume density function $\omega(m,p)$ is (up to a factor r^{1-n}) an elementary symmetric function of the eigenvalues of the endomorphisms $A(r)$, namely its determinant. So, it is quite natural to consider also some other elementary symmetric functions of $A(r)$ or some other endomorphisms. This has been done in [WI1], [WEH] in the framework of the study of harmonic spaces. The concepts of a $\mathcal{TC}$- and $\mathcal{SC}$-space also belong to this field of study. See also [WI3].

Depending on the endomorphism function in question, one can introduce several kinds of *D'Atri spaces of type* k where $k \in \mathbb{N}$ and $1 \leq k \leq n$. But we mention already that it is an open problem whether all these different notions are equivalent or not.

8.2. We start by considering the shape operator of small geodesic spheres.

Definition. An n-dimensional Riemannian manifold is called a *D'Atri space of type* k if the k-th elementary symmetric function $(1 \leq k \leq n-1)$ of the eigenvalues of the shape operator of all geodesic spheres of sufficiently small radius is left-centrally symmetric.

From 3.2 (iv) we see that the property "D'Atri of type 1" is equivalent to the D'Atri property in the usual sense. Moreover, an $\mathcal{SC}$-space (see 4.9) is a Riemannian manifold which is D'Atri of type k for all possible k. Now, it is known that a DR-space is an $\mathcal{SC}$-space if and only if it is symmetric [BTV1], [BTV2]. Hence, a non-symmetric DR-space is never a D'Atri space of type k for all k, that is, the D'Atri property of type $k = 1$ does not imply such a property of $k = 2, 3, \ldots$ and this example justifies our new definition.

8.3. These elementary symmetric functions have also been used in [TV4]. There, a Riemannian manifold (M, g) is said to be *k-harmonic* if the k-th elementary symmetric functions of the principal curvatures of sufficiently small geodesic spheres $G_m(r)$ are radial functions, that is, constant on $G_m(r)$, for all $m \in M$. Here 1-harmonic means harmonic in the usual sense (see 4.8). Moreover, from the results in [CV] it follows that 2-harmonic is equivalent to 1-harmonic. The non-symmetric DR-spaces provide examples of harmonic spaces which are not 3-harmonic [TV4].

In this context the following conjecture has been formulated in [TV4] : *A Riemannian manifold is locally isometric to a two-point homogeneous space if and only if it is k-harmonic for all possible k.* Despite the many positive answers given in [GSV], the general answer to the conjecture is still unknown.

We also refer to [GSV] for an intrinsic analog of this conjecture introduced by using the Ricci operator of small geodesic spheres.

8.4. Now we shall introduce a second notion of a D'Atri space of type k.

Let $(x^1, ..., x^n)$ be a normal coordinate system centered at $m \in (M, g)$ and let (g^{ij}), respectively (g_{ij}), be the corresponding contravariant, repectively covariant, matrices of g. We denote by $\tilde{\mu}_k(m,p)$, respectively $\mu_k(m,p)$, the associated k-th elementary symmetric functions of the eigenvalues. Obviously, $\mu_k(m,p)$ and $\tilde{\mu}_k(m,p)$ are smooth two-point functions. It is shown in [V3] that they are symmetric functions for all $k = 1, ..., n$.

Note that $\tilde{\mu}_n^{-1}(m,p) = \mu_n(m,p) = \omega^2(m,p)$ and hence a D'Atri space is characterized by the property that $\mu_n(m,p)$ or $\tilde{\mu}_n(m,p)$ are left-centrally symmetric. This motivates the following

Definition. (M, g) is said to be a *D'Atri space of type* k for $k \in \mathbb{N}, 1 \leq k \leq n$, if $\mu_k(m,p)$ (respectively $\tilde{\mu}_k(m,p)$) is a left-centrally symmetric function for any $m \in M$.

So, D'Atri spaces are D'Atri spaces of type n and conversely.

In [KV7] it is proved that (M, g) is a D'Atri space if and only if $\tilde{\mu}_1(m,p) = \mathrm{tr}\,(g^{ij})$ is left-centrally symmetric.

8.5. For special classes of Riemannian manifolds we get stronger results [BTV2], [KV2], [KV5]. Indeed, we have : Let (M, g) be a g.o. space, a commutative space or a generalized Heisenberg group. Then (M, g) is a D'Atri space of type k in the sense of 8.4 for all possible k.

It is worthwhile to note once again that the non-symmetric DR-spaces cannot be D'Atri spaces of type k for all k in the sense of 8.4.

8.6. Finally, we return to the notation of Section 2. In [SZ4] the author considers the endomorphism $\tilde{A}(m,p) = \frac{1}{r}A(m,p)$. Then we have that

$\omega(m,p) = \det\tilde{A}(m,p)$ and hence a D'Atri space is characterized by the property "$\det\tilde{A}(m,p)$ is a left-centrally symmetric function".

Based on this result, one considers in [SZ4] the following symmetric two-point functions :

$$S_i(m,p) = \frac{(\alpha_i(m,p))^2}{\omega(m,p)}$$

where $\alpha_i(m,p) = \mathrm{tr}(\tilde{A}(m,p))^i, i = 0, 1, ..., n$. Then an (M,g) is said to be α_i-*symmetric* for some $i \in \mathbb{N}, 0 \leq i \leq n$, if $S_i(m,p)$ is left-centrally symmetric.

From the results in [KV7] and [BTV2] it follows that g.o. spaces, commutative spaces and generalized Heisenberg groups are α_i-symmetric for all possible i.

Finally, as in [SZ4] we introduce the differential operators $\Box_H^{(k)} : \varphi \mapsto \tilde{\Delta}^{(k)}(H(m,\cdot)\varphi)$ for $m \in M$ and a suitable two-point function H. If we choose $H(m,p) = \alpha_i(m,p)$, then we have the commutativity $\Delta\Box_{\alpha_i}^{(k)} = \Box_{\alpha_i}^{(k)}\Delta$ for all $k \in \mathbb{N}$ if and only if the space is α_i-symmetric and satisfies the ultra-hyperbolic equation

$$\Delta_m \frac{\alpha_i}{\omega}(m,p) = \Delta_p \frac{\alpha_i}{\omega}(m,p)$$

[SZ4].

9. D'Atri spaces in special geometries

Now we shall consider D'Atri spaces endowed with some additional structure. We shall mainly concentrate on two cases : (M,g) is an almost Hermitian manifold or there exists a unit Killing vector field on (M,g), respectively. This last case arises from considerations about contact geometry.

9.1. First, let (M,g,J) be an *almost Hermitian manifold*, that is, (M,g) is equipped with a (1,1)-tensor field J such that

$$J^2 = -I, \qquad g(JX,JY) = g(X,Y)$$

for all vector fields X, Y. Let Ω denote the corresponding Kähler form defined by

$$\Omega(X,Y) = g(X,JY)$$

for all X, Y. Note that dim $M = 2n$. Further, (M, g, J) is said to be a *Kähler manifold* if J is parallel, that is $\nabla J = 0$. In this case Ω is closed.

In [J] the author started the study of Kähler manifolds which satisfy the additional condition that all local geodesic symmetries preserve the symplectic form Ω, that is, $s_m^*\Omega = \Omega$ for all $m \in M$. Then s_m is said to be a *symplectic geodesic symmetry*. Because $\Omega^n = \Omega \wedge \ldots \wedge \Omega = (n!)\, v_g$ where v_g is the canonical volume form of (M, g), it follows that each s_m is volume-preserving (up to sign).

Examples of such manifolds are provided by all the Hermitian symmetric spaces. Note that on these spaces also g and J are preserved by all s_m. This fact led to the following conjecture of Kobayashi : A compact homogeneous Kähler manifold all of whose geodesic symmetries are symplectic is a Hermitian symmetric space. The main purpose of [J] was to prove this conjecture but this was done only in some special cases. The first general proof of the conjecture was published in [SV3]. This proof (using the method of Jacobi vector fields) works even for a much broader class of spaces, namely for all almost Hermitian manifolds (M, g, J). Here the s_m are also said to be symplectic if they preserve the Kähler form.

Finally, the authors of [SV3] considered also the case of *holomorphic* geodesic symmetries, that is, s_m satisfies

$$s_{m*} \circ J = J \circ s_{m*}$$

and proved that an almost Hermitian manifold all of whose geodesic symmetries are holomorphic is necessarily Kählerian and locally symmetric. Note that the converse also holds.

We refer to [V4], [V6], [V7] for surveys and related material and to [MV] for a similar study in *quaternionic geometry*.

9.2. Next, we turn to the second class of manifolds. Let (M, g) be an n-dimensional, smooth, connected Riemannian manifold. A tangentially oriented foliation of dimension one on (M, g) is called a flow [TO]. We shall denote by $\mathcal{F}_\xi$ an isometric flow generated by a unit Killing vector field ξ. Let η be the dual one-form of ξ. Then $\mathcal{F}_\xi$ is said to be a contact flow if η is a contact form, that is, $\eta \wedge (d\eta)^n \neq 0$ on M. In this case, n is necessarily odd.

$(M, g, \mathcal{F}_\xi)$ is called a *locally Killing-transversally symmetric space* (briefly, a *locally* KTS-*space*) if and only if the local reflections with respect to all flow lines are isometries. Moreover, $(M, g, \mathcal{F}_\xi)$ is said to be a *globally* KTS-*space* if ξ is complete and if the local reflections can be extended to global ones. These manifolds have been introduced and studied in [GGV1]-[GGV5], [GV1], [GV2], [G1], [G2]. They are special cases of the *Killing-transversally symmetric foliations* treated in [TOV]. We will see below that there are many examples in contact geometry. These last ones provided the motivation for the extensions.

We refer to the cited papers for more information and restrict ourselves here to some of the relevant results relating to D'Atri spaces.

First, we note that any simply connected, complete locally KTS-space is a globally KTS-space. Further, any locally KTS-space is locally homogeneous and is equipped with a naturally reductive structure and hence is a D'Atri space.

The class of *locally* or *globally* φ*-symmetric spaces* form a proper subclass of the class of KTS-spaces. These manifolds have been introduced in [T] for Sasakian geometry. The Sasakian manifolds play an important role in contact geometry. For this reason we will now mention some aspects of it. See [BL] for more details.

A C^∞ Riemannian manifold (M, g) (necessarily of odd dimension $n = 2m+1$) is said to be an *almost contact metric manifold* if it admits a tensor field φ of type $(1,1)$, a vector field ξ and a one-form η satisfying

$$\eta(\xi) = 1, \quad \varphi^2 = -I + \eta \otimes \xi, \quad g(\varphi X, \varphi Y) = g(X, Y) - \eta(X)\eta(Y)$$

for all tangent vector fields X, Y. $(M, g, \varphi, \xi, \eta)$ is called a *Sasakian manifold* if

$$(\nabla_X \varphi)Y = g(X, Y)\xi - \eta(Y)X.$$

In this case ξ is a unit Killing vector field.

Now, the Sasakian manifold is said to be a *locally* φ*-symmetric space* if and only if

$$\varphi^2(\nabla_V R)_{XY} Z = 0$$

for all vector fields V, X, Y, Z orthogonal to ξ. This is equivalent to the condition that all the reflections with respect to the flow lines of ξ are isometries. These reflections are called φ*-geodesic symmetries*. The Sasakian manifold $(M, g, \varphi, \xi, \eta)$ is said to be a *globally* φ*-symmetric space* if any φ-geodesic symmetry is extendable to a global automorphism of the structure (g, φ, ξ, η) and the Killing vector field ξ generates a global one-parameter subgroup of isometries. We refer to [BLV1] - [BLV3], [BU] for other characterizations and more information. See also [WA], [WF].

A particular class of locally φ-symmetric spaces is formed by the *Sasakian space forms*, that is, by the Sasakian manifolds of constant φ-sectional curvature. This means that the sectional curvature of sections

$\{X, \varphi X\}$, X orthogonal to ξ, is independent of X. For dim $M \geq 5$ it is then automatically globally constant. For dim $M = 3$ this property is included in the definition of Sasakian space form. We refer to [BL] for examples.

The simply connected three-dimensional globally φ-symmetric spaces are classified in [BLV2]. We just have the simply connected complete Sasakian space forms. Five-dimensional simply connected globally φ-symmetric spaces are explicitly classified in [KW] and it follows that they are not all Sasakian space forms.

The classification for arbitrary (odd) dimensions is given in [JK] where all simply connected φ-symmetric spaces are constructed as one-dimensional fibre bundles over some Hermitian symmetric spaces. It was proved that if the de Rham decomposition of the Hermitian symmetric space involves at least two factors of compact type, then there exist "rationality obstructions" connected with the curvatures for constructing a φ-symmetric space over it; otherwise there is no such obstruction.

From the results above we may conclude that all locally φ-symmetric spaces are locally homogeneous D'Atri spaces but for Sasakian space forms we can say more. Indeed, in [BPV] it is proved that a Riemannian manifold which is locally homothetic to a Sasakian space form is a $\mathcal{TC}$-space. In the global case this follows also from [BV4] where it is proved that complete, simply connected Sasakian space forms are weakly symmetric spaces, even after a homothetic change of metric.

We finish by noting that we do not know if all locally φ-symmetric spaces are $\mathcal{TC}$-spaces.

9.3. Remark

The cases we considered above do not exhaust the list of results and it would be possible to consider other special Riemannian manifolds. We only want to mention one result because it is related to the curvature condition $\nabla_\xi \rho_{\xi\xi} = 0$.

(M, g) is said to be semi-symmetric if $R_{XY} \cdot R = 0$ for all X, Y. For these spaces it is proved in [BO] that such manifolds are locally symmetric when ρ is cyclic-parallel. This implies that semi-symmetric D'Atri spaces are locally symmetric. See also [CO].

10. Volume of tubes about curves in D'Atri spaces

Many papers have been devoted to the study of volumes of tubes in Riemannian manifolds (see, for example, [GR2], [GRV 2,3], [V5], [VW2]), which was inspired by the remarkable results of H. Weyl for tubes in a space of constant curvature [WE]. In this section we focus our attention on a particular behavior of the D'Atri spaces as concerns volumes of tubes about *curves*. A result in this direction was proved in [KV4] based on basic material from [VW2].

Let (M, g) be a smooth Riemannian manifold and $\gamma : [a, b] \to M$ a unit speed curve of finite length which is topologically embedded in M. The total space $N_\sigma \subset TM$ of the normal bundle of $\gamma([a, b])$ in M is naturally diffeomorphic to the direct product $[a, b] \times \mathbb{R}^{n-1}$ via the parallel translation τ along γ with respect to the induced normal connection. Thus, it can be equipped with the natural flat Riemannian product g_γ. Let $N_\gamma(r) \subset N_\gamma$ denote the open solid tube of radius r about the null section of N_γ. If $r > 0$ is small enough, the map

$$\exp_\gamma : \eta \in N_{\gamma(t)}(r) \mapsto \exp_{\gamma(t)} \eta$$

is a diffeomorphism of class C^∞ of $N_\gamma(r)$ onto the open solid tube $U_\gamma(r)$ in M of radius r about γ. Since $U_\gamma(r)$ is an n-cell, we can choose an orientation and the corresponding volume form ω of (M, g) on $U_\gamma(r)$. Let ω_γ denote the volume form of the coherently oriented total space of the normal bundle of γ with its flat metric. Then we have

$$\exp_\gamma^* \omega = \theta_\gamma \omega_\gamma$$

for some positive smooth function θ_γ defined on $N_\gamma(r)$. This function is called the volume density function of $\exp_\gamma$. If $V_\gamma(r)$ denotes the volume of the tube $U_\gamma(r)$, then we have

$$\begin{aligned} V_\gamma(r) &= \int_{N_\gamma(r)} \theta_\gamma(\eta) d(\mathrm{vol} g_\gamma(\eta)) \\ &= \int_a^b \int_0^r s^{n-2} \int_{S^{n-2}_{\gamma(a)}(1)} \theta_\gamma(s\tau_t(\xi)) d\xi ds dt \end{aligned}$$

where $S^{n-2}_{\gamma(a)}(1)$ is the unit sphere with center $\gamma(a)$ in the normal space of γ at $\gamma(a)$.

Now, let ω_m denote the normal volume density function defined in Section 2. For $\eta \in N_\gamma(r)$ we adopt the following notations : m is the initial point of $\eta, m = \gamma(t)$ where $t \in [a, b], p = \exp_m \eta, T_p$ is the shape operator of the geodesic sphere with center p and radius $s = \|\eta\|$ evaluated at m. Then we have

$$\theta_\gamma(\eta) = -\{g(T_p\gamma'(t), \gamma'(t)) + \kappa_\eta\} s\omega_m(p)$$

and $\kappa_\eta = g(\gamma''(t), \|\eta\|^{-1}\eta)$.

For a D'Atri space the formula for $V_\gamma(r)$ becomes

$$V_\gamma(r) = -\int_a^b \int_0^r s^{n-2} \int_{S^{n-2}_{\gamma(a)}(1)} g(T_p\gamma'(t), \gamma'(t))\omega_{\gamma(t)}(s\tau_t\xi)d\xi ds dt.$$

So, $V_\gamma(r)$ does not involve the curvature of γ and depends only on first derivatives of γ.

In [KV4] it is proved that the volume of a tube about a curve in (M, g) remains unchanged under a G-deformation of the *second* order of this curve, where G is the group of all isometries of (M, g).

If (M, g) is a D'Atri space, then the volume is invariant even under *first* order deformations of the initial curve. More precisely, let $k \in \mathbb{N}$. Then two parametrized curves $\gamma_1, \gamma_2 : [a, b] \to M$ are said to be locally k-equivalent if for each $t \in [a, b]$ there is a local isometry D_t of M such that $j_t^k\gamma_1 = j_t^k(D_t \circ \gamma_2)$. Here j_t^k is the k-jet of a map $[a, b] \to M$ at the value t.

Here we have the following result. Let $\gamma_1, \gamma_2 : [a, b] \to M$ be two smooth regular parametrized curves of finite length which are topologically embedded in a D'Atri space M and let $P_{\gamma_1}(r), P_{\gamma_2}(r)$ respectively, denote the tubes of sufficiently small radius $r > 0$ about γ_1 and γ_2, respectively. If γ_1 and γ_2 are locally 1-equivalent, then γ_1 and γ_2 have the same length and $P_{\gamma_1}(r)$ and $P_{\gamma_2}(r)$ have the same volume for any sufficiently small radius $r > 0$.

We remark that in a homogeneous manifold there exist in general infinitely many curves with the same initial vector which are 1-equivalent but not congruent. We also notice that in a two-point homogeneous space every two curves of the same length are locally 1-equivalent and hence the volume of a tube about a curve only depends on the length of the curve and the radius of the tube. This reproves a result from [GRV2] and for spaces of constant curvature we obtain a well-known fact proved by H. Weyl. See [KV4] for more details and proofs.

11. Generalizations

11.1. Up to now we considered only (proper) Riemannian manifolds but the D'Atri property may also be considered on *pseudo-Riemannian spaces.* This has already been done in [D], [DN2], [VW1]. With some evident modification (for example, instead of the distance function one uses the function $\frac{1}{2}ed^2(m, p)$ where e is the signature of the geodesic from p to m)

the characterizations given in Section 3 also hold in this case. Furthermore, in [D], [DN2] it it proved that natural reductive spaces of any signature are D'Atri spaces.

11.2. Some pseudo-Riemannian examples are obtained in [VW1]. We shall describe them now. So, let M be an n-dimensional manifold of class C^ω endowed with the torsionfree affine connection ∇. The definition of a *Riemannian extension* of (M, ∇) was introduced in [PW]. It is the pseudo-Riemannian manifold $(T^*M, \tilde{g})$ where T^*M denotes the cotangent bundle of M and $\tilde{g}$ is the pseudo-Riemannian metric defined in the following way. Let (U, x) be a coordinate chart of M with $x(p) = (x^1, ..., x^n), p \in U$. Further, let Γ^k_{ij} denote the Christoffel symbols of ∇ with respect to the coordinate system (U, x). The points in the bundle over U can be represented by (x^i, ξ_j) where ξ_j are the components of a covariant vector of T^*M relative to the chart (U, x). Then we put

$$(\tilde{g}_{ab}(x^i, \xi_j)) = \begin{pmatrix} g_{\alpha\beta} & I \\ I & 0 \end{pmatrix}$$

$a, b = 1, ..., 2n$, where I denotes the unit matrix of type (n, n) and $g_{\alpha\beta} = -2\Gamma^h_{\alpha\beta}\xi_h$.

In [VW1] it is proved that the Riemannian extension with respect to the Levi Civita connection of a pseudo-Riemannian manifold M is a D'Atri space if and only if M is a D'Atri space.

This construction makes it possible to extend the primarily metric concept of a D'Atri space to *affine manifolds*: A manifold with a torsionfree affine connection is said to be an *affine D'Atri space* if its Riemannian extension is a pseudo-Riemannian D'Atri space.

11.3. We finish this paper with the notion of an *asymptotic D'Atri space* introduced for complete manifolds of non-positive curvature in [H2].

Definition. A complete Riemannian manifold of non-positive curvature is called an *asymptotic D'Atri space* if for every unit speed geodesic $\gamma_\xi : r \mapsto exp_m(r\xi)$ we have

$$\lim_{r\to\infty} \frac{\ln \det A_\xi(r)}{\ln \det A_{-\xi}(r)} = 1$$

where $A_\xi(r)$ is the endomorphism-valued function defined in Section 2.

Note that for the manifolds of non-positive curvature the Rauch comparison theorem gives that both the numerator and the denominator in the limit formula have at least logarithmic growth and so the definition looks natural.

In [H2] one proves the following result : A homogeneous Hadamard manifold is an asymptotic D'Atri space if and only if each of its de Rham factors is either flat, or a higher rank symmetric space or has algebraic rank equal to one.

Up to now we do not have any explicit examples of asymptotic D'Atri spaces which are not D'Atri spaces.

References

[AL] D.V. Alekseevskii, Homogeneous Riemannian spaces of negative curvature, *Math. USSR* 25 (1975), 87-109.

[AW1] R. Azencott and E. Wilson, Homogeneous manifolds with negative curvature I, *Trans. Amer. Math. Soc.* 215 (1976), 323-362.

[AW2] R. Azencott and E. Wilson, *Homogeneous manifolds with negative curvature* II, Mem. Amer. Math. Soc. 178, 1976.

[BKV] J. Berndt, O. Kowalski and L. Vanhecke, Geodesics in weakly symmetric spaces, preprint 1995.

[BPV] J. Berndt, F. Prüfer and L. Vanhecke, Symmetric-like Riemannian manifolds and geodesic symmetries, *Proc. Roy. Soc. Edinburgh Sect. A*, to appear.

[BRV] J. Berndt, F. Ricci and L. Vanhecke, Weakly symmetric generalized Heisenberg groups, in preparation.

[BTOV] J. Berndt, Ph. Tondeur and L. Vanhecke, Examples of weakly symmetric spaces in contact geometry, preprint 1995.

[BTV1] J. Berndt, F. Tricerri and L. Vanhecke, Geometry of generalized Heisenberg groups and their Damek-Ricci harmonic extensions, *C.R. Acad. Sci. Paris Sér. I Math.* 318 (1994), 471-476.

[BTV2] J. Berndt, F. Tricerri and L. Vanhecke, *Generalized Heisenberg groups and Damek-Ricci harmonic spaces*, Lecture Notes in Mathematics 1598, Springer-Verlag, Berlin, Heidelberg, New York, 1995.

[BV1] J. Berndt and L. Vanhecke, Two natural generalizations of locally symmetric spaces, *Diff. Geom. Appl.* 2 (1992), 57-80.

[BV2] J. Berndt and L. Vanhecke, Naturally reductive Riemannian homogeneous spaces and real hypersurfaces in complex and quaternionic space forms, *Differential Geometry and Its Applications*, Proc. Conf. Opava,

Czechoslovakia 1992 (Eds. O. Kowalski and D. Krupka), Silesian Univ. Opava and Open Education and Sciences, 1993, 415-426.

[BV3] J. Berndt and L. Vanhecke, Geodesic spheres and generalizations of symmetric spaces, *Boll. Un. Mat. Ital. A* (7) 7 (1993), 125-134.

[BV4] J. Berndt and L. Vanhecke, Geometry of weakly symmetric spaces, preprint, 1994.

[BE] A.L. Besse, *Manifolds all of whose geodesics are closed*, Ergeb. Math. Grenzgeb. 93, Springer-Verlag, Berlin, Heidelberg, New York, 1978.

[BCG] G. Besson, G. Courtois and S. Gallot, Volumes, entropies et rigidités des espaces localement symétriques de courbure strictement négative, *C.R. Acad. Sci. Paris Sér. I Math.* 319 (1994), 81-84.

[BI] L. Bieszk, On natural reductivity of five-dimensional commutative spaces, *Note Mat.* 8 (1988), 13-43.

[BL] D.E. Blair, *Contact manifolds in Riemannian geometry*, Lecture Notes in Mathematics 509, Springer-Verlag, Berlin, Heidelberg, New York, 1976.

[BLV1] D.E. Blair and L. Vanhecke, New characterizations of φ-symmetric spaces, *Kodai Math. J.* 10 (1987), 102-107.

[BLV2] D.E. Blair and L. Vanhecke, Symmetries and φ-symmetric spaces, *Tôhoku Math. J.* 39 (1987), 373-383.

[BLV3] D.E. Blair and L. Vanhecke, Volume-preserving φ-geodesic symmetries, *C.R. Math. Rep. Acad. Sci. Canada* 9 (1987), 31-36.

[BO] E. Boeckx, Einstein-like semi-symmetric spaces, *Arch. Math. (Brno)* 29 (1993), 235-240.

[BU] P. Bueken, *Reflections and rotations in contact geometry*, doctoral dissertation, Catholic University Leuven, 1992.

[CA] J. Cao, Rigidity for non-compact surfaces of finite area and certain Kähler manifolds, *Ergodic Theory Dynamical Systems*, to appear.

[CV] B.Y. Chen and L. Vanhecke, Differential geometry of geodesic spheres, *J. Reine Angew. Math.* 325 (1981), 28-67.

[CO] J.T. Cho, Natural generalizations of locally symmetric spaces, *Indian J. Pure Appl. Math.* 24 (1993), 231-240.

[DR] E. Damek and F. Ricci, A class of nonsymmetric harmonic Riemannian spaces, *Bull. Amer. Math. Soc.* 27 (1992), 139-142.

[D] J.E. D'Atri, Geodesic spheres and symmetries in naturally reductive spaces, *Michigan Math. J.* 22 (1975), 71-76.

[DN1] J.E. D'Atri and H.K. Nickerson, Divergence-preserving geodesic symmetries, *J. Differential Geom.* 3 (1969), 467-476.

[DN2] J.E. D'Atri and H.K. Nickerson, Geodesic symmetries in spaces with special curvature tensors, *J. Differential Geom.* 9 (1974), 251-262.

[DZ] J.E. D'Atri and W. Ziller, *Naturally reductive metrics and Einstein metrics on compact Lie groups*, Mem. Amer. Math. Soc. 215 (1979).

[DK] D.M. DeTurck and J.L. Kazdan, Some regularity theorems in Riemannian geometry, *Ann. Scient. Ec. Norm. Sup.* 14 (1981), 249-260.

[GI] P.B. Gilkey, Spectral geometry of the higher order Laplacian, *Duke Math. J.* 47 (1980), 511-528.

[GSV] P. Gilkey, A. Swann and L. Vanhecke, Isoparametric geodesic spheres and a conjecture of Osserman concerning the Jacobi operator, *Quart. J. Math. Oxford Ser. (2)*, to appear.

[GGV1] J.E. González-Dávila, M.C. González-Dávila and L. Vanhecke, Killing-transversally symmetric spaces, *Proc. Workshop Recent Topics in Differential Geometry, Puerto de la Cruz* 1990 (Eds. D. Chinea and J.M. Sierra), Secret. Publ. Univ. de La Laguna, Serie Informes 32 (1991), 77-78.

[GGV2] J.C. González-Dávila, M.C. González-Dávila and L. Vanhecke, Reflections and isometric flows, *Kyungpook Math. J.*, to appear.

[GGV3] J.C. González-Dávila, M.C. González-Dávila and L. Vanhecke, Classification of Killing-transversally symmetric spaces, *Tsukuba J. Math.*, to appear.

[GGV4] J.C. González-Dávila, M.C. González-Dávila and L. Vanhecke, The Gelfand theorem in flow geometry, *C.R. Math. Rep. Acad. Sci. Canada* 15 (1993), 281-285.

[GGV5] J.C. González-Dávila, M.C. González-Dávila and L. Vanhecke, Normal flow forms and their classification, preprint 1995.

[GV1] J.C. González-Dávila and L. Vanhecke, Geodesic spheres and isometric flows, *Colloq. Math.* 67 (1994), 223-240.

[GV2] J.C. González-Dávila and L. Vanhecke, Geometry of tubes and isometric flows, *Math. J. Toyama Univ.*, to appear.

[GV3] J.C. González-Dávila and L. Vanhecke, New examples of weakly symmetric spaces, preprint 1995.

[G1] M.C. González-Dávila, *Espacios transversalmente simétricos de tipo Killing*, doctoral dissertation, Universidad de La Laguna, 1992.

[G2] M.C. González-Dávila, KTS-spaces and natural reductivity, *Nihonkai Math. J.*, to appear.

[GO] C.S. Gordon, Naturally reductive homogeneous Riemannian manifolds, *Can. J. Math.* 37 (1985), 467-487.

[GP] P. Günther and F. Prüfer, D'Atri spaces are ball-homogeneous, in preparation.

[GR1] A. Gray, Riemannian manifolds with geodesic symmetries of order 3, *J. Differential Geom.* 7 (1972), 343-369.

[GR2] A. Gray, *Tubes*, Addison-Wesley Publ. Co. Reading, 1990.

[GRV1] A. Gray and L. Vanhecke, The volume of tubes in a Riemannian manifold, *Acta Math.* 142 (1972) 157-198.

[GRV2] A. Gray and L. Vanhecke, The volumes of tubes in a Riemannian manifold, *Rend. Sem. Mat. Univ. Politec. Torino* 39 (1981), 1-50.

[GRV3] A. Gray and L. Vanhecke, The volume of tubes about curves in a Riemannian manifold, *Proc. London Math. Soc.* 44 (1982), 215-243.

[GW] A. Gray and T.J. Willmore, Mean-value theorems for Riemannian manifolds, *Proc. Roy. Soc. Edinburgh Sect. A* 92 (1982), 343-364.

[H1] J. Heber, On the geometric rank of homogeneous spaces of nonpositive curvature, *Invent. Math.* 112 (1993), 151-170.

[H2] J. Heber, Homogeneous spaces of nonpositive curvature and their geodesic flow, *Int. J. Math.*, to appear.

[HE] E. Heintze, On homogeneous manifolds of negative curvature, *Math. Ann.* 211 (1974), 23-34.

[J] J.P. Jacob, Geodesic symmetries of homogeneous Kähler manifolds, *Geom. Dedicata* 10 (1981), 223-259.

[JE] G.R. Jensen, Homogeneous Einstein spaces of dimension four, *J. Differential Geom.* 3 (1969), 309-349.

[JI1] J.A. Jiménez, Existence of Hermitian n-symmetric spaces and of non-commutative naturally reductive spaces, *Math. Z.* 196 (1987), 133-139.

[JI2] J.A. Jiménez, Addendum to "Existence of Hermitian n-symmetric spaces and of non-commutative naturally reductive spaces, *Math. Z.* 197 (1988), 455-456.

[JI3] J.A. Jiménez, Non-commutative naturally reductive spaces of odd dimension, preprint.

[JI4] J.A. Jiménez, Stiefel manifolds and non-commutative φ-symmetric spaces, preprint.

[JK] J.A. Jiménez and O. Kowalski, The classification of φ-symmetric Sasakian manifolds, *Monatsh. Math.* 115 (1993), 83-98.

[K1] A. Kaplan, Fundamental solutions for a class of hypoelliptic PDE generated by composition of quadratic forms, *Trans. Amer. Math. Soc.* 258 (1980), 147-153.

[K2] A. Kaplan, On the geometry of groups of Heisenberg type, *Bull. London Math. Soc.* 15 (1983), 35-42.

[K3] A. Kaplan, Composition of quadratic forms in geometry and analysis : Some recent applications, *Proc. Quadratic Forms and Hermitian K-Theory*, McMasters Univ., 1983.

[KA] J.L. Kazdan, personal communication with the third author.

[KN] S. Kobayashi and K. Nomizu, *Foundations of the differential geometry* II, Interscience Publ., New York, 1969.

[KO1] O. Kowalski, Some curvature identities for commutative spaces, *Czech. Math. J.* 32 (1982), 389-396.

[KO2] O. Kowalski, The second mean-value operator on Riemannian manifolds, *Proc. of the ČSSR-GDR-Polish conference on Differential Geometry and Its Applications, Nove Město* 1980, Univerzita Karlova, 1982, 33-45.

[KO3] O. Kowalski, Spaces with volume-preserving symmetries and related classes of Riemannian manifolds, *Rend. Sem. Mat. Univ. Politec. Torino*, Fascicolo Speciale 1983, 131-158.

[KO4] O. Kowalski, An explicit classification of 3-dimensional Riemannian spaces satisfying $R(X,Y) \cdot R = 0$, preprint 1992.

[KO5] O. Kowalski, On strictly locally homogeneous Riemannian manifolds, preprint 1995.

[KOM] O. Kowalski and R.A. Marinosci, Weakly symmetric spaces in dimension five, preprint 1994.

[KP] O. Kowalski and F. Prüfer, On probabilistic commutative spaces, *Monatsh. Math.* 107 (1989), 57-68.

[KV1] O. Kowalski and L. Vanhecke, Ball-homogeneous and disk-homogeneous Riemannian manifolds, *Math. Z.* 180 (1982), 429-444.

[KV2] O. Kowalski and L. Vanhecke, Opérateurs différentiels invariants et symétries géodésiques préservant le volume, *C.R. Acad. Sc. Paris Sér. I Math.* 296 (1983), 1001-1003.

[KV3] O. Kowalski and L. Vanhecke, Four-dimensional naturally reductive homogeneous spaces, *Rend. Sem. Mat. Univ. Politec. Torino*, Fascicolo Speciale 1983, 222-232.

[KV4] O. Kowalski and L. Vanhecke, G-deformations of curves and volumes of tubes in Riemannian manifolds, *Geom. Dedicata* 15 (1983), 125-135.

[KV5] O. Kowalski and L. Vanhecke, A generalization of a theorem on naturally reductive homogeneous spaces, *Proc. Amer. Math. Soc.* 91 (1984), 433-435.

[KV6] O. Kowalski and L. Vanhecke, Classification of four-dimensional commutative spaces, *Quart. J. Math. Oxford (2)* 35 (1984), 281-291.

[KV7] O. Kowalski and L. Vanhecke, Two-point functions on Riemannian manifolds, *Ann. Global Anal. Geom.* 3 (1985), 95-119.

[KV8] O. Kowalski and L. Vanhecke, Classification of five-dimensional naturally reductive spaces, *Math. Proc. Cambridge Philos. Soc.* 97 (1985), 445-463.

[KV9] O. Kowalski and L. Vanhecke, The Gelfand theorem and its converse for Kähler manifolds, *Proc. Amer. Math. Soc.* 102 (1988), 150-152.

[KV10] O. Kowalski and L. Vanhecke, The Gelfand theorem and its converse in Sasakian geometry, *Kodai Math. J.* 11 (1988), 70-77.

[KV11] O. Kowalski and L. Vanhecke, Riemannian manifolds with homogeneous geodesics, *Boll. Un. Mat. Ital. B (7)* 5 (1991), 189-246.

[KW] O. Kowalski and S. Węgrzynowski, A classification of five-dimensional φ-symmetric spaces, *Tensor (N.S.)* 46 (1987), 379-386.

[KOZ] S. Kôzaki, On mean value theorems for small geodesic spheres in Riemannian manifolds, *Czech. Math. J.* 42 (1992), 519-547.

[KR] K. Krämer, Eine Klassifikation bestimmter Untergruppen kompakter zusammenhängender Liegruppen, *Comm. in Algebra* 3 (8) (1975), 691-737.

[MA] O.V. Manturov, Homogeneous Riemannian manifolds with irreducible isotropy group, *Trudy Sem. Vector and Tensor Analysis* 13 (1966), 68-145.

[MV] M.D. Monar and L. Vanhecke, Locally symmetric quaternionic Kähler manifolds, *Diff. Geom. Appl.* 4 (1994), 127-149.

[MO] A. Montesinos, Flot géodésique et espaces $\mathcal{A}$ de A. Gray, *C.R. Acad. Sci. Paris Sér. I Math.* 305 (1987), 789-791.

[N] S. Nagai, Naturally reductive Riemannian homogeneous structure on a homogeneous real hypersurface in complex space forms, preprint, 1993.

[PW] E.M. Patterson and A.G. Walker, Riemannian extensions, *Quart. J. Math. Oxford* 3 (1952), 19-28.

[P1] F. Prüfer, On compact Riemannian manifolds with volume-preserving symmetries, *Ann. Global Anal. Geom.* 7 (1989), 133-140.

[P2] F. Prüfer, *Mittelwertmethoden in D'Atri-Räumen unter besonderer Berücksichtigung der Herleitung von Poissonschen Summationsformeln für Räume konstanter positiver Krümmung*, Habilitationsschrift, Karl-Marx-Universität Leipzig, 1990.

[R] F. Ricci, Commutative algebras of invariant functions on groups of Heisenberg type, *J. London Math. Soc.* (2) 32 (1985), 265-271.

[RI1] C. Riehm, The automorphism group of a composition of quadratic forms, *Trans. Amer. Math. Soc.* 269 (1982), 403-414.

[RI2] C. Riehm, Explicit spin representations and Lie algebras of Heisenberg type, *J. London Math. Soc.* (2) 29 (1984), 49-62.

[ROU] P.H. Roberts and H.D. Ursell, Random walk on a sphere and on a Riemannian manifold, *Philos. Trans. Roy. Soc. London A* 252 (1960), 317-356.

[RU] H.S. Ruse, On commutative Riemannian manifolds, *Tensor (N.S.)* 26 (1972), 180-184.

[RWW] H.S. Ruse, A.G. Walker and T.J. Willmore, *Harmonic spaces*, Cremonese, Roma, 1961.

[S] R. Schneider, Curvature measures and integral geometry of convex bodies, *Rend. Sem. Mat. Univ. Politec. Torino* 38 (1980), 79-98.

[SV1] K. Sekigawa and L. Vanhecke, Volume-preserving geodesic symmetries on four-dimensional Kähler manifolds, *Differential Geometry*, Proc. 2nd Symp. Peñiscola/Spain 1985 (Eds. A.M. Naveira, A. Ferrández and F. Mascaró), Lecture Notes in Mathematics 1209 (1986), 275-291.

[SV2] K. Sekigawa and L. Vanhecke, Volume-preserving geodesic symmetries on four-dimensional 2-stein spaces, *Kodai Math. J.* 9 (1986), 215-224.

[SV3] K. Sekigawa and L. Vanhecke, Symplectic geodesic symmetries on Kähler manifolds, *Quart. J. Math. Oxford* (2) 37 (1986), 95-103.

[SE] A. Selberg, Harmonic analysis and discontinuous groups in weakly symmetric Riemannian spaces with applications to Dirichlet series, *J. Indian Math. Soc. (N.S.)* 20 (1956), 47-87.

[SZ1] Z.I. Szabó, Higher order Laplacians I, Harmonic and two-point homogeneous spaces, *Preprint Max Planck Inst. Math. Bonn* 65 (1989).

[SZ2] Z.I.Szabó, Higher order Laplacians II, Laplacians commuting with the higher orders, Spectral theorems, *Preprint Max Planck Inst. Math. Bonn* 66 (1989).

[SZ3] Z.I. Szabó, The Lichnerowicz conjecture on harmonic manifolds, *J. Differential Geom.* 31 (1990), 1-28.

[SZ4] Z.I. Szabó, Spectral theory for operator families on Riemannian manifolds, *Proc. Sympos. Pure Math.* Part 3, 54 (1993), 615-665.

[T] T. Takahashi, Sasakian φ-symmetric spaces, *Tôhoku Math. J.* 29 (1977), 91-113.

[TO] Ph. Tondeur, *Foliations on Riemannian manifolds*, Universitext, Springer-Verlag, Berlin, Heidelberg, New York, 1988.

[TOV] Ph. Tondeur and L. Vanhecke, Transversally symmetric Riemannian foliations, *Tôhoku Math. J.* 42 (1990), 307-317.

[TV1] F. Tricerri and L. Vanhecke, *Homogeneous structures on Riemannian manifolds*, London Math. Soc. Lecture Note Series 83, Cambridge Univ. Press, Cambridge, 1983.

[TV2] F. Tricerri and L. Vanhecke, Naturally reductive homogeneous spaces and generalized Heisenberg groups, *Compositio Math.* 52 (1984), 389-408.

[TV3] F. Tricerri and L. Vanhecke, Geodesic spheres and naturally reductive homogeneous spaces, *Riv. Mat. Univ. Parma (4)* 10 (1984), 123-131.

[TV4] F. Tricerri and L. Vanhecke, Geometry of a class of non-symmetric harmonic manifolds, *Differential Geometry and Its Applications*, Proc. Conf. Opava, Czechoslovakia 1992 (Eds. O. Kowalski and D. Krupka), Silesian Univ. Opava and Open Education and Sciences, 1993, 415-426.

[TV5] F. Tricerri and L. Vanhecke, The geometry of the Damek-Ricci harmonic spaces, in preparation.

[V1] L. Vanhecke, A note on harmonic spaces, *Bull. London Math. Soc.* 13 (1981), 545-546.

[V2] L. Vanhecke, Some solved and unsolved problems about harmonic and commutative spaces, *Bull. Soc. Math. Belg. Sér. B* 34 (1982), 1-24.

[V3] L. Vanhecke, The canonical geodesic involution and harmonic spaces, *Ann. Global Anal. Geom.* 1 (1983), 131-136.

[V4] L. Vanhecke, Symmetries and homogeneous Kähler manifolds, *Differential Geometry and Its Applications* (Eds. D. Krupka and A. Sveč), Reidel Publ. Co., Dordrecht, 1987, 339-357.

[V5] L. Vanhecke, Geometry in normal and tubular neighborhoods, *Rend. Sem. Fac. Sci. Univ. Cagliari*, Supplemento al Vol. 58 (1988), 73-176.

[V6] L. Vanhecke, The geometry of reflections on Riemannian manifolds, *Proc. International Conference on Diff. Geom. and Its Appl., Dubrovnik* 1988 (Eds. N. Bokan, I. Čomić, J. Nikić, M. Prvanović), Univ. Belgrade and Novi Sad, 1989, 387-400.

[V7] L. Vanhecke, Geometry and symmetry, *Advances in Differential Geometry and Topology* (Eds. I.S.I. and F. Tricerri), World Scientific Publ. Co., Singapore, 1990, 115-129.

[VW1] L. Vanhecke and T.J. Willmore, Riemannian extensions of D'Atri spaces, *Tensor (N.S.)* 38 (1982), 154-158.

[VW2] L. Vanhecke and T.J. Willmore, Interaction of tubes and spheres, *Math. Ann.* 263 (1983), 31-42.

[W] A.G. Walker, On Lichnerowicz's conjecture for harmonic 4-spaces, *J. London Math. Soc.* 24 (1948-49), 317-329.

[WA] Y. Watanabe, Geodesic symmetries in Sasakian locally φ-symmetric spaces, *Kodai Math. J.* 3 (1980), 48-55.

[WF] Y. Watanabe and H. Fujita, A family of homogeneous Sasakian structures on $S^2 \times S^3$, *C.R. Math. Rep. Acad. Sci. Canada* 10 (1988), 57-61.

[WE] H. Weyl, On the volume of tubes, *Amer. J. Math.* 61 (1939), 461-472.

[WI1] T.J. Willmore, 2-point invariant functions and k-harmonic manifolds, *Rev. Roumaine Math. Pures Appl.* 13 (1968), 1051-1057.

[WI2] T.J. Willmore, Commutative and related metrics, *Astérisque* 107-108 (1983), 219-228.

[WI3] T.J. Willmore, *Riemannian Geometry*, Oxford Science Publications, Clarendon Press, Oxford, 1993.

[WEH] T.J. Willmore and K. El Hadi, k-harmonic symmetric manifolds, *Rev. Roumaine Math. Pures Appl.* 15 (1970), 1573-1577.

[WO] J. Wolf, The geometry and structure of isotropy irreducible homogeneous spaces, *Acta Math.* 120 (1968), 59-148. Correction, *Acta Math.* 152 (1984), 141-142.

[Z] W. Ziller, Homogeneous Einstein metrics on spheres and projective spaces, *Math. Ann.* 259 (1982), 351-358.

Charles University, Faculty of Mathematics and Physics, Sokolovská 83, 18600 Praha, Czech Republic

Universität Leipzig, Fakultät für Mathematik und Informatik, Mathematisches Institut, Augustusplatz 10, 04109 Leipzig, Germany

Katholieke Universiteit Leuven, Department of Mathematics, Celestijnenlaan 200B, 3001 Leuven, Belgium

Revised December 1995; corrections January 1996

Multiple Point Blowup Phenomenon in Scalar Curvature Equations on Spheres of Dimension Greater Than Three

YanYan Li *

§0.

Let $(\mathbf{S}^n, g_0)$ be the standard n-sphere. The following question was raised by L. Nirenberg. Which function K on $\mathbf{S}^2$ is the Gauss curvature of a metric g on $\mathbf{S}^2$ conformally equivalent to g_0? Naturally one may ask a similar question in higher dimensional case, namely, which function K on $\mathbf{S}^n$ is the scalar curvature of a metric g on $\mathbf{S}^n$ conformally equivalent to g_0?

For $n \geq 3$, we write $g = v^{\frac{4}{n-2}} g_0$, the problem is equivalent to finding a function v on $\mathbf{S}^n$ which satisfies the following equation

$$(1) \qquad -\Delta_{g_0} v + c(n) R_0 v = c(n) K v^{\frac{n+2}{n-2}}, \quad v > 0, \quad \text{on } \mathbf{S}^n,$$

where $c(n) = \frac{n-2}{4(n-1)}$, $R_0 = n(n-1)$ is the scalar curvature of g_0 and Δ_{g_0} denotes the Laplace-Beltrami operator associated with the metric g_0.

For $n = 2$, we write $g = e^{2v} g_0$, the problem is equivalent to finding a function v on $\mathbf{S}^2$ which satisfies the following equation

$$(2) \qquad -\Delta_{g_0} v + 1 = K e^{2v}.$$

There has been much work devoted to (1) and (2). The purpose of this paper is to briefly outline the proof of the main results in [L2], and therefore we will not give references to many earlier results mentioned below. For such information and other work on the problem, please see the introductions and the references in [L1] and [L2].

For $P \in \mathbf{S}^n, 0 < t < \infty$, we define a centered dilation conformal transformation $\varphi_{P,t} : \mathbf{S}^n \to \mathbf{S}^n$, by $y \mapsto ty$, where $y \in \mathbf{R}^n$ is the stereographic projection coordinates of points on $\mathbf{S}^n$ while the stereographic projection is performed with P as the north pole to the equatorial plane of $\mathbf{S}^n$.

For a conformal transformation $\varphi : \mathbf{S}^n \to \mathbf{S}^n$, we associate it with a natural transformation on functions:

$$T_\varphi u = u \circ \varphi |\det d\varphi|^{\frac{n-2}{2n}}.$$

---

*Partially supported by the Alfred P. Sloan Foundation Research Fellowship and NSF grant DMS-9401815.

For a function K on $\mathbf{S}^n$, let

$$\mathcal{M}_K = \{v \in C^2(\mathbf{S}^n) \mid v \text{ satisfies (1) or (2) }\}$$

denote the set of solutions. For $K \equiv R_0$, it is well known that for the obvious choice of $a = a(n)$,

$$\mathcal{M}_{R_0} = \{T_{\varphi_{P,t}} a \mid P \in \mathbf{S}^n, 1 \le t < \infty\}.$$

It is clear that $\mathcal{M}_{R_0}$ is not compact even in $C^0(\mathbf{S}^n)$. However it is compact modulo $T_{\varphi_{P,t}}$. This turns out to be a general fact for dimension $n = 2$ and 3, namely, for any $K \in C^2(\mathbf{S}^n)^+ = \{K \in C^2(\mathbf{S}^n) \mid K > 0 \text{ on } \mathbf{S}^n\}$ $(n = 2, 3)$, any $K_i \to K$ in $C^2(\mathbf{S}^n)$, any $v_i \in \mathcal{M}_{K_i}$, after passing to a subsequence,

$$\text{(3)} \qquad \{T_{\varphi_{P_i,t_i}} v_i\} \text{ is compact in } C^{1,\alpha}(\mathbf{S}^n)\ \forall\, 0 < \alpha < 1,$$

where $P_i \in \mathbf{S}^n$ and $1 \le t_i < \infty$.

In particular, $\{v_i\}$ has at most one (isolated simple) blow up point on $\mathbf{S}^n$. (3) plays an important role in the existence theory. In particular, a Leray-Schauder degree approach to the existence results is initiated in [CGY]. Naturally, one would like to know whether (3) is true in higher dimensions. It is proved in [L1] that (3) still holds provided $\{\nabla K_i\}$ are suitably flat near critical points of K. This flatness hypothesis automatically holds when $n = 3$. More precisely, we have

Theorem 0.1 ([L1]). *Let $K \in C^{n-2}(\mathbf{S}^n)^+$ $(n \ge 3)$, and $K_i \to K$ in $C^{n-2}(\mathbf{S}^n)$ satisfy*

$$(*)_\beta \qquad |\nabla^\alpha K_i| \le C|\nabla K_i|^{\frac{\beta-|\alpha|}{\beta-1}} \quad on\ \mathbf{S}^n \qquad \forall\, 1 \le |\alpha| \le n-2,$$

where $\beta > n-2$ and $C > 0$ are independent of i. Then for any $v_i \in \mathcal{M}_{K_i}$, we have (after passing to a subsequence) (3). In particular, $\{v_i\}$ has at most one (isolated simple) blow up point on $\mathbf{S}^n$.

Using Theorem 0.1, certain existence results for $n \ge 4$ are established in [L1] through the Leray-Schauder degree approach.

It is proved in [L2] that (3) is *false* in general when $n \ge 4$. Multiple point blow up indeed occurs for $n \ge 4$. In fact, we have such examples satisfying $(*)_{n-2}$. This in particular shows that Theorem 0.1 is sharp with respect to the range of β. This causes difficulties in understanding the problem in higher dimensions. Due to the results in [L2], in dimension $n \ge 4$, any *finite number* of point blow up may occur. For $n = 4$, this is the only possible blow up. The following question remains open.

Question. *For $n \geq 5$, does there exist $K \in C^2(\mathbf{S}^n)^+$, $K_i \to K$ in $C^2(\mathbf{S}^n)$, and $v_i \in \mathcal{M}_{K_i}$ such that $\{v_i\}$ blows up on a set consisting of infinitely many points?*

One of the main results in [L2] concerns dimension $n = 4$, where multiple point blow up has to be handled. For $n = 4$, (1) becomes

$$-\Delta_{g_0} v + 2v = \frac{1}{6} K v^3, \quad v > 0, \quad \text{on } \mathbf{S}^4. \tag{4}$$

For $P \in \mathbf{S}^4$, let $G_P(q)$ denote the Green's function of $-L_{g_0} \equiv -\Delta_{g_0} + 2$ on $\mathbf{S}^4$. It is well known that G_P satisfies

$$\begin{cases} G_P(q) > 0, \quad -L_{g_0} G_P(q) = 0, & \text{for all} \quad q \in \mathbf{S}^4 \setminus \{P\}, \\ -\dfrac{1}{|\mathbf{S}^4|} \displaystyle\int_{\mathbf{S}^4} G_P L_{g_0} \psi dV_{g_0} = \psi(P), & \text{for all } \psi \in C^\infty(\mathbf{S}^4). \end{cases}$$

For $K \in C^2(\mathbf{S}^4)^+$, we set

$$\mathcal{K} = \{q \in \mathbf{S}^4 \mid \nabla_{g_0} K(q) = 0\}, \quad \mathcal{K}^+ = \{q \in \mathbf{S}^4 \mid \nabla_{g_0} K(q) = 0, \Delta_{g_0} K(q) > 0\},$$
$$\mathcal{K}^- = \{q \in \mathbf{S}^4 \mid \nabla_{g_0} K(q) = 0, \Delta_{g_0} K(q) < 0\}.$$

We also associate any k ($k \geq 1$) distinct points $q^{(1)}, \cdots, q^{(k)} \in \mathcal{K} \setminus \mathcal{K}^+$ with a $k \times k$ symmetric matrix $M = M(K; q^{(1)}, \cdots, q^{(k)})$ defined by

$$M_{ij} = \begin{cases} -\dfrac{\Delta_{g_0} K(q^{(i)})}{K(q^{(i)})^2}, & i = j, \\ -\dfrac{48|\mathbf{S}^3|}{|\mathbf{S}^4|} \dfrac{G_{q^{(i)}}(q^{(j)})}{\sqrt{K(q^{(i)}) K(q^{(j)})}}, & i \neq j. \end{cases}$$

Let $\mu(M)$ denote the least eigenvalue of M.

Set

$\mathcal{A} =$

$$\{K \in C^2(\mathbf{S}^4)^+ \mid \mu(M(K; q^{(1)}, \cdots, q^{(k)})) \neq 0, \forall\, q^{(1)}, \cdots, q^{(k)} \in \mathcal{K} \setminus \mathcal{K}^+, k \geq 1\}.$$

Lemma 0.1. *$\mathcal{A}$ is an open and dense subset of $C^2(\mathbf{S}^4)^+$ with respect to the C^2 topology.*

The proof of Lemma 0.1 is elementary, see [L2]. It is clear that $\mathcal{A}$, equipped with the C^2 norm, is a Banach space. It turns out that we can define an *integer valued, continuous* function (hence locally constant) Index: $\mathcal{A} \to \mathbf{Z}$ with the following property: For any Morse function $K \in \mathcal{A}$ with $\mathcal{K}^- = \{q^{(1)}, \cdots, q^{(m)}\}$,

$$\text{Index}\,(K) = -1 + \sum_{k=1}^{m} \sum_{\substack{M(K; q^{(i_1)}, \cdots, q^{(i_k)}) > 0 \\ 1 \leq i_1 < \cdots < i_k \leq m}} (-1)^{k-1+\sum_{j=1}^{k} i(q^{(i_j)})}, \tag{5}$$

where $i(q^{(i_j)})$ denotes the Morse index of K at $q^{(i_j)}$.

Theorem 0.2 ([L2]).

(a) For any $K \in \mathcal{A}$, there exists some positive constant $C = C(K)$ such that for any $K_i \to K$ in $C^2(\mathbf{S}^4)$, and any $v_i \in \mathcal{M}_{K_i}$,

$$C^{-1} \le \liminf_{i\to\infty}(\min_{\mathbf{S}^4} v_i) \le \limsup_{i\to\infty}(\max_{\mathbf{S}^4} v_i) \le C.$$

(b) For any $K \in C^2(\mathbf{S}^4)^+ \setminus \mathcal{A} = \partial\mathcal{A}$, there exists $K_i \to K$ in $C^2(\mathbf{S}^4)$, and $v_i \in \mathcal{M}_{K_i}$ such that

$$\lim_{i\to\infty}(\max_{\mathbf{S}^4} v_i) = \infty, \quad \lim_{i\to\infty}(\min_{\mathbf{S}^4} v_i) = 0. \tag{6}$$

Let

$$R_{K,\alpha} = \inf\{R \mid 1/R < v < R, \text{ on } \mathbf{S}^4, \ \|v\|_{C^{2,\alpha}(\mathbf{S}^4)} < R, \ \forall\, v \in \mathcal{M}_K\},$$

and

$$\mathcal{O}_R = \{v \in C^{2,\alpha}(\mathbf{S}^4) \mid 1/R < v < R, \|v\|_{C^{2,\alpha}(\mathbf{S}^4)} < R\}.$$

It follows from Theorem 0.2 and standard elliptic estimates that $\mathcal{M}_K$ is precompact in $C^{2,\alpha}(\mathbf{S}^4)$ for any $K \in \mathcal{A}$ and $0 < \alpha < 1$; moreover $1 \le R_{K,\alpha} < \infty$.

Theorem 0.3 ([L2]). *For any $K \in \mathcal{A}$ and $0 < \alpha < 1$, we have for all $R > R_{K,\alpha}$,*

$$\deg\left(v - \frac{1}{6}(-\Delta_{g_0} + 2)^{-1}(Kv^3), \mathcal{O}_R, 0\right) = \mathit{Index}\,(K). \tag{7}$$

In particular, (4) *has at least one solution when $\mathit{Index}(K) \ne 0$.*

Throughout this paper, deg denotes the Leray-Schauder degree with respect to $C^{2,\alpha}(\mathbf{S}^4)$.

§1.

In this section we prove Theorem 0.2(b), by using Theorem 0.2(a) and Theorem 0.3.

To prove Theorem 0.2(b), we first prove without much difficulty that Morse functions in $C^2(\mathbf{S}^4)^+ \setminus \mathcal{A} = \partial\mathcal{A}$ are dense in $C^2(\mathbf{S}^4)^+ \setminus \mathcal{A} = \partial\mathcal{A}$. Therefore we only need to establish Theorem 0.2(b) for $K \in \mathcal{A}$ being a Morse function. Let $\mathcal{K} \setminus \mathcal{K}^+ = \{q^{(1)}, \cdots, q^{(m)}\}$, it follows from the definition that there exists $1 \le i_1 < \cdots < i_k \le m$ $(k \ge 1)$ such that $\mu(M(K; q^{(i_1)}, \cdots, q^{(i_k)})) = 0$. By making small C^2 perturbations of K, we can assume without loss of generality that there is only one such $(i_1, \cdots, i_k)$. We can easily produce a smooth one parameter family of Morse functions $\{K_t\}$ $(-1 \le t \le 1)$ with the following properties:

(i) For $-1 \le t \le 1$, $\{K_t\}$ have the same critical points with the same Morse index as K, $K_0 = K$, and $\{K_t\}$ are identically the same as K except in some small balls around $q^{(i_1)}, \cdots, q^{(i_k)}$,

(ii) $K_t \in \mathcal{A}$ for $t \neq 0$,

(iii) For any $1 \le j_1 < \cdots < j_l \le m, (j_1, \cdots, j_l) \neq (i_1, \cdots, i_k)$, $\mu(M(K_t; q^{(j_1)}, \cdots, q^{(j_l)}))$ have the same sign for $-1 < t < 1$.

(iv) $\mu(M(K_t; q^{(i_1)}, \cdots, q^{(i_k)})) < 0$ for $-1 < t < 0$, but $\mu(M(K_t; q^{(i_1)}, \cdots, q^{(i_k)})) > 0$ for $0 < t < 1$.

The above can be achieved easily. The idea is to perturb the function K near $q^{(i_1)}, \cdots, q^{(i_k)}$ to change the Hessian of K at $q^{(i_1)}, \cdots, q^{(i_k)}$.

Using (5), we see that

$$\text{Index}(K_1) = \text{Index}(K_{-1}) + (-1)^{k-1+\sum_{j=1}^{k} i(q^{(i_j)})} \neq \text{Index}(K_{-1}). \tag{8}$$

It follows from Theorem 0.3, (8), and the homotopy invariance of the Leray-Schauder degree that there exists t_i and $v_i \in \mathcal{M}_{K_{t_i}}$ such that

$$\lim_{i\to\infty} \|v_i\|_{C^{2,\alpha}(\mathbf{S}^4)} = \infty, \qquad \text{or} \qquad \lim_{i\to\infty} (\min_{\mathbf{S}^4} v_i) = 0.$$

It follows from the above, Harnack inequality and standard elliptic estimates that (6) holds. Using Theorem 0.2(a) and (ii), we know that $t_i \to 0$. In fact, we know that $\{v_i\}$ blows up exactly at the k points $q^{(i_1)}, \cdots, q^{(i_k)}$.

§2.

In this section we outline the proof of Theorem 0.3 by using a more general form of Theorem 0.2(a). We introduce approximate equations to (4) for $K \in \mathcal{A}$:

$$-\Delta_{g_0} u + 2u = \frac{1}{6} K(x) u^{3-\tau}, \quad u > 0, \quad \text{on } \mathbf{S}^4, \tag{9}$$

where $\tau > 0$. We will analyze (9) for $\tau > 0$ very small and calculate the total degree contribution of solutions of (9) in terms of Index(K) and the total degree contribution of solutions of (4). The total degree contribution of solutions of (9) is known to be -1. It is shown that as $\tau \to 0^+$, solutions of (9) either stay bounded and subconverge in $C^{2,\alpha}$ norm to solutions of (4) or become unbounded and blow up at finite points. The degree contribution of the bounded solutions of (9) is equal to the degree contribution of all solutions of (4), while the degree contribution of the unbounded solutions of (9) can be shown equal to Index(K)+1. This yields (7). The following theorem, which is more general than Theorem 0.2(a), is proved in [L2].

Theorem 1.1 ([L2]). *For any $K \in C^2(\mathbf{S}^4)^+$, there exists $\delta^* = \delta^*(K) > 0$ with the following property. Let $\tau_i \geq 0$, $\tau_i \to 0$, $K_i \to K$ in $C^2(\mathbf{S}^4)$, $\{v_i\}$ satisfy*

$$-\Delta_{g_0} v_i + 2v_i = \frac{1}{6} K_i v_i^{3-\tau_i}, \quad v_i > 0, \quad on\ \mathbf{S}^4,$$

and

$$\lim_{i\to\infty} (\max_{\mathbf{S}^4} v_i) = \infty.$$

Then after passing to some subsequence we have

(i) *$\{v_i\}$ has only (isolated simple) blow up points $q^{(1)}, \cdots, q^{(k)} \in \mathcal{K} \setminus \mathcal{K}^+$ $(k \geq 1)$ with $|q^{(j)} - q^{(l)}| \geq \delta^*$ $\forall\, j \neq l$ and $\mu\left(M(K; q^{(1)}, \cdots, q^{(k)})\right) \geq 0$. Furthermore $q^{(1)}, \cdots, q^{(k)} \in \mathcal{K}^-$ if $k \geq 2$.*

(ii) *$\lambda_j := K(q^{(j)})^{-1/2} \lim_{i\to\infty} v_i(q_i^{(1)}) v_i(q_i^{(j)})^{-1} \in (0, \infty)$, $\mu^{(j)} := \lim_{i\to\infty} \tau_i v_i(q_i^{(j)})^2 \in [0, \infty)$, $\forall\, 1 \leq j \leq k$. Here $q_i^{(j)} \to q^{(j)}$ is the local maximum of v_i.*

(iii) *When $k = 1$,*

$$\mu^{(1)} = -24 K(q^{(1)})^{-2} \Delta_{g_0} K(q^{(1)}),$$

When $k \geq 2$,

$$\sum_{\ell=1}^{k} M_{\ell j}(K; q^{(1)}, \cdots, q^{(k)}) \lambda_\ell = \frac{1}{24} \lambda_j \mu^{(j)}, \qquad \forall\, 1 \leq j \leq k.$$

(iv) *$\mu^{(j)} \in (0, \infty)$ $\forall\, 1 \leq j \leq k$ if and only if $\mu(M(K; q^{(1)}, \cdots, q^{(k)}) > 0$.*

The above theorem characterizes the blow up behavior of solutions of (9). The following lemma follows easily from the Sobolev embedding theorem and the strong maximum principle.

Lemma 1.1. *There exists $C^* = C^*(n, \|K\|_{L^\infty(\mathbf{S}^4)}) > 0$ such that for all $0 \leq \tau \leq 2$ and any nontrivial solution $u \in H^1(\mathbf{S}^4)$ of*

$$-\Delta_{g_0} u + 2u = \frac{1}{6} K |u|^{2-\tau} u, \quad on\ \mathbf{S}^4$$

with $\|u^-\|_{L^4} \leq 1/C^*$, we have $u > 0$ on $\mathbf{S}^4$, hence a solution of (9).

For any Morse function $K \in \mathcal{A}$, let $\mathcal{K}^- = \{q^{(1)}, \cdots, q^{(m)}\}$. It follows from Theorem 0.2(a) and standard elliptic estimates that there exists some $R = R(K) > 0$ such that

$$v \in \mathcal{O}_R, \qquad \forall\, v \in \mathcal{M}_K.$$

Set

$$\mathcal{O}_{R,\delta} = \{u \in H^1(\mathbf{S}^4) \mid \inf_{w \in \mathcal{O}_R} \|u - w\|_{H^1} < \delta \}.$$

We fix $\delta = \delta(K) > 0$ to be small so that $\overline{\mathcal{O}_{R,\delta}} \subset \{u \in H^1(\mathbf{S}^4) \mid \|u^-\|_{L^4} < 1/C^*\}$.

For any $1 \le i_1 < \cdots < i_k \le m$, and $\tau > 0$ small, we can apply Theorem 1.1 to define a bounded open set

$$\Sigma_\tau(q^{(i_1)}, \cdots, q^{(i_k)}) \subset H^1(\mathbf{S}^4),$$

with the following properties.

(i) $\mathcal{O}_{R,\delta}, \{\Sigma_\tau(q^{(i_1)}, \cdots, q^{(i_k)}) \mid 1 \le i_1 < \cdots < i_k \le m\}$ are pairwise disjoint,

(ii) For all $u \in H^1(\mathbf{S}^4), u > 0$ a.e., u satisfies (9), we have

$$u \in \mathcal{O}_R \bigcup \{ \cup_{1 \le i_1 < \cdots < i_k \le m, M(K; q^{(i_1)}, \cdots, q^{(i_k)}) > 0} \Sigma_\tau(q^{(i_1)}, \cdots, q^{(i_k)})\}.$$

(iii) $\overline{\Sigma_\tau(q^{(i_1)}, \cdots, q^{(i_k)})} \subset \{u \in H^1(\mathbf{S}^4) \mid \|u^-\|_{L^4} < 1/C^*\}$ $\forall\, 1 \le i_1 < \cdots < i_k \le m$.

Due to the above we write the degree contribution of all solutions of (9) as the sum of the degree contribution of solutions of (9) in $\mathcal{O}_R$ and $\cup_{1 \le i_1 < \cdots < i_k \le m, M(K; q^{(i_1)}, \cdots, q^{(i_k)}) > 0} \Sigma_\tau(q^{(i_1)}, \cdots, q^{(i_k)})$.

It is known that for $\tau > 0$ small, there exists $C = C(\tau, K)$ such that all solutions of (9) are contained in $\mathcal{V} = \{ u \in H^1(\mathbf{S}^4) \mid \|u^-\|_{L^4} < 1/C^*,\ 1/C < \|u\|_{H^1} < C\}$.

Set

$$T_\tau(u) = \frac{1}{6}(-\Delta_{g_0} + 2)^{-1}(K|u|^{2-\tau}u).$$

It is a quite general fact that

$$\deg_{H^1}(u - T_\tau(u), \mathcal{V}, 0) = -1, \tag{10}$$

where $\deg_{H^1}$ denotes the Leray-Schauder degree with respect to $H^1(\mathbf{S}^4)$, which is well defined since $\tau > 0$.

It follows from (i), (ii), (10) and the properties of Leray-Schauder degree that for $\tau > 0$ small we have

$$\begin{aligned} & -1 \\ = & \deg_{H^1}(u - T_\tau(u), \mathcal{V}, 0) \\ = & \sum_{\substack{1 \le i_1 < \cdots < i_k \le m, \\ M(K; q^{(i_1)}, \cdots, q^{(i_k)}) > 0}} \deg_{H^1}\left(u - T_\tau(u), \Sigma_\tau(q^{(i_1)}, \cdots, q^{(i_k)}), 0\right) \\ & + \deg_{H^1}(u - T_\tau(u), \mathcal{O}_{R,\delta}, 0) \end{aligned} \tag{11}$$

Fix $0 < \alpha < 1$, it follows from Theorem B.2 of [L1] and Theorem 1.1 that

$$\deg_{H^1}(u - T_\tau(u), \mathcal{O}_{R,\delta}, 0) = \deg(u - T_\tau(u), \mathcal{O}_R, 0). \tag{12}$$

Finally, *a quite lengthy calculation* in [L2] shows that

$$\deg_{H^1}\Big(u - T_\tau(u), \Sigma_\tau(q^{(i_1)}, \cdots, q^{(i_k)}), 0\Big) = (-1)^{k+\sum_{j=1}^k i(q^{(i_j)})}. \tag{13}$$

In fact we have stronger information: There exists a unique solution of (9) in $\Sigma_\tau(q^{(i_1)}, \cdots, q^{(i_k)})$ which is nondegenerate with Morse index $5k - \sum_{j=1}^k i(q^{(i_j)})$. Theorem 0.3 follows from (11), (12), (13) and the definition of Index(K).

References

[A] T. Aubin, Nonlinear analysis on manifolds. Monge-Ampère equations. Springer-Verlag, New York, 1982.

[BC] A. Bahri, J.M. Coron, The scalar-curvature problem on standard three-dimensional sphere, J. of Func. Anal., 95 (1991), 106-172.

[BLR] A. Bahri, Y.Y. Li, O. Rey, On a variational problem with lack of compactness: the topological effect of the critical points at infinity, Calculus of Variations and PDE's 3 (1995), 67-93.

[CGS] L. Caffarelli, B. Gidas, J. Spruck, Asymptotic symmetry and local behavior of semilinear elliptic equations with critical Sobolev growth, Comm. Pure Appl. Math. 42 (1989), 271-297.

[CGY] S.Y. Chang, M.J. Gursky, P. Yang, The scalar curvature equation on 2- and 3-sphere, Calculus of Variations and PDE's 1 (1993), 205-229.

[CY] S.Y. Chang, P. Yang, Conformal deformations of metrics on $\mathbf{S}^2$, J. Diff. Geom. 27 (1988), 256-296.

[CD] W.X. Chen, W. Ding, Scalar curvature on $\mathbf{S}^2$, Trans. Amer. Math. Soc. 303 (1987), 365-382.

[ES] J. Escobar, R. Schoen, Conformal metrics with prescribed scalar curvature, Invent. Math. 86 (1986), 243-254.

[H] Z.C. Han, Prescribing Gaussian curvature on $\mathbf{S}^2$, Duke Math. J. 61 (1990), 679-703.

[KW] J. Kazdan, F. Warner, Existence and conformal deformation of metrics with prescribed Gaussian and scalar curvature, Ann. of Math. 101 (1975), 317-331.

[L1] Y.Y. Li, Prescribing scalar curvature on $\mathbf{S}^n$ and related problems, Part I, J. Differential Equations, 120 (1995), 319–410.

[L2] Y.Y. Li, Prescribing scalar curvature on $\mathbf{S}^n$ and related problems, Part II: existence and compactness, Comm. Pure Appl. Math., 49 (1996).

[M] J. Moser, On a nonlinear problem in differential geometry, Dynamical systems (M. Peixoto, ed.) Academic Press, New York, 1973.

[S] R. Schoen, On the number of constant scalar curvature metrics in a conformal class, Differential Geometry: A symposium in honor of Manfredo Do Carmo (H.B. Lawson and K. Tenenblat, eds), Wiley, 1991, 311-320.

[Z] D. Zhang, New results on geometric variational problems, thesis, Stanford University, 1990.

Department of Mathematics
Rutgers University
New Brunswick, NJ 08903

Received February 1995

The Harish-Chandra Realization for Non-Symmetric Domains in $\mathbb{C}^n$

R. Penney

Let $\mathcal{D} \subset \mathbb{C}^n$ be a bounded, homogeneous domain. This means that the group Aut $(\mathcal{D})$ of bi-holomorphisms of $\mathcal{D}$ acts transitively on $\mathcal{D}$. Let G be the component of the identity in Aut $(\mathcal{D})$. It is known that G is a Lie group.

We shall consider $\mathcal{D}$ as a Kähler manifold under the Bergman metric. If $\mathcal{D}$ is a symmetric space, then $\mathcal{D}$ is a Hermitian symmetric space. Harish-Chandra has given a *canonical* bounded realization of $\mathcal{D}$ defined completely in terms of the Lie algebra of G ([HC]). In this work we present a remarkably simple generalization of this theory to arbitrary (non-symmetric) bounded homogeneous domains. We also give a quite pretty description of the Cayley transformation for the general such domain. Our formula generalizes the Cayley transform defined via Jordan algebras. Our Harish-Chandra formula, in fact, provides canonical realizations of unbounded domains as well, although this will not be discussed here.

The main results which we are aware of concerning canonical realizations of non-symmetric domains are due to Dorfmeister [Do]. In the case of a quasi-symmetric domain, Dorfmeister writes down a Cayley transformation which, we suspect, is equivalent to ours. We should also mention that in [Ku], Kaneyuki defines a 'Cayley' transform. This transformation, however, is different from ours and does not seem to yield bounded realizations.

To describe our bounded realization, let $\omega_o \in \mathcal{D}$ be a fixed base point and let K be the isotropy subgroup. Let $\mathcal{G}$ and $\mathcal{K}$ be the respective Lie algebras of G and K. Then the tangent space of $\mathcal{D}$ at ω_o is $\mathcal{G}/\mathcal{K}$. Let $\mathcal{G}_c$ and $\mathcal{K}_c$ be the corresponding complexifications so that the complex tangent space at ω_o is $\mathcal{G}_c/\mathcal{K}_c$. Since G/K is a complex manifold, $\mathcal{G}/\mathcal{K}$ possesses a natural complex structure J. The $\pm i$ eigenspaces of J define, respectively, subspaces $\mathcal{Q}$ and $\mathcal{P}$ of $\mathcal{G}_c$, both of which contain $\mathcal{K}_c$, such that

$$\mathcal{G}_c/\mathcal{K}_c = \mathcal{P}/\mathcal{K}_c \oplus \mathcal{Q}/\mathcal{K}_c. \tag{1}$$

Furthermore, $\mathcal{Q} = \overline{\mathcal{P}}$. Due to the integrality condition for G/K, both $\mathcal{P}$ and $\mathcal{Q}$ are complex subalgebras of $\mathcal{G}_c$.

It is known (and will be shown below) that the center of G is contained in K. Since G acts effectively on $\mathcal{D}$, this implies that the center is in fact trivial. In particular, G is isomorphic (under the adjoint representation) with an algebraic subgroup of $\mathrm{Gl}(q)$ for some q. Let G_c be the complex algebraic closure of G in $\mathrm{Gl}(q, \mathbb{C})$. Let P and Q be the connected subgroups of G_c corresponding (under the exponential map) to $\mathcal{P}$ and $\mathcal{Q}$ respectively. Then $P \cap G = K$ (see Lemma 5 below). This produces the Borel embedding of $\mathcal{D}$:

$$\mathcal{D} = G/(P \cap G) = GP/P \subset G_c/P. \tag{2}$$

This embedding is useful because it describes the complex structure of $\mathcal{D}$. Explicitly, GP is an open subset of G_c and the complex structure of $\mathcal{D}$ is the complex structure on GP/P as an open subset of G_c/P. A useful consequence of this is the following observation:

Proposition 1. *The mapping $g \to g\omega_o$ of G into $\mathbb{C}^n$ extends holomorphically to a mapping of an open subset of G_c which contains G and P. On this set, P is the isotropy subgroup of ω_o.*

Next we need to bring some geometry into play. Since $\mathcal{D}$ has a G-invariant Kähler structure, it also has an invariant symplectic structure. To describe this, let $J : \mathcal{G} \to \mathcal{G}$ be some mapping which projects to the complex structure on $\mathcal{G}/\mathcal{K}$. Following Koszul ([Kl], Formula 4.5) we define a functional $\beta \in \mathcal{G}^*$ by:

$$\beta(X) = c \operatorname{Tr}_{\mathcal{G}/\mathcal{K}}(\operatorname{ad} JX - J \operatorname{ad} X). \tag{3}$$

where c is an arbitrary positive constant. Then the real, alternating bilinear form B_β on $\mathcal{G}$ defined by

$$B_\beta(X,Y) = \beta([X,Y])$$

has $\mathcal{K}$ as its radical and projects to the Kähler form on $\mathcal{G}/\mathcal{K}$. (In the Hermitian symmetric case, we may take $J = \operatorname{ad} Z$ where Z is a certain element of the center of $\mathcal{K}$. In this case, it follows that $\beta(X) = k(Z,X)$ where k is the Killing form.) Since the Kähler form is J-invariant,

$$\beta([\mathcal{Q},\mathcal{Q}]) = \{0\}. \tag{4}$$

It also follows that the invariant metric in $\mathcal{D}$ is defined by projecting the form

$$g_\beta(X,Y) = \beta([JX,Y]) \tag{5}$$

to the quotient. The coadjoint orbit of β under G is in fact identifiable with $\mathcal{D}$ since it is known (see Lemma 5 below) that:

$$\operatorname{Ad}^*_G(x)\beta = \beta \qquad \text{if and only if } x \in K. \tag{6}$$

Lemma 2. *Formula 6 holds with* G_c *and* K_c *in place of* G *and* K.

Proof Let (for the moment) $\overline{K}$ denote the set of x in G_c for which the analogue of Formula 6 holds. Then $\overline{K}$ is an algebraic subgroup of G_c which is defined over $\mathbb{R}$ and satisfies

$$G \cap \overline{K} = K.$$

In fact, since $\overline{K}$ is defined over $\mathbb{R}$, it is the algebraic closure of K. It follows that $\overline{K}$ is connected. Hence, $\overline{K} = K_c$ as claimed.

We are now ready to describe our 'Harish-Chandra embedding'. Our first main result is:

Theorem 3. $G \subset QP$.

The Borel embedding (Formula 2) allows us to realize $\mathcal{D}$ in $QP/P = Q/Q \cap P = Q/K_c$. This is not yet our bounded realization. However, for each $q \in Q$, let

$$\psi(q) = \operatorname{Ad}^*_{G_c}(q)(\beta) - \beta.$$

where $\operatorname{Ad}^*_{G_c}$ is the coadjoint representation of G_c in $\mathcal{G}_c^*$.

From Formula 4 above, $\psi(q)$ equals 0 on $\mathcal{Q}$. Hence, $\psi(q)$ may be identified with an element of $(\mathcal{G}_c/\mathcal{Q})^* = (\mathcal{P}/\mathcal{K}_c)^*$. From Formula 6, ψ defines a biholomorphism of Q/K_c onto an open subset of $(\mathcal{P}/\mathcal{K}_c)^*$. The main result of this work is the following:

Theorem 4. *The image of* $\mathcal{D}$ *under* ψ *is a bounded domain in* $(\mathcal{P}/\mathcal{K}_c)^*$.

We refer to $\psi(\mathcal{D})$ as the Harish-Chandra realization of $\mathcal{D}$. The mapping which maps $\mathcal{D}$ onto its Harish-Chandra realization is called the Cayley transformation. It depends, apparently, upon the choice of the transitive subgroup G of Aut $(\mathcal{D})$ and upon the choice of base point in $\mathcal{D}$. It is easily seen, however, that any two such embeddings are related by a complex linear transformation. We should note also that by applying this result to the full automorphism group of the domain, we obtain an embedding which depends only upon the choice of base point in $\mathcal{D}$.

The relation between our 'Harish-Chandra' embedding and the classical version is simple. Assume for the moment that $\mathcal{D}$ is symmetric so that G is semi-simple. We assume that $J =$ ad Z as above. Let $\mathcal{G} = \mathcal{P}^o \oplus \mathcal{K}$ be a Cartan decomposition. Then we may write

$$\mathcal{P}^o_c = \mathcal{P}^+ \oplus \mathcal{P}^-$$

where $\mathcal{P}^\pm$ are the $\pm i$ eigenspaces of ad Z on $\mathcal{P}^o_c$. Then, in the above notation, $\mathcal{P} = \mathcal{P}^- + \mathcal{K}_c$ and $\mathcal{Q} = \mathcal{P}^+ + \mathcal{K}_c$. The classical Harish-Chandra embedding realizes the domain in $\mathcal{P}^+$ by noting that the exponential map defines a bi-holomorphism of $\mathcal{P}^+$ with Q/K_c. On the other hand, using the Killing form, we may identify $\mathcal{P}^+$ with $(\mathcal{P}/\mathcal{K}_c)^*$. It is a simple exercise to verify that under this identification, our realization is essentially the same as the Harish-Chandra realization.

All bounded homogeneous domains may be realized as a Siegel domain of type I or type II. (See below.) There are some beautiful formulas which explicitly describe the Cayley transformation if the domain is so realized. The formula for domains of type one is particularly nice. To describe it we need to recall some basic facts concerning Siegel domains.

Let $\mathcal{V} \subset \mathbb{R}^m$ be an open, convex cone which does not contain straight lines. The Siegel domain of type I associated with $\mathcal{V}$ is the domain in $\mathbb{C}^m$ defined by

$$\mathcal{E} = \mathbb{R}^m + i\mathcal{V}.$$

Let $\mathcal{V}^*$ be the dual cone in $(\mathbb{R}^m)^*$. (This is the cone of all functionals which are strictly positive on $\overline{\mathcal{V}} - \{0\}$.) For all $x \in \mathcal{V}$ we define

$$\phi_{\mathcal{V}}(x) = \int_{\mathcal{V}^*} e^{-<\lambda,x>} d\lambda. \tag{7}$$

where $d\lambda$ is Lebesgue measure on $\mathbb{R}^m$. In [Vin], this function is referred to as the characteristic function of the cone, although it plays the role of the determinant function in Jordan algebras. The differential $D\phi_{\mathcal{V}}$ of $\phi_{\mathcal{V}}$ is the mapping of $\mathcal{V}$ into $(\mathbb{R}^m)^*$ defined by:

$$D\phi_{\mathcal{V}}(x) = \int_{\mathcal{V}^*} - < \lambda, \cdot > e^{-<\lambda,x>} d\lambda. \tag{8}$$

We define $I : \mathcal{V} \to (\mathbb{R}^m)^*$ by $I(x) = -D\phi_{\mathcal{V}}(x)/\phi_{\mathcal{V}}(x)$, the logarithmic derivative of $\phi_{\mathcal{V}}$. Vinberg shows that I maps $\mathcal{V}$ injectively onto $\mathcal{V}^*$. We refer to $I(x)$ as the 'pseudo-inverse' of x. (If $\mathcal{V}$ is the cone of squares in a Euclidian Jordan algebra, then I is essentially the inverse function.)

Now, it is is easily seen from Vinberg's arguments that both $\phi_{\mathcal{V}}$ and $D\phi_{\mathcal{V}}$ extend holomorphically to $-i\mathcal{E} = \mathcal{V} + i\mathbb{R}^m$. Thus, I extends at least meromorphically to a mapping of $-i\mathcal{E}$ into $(\mathbb{C}^m)^*$. We prove that for a homogeneous domain, I extends holomorphically to $-i\mathcal{E}$.

Now, let $c \in \mathcal{V}$ be a fixed base point. We shall assume that Lebesgue measure on $\mathbb{R}^m$ is normalized so that $\phi_{\mathcal{V}}(c) = 1$. Let $c^* = I(c) \in \mathcal{V}^*$. We define

$$C(z) = I(c - iz)$$

for all $z \in \mathcal{E}$. We show that the Harish-Chandra realization of $\mathcal{E}$ is identifiable with the image of $\mathcal{E}$ in $(\mathbb{C}^m)^*$ under the biholomorphism C. In particular, this mapping has bounded image on $\mathcal{E}$. In the Jordan algebra case, our C is a translate of the standard Cayley transformation.

There is a generalization of this to homogeneous Siegel domains of Type II. To describe this, we need to recall the definition of the Type II domains. Suppose that we are given a cone $\mathcal{V} \subset \mathbb{R}^m$ as above. Suppose further that we are given a complex vector space $\mathcal{Z}$ and a Hermitian symmetric, bi-linear mapping $H : \mathcal{Z} \times \mathcal{Z} \to \mathbb{C}^m$. We shall assume that

$$(a)\, H(z,z) \in \overline{\mathcal{V}} \text{ for all } z \in \mathcal{Z}$$

$$(b)\, H(z,z) = 0 \text{ implies } z = 0$$

The Siegel domain associated with this data is defined as the set $\mathcal{E}$ of points (z, w) in $\mathcal{Z} \times \mathbb{C}^m$ described by the equation

$$\text{im } w - H(z,z) \in \mathcal{V}.$$

Let $\mathcal{E} = \mathbb{R}^m + i\mathcal{V}$ be the tube domain associated with $\mathcal{V}$. Then, for $(z, w) \in \mathcal{D}$, $w \in \mathcal{E}$. Hence $C(w)$ is defined as an element of $(\mathbb{C}^m)^*$. Let $\mathcal{Z}^c$ be the *conjugate* dual space of $\mathcal{Z}$. (i.e. the space of conjugate linear functionals on $\mathcal{Z}$.) We define $D : \mathcal{D} \to \mathcal{Z}^c \times (\mathbb{C}^m)^*$ by

$$D(z,w)(\cdot) = (< H(z,\cdot), C(w) >, C(w)). \tag{9}$$

Then, it turns out that the Harish-Chandra realization is identifiable with the image of D.

The proofs of Theorem 3 and Theorem 4 are inductive and utilize the theorem that any bounded homogeneous domain is realizable as a Siegel domain of type I or II [GPV].

Section 1.

In this section, we present the main results needed pertaining to homogeneous cones in $\mathbb{R}^m$. Our results here are not new. We could, in fact, obtain all that we require by reference to the structure theory of $\mathcal{T}$-algebras as presented in [Vin]. However, we only require a small body of material from this reference. In the interest of keeping this work relatively self contained, we have chosen to prove what we need directly. We shall, however, make free use of a number of results from [Vin] which do not refer to $\mathcal{T}$-algebras. *We shall uniformly use upper case script letters to denote Lie algebras and the corresponding upper case Roman letter to denote the corresponding Lie group.*

We shall continue the notation from the introduction. Specifically, $\mathcal{V}$, $\mathcal{Z}$ and H will have the same meaning as above. The cone $\mathcal{V}$ is said to be homogeneous if there is an algebraic subgroup S of $\text{Gl}(n, \mathbb{R})$ which acts transitively on $\mathcal{V}$ via the usual action of $\mathrm{Gl}(n)$ on $\mathbb{R}^m$. It is a result of [Vin] that S may be taken to be a triangular subgroup which acts simply transitively on $\mathcal{V}$. We may also assume that S contains the dilations $v \to tv$ for all $t \in \mathbb{R}^+$. Suppose further that we are given a complex linear algebraic representation ρ of S in $\mathcal{Z}$ such that

$$H(\rho(s)z, \rho(s)w) = sH(z,w) \text{ for all } z, w \in \mathcal{Z}. \tag{10}$$

The group S acts on $\mathcal{D}$ by

$$s(z,w) = (\rho(s)z, sw). \tag{11}$$

We let $\mathbb{R}^m$ act on $\mathcal{D}$ by translation:

$$t(z,w) = (z, w+t). \tag{12}$$

Finally, we let $\mathcal{Z}$ act by

$$z_o(z,w) = (z + z_o, w + 2iH(z, z_o) + iH(z_o, z_o)). \tag{13}$$

These actions generate a group which acts simply transitively on $\mathcal{D}$. Algebraically, this group may be described as follows. Let $\phi =$ im H. We let $\mathcal{L} = \mathcal{Z} \times \mathbb{R}^m$ with the Lie structure

$$[(z_1, t_1), (z_2, t_2)] = (0, 4\phi(z_1, z_2)).$$

The corresponding group is $\mathcal{L}$ with the product

$$(z_1, t_1) \cdot (z_2, t_2) = (z_1 + z_2, t_1 + t_2 + 2\phi(z_1, z_2)).$$

However, following our convention of denoting Lie groups by upper case Roman letters, we shall denote this space by L when it is considered as a group.

Let $G = L \times_s S$ where $s(z,t)s^{-1} = (\rho(s)z, st)$. Then, G is a completely solvable group which acts simply transitively on $\mathcal{D}$. From the given transformation formulas, the identification of G and $\mathcal{D}$ is defined by

$$((z,t), s) \to (z, t + isc + iH(z,z)). \tag{14}$$

The existence of G_o allows us to prove several facts which were mentioned in the introduction.

Lemma 5. *The subgroup K is connected and is the isotropy subgroup of β under the coadjoint representation. Furthermore, $P \cap G = K$.*

Proof The connectedness of K is obvious since $\mathcal{D}$ is simply connected. To prove the rest of the lemma, we note first that Koszul [Kl] proved that K and the isotropy subgroup of β have the same Lie algebra.

Since G_o is a subgroup of G which acts simply transitively on $\mathcal{D}$, then$G_oK = G$ and $K \cap G_o = \{e\}$. Let K_o be the isotropy subgroup of $\beta|\mathcal{G}_o$ in G_o relative to the co-adjoint representation of G_o. The full isotropy group of β is a subgroup of K_oK.

Since G_o is completely solvable, K_o must be connected. (See [B]). From the comments in the first paragraph, K_o is trivial, proving the second statement of the lemma. The final statement also follows since it is clear from formula 4 and formula 1, that $P \cap G$ stabilizes β. This finishes the proof.

We shall let $T = \mathbb{R}^m \times_s S \subset G$. Note that T is the group of the Type I domain $\mathcal{E} = \mathbb{R}^m + i\mathcal{V}$. Of course, if $\mathcal{D}$ is type one, then $T = G$. In this case, we shall prefer to call the group T rather than G. We shall also refer to the domain as $\mathcal{E}$.

If the cone $\mathcal{V}$ is one dimensional, then (as is well known) $\mathcal{D}$ is the standard unbounded realization of the unit ball in $\mathbb{C}^{d+1}$ where $d = \dim \mathcal{Z}$. In this case, we shall refer to $\mathcal{G}$ as a 'unit ball algebra'. We will denote it by $\mathcal{G}_b$. The element $((0,1),0) \in \mathcal{L} \times \mathbb{R}$ is denoted by X_1 and the element $((0,0),1)$ is denoted E.

The main structural result is the following:

Theorem 6. *There is a J-invariant decomposition*

$$\mathcal{G}_o = \mathcal{G}_b \oplus \tilde{\mathcal{G}}$$

where $\mathcal{G}_b$ is a unit ball algebra which is also an ideal and $\tilde{\mathcal{G}}$ is the algebra of a Siegel domain. Furthermore, $ad^*(X)\beta|\mathcal{G}_b = 0$ for all $X \in \tilde{\mathcal{G}}$.

Before proving this result, we shall pause to give and example of this decomposition which will be used in the proof of Theorem 3.

Example 1: Let M be the space of all $p \times p$ matricies with real entries. We shall write the elements of M in block form

$$\begin{bmatrix} m_{11} & m_{12} \\ m_{21} & M_{22} \end{bmatrix}$$

where m_{11} is 1×1 and M_{22} is $(p-1) \times (p-1)$. We let $\mathcal{M}$ be the space of all elements A of M which satisfy $A^t = A$. We identify $\mathcal{M}$ with $\mathbb{R}^m$ where $m = p(p+1)/2$.

Let S be the set of invertible, upper triangular elements of M which have positive diagonal and let $\mathcal{V}$ be the cone of positive definite elements of $\mathcal{M}$. Then S acts simply transitively on $\mathcal{V}$ where the action is defined by

$$\tau(s)X = sXs^t.$$

The Lie algebra $\mathcal{S}$ is the upper-triangular elements of M and the differentiated action is

$$\tau(S)X = SX + XS^t.$$

We let $\mathcal{T} = \mathcal{M} \times_s \mathcal{S}$. This is a 'tube algebra'.

Let $X_1 \in M$ have $(X_1)_{11} = 1$ and all other entries 0 and let $E = X_1/2$. We shall consider X_1 as an element of $\mathcal{M} \subset \mathcal{T}$ and E as an element of $\mathcal{S} \subset \mathcal{T}$. We define subalgebras $\mathcal{S}_0$ and $\mathcal{S}_1$ of $\mathcal{S}$ to be, respectively, the sets of all elements of $\mathcal{S}$ having the respective forms

$$\begin{bmatrix} 0 & 0 \\ 0 & S \end{bmatrix} \text{ and } \begin{bmatrix} 0 & s \\ 0 & 0 \end{bmatrix}.$$

We define subspaces $\mathcal{M}_i$ of $\mathcal{M}$ to be the set of all elements of $\mathcal{M}$ of the form $A + A^t$ where $A \in \mathcal{S}_i$ where $i = 0, 1$. Finally, we set

$$\tilde{\mathcal{T}} = \mathcal{M}_0 \times \mathcal{S}_0 \text{ and } \mathcal{T}_b = \mathcal{M}_1 \times \mathcal{S}_1 + \mathbb{R}X_1 + \mathbb{R}E.$$

More explicitly, $\mathcal{T}_b$ is the space of pairs matrices of the form

$$B = \left(\begin{bmatrix} a & m^t \\ m & 0 \end{bmatrix}, \begin{bmatrix} b & s^t \\ 0 & 0 \end{bmatrix} \right)$$

wmere m and s are $(p-1) \times 1$ matrices and a and b are real numbers and $\tilde{\mathcal{T}}$ is the space of pairs of matrices of the form

$$A = \left(\begin{bmatrix} 0 & 0 \\ 0 & M \end{bmatrix}, \begin{bmatrix} 0 & 0 \\ 0 & S \end{bmatrix} \right)$$

where M and S are $(p-1) \times (p-1)$ matrices and $M = M^t$.

For A and B as above,

$$[A, B] = \left(\begin{bmatrix} 0 & (Sm)^t - (Ms)^t \\ Sm - Ms & 0 \end{bmatrix}, \begin{bmatrix} 0 & -(S^t s)^t \\ 0 & 0 \end{bmatrix} \right)$$

Thus, $\mathcal{T}_b$ is an ideal. Clearly $\mathcal{T} = \mathcal{T}_b + \tilde{\mathcal{T}}$ is the decomposition stated in Theorem 6 in this case.

There is an "obvious" identificaltion of $\mathcal{T}_b$ with $\mathbb{R}^p \times \mathbb{R}^p$. If we reorder the standard basis, (i.e. reverse the standard basis on $\mathcal{S}_1 + \mathbb{R}E$), we see that all of the elements of the image of ad_b are described by upper-triangular matrices. This image is, of courrse, a subalgebra of $Der(\mathcal{T}_b)$. We shall require the following description of this image in the proof of Theorem 3.

Lemma 7. *The image of ad $_b$ is precisely the set of all elements of $Der(\mathcal{T}_b)$ which are upper-triangular, map $\mathcal{M}_1+\mathcal{S}_1$ into itself, and annihilate X_1 and E.*

Proof We identify $\mathcal{M}_1+\mathcal{S}_1$ with $\mathbb{R}^{p-1}\times\mathbb{R}^{p-1}$. On this space let ϕ be the bi-linear form

$$\phi((x_1,y_1),(x_2,y_2)) = < x_1, y_2 > - < x_2, y_1 >$$

where $<\cdot,\cdot>$ is the usual scalar product on $\mathbb{R}^{p-1}$.

The derivations described in the lemma are determined by upper-triangular linear mappings U of $\mathbb{R}^{p-1}\times\mathbb{R}^{p-1}$ into itself which satisfy

$$\phi(Uz,w) = -\phi(w,Uz). \tag{15}$$

Let U be written in block form as

$$U = \begin{bmatrix} A & B \\ 0 & C \end{bmatrix}$$

where A, B, and C are all $(p-1)\times(p-1)$ with A and C upper-tiangular.. It is easily seen that formula 15 is equivalent with $B = B^t$ and $A = -C$. That such transformations are described by appropriate elements of $\tilde{\mathcal{T}}$ is left to the reader.

Proof of Theorem 6 Since G_o acts transitively on $\mathcal{D}$, we may define the complexification $(G_o)_c$ and the subgroups P' and Q' in $(G_o)_c$ just as we defined G_c, P and Q above. Let us first assume that $\mathcal{D}=\mathcal{E}$ is a Type I domain. In this case, our algebra is denoted $\mathcal{T}$, as explained above. Our first goal is to explicitly describe the group $P'\subset T_c$.

Clearly, $T_c = \mathbb{C}^m \times_c S_c$ where S_c acts as a group of complex matricies on $\mathbb{C}^m$. (S_c is the complex algebraic closure of S.) T_c acts on $\mathbb{C}^m$ as in formula 11 and formula 12. (There is no z-variable in this case.) Let $c\in\mathcal{V}$ be a fixed base point and let $\omega_o = ic\in\mathcal{D}$. We also define $\tilde{c}=(ic,e)\in T_c$. Then $\tilde{c}^{-1}ic = 0$. The isotropy of 0 in T_c is S_c. It follows from Corollary 1 that $P' = \tilde{c}S_c\tilde{c}^{-1}$. Thus, we have:

Lemma 8. $\mathcal{P}' = \ Ad\ \tilde{c}(\mathcal{S}_c) = \{(-iYc, Y) | Y\in\mathcal{S}_c\}$.

Corollary 9. *The mapping $J:\mathcal{T}\to\mathcal{T}$ is described by $J:(Xc,Y)\to(-Yc,X)$ where X and Y belong to $\mathcal{S}$.*

Now, since $\mathcal{S}$ is triangular, there is a $X_1\in\mathbb{R}^m$ and a functional $\gamma\in\mathcal{S}^*$ such that $YX_1 = \gamma(Y)X_1$ for all $Y\in\mathcal{S}$. Let $E\in\mathcal{T}$ be defined by $E = J(X_1,0)$. Then E may be identified with an element E of $\mathcal{S}$ where $Ec = X_1$. It follows from formula 5, that

$$0 < \beta([E,X_1]) = \gamma(E)\beta(X_1).$$

By proper choice of c and of X_1, we may normalize these factors so that $\gamma(E) = 1$ and $\beta(X_1) = 1$. For $t \in \mathbb{R}$, we set

$$\beta_t = \operatorname{Ad}^*(\exp_T(tE))(\beta).$$

The following proposition contains most of the structural information required.

Proposition 10. $\beta_t = \beta - (1 - e^{-t})\gamma \circ J$.

Proof Let $\mathcal{P}_\gamma = \mathcal{P}' \cap \ker \gamma$. Then

$$\mathcal{T}_c = \mathcal{P}_\gamma + \overline{\mathcal{P}}_\gamma + \mathbb{C}E + \mathbb{C}X_1. \tag{16}$$

It suffices to prove our lemma on each of the four factors above. We consider $\mathcal{P}_\gamma$ first. Let $Z = E - iX_1$. Then, $Z \in \mathcal{P}'$ and, on $\mathcal{P}_\gamma$, $\operatorname{ad} Z = \operatorname{ad} E$. It then follows from formula 4 that $\beta_t = \beta$ on $\mathcal{P}_\gamma$. This is consistent with our lemma. Since β_t is real, this also establishs the lemma on $\overline{\mathcal{P}}_\gamma$. The formula is easily checked on the other two factors.

Next, we prove the following:

Proposition 11. *E is a semi-simple element of $\mathcal{T}$ which only has eigenvalues 0, $1/2$ and 1. The eigenspace corresponding to 1 is one-dimensional and is spanned by X_1. Furthermore, the eigenspaces corresponding to 0 and $1/2$ are orthogonal under g_β.*

Proof The decomposition in formula 16 above is $\operatorname{ad} E$ invariant. Thus, it suffices to prove semi-simplicity in each factor. Clearly, the only question is $\mathcal{P}_\gamma$. Let $Z \in \mathcal{P}_\gamma$ be a generalized eigenvector corresponding to the value α. Note that α is real, since S is triangular. Then

$$\operatorname{Ad}(\exp_T(-tE))(Z) = e^{-\alpha t} Z(t)$$

for some $\mathcal{P}_\gamma$ valued polynomial function $Z(t)$. Then

$$\beta_t([Z, \overline{Z}]) = \beta([e^{-\alpha t} Z(t), e^{-\alpha t} Z(t)]) = e^{-2\alpha t} \beta([Z(t), \overline{Z}(t)]).$$

On the other hand

$$\beta_t([Z, \overline{Z}]) = \beta([Z, \overline{Z}]) - (1 - e^{-t})\gamma(J[Z, \overline{Z}]).$$

The term $\beta([Z, \overline{Z}])$ is, up to constant factor, the Hermitian scalar product. Thus, all such terms in the above expression are non-zero. It follows that these two formulas will be inconsistent unless $\alpha = 0$ or $\alpha = 1/2$. In either case, it follows that $Z(t)$ has constant norm relative to the Hermitian scalar product. This forces $Z(t)$ to be constant, proving that Z is in fact an eigenvector. This shows semi-simplicity, as well as the fact that $\mathbb{R}X_1$ is the only elements with eigenvalue 1.

A similar argument will show that eigenspaces corresponding to values α_1 and α_2 will be orthogonal unless $\alpha_1 + \alpha_2$ is 1 or 0. This proves the desired orthogonality.

Let $\mathcal{T}_\alpha$ be the eigenspace of $\operatorname{ad} E$ in $\mathcal{T}$ corresponding to α. Then $[\mathcal{T}_\alpha, \mathcal{T}_\beta] \subset \mathcal{T}_{\alpha+\beta}$. Let $\mathcal{T}_b = \mathcal{T}_{1/2} + \mathbb{R}X_1 + \mathbb{R}E$ and let $\tilde{\mathcal{T}} = \mathcal{T}_0 \cap \ker\gamma$. Then $\mathcal{T}_b$ is an ideal in $\mathcal{T}$ and

$$\mathcal{T} = \mathcal{T}_b \oplus \tilde{\mathcal{T}}.$$

Lemma 12. *The above decomposition is J invariant.*

Proof Since $\operatorname{ad} E$ leaves $\mathcal{P}_\gamma$ and $\overline{\mathcal{P}}_\gamma$ invariant, it follows that $\operatorname{ad} E$ commutes with J on $\mathcal{P}_\gamma + \overline{\mathcal{P}}_\gamma$. The J-invariance follows easily.

Next, we must prove that $\tilde{\mathcal{T}}$ is the algebra of a Siegel domain and $\mathcal{T}_b$ is a unit ball algebra. We shall do the former first.

Since $\mathcal{S}$ and $\mathbb{R}^m$ are both $\operatorname{ad} E$ invariant, the above decomposition restricts to these spaces as

$$\mathcal{S} = \mathcal{S}_b \oplus \tilde{\mathcal{S}} \text{ and } \mathbb{R}^m = \mathcal{M}_b \oplus \tilde{\mathcal{M}}.$$

The space $\mathcal{M}_b$ is S invariant and $\tilde{\mathcal{M}} = \mathbb{R}^m/\mathcal{M}_b$. Our cone $\mathcal{V}$ projects to an open cone $\tilde{\mathcal{V}} \subset \tilde{\mathcal{M}}$.

Lemma 13. $\tilde{\mathcal{V}}$ *is homogeneous under* $\tilde{S}$. *Furthermore,* $\tilde{\mathcal{V}}$ *contains no straight lines and* $\tilde{S}$ *acts simply-transitively on* $\tilde{\mathcal{V}}$.

Proof Let us first note that $\mathcal{S}_b$ acts trivially on $\mathbb{R}^m/\mathcal{M}_b$. It follows that $\tilde{\mathcal{V}}$ is transitive under $\tilde{S}$.

Let c be the base point for $\mathcal{V}$ as before. Let $c = c_0 + c_{1/2} + c_1$ where $c_i \in \mathcal{T}_i$. From $Ec = X_1$, we see immediately that $c = c_0 + X_1$. Then the $\tilde{S}$ orbit of c is the set of points $sc_0 + X_1$ where $s \in \tilde{S}$. We conclude that $X_1 + \tilde{\mathcal{V}} \subset \mathcal{V}$. This shows that $\tilde{\mathcal{V}}$ contains no lines. We also see that $\tilde{S}$ acts simply transitively, as desired.

The above proves that $\tilde{\mathcal{S}}$ is a Siegel algebra (Type I in fact). The fact that $\mathcal{S}_b$ is a unit ball algebra is simple and is left to the reader.

Next, we consider the case of a general Siegel domain. We continue the above notation, relative to the subalgebra $\mathcal{T} \subset \mathcal{G}_o$. We differentiate ρ to produce a representation of the Lie algebra $\mathcal{S}$ (still denoted ρ) on $\mathcal{Z}$. Since ρ is algebraic and E is semi-simple, $\rho(E)$ is semi-simple with only real eigenvalues. Let $Z \in \mathcal{Z}$ be an eigenvector for $\rho(E)$ with eigenvalue α.

Then, $H(Z, Z)$ is an eigenvector for E with eigenvalue 2α. Hence, α can only assume the values 0, 1/4, or 1/2.

Lemma 14. *The eigenvalue 1/4 does not occur.*

Proof Let $Z \in \mathcal{Z}_{1/4}$ and let $u = H(Z, \overline{Z})$. Then $u \in \mathbb{R}^m \cap \mathcal{T}_{1/2}$. Let $X = Ju$, $X \in \mathcal{S}_{1/2} = \mathcal{S} \cap \mathcal{T}_{1/2}$. Since there is no eigenvalue 3/4, we see that $\rho(X)Z = 0$. Then, from formula 10, $Xu = 0$. This, however, implies that $[Ju, u] = 0$ which implies that $u = 0$ and hence that $Z = 0$. This proves the lemma.

We decompose $\mathcal{Z}$ as $\mathcal{Z} = \mathcal{Z}_0 + \mathcal{Z}_{1/2}$ where $\mathcal{Z}_i$ is the eigenspace for $\rho(E)$ with eigenvalue i. We then let

$$\tilde{\mathcal{G}} = \mathcal{Z}_0 + \tilde{\mathcal{T}} \subset \mathcal{G}_o \text{ and } \mathcal{G}_b = \mathcal{Z}_{1/2} + \mathcal{T}_b \subset \mathcal{G}_o.$$

Clearly, $\tilde{\mathcal{G}}$ is the Lie algebra of a Siegel domain while $\mathcal{G}_b$ is an ideal which is a unit ball algebra. This proves Theorem 6.

Section 2.

In this section, we prove our main results. We shall first prove them in the case of a type I domain. The extension to the type II case requires only minor modification of the basic argument. We shall let G_o be as in the previous section. We define $\beta' = \beta|\mathcal{G}_o$. Notice that Theorems 3 and 4 both have meaning for G_o reletive tp P', Q' and β'.

Proposition 15. *Suppose Theorems 3 and 4 are true for G_o. Then they are true for G.*

Proof This is a trivial consequence of the facts that $G_oK = G$ and $\mathcal{P}' + \mathcal{K}_c = \mathcal{P}$, both of which follow from transitivity. We leave the details to the reader.

Due to the above result, we may prove our theorems with G_o in place of G. We begin with the proof of Theorem 3. We consider first the special case that $\mathcal{G}_o$ is a unit ball algebra. We shall, in fact, assume first that $\mathcal{Z} = 0$, so that

$$G_o = T = \mathbb{R} \times_s \mathbb{R}^+$$

where the $\mathbb{R}^+$ action is simply multiplication. The complexification is $\mathbb{C} \times \mathbb{C}^*$ In the cone $\mathcal{V}$, we choose $c = 1$ as the base point. Then $P' = \{((1-a)i, a) | \, a \in \mathbb{C}^*\}$. Q' is described similarly, with i replaced by $-i$. The following lemma (which the reader may check for himself) establishes Theorem 3 in this case:

Lemma 16. *Let $(b,a) \in T_c$. Then $(b,a) \in Q'P'$ if and only if $b \neq -i(1+a)$.*

Next, we consider the general unit ball algebra. Now, we write $\mathcal{G}_o = \mathcal{G}_b$. In this case we shall require a slightly more general result. First, however, we need some notation. Let $\mathcal{Z}$, H, and ϕ be as above. We consider $\mathcal{Z}$ as a real vector space and consider the complexification $\mathcal{Z}_c$. Let $\mathcal{R} \subset \mathcal{Z}_c$ be the $-i$ eigenspace of J where J is the complex structure of $\mathcal{Z}$. We extend ϕ via complex linearity to a bi-linear form on $\mathcal{Z}_c$. It is easily verified (from the J-invariance of ϕ) that ϕ is zero on $\mathcal{R} \times \mathcal{R}$. Furthermore, from the positivity of H,

$$\text{im } \phi(Z, \overline{Z}) > 0$$

for all non-zero $Z \in \mathcal{R}$.

In general, we shall say that a complex subspace $\mathcal{R}_1 \subset \mathcal{Z}_c$ is a totally complex polarization provided

(1) The above inequality holds relative to $\mathcal{R}_1$.
(2) ϕ is zero on $\mathcal{R}_1 \times \mathcal{R}_1$.
(3) $\mathcal{R}_1 + \overline{\mathcal{R}}_1 = \mathcal{Z}_c$

Now, let $\mathcal{R}_1$ and $\mathcal{R}_2$ be two totally complex polarizations. We identify $\mathcal{R}_i$ with the subalgebras $\mathcal{R}_i \times \{0\}$ in $\mathcal{L}$. Let $\mathcal{P}_i \subset (\mathcal{G}_b)_c$ be defined by

$$\mathcal{P}_i = \mathcal{R}_i + \mathbb{C}(E - iX_1)$$

for $i = 2, 3$. This is clearly a subalgebra of $\mathcal{G}_b$. Let $\mathcal{Q}_i = \overline{\mathcal{P}}_i$. The following proposition implies Theorem 3 in the ball case.

Proposition 17. $G_b \subset Q_1 P_2$.

Proof As in the discussion preceding Theorem 6, we write $G_b = L \times_s \mathbb{R}^+$. The $\mathbb{R}^+$ action on L is defined by

$$s(Z,t)s^{-1} = (s^{1/2}Z, st).$$

The subgroup $T = (\{0\} \times \mathbb{R}) \times_s \mathbb{R}^+$ is essentially the same as the group considered in Lemma 16. Furthermore $\mathcal{P}_i \cap \mathcal{T}_c = \mathcal{P}_T$ where $\mathcal{P}_T$ is the algebra $\mathcal{P}$ of Lemma 16. It is easily seen that P_T normalizes both R_i and $\overline{R}_i$.

We may write the general element g of G_b as

$$g = ((Z,t),s) = ((Z,0),1)((0,t),s).$$

Lemma 18. $\overline{\mathcal{R}}_1 + \mathcal{R}_2 = \mathcal{Z}_c$. *Furthermore, If $Z \in \mathcal{Z}$ is written $Z = q + p$ relative to this decomposition, then im $\phi(q,p) < 0$.*

Let us assume this for the moment and continue with our proof. We write $Z = q + p$ relative to this decomposition. Then, in L,

$$(Z,0) = (q,0)(p,0)(0,\alpha)$$

where $\alpha = -2\phi(q,p)$. Then $g = (q,0)(p,0)((0,t+\alpha),s)$. Since im $(t+\alpha) > 0$, it follows from Lemma 16 that $((0,t+\alpha),s) \in Q_T P_T$. Proposition 17 follows from the fact that Q_T normalizes R_2.

Thus, we need only prove Lemma 18. For the decomposition, we only need show that $\overline{\mathcal{R}}_1 \cap \mathcal{R}_2 = \{0\}$. However, if Z is a non-zero element of this intersection, then

$$\text{im } \phi(Z,\overline{Z}) > 0 \text{ and } \text{im } \phi(\overline{Z},Z) > 0.$$

This is impossible since ϕ is alternating.

Now suppose $Z \in \mathcal{Z}$ and $Z = q + p$ as before. Since Z is real (relative to $\mathcal{Z}_c$), we also have $Z = \overline{p} + \overline{q}$. Then

$$\phi(q,p) = \phi(\overline{q} + \overline{p} - p, p) = \phi(\overline{p},p) + \phi(\overline{q}, \overline{p} + \overline{q} - q).$$

We conclude that

$$\phi(q,p) - \phi(\overline{q},\overline{p}) = \phi(\overline{p},p) - \phi(\overline{q},q).$$

This proves the lemma.

Finally, we are ready to prove Theorem 3 in the general case. We shall reason by induction on the dimension of $\mathcal{V}$. The dimension one case is the unit ball case which was done above. Thus, we assume that it is true for all cones of smaller dimension than $\mathcal{V}$. Let us decompose $\mathcal{G}_o$ as in Theorem 6 as

$$\mathcal{G}_o = \mathcal{G}_b + \tilde{\mathcal{G}}.$$

Since this decomposition is J-invariant, $\mathcal{P}'$ and $\mathcal{Q}'$ decompose relative to this decomposition as $\mathcal{P}' = \mathcal{P}_b + \tilde{\mathcal{P}}$ and $\mathcal{Q}' = \mathcal{Q}_b + \tilde{\mathcal{Q}}$.

Now, let $g = g_b t \in G$ where $g_b \in G_b$ and $t \in \tilde{G}$. Let $P_t = tP_b t^{-1}$. It is clear from Theorem 6 that P_t and Q_b satisfy the hypotheses of Proposition 17. (Note that by construction, $\tilde{\mathcal{G}}$ centralizes E and X_1.) Thus, we may write $g_b = q(t)p(t)$ where $q(t) \in Q_b$ and $p(t) \in P_t$. It follows from the inductive hypothesis that we may also write $t = \tilde{q}\tilde{p}$ where $\tilde{q} \in \tilde{Q}$ and $\tilde{p} \in \tilde{P}$. Then

$$g = g_b t = q(t)p(t)t = q(t)\tilde{q}\tilde{p}t^{-1}p(t)t \in Q'P'. \tag{17}$$

This finishes the proof of Theorem 3.

Next we turn to the proof of Theorem 4. Again, we shall prove the result for G_o.

Let $\mathcal{R}^{\pm}$ be the $\pm i$ eigenspaces of J in $\mathcal{Z}_c$. As noted in §2, the bi-linear form im ϕ is zero on each of these spaces. Thus, $\mathcal{R}^{\pm}$ defines a *subgroup* of L. We denote this space by $R^{\pm}$ when thought of as a group. Since S_c normalizes $R^{\pm}$, we may form the subgroup $R_G^{\pm} = S_c R^{\pm}$. From the argument above in Lemma 8, $P = CR_G^- C^{-1}$ and $Q = C^{-1} R_G^+ C$.

Let $p \in P$ be given. Then Ad (p) projects to a mapping of $(\mathcal{G}_o)_c/\mathcal{P}' \to (\mathcal{G}_o)_c/\mathcal{P}'$. We define $\chi_P(p)$ to be the determinant of this mapping. We similarly define $\chi_Q(q)$ for $q \in Q$. Both χ_P and χ_Q are holomorphic characters. We then define a function Φ on G by

$$\Phi(g) = \chi_Q(q^{-1})\chi_P(p) \tag{18}$$

where $g = qp$ as in Theorem 3. We refer to Φ as the 'characteristic function' of the domain. The relevance of this to Theorem 4 is based upon the following simple proposition. For this proposition, $X \in \mathcal{G}_o$ is considered as a *right* invariant vector field.

Proposition 19. *Let $g \in G_o$ be written in the form $g = qp$ as above. Then for all $X \in \mathcal{G}_o$, $X\Phi(g) = \frac{i}{2c}$ Ad $^*(q)\beta(X)\Phi(g)$.*

Proof Let $d\chi_P \in (\mathcal{P}')^*$ and $d\chi_Q \in (\mathcal{Q}')^*$ be the differentials of χ_P and χ_Q at the identity. Then, for $Z \in \mathcal{P}'$,

$$d\chi_P(Z) = \text{ Tr }_{(\mathcal{G}_o)_c/\mathcal{P}'} \text{ ad } Z.$$

On the otherhand, ad $JZ - J$ ad Z is zero on $\mathcal{P}'$ and is congruent to $-2i$ ad Z modulo $\mathcal{P}'$. We conclude from formula 3 that $\beta(Z) = -2icd\chi_P(Z)$ for $Z \in \mathcal{P}'$. Similarly, $\beta(W) = 2icd\chi_Q(W)$ for $W \in \mathcal{Q}'$.

Let $g = qp$ as in the statement of the proposition. Let $\gamma(t) = q^{-1}(\exp_G tX)q$. For t near zero, we may write

$$\gamma(t) = \gamma_Q(t)\gamma_P(t)$$

where $\gamma_P(t) \in P'$ and $\gamma_Q(t) \in Q'$. Clearly,

$$\Phi((\exp tX)g) = \chi_Q(\gamma_Q(t)^{-1})\chi_P(\gamma_P(t))\Phi(g).$$

Hence

$$X\Phi(g) = (-d\chi_Q(\gamma_Q'(0)) + d\chi_P(\gamma_P'(0)))\Phi = \frac{i}{2c}\beta(\text{ Ad } q^{-1}(X))\Phi.$$

This proves the lemma.

It follows that Theorem 4 is equivalent with proving that $X\Phi/\Phi$ is bounded for all $X \in \mathcal{G}_o$. We prove this by constructing an algorithm which inductively computes Φ in terms of the corresponding function for lower dimensional domains.

Let $g \in G$. From Theorem 6, we may write $g = tx$ for $t \in \tilde{G}$ and $x \in G_b$. We write $x = q(t)p(t)$ where $q(t) \in t^{-1}Q_b t$ and $p(t) \in P$. From formula 18, we see

$$\Phi(g) = \Phi(t)\psi(t, x)$$

where

$$\psi(t, x) = \chi_Q(tq(t)^{-1}t^{-1})\chi_P(p(t)).$$

It is clear that $\Phi|\tilde{G}$ is the characteristic function $\tilde{\Phi}$ for the domain corresponding to $\tilde{G}$. Furthermore, for $X \in \tilde{\mathcal{G}}$,

$$X\Phi/\Phi = X\tilde{\Phi}/\tilde{\Phi} + X_t\psi/\psi.$$

where the subscript indicates that X acts in the t variable. For $X \in \mathcal{G}_b$

$$X\Phi/\Phi = (\text{ Ad } (t^{-1})X)_x\psi/\psi.$$

By induction, the term $X\tilde{\Phi}/\tilde{\Phi}$ may be assumed bounded for fixed X. Thus, our theorem is equivalent with the boundedness of both $X_t\psi/\psi$, $X \in \tilde{\mathcal{G}}$, and $(\text{ Ad } (t^{-1})X)_x\psi/\psi$, $X \in \mathcal{G}_b$. To prove this, we use Lemma 7. Explicitly, H defines a Hermitian scalar product on $\mathcal{Z}_{1/2}$. (Note that on $\mathcal{Z}_{1/2}$, H is valued in $\mathbb{C}X_1$.) Let $B = \{Z_1, \ldots, Z_m\}$ be an orthonormal $\mathbb{C}$-basis for $\mathcal{Z}_{1/2}$. Let $\mathcal{S}_z$ be the span over $\mathbb{R}$ of B and let $\mathcal{M}_z = J(\mathcal{S}_z)$ where J is the complex multiplication on $\mathcal{Z}_{1/2}$. Then, we may write

$$\mathcal{G}_b = (\mathcal{S}_z + \mathcal{S}_b) + (\mathcal{M}_z + \mathcal{M}_b) + \mathbb{R}X_1 + \mathbb{R}E.$$

We may identify $\mathcal{G}_b$ with the algebra $\mathcal{T}_b$ of Example 1 above Lemma 7 in such a way that $\mathcal{S}_z + \mathcal{S}_b$ is identified with $\mathcal{S}_1$ and $\mathcal{M}_z + \mathcal{M}_b$ is identified with $\mathcal{M}_1$. These identifications may be carried out in a manner that the adjoint action of $\tilde{\mathcal{G}}$ on $\mathcal{G}_b = \mathcal{T}_b$ triangulizes.

Now we come to the main idea. Theorem 4 is known in the context of Example 1. (One could apply the symmetric space Harish-Chandra embedding, although it also is simple to prove directly.) Let ψ_T be the function on $\tilde{T} \times T_b$ defined in the same manner as ψ was. The adjoint action of $\tilde{G}$ on G_b maps $\tilde{G}$ into the automorphism group of $G_b = T_b$. According to Lemma 7, for all $g \in \tilde{G}$, there exists a $g' \in \tilde{T}$ such that $\text{Ad }(g)|T_b = \text{ Ad }(g')|T_b$. Let $G' \subset \tilde{T}$ be the set of all such g'. The restriction of ψ_T to $G' \times T_b$ may be identified with ψ. Since Theroem 4 is true for both T and $\tilde{T}$, all of the required derivatives of ψ_T are known to be bounded on $\tilde{T}$ and hence on G'. This proves the boundedness of ψ, as desired. This also proves Theorem 4.

Section 3.

Explicit Embeddings

In this final section, we shall use the Harish-Chandra embedding to prove that formula 9 does indeed define a bounded realization of $\mathcal{D}$. We continue the notation established in the previous sections. Let Φ be the function on G_o defined in formula 18. By definition, Φ extends holomorphically to $P'Q' \subset (G_o)_c$. From the discussion above Lemma 8, $P' = CS_cC^{-1}R^-$ and $Q' = C^{-1}S_cCR^+$. It follows that for $s \in S_c$,

$$\chi_P(CsC^{-1}) = \det s \quad \text{ and } \chi_Q(C^{-1}sC) = \det s.$$

Let

$$\Phi_{\mathcal{V}}(z) = \Phi(C^{-1}zC^{-1}).$$

The domain of $\Phi_{\mathcal{V}}$ is $S_cR^+C^2R^-S_c$ and on this domain

$$\Phi_{\mathcal{V}}(s_1xs_2) = \det s_1^{-1} \det s_2 \Phi_{\mathcal{V}}(x).$$

for $s_i \in S$. Note that since S acts transitively on $\mathcal{V}$, C^2 may be replaced by any element of $i\mathcal{V} \times \{e\}$ in the description of the domain. (e is the identity element of S.)

For the next lemma, note that the domain of $\Phi_{\mathcal{V}}$ contains the set $sC^2s^{-1} = i\mathcal{V} \times \{e\} \subset T_c \subset G_c$.

Lemma 20. *For all $v \in \mathcal{V}$,*

$$\Phi_{\mathcal{V}}((iv, e)) = 4^m \phi_{\mathcal{V}}(v)^2.$$

where $\phi_{\mathcal{V}}$ is as in formula 7.

Proof For $v \in \mathcal{V}$,

$$\Phi_{\mathcal{V}}((isc, e)) = \Phi_{\mathcal{V}}(s(ic, e)s^{-1}) = \det s^{-2}\Phi(C^{-1}).$$

We may evaluate $\Phi(C^{-1})$ as follows:

Let $s : \mathcal{V} \to \mathcal{V}$ be defined by $sx = x/2$. Then $s \in S$ and

$$C^{-1} = C^{-1}sCCs^{-1}C^{-1} \in QP.$$

Hence,

$$\Phi(C^{-1}) = \det s^{-2} = 4^m.$$

On the other hand, according to Formula 2, p. 348 of [Vin],

$$\phi_{\mathcal{V}}(sc) = \det s^{-1}\phi_{\mathcal{V}}(c) = \det s^{-1}.$$

This proves the lemma.

Now, let $((z,m),s) \in G_o \subset Q'P'$ where $m \in \mathbb{R}^m$, $z \in \mathcal{Z}$ and $s \in S$. Consider the point $y = C((z,m),s)C(0,s^{-1}) = ((z, m + isc + ic), e)$. Then

$$\Phi_{\mathcal{V}}(y) = \Phi(C^{-1}yC^{-1}) = \Phi((m,z),s)\det s^{-1}.$$

On the other hand, let J_Z denote the complex structure of $\mathcal{Z}$. Let $p \in \mathcal{Z}_c$ be $p = (z + iJ_Z z)/2$ and $q = (z - iJ_Z z)/2$. Then, in L,

$$(z,0) = (0, -2\phi(q,p))(q,0)(p,0) = (0, iH(z,z))(q,0)(p,0).$$

It follows that

$$\Phi_{\mathcal{V}}(y) = \Phi_{\mathcal{V}}((m + isc + ic + iH(z,z), e)) = 4^m \phi_{\mathcal{V}}^2(c - iW) \tag{19}$$

where $W = m + isc + iH(z,z)$. Note that under the identification of G with $\mathcal{D}$ stated in formula 14, W is the point in $\mathcal{D}$ identified with $((z,m),s) \in G_o$.

From Proposition 20, the Harish-Chandra embedding is equivalent to the embedding obtained by restricting the logarithmic derivative of Φ to $\mathcal{P}'$. (We consider two embeddings which are related by translation and multiplication by scalars to be equivalent.) Note, however, that $(\mathcal{G}_o)_c/\mathcal{Q}'$ is isomorphic with $\mathcal{W} = \mathcal{R}^- + \mathbb{C}^m$. Thus, we may also obtain an embedding by restricting to $\mathcal{W}$. This will map $\mathcal{D}$ into $\mathcal{W}^*$. Furthermore, $\mathcal{R}^-$ is anti-isomorphic with $\mathcal{Z}$ under the conjugate linear mapping $X \to (X + iJ_Z X)/2$.

More explicitly, let $M \in \mathbb{R}^m$. Taking the logarithmic derivative of formula 19 in the M direction, we see:

$$-2i < M, I(c - iW) >= \frac{(0,M)\Phi}{\Phi}((z,m),s))$$

where I is the mapping of $-i\mathcal{D}$ into $(\mathbb{C}^m)^*$ defined below formula 8. Since both Φ and I are holomorphic, this equality is true for all $M \in \mathbb{C}^m$.

Next, consider $X \in \mathcal{Z}$. Let $Z = (X + iJ_Z X)/2$. We extend both $\phi = \operatorname{im} H$ and H to complex-bilinear forms on $\mathcal{Z}_c \times \mathcal{Z}_c$. Then, for all $z \in \mathcal{Z}$, $H(Z,z) = 0$ and $H(z,Z) = H(z,X)$. In L_c, $\exp tZ = tZ$ and $tZ(z,m) = (z + tZ, m + 2t\phi(Z,z)) = (z + tZ, m + 2itH(z,X))$. It follows from formula 19 that

$$4 < H(z,X), I(c - iW) >= \frac{(Z,0)\Phi}{\Phi}((z,m),s)).$$

This establishes our formula.

References

[B] P. Bernat et al., *Représentations des groupes de Lie résoluble*, Dunod Paris, 1972

[Do] S. Dorfmeister, Quasi-symmetric Siegel domains and the automorphisms of homogeneous Siegel domains, *Amer. J. Math* **102** (1980), 537–563

[HC] Harish-Chandra, Representations of semi-simple Lie groups, IV, *Amer. J. Math.* **77** (1955), 743–777

[Ku] S. Kaneyuki, Homogeneous Bounded Domains and Siegel Domains, *Lecture Notes in Mathematics* **241** (1971), Springer-Verlag, Berlin

[Kl] J. Koszul, Sur la forme hermitienne canonique des espaces homogènes complexes, *Canad. J. Math.* **7** (1955), 562–576

[GPV] S. Gindikin, I. Pjatecki-Shapiro, E. Vinberg, Classification and canonical realization of complex bounded homogeneous domains, *Trans. Moscow Math. Soc.* 1963, 404–437

[Vin] E. Vinberg, Theory of homogeneous convex cones *Trudy Moskva Math. Obsc.* **12** (1963) 303–358; *Trans. Moscow Math. Soc.* 1963, 340–403

Department of Mathematics
Purdue University
West Lafayette, IN 47907

Received August 1995

How Many Lorentz Surfaces Are There?

Robert Smyth and Tilla Weinstein[1]

1. Introduction

Lorentz surfaces are the indefinite metric analogs of Riemann surfaces. While there are just three distinct simply connected Riemann surfaces (the round sphere, the Euclidean plane and the Euclidean disc), there are infinitely many distinct simply connected Lorentz surfaces. This was first noted in a landmark paper [2] by Kulkarni which inspired the systematic study of Lorentz surfaces. Though the full complexity of simply connected Lorentz surfaces is best understood by working with the conformal boundary introduced in [2] (see [7] and [10]), this paper will use a series of elementary examples to give some feeling for the great variety of structures involved.

Expanding on results in [8], we describe uncountably many C^j conformally equivalent simply connected Lorentz surfaces any two of which are C^{j+1} conformally distinct for any $j = 0, 1, \dots$, and uncountably many C^∞ conformally equivalent simply connected Lorentz surfaces any two of which are C^ω (that is, real analytically) conformally distinct. For a particular family of C^0 conformally equivalent triangular regions in $E^{2,1}$, we describe a real valued C^1 conformal invariant, to show that uncountably many of these triangular regions are C^1 conformally distinct. Exploiting a construction from [7], we describe uncountably many C^0 conformally distinct simply connected Lorentz surfaces, each C^0 conformally distinct from every subset of the Minkowski plane $E^{2,1}$. Similarly, we describe uncountably many C^0 conformally distinct simply connected Lorentz surfaces, each C^0 conformally equivalent to a subset of $E^{2,1}$, and each C^1 conformally distinct from every subset of $E^{2,1}$. Finally, we outline arguments from [6] which show there are uncountably many C^0 conformally distinct rectangular regions in $E^{2,1}$.

This uncanny abundance of simply connected Lorentz surfaces suggests the usefulness of theorems which identify conformal type, singling out one structure from among uncountably many. As samples, we cite without proof these two results from [4].

The Hilbert-Holmgren Theorem for Harmonic Maps. *Suppose* f :

[1]Partially supported under NSF grant DMS 94-01825

$(S,h) \to (\mathcal{M},\mathcal{G})$ is a C^∞ harmonic map from a surface S with indefinite prescribed metric into a manifold $\mathcal{M}$ of dimension $n \geq 2$ with pseudo-Riemannian metric $\mathcal{G}$. If the induced metric $g = f^*\mathcal{G}$ on S is complete and Riemannian,

(i) the intrinsic curvature of g cannot be bounded away from zero, and
(ii) the universal cover $\tilde{S}$ of S with the lift $\tilde{h}$ of h is C^∞ conformally equivalent to $E^{2,1}$.

A Conformal Bernstein Theorem. *Every entire timelike minimal surface in Minkowski 3-space $E^{3,1}$ is C^∞ conformally equivalent to $E^{2,1}$.*

For a description of the varied shapes possible for an entire timelike minimal surface in $E^{3,1}$, see [5].

2. Definitions and Preliminaries

By a surface S, we mean an oriented, connected, paracompact C^∞ 2-manifold. The metrics used on S are all C^∞ and non-degenerate, sometimes Riemannian (positive definite) and sometimes Lorentzian (indefinite). Two metrics h and $\hat{h}$ on S are conformally equivalent (written $h \sim \hat{h}$) if and only if h is the multiple of $\hat{h}$ by a *positive* C^∞ function. Thus when $h \sim \hat{h}$, h is Riemannian (resp. Lorentzian) if and only if $\hat{h}$ is Riemannian (resp. Lorentzian).

Given a Lorentzian metric h on S and the tangent space S_p at any point p on S, the causal character of a vector $\vec{t} \neq \vec{0}$ in S_p is spacelike if $h_p(\vec{t},\vec{t}) > 0$, timelike if $h_p(\vec{t},\vec{t}) < 0$ and null if $h_p(\vec{t},\vec{t}) = 0$. The zero vector $\vec{0}$ is taken to be spacelike. While conformally equivalent Riemannian metrics assign the same measure to any oriented angle on S, conformally equivalent Lorentzian metrics assign the same causal character to any tangent vector on S.

At any point p on S, a 1-dimensional linear subspace of S_p is called a direction at p. For a fixed Lorentzian metric h on S, the causal character of a direction is the causal character of any non-zero vector in it. By Lemma 1 in [10], there are C^∞ null direction fields X and Y on S, naturally ordered so that at each point p on S, sufficiently small rotations of X_p (resp. Y_p) in the positive sense carry X_p (resp. Y_p) into a spacelike (resp. timelike) position. Conformally equivalent Lorentzian metrics h and $\hat{h}$ determine the same ordered pair of null direction fields X, Y on S. Thus they also determine the same collection $\mathcal{X}$ (resp. $\mathcal{Y}$) of maximal C^∞ integral curves for X (resp. Y).

A Riemann surface $\mathcal{R} = (S, [h])$ is the pairing of S with a conformal equivalence class of Riemannian metrics. Coordinates x, y are h-isothermal for a Riemannian metric h on S if and only if $h \sim dx^2 + dy^2$. It follows that h-isothermal coordinates are $\hat{h}$-isothermal for any $\hat{h} \sim h$, allowing one to speak of isothermal coordinates on $\mathcal{R} = (S, [h])$. The use of complex parameters $z = x + iy$ for all isothermal coordinates x, y on $\mathcal{R}$ makes $\mathcal{R}$ into a 1-complex dimensional manifold.

A Lorentz surface $\mathcal{L} = (S, [h])$ is the pairing of S with a conformal equivalence class of Lorentzian metrics. Coordinates x, y are proper h-null for a Lorentzian metric h on S if and only if $h \sim dxdy$. It follows that proper h-null coordinates are proper $\hat{h}$-null for any $\hat{h} \sim h$, allowing one to speak of proper null coordinates on $\mathcal{L} = (S, [h])$. While the local existence of isothermal coordinates on a Riemann surface is quite difficult to prove, (see pp.445-500 in [9]), the existence of proper null coordinates on a Lorentz surface is easily obtained (see Lemma 2 in [10]). Generally, there is no structure canonically determined on a Lorentz surface $\mathcal{L} = (S, [h])$ which is smoother than the C^∞ structure on the underlying surface S.

Just as the most basic Riemann surface is the Euclidean plane

$$E^2 = (u, v\text{-plane}, [du^2 + dv^2]),$$

the most basic Lorentz surface is the Minkowski plane

$$E^{2,1} = (u, v\text{-plane}, [du^2 - dv^2]).$$

Because it is convenient to work with horizontal X directions and vertical Y directions, the proper null coordinates $x = u - v$, $y = u + v$ are generally used on $E^{2,1}$, giving

$$E^{2,1} = (x, y\text{-plane}, [dxdy]).$$

Note that there is a natural C^ω (that is, real analytic) structure determined by x, y (or u, v) on $E^{2,1}$ in terms of which $dxdy$ (or $du^2 - dv^2$) is a C^ω metric. If the charts associated with some well-defined selection of proper null coordinates on $\mathcal{L} = (S, [h])$ determine a C^ω atlas on S in terms of which at least one metric in $[h]$ is C^ω, the Lorentz surface $\mathcal{L} = (S, [h])$ is said to have a C^ω structure. For any open, connected, non-empty subset U of the x, y-plane, the Lorentz subsurface $\mathcal{L} = (U, [dxdy])$ of $E^{2,1}$ has a C^ω structure.

The definition of C^∞ conformal equivalence between a surface S with metric h and a surface $\hat{S}$ with metric $\hat{h}$ is identical in the Riemannian and Lorentzian cases. But the implications of C^∞ conformal equivalence are drastically different in the two cases. A C^∞ conformal diffeomorphism $F : (S, h) \to (\hat{S}, \hat{h})$ is an orientation preserving C^∞ diffeomorphism for which $F^*\hat{h} \sim h$. When such a C^∞ conformal diffeomorphism F exists, one says that (S, h) is C^∞ conformally equivalent to $(\hat{S}, \hat{h})$, writing $(S, h) \sim_\infty (\hat{S}, \hat{h})$.

If h is Riemannian, a C^∞ conformal diffeomorphism $F : (S, h) \to (\hat{S}, \hat{h})$ preserves the measure of oriented angles, and is actually a complex analytic map $F : \mathcal{R} \to \hat{\mathcal{R}}$ between the Riemann surfaces $\mathcal{R} = (S, [h])$ and $\hat{\mathcal{R}} = (\hat{S}, [\hat{h}])$. Similarly, if $F : (S, h) \to (\hat{S}, \hat{h})$ is a C^j orientation preserving diffeomorphism for $j = 1, 2, \ldots$ which preserves the measure of oriented angles, or is just a 1-quasiconformal homeomorphism, F is really a complex analytic map $F : \mathcal{R} \to \hat{\mathcal{R}}$. (See [1].) In short, any effort to reduce smoothness and keep conformality is doomed in the definite metric case.

If h is Lorentzian, a C^∞ conformal diffeomorphism $F : (S, h) \to (\hat{S}, \hat{h})$ preserves the causal character of any tangent vector, and takes the ordered pair of null direction fields X, Y for h onto the ordered pair of null direction fields $\hat{X}, \hat{Y}$ for $\hat{h}$. Thus F also takes the collection $\mathcal{X}$ (resp. $\mathcal{Y}$) of maximal C^∞ integral curves for X (resp. Y) onto the collection $\hat{\mathcal{X}}$ (resp. $\hat{\mathcal{Y}}$) of maximal C^∞ integral curves for $\hat{X}$ (resp. $\hat{Y}$). Locally, F is given by C^∞ functions $\hat{x} = f(x)$ and $\hat{y} = g(y)$ with $f'(x)g'(y) > 0$ connecting proper null coordinates $\hat{x}, \hat{y}$ on $\hat{\mathcal{L}} = (\hat{S}, [\hat{h}])$ to proper null coordinates x, y on $\mathcal{L} = (S, [h])$. Here conformality forces no greater smoothness on F than we assumed F to have. This may seem disappointing at first, but it opens the way for separate notions of conformality for each degree of smoothness.

Suppose h is a Lorentzian metric on S. For each $j = 1, 2, \ldots$, a C^j conformal diffeomorphism $F : (S, h) \to (\hat{S}, \hat{h})$ is an orientation preserving C^j diffeomorphism which takes each C^∞ parameterized curve in $\mathcal{X}$ (resp. $\mathcal{Y}$) to a C^j reparameterization of a curve in $\hat{\mathcal{X}}$ (resp. $\hat{\mathcal{Y}}$). A conformal homeomorphism $F : (S, h) \to (\hat{S}, \hat{h})$ is an orientation preserving homeomorphism which takes each C^∞ parameterized curve in $\mathcal{X}$ (resp. $\mathcal{Y}$) to a C^0 reparameterization of a curve in $\hat{\mathcal{X}}$ (resp. $\hat{\mathcal{Y}}$). In case $\mathcal{L} = (S, [h])$ and $\hat{\mathcal{L}} = (\hat{S}, [\hat{h}])$ have C^ω structures, then in terms of these structures, a C^ω conformal diffeomorphism is an orientation preserving C^ω diffeomorphism which takes each C^ω parameterization of a curve in $\mathcal{X}$ (resp. $\mathcal{Y}$) to a C^ω parameterization of a curve in $\hat{\mathcal{X}}$ (resp. $\hat{\mathcal{Y}}$). If for $j = 1, 2, \ldots; \infty; \omega$ (resp. $j = 0$), there exists a C^j conformal diffeomorphism (resp. a conformal homeomorphism) $F : (S, h) \to (\hat{S}, \hat{h})$, one says that (S, h) and $(\hat{S}, \hat{h})$ are C^j conformally equivalent, writing $(S, h) \sim_j (\hat{S}, \hat{h})$. One also says that $\mathcal{L} = (S, [h])$ is C^j conformally equivalent to $\hat{\mathcal{L}} = (\hat{S}, [\hat{h}])$, writing $\mathcal{L} \sim_j \hat{\mathcal{L}}$.

If $F : (S, h) \to (\hat{S}, \hat{h})$ is a C^j conformal diffeomorphism for $j =$

$1, 2, \ldots; \infty$ (resp. a conformal homeomorphism), then F is given locally by C^j (resp. C^0) functions $\hat{x} = f(x)$ and $\hat{y} = g(y)$ relating C^∞ null coordinates $\hat{x}, \hat{y}$ on $\hat{\mathcal{L}}$ to C^∞ null coordinates x, y on $\mathcal{L}$, with $f'(x)g'(y) > 0$ (resp. with f and g both strictly increasing or both strictly decreasing). In case $\mathcal{L} = (S, [h])$ and $\hat{\mathcal{L}} = (\hat{S}, [\hat{h}])$ have C^ω structures, a C^ω conformal diffeomorphism $F : (S, h) \to (\hat{S}, \hat{h})$ is given locally by C^ω functions $\hat{x} = f(x)$ and $\hat{y} = g(y)$ relating C^ω null coordinates $\hat{x}, \hat{y}$ on $\hat{\mathcal{L}}$ to C^ω null coordinates x, y on $\mathcal{L}$, with $f'(x)g'(y) > 0$.

3. Results

Before constructing conformally distinct Lorentz surfaces, it is helpful to note the different conformally equivalent guises in which a single Lorentz surface can appear. Any region of the form $(a, b) \times (c, d)$ in the x, y-plane for $-\infty \leq a < b \leq \infty$ and $-\infty \leq c < d \leq \infty$ is called a grid box. Our first observation is that any grid box with the metric $dxdy$ can be used as a conformal model for $E^{2,1}$.

Lemma 1. *$E^{2,1} \sim_\omega ((a, b) \times (c, d), [dxdy])$, where $-\infty \leq a < b \leq \infty$ and $-\infty \leq c < d \leq \infty$.*

Proof. Modulo translation, 180° rotation and separate uniform scaling of x and/or y, one can assume that $(a, b) \times (c, d)$ is one of the grid boxes $(-\infty, \infty) \times (-\infty, \infty)$, $(-\infty, \infty) \times (-1, 1)$, $(-\infty, \infty) \times (0, \infty)$, $(-1, 1) \times (-\infty, \infty)$, $(-1, 1) \times (-1, 1)$, $(-1, 1) \times (0, \infty)$, $(0, \infty) \times (0, \infty)$, $(0, \infty) \times (-1, 1)$ or $(-\infty, 0) \times (0, \infty)$. To define a C^ω conformal diffeomorphism from $E^{2,1}$ onto each of these grid boxes, take (x, y) to $(\hat{x}, \hat{y})$ where $\hat{x}(x) = x$, $\tanh x$, e^x or $-e^{-x}$ as needed, and $\hat{y}(y) = y$, $\tanh y$ or e^y as needed. □

While it is not unusual for a Lorentz surface to be conformally equivalent to a proper subset of itself, the situation in Lemma 1 (where $E^{2,1}$ is conformally equivalent to "arbitrarily small" subsets of itself) is easily used to characterize those Lorentz surfaces which are C^j conformally equivalent to a subset of $E^{2,1}$. To be specific, we cite without proof a result from [10], in which $\mathcal{L}$ need not be simply connected.

The Cloning Lemma. *A Lorentz surface $\mathcal{L} = (S, [h])$ is C^j conformally equivalent to a subset of $E^{2,1}$ for a fixed $j = 0, 1, \ldots; \infty; \omega$ if and only if for any non-empty open subset V of S, there is a subset U of V so that $\mathcal{L} \sim_j (U, [h])$.*

The next two results from [8] are central to many of our constructions. In both lemmas, $f : [0,1] \to [0,1]$ is assumed to be continuous and strictly increasing with $f(0) = 0$ and $f(1) = 1$. In addition, $\mathcal{L} = (U, [dxdy])$ where U is the region enclosed by the lines $x = 1$, $y = 0$ and the graph of f, while $\mathcal{T} = (T, [dxdy])$ where T is the triangular region with vertices at $(0,0)$, $(1,0)$ and $(1,1)$.

The Flattening Lemma. *$\mathcal{L} \sim_0 \mathcal{T}$ and if f is C^j with $f'(x) > 0$ on $(0,1)$ for a fixed $j = 1, 2, ...; \infty; \omega$, then $\mathcal{L} \sim_j \mathcal{T}$.*

Proof. Take (x, y) on $\mathcal{L}$ to $(f(x), y)$ on $\mathcal{T}$. □

The Breaking Lemma. *$\mathcal{L} \sim_j \mathcal{T}$ fails if f is not C^j with $f'(x) > 0$ on $(0,1)$ for a fixed $j = 1, 2, ...; \infty; \omega$.*

Proof. We outline the argument from [8]. Suppose $F : \mathcal{L} \to \mathcal{T}$ is a C^j conformal diffeomorphism. Then F takes any horizontal (resp. vertical) line segment in $\mathcal{L}$ onto a horizontal (resp. vertical) line segment in $\mathcal{T}$, and any grid box in $\mathcal{L}$ onto a grid box in $\mathcal{T}$. It follows that $F : U \to T$ extends to a homeomorphism $\bar{F} : \bar{U} \to \bar{T}$ fixing the points $(0,0)$, $(1,0)$ and $(1,1)$. If $\bar{F}$ takes $(x, 0)$ to $(\alpha(x), 0)$ for $0 \le x \le 1$ and $(1, y)$ to $(1, \beta(y))$ for $0 \le y \le 1$, then F takes any (x, y) in $\mathcal{L}$ to $(\alpha(x), \beta(y))$ in $\mathcal{T}$. Thus $\alpha : (0,1) \to (0,1)$ and $\beta : (0,1) \to (0,1)$ are C^j and onto with positive first derivatives. Since $\alpha(x) = \beta(y)$ for $y = f(x)$, one gets

$$\alpha(x) = \beta(f(x)),$$

making $f = \beta^{-1} \circ \alpha$ on $(0,1)$ the composite of C^j functions with positive first derivative, a contradiction. □

The next result shows that monotonicity of f is essential in the Breaking Lemma. It also makes clear that relatively compact Lorentz subsurfaces of $E^{2,1}$ can be C^ω conformally equivalent, even when the boundary of one is C^ω smooth, and the boundary of the other fails to be C^1 smooth.

Lemma 2. $(\{|x| + |y| < 1\}, [dxdy]) \sim_\omega (\{x^2 + y^2 < 1\}, [dxdy])$.

Proof. Take (x, y) in the square to $(\sin(\pi x/2), \sin(\pi y/2))$ in the disc. (This map was found by Senchun Lin.) □

For some families of Lorentz surfaces, there is a numerical C^j invariant which must match if two members of the family are C^j conformally equivalent. To illustrate this phenomenon, let D (resp. $\hat{D}$) be the triangular region enclosed by the lines $y = \lambda x$, $y = \mu x$ and $y = -1$ with $\mu < 0 < \lambda$ (resp. $y = \hat{\lambda} x$, $y = \hat{\mu} x$ and $y = -1$ with $\hat{\mu} < 0 < \hat{\lambda}$). The C^1 conformal invariant λ/μ in Lemma 3 was first described in [8] for a more complicated family of examples. Kulkarni suggested that we display this invariant using the simpler family of triangular regions just defined.

Lemma 3. *Suppose* $\mathcal{L} = (D, [dxdy])$ *and* $\hat{\mathcal{L}} = (\hat{D}, [dxdy])$ *then* $\mathcal{L} \sim_0 \hat{\mathcal{L}}$ *is always valid, while* $\mathcal{L} \sim_j \hat{\mathcal{L}}$ *for any fixed* $j = 1, 2, \ldots; \infty; \omega$ *if and only if* $\lambda\hat{\mu} = \hat{\lambda}\mu$.

Proof. We outline an argument from [10]. The map taking (x, y) in $\mathcal{L}$ to $(\lambda x/\hat{\lambda}, y)$ in $\hat{\mathcal{L}}$ for $x \leq 0$ and to $(\mu x/\hat{\mu}, y)$ in $\hat{\mathcal{L}}$ for $x \geq 0$ is a conformal homeomorphism. When $\lambda\hat{\mu} = \hat{\lambda}\mu$, this map is a C^ω conformal diffeomorphism.

If $\mathcal{L} \sim_j \hat{\mathcal{L}}$ for $j = 1, 2, \ldots; \infty; \omega$, there is a C^j conformal diffeomorphism $F : \mathcal{L} \to \hat{\mathcal{L}}$ which takes any horizontal (resp. vertical) line segment in $\mathcal{L}$ onto a horizontal (resp. vertical) line segment in $\hat{\mathcal{L}}$, and any grid box in $\mathcal{L}$ onto a grid box in $\hat{\mathcal{L}}$. It follows that $F : D \to \hat{D}$ extends to a homeomorphism $\bar{F}$ from the closure of D onto the closure of $\hat{D}$, taking $(-1/\lambda, -1)$ to $(-1/\hat{\lambda}, -1)$, $(-1/\mu, -1)$ to $(-1/\hat{\mu}, -1)$ and fixing the origin. If $\bar{F}$ takes $(x, -1)$ to $(\alpha(x), -1)$ for $-1/\lambda \leq x \leq -1/\mu$ and $(0, y)$ to $(0, \beta(y))$ for $-1 \leq y \leq 0$, then F takes any (x, y) in $\mathcal{L}$ to $(\alpha(x), \beta(y))$ in $\hat{\mathcal{L}}$. Thus the functions $\alpha : (-1/\lambda, -1/\mu) \to (-1/\hat{\lambda}, -1/\hat{\mu})$ and $\beta : (-1, 0) \to (-1, 0)$ are C^j and onto with positive first derivatives. In particular, $\alpha'(0) > 0$. Since $\beta(y) = \hat{\lambda}\alpha(x)$ if $y = \lambda x$ for $-1/\lambda \leq x \leq 0$,

$$\beta(\lambda x) = \hat{\lambda}\alpha(x)$$

when $-1/\lambda \leq x \leq 0$. Similarly, since $\beta(y) = \hat{\mu}\alpha(x)$ if $y = \mu x$ for $0 \leq x \leq -1/\mu$,

$$\beta(\mu x) = \hat{\mu}\alpha(x)$$

when $0 \leq x \leq -1/\mu$. By the chain rule,

$$\hat{\lambda}\alpha'(x) = \lambda\beta'(\lambda x)$$

when $-1/\lambda < x < 0$, while

$$\hat{\mu}\alpha'(x) = \mu\beta'(\mu x)$$

when $0 < x < -1/\mu$. Thus

$$\lim_{y \to 0^-} \beta'(y) = \frac{\hat{\lambda}}{\lambda}\alpha'(0) = \frac{\hat{\mu}}{\mu}\alpha'(0)$$

gives $\hat{\lambda}\mu = \lambda\hat{\mu}$. □

Given the Lorentz surface $\mathcal{L} = (D, [dxdy])$ in Lemma 3, small variations in λ/μ produce uncountably many conformally homeomorphic Lorentz surfaces, no two of which are C^1 conformally diffeomorphic. It is shown in [8] that C^j conformally equivalent simply connected Lorentz surfaces need not be C^{j+1} conformally equivalent for any $j = 0, 1, \ldots$. Here we prove the following stronger result.

Theorem 1. *For any $j = 0, 1, \ldots$, there are uncountably many C^j conformally equivalent simply connected Lorentz surfaces, no two of which are C^{j+1} conformally equivalent. Moreover, there are uncountably many C^∞ conformally equivalent simply connected Lorentz surfaces with C^ω structures, no two of which are C^ω conformally equivalent.*

Proof. We gave the argument above for $j = 0$, so take a fixed value for $j = 2, 3, \ldots; \infty$. In case $j = \infty$, interpret $j + 1$ to mean ω. For each $k = 1, 2, \ldots$, take $a_k = 1-(1/2)^{k-1}$, define the intervals $I_k = [a_{2k}, a_{2k+1}]$ and $J_k = (a_{2k-1}, a_{2k})$, and pick a fixed closed subinterval $\mathcal{J}_k$ of J_k. Represent each real number r in $(0, 1)$ by a binary expansion $r = .r_1 r_2 \cdots$ so that $r_k = 0$ or $r_k = 1$ for each $k = 1, 2, \ldots$. Given a fixed r in $(0, 1)$, construct a continuous function $f_r : [0, 1] \to [0, 1]$ such that

(i) $f_r(0) = 0$ and $f_r(1) = 1$,
(ii) $f_r(x) = 1 - (1/2)^k$ on I_k for any $k = 1, 2, \ldots$,
(iii) f_r is C^j with $f_r'(x) > 0$ on J_k, while $f_r'(x) = 1$ for all x in J_k which lie outside $\mathcal{J}_k$ for any $k = 1, 2, \ldots$, and
(iv) f_r is C^{j+1} (resp. *not* C^{j+1}) on J_k if $r_k = 0$ (resp. if $r_k = 1$).

Denote by Ω_r the region beneath the graph of $y = f_r(x)$ over the interval $(0, 1)$, and define the Lorentz surface $\mathcal{L}_r = (\Omega_r, [dxdy])$. We will show that $\mathcal{L}_r \sim_j \mathcal{L}_{r'}$ for every r and r' in $(0, 1)$, while $\mathcal{L}_r \sim_{j+1} \mathcal{L}_{r'}$ fails unless $r = r'$.

To construct a C^j conformal diffeomorphism $F : \mathcal{L}_r \to \mathcal{L}_{r'}$, let F take (x, y) in Ω_r to $(\hat{x}, \hat{y})$ in $\Omega_{r'}$ with $\hat{x} = x$ and $\hat{y} = g(y)$, setting $g(y) = y$ if $y = 1 - (1/2)^k$ for some $k = 1, 2, \ldots$ and taking

$$g(y) = (f_{r'} \circ f_r^{-1})(y)$$

for any other y in $(0,1)$. Then $\hat{y} = g(y)$ is C^j in each interval $(1-(1/2)^k, 1-(1/2)^{k+1})$ since f_r and $f_{r'}$ are each C^j with positive first derivative on J_k. In addition, $\hat{y} = g(y) = y$ in some neighborhood of $y = 1-(1/2)^k$ for $k = 1, 2, \ldots$. It follows that $g(y)$ is C^j with $g'(y) > 0$ on $(0,1)$, making $F : \mathcal{L}_r \to \mathcal{L}_{r'}$ a C^j conformal diffeomorphism.

Suppose next that $\hat{F} : \mathcal{L}_r \to \mathcal{L}_{r'}$ is a C^{j+1} conformal diffeomorphism. Then $\hat{F}$ takes horizontal (resp. vertical) line segments in Ω_r onto horizontal (resp. vertical) line segments in $\Omega_{r'}$, and any grid box in Ω_r onto a grid box in $\Omega_{r'}$. To see that $\hat{F}$ takes the portion of Ω_r over J_k (resp. I_k) onto the portion of $\Omega_{r'}$ over J_k (resp. I_k), argue as follows. Call a grid box in Ω_r (resp. $\Omega_{r'}$) maximal if and only if it is contained in no strictly larger grid box in Ω_r (resp. $\Omega_{r'}$). Any maximal grid box in Ω_r (resp. $\Omega_{r'}$) has its bottom edge along $y \equiv 0$, its right edge along $x \equiv 1$, and its left top vertex on the graph of $y = f_r(x)$ (resp. $y = f_{r'}(x)$), where x lies in J_k or at the right endpoint a_{2k} of J_k for some $k = 1, 2, \ldots$. A maximal grid box of the form

$$R_k = (a_{2k}, 1) \times (0, 1-(1/2)^k)$$

in Ω_r (or $\Omega_{r'}$) is called special, since no other kind of maximal grid box has an interval along its upper edge outside of Ω_r (or $\Omega_{r'}$). Thus $\hat{F}$ takes each special maximal grid box in Ω_r onto a special maximal grid box in $\Omega_{r'}$ so as to establish a one-one correspondence between the R_k in Ω_r and the R_k in $\Omega_{r'}$ which preserves the order in the nested sequence of their lower edges for $k = 1, 2, \ldots$. It follows that $\hat{F}$ takes R_k onto itself for each $k = 1, 2, \ldots$, so the portion of the line $x = a_{2k}$ in Ω_r is taken onto itself. Within every R_k there is a largest grid box

$$\hat{R}_k = (a_{2k}, a_{2k+1}) \times (0, 1-(1/2)^k)$$

whose upper edge lies on the graphs of $y = f_r(x)$ and $y = f_{r'}(x)$. Thus $\hat{F}$ takes $\hat{R}_k$ onto itself for each $k = 1, 2, \ldots$, so the portion of the line $x = a_{2k+1}$ in Ω_r is taken onto itself. Since $\hat{F}$ takes the portion of R_k outside of R_{k+1} onto itself, we conclude that $\hat{F}$ takes the portion of Ω_r over J_k (resp. I_k) onto the portion of $\Omega_{r'}$ over J_k (resp. I_k). Moreover, $\hat{F}$ takes the portion T_r^k of Ω_r over J_k on which $y > 1-(1/2)^{k-1}$, C^{j+1} conformally onto the portion $T_{r'}^k$ of $\Omega_{r'}$ over J_k on which $y > 1-(1/2)^{k-1}$. Modulo uniform rescaling of x and of y (and translation), each region T_r^k and $T_{r'}^k$ can be taken as the region U in the Breaking Lemma.

But if $r \neq r'$, one of the functions f_r or $f_{r'}$ is C^{j+1} and the other *not* C^{j+1} over J_k for at least one value of $k = 1, 2, \ldots$. By the Breaking Lemma,

$T_r^k \sim_{j+1} T_{r'}^k$ is impossible for such a value of k. This contradiction proves the theorem. □

Suppose now that S^∞ is the universal cover of the x, y-plane punctured at the origin. To construct S^∞, cut a copy S_k of the x, y-plane along the non-positive x-axis for each integer k, leaving the neqative x-axis along the top edge of the cut, and discarding the origin. Then identify the upper edge of the cut on S_k with the lower edge of the cut on S_{k+1}. The local coordinates x, y on S^∞ are proper h-null for the lift h of the metric $dxdy$ to S^∞, giving the Lorentz surface $\mathcal{L}_\infty = (S^\infty, [h])$ a C^ω structure. It is shown in [10] that $\mathcal{L}_\infty$ is C^ω conformally equivalent to the u, v-plane with the metric

$$h = \sin u(dv^2 - du^2) - 2\cos u du dv.$$

However, $\mathcal{L}_\infty$ is not even C^0 conformally equivalent to any subset of $E^{2,1}$. To see this, fix any integer k, and take $m = 0, 1, 2$ or 3. Let γ_{4k+m} be the copy on $S_k \subset S^\infty$ of the negative y-axis if $m = 0$, the positive x-axis if $m = 1$, the positive y-axis if $m = 2$, and the negative x-axis if $m = 3$. Use any proper h-null coordinates x, y on S^∞ to define grid boxes on S^∞. If F is a conformal homeomorphism taking S^∞ onto a subset of $E^{2,1} = (x, y\text{-plane}, [dxdy])$, then F takes each grid box on S^∞ onto a grid box in $E^{2,1}$. Thus F takes γ_j and γ_{j+1} for each integer j onto open line segments (one horizontal and one vertical) with a common endpoint p. It follows that F takes all γ_j to open horizontal or vertical line segments with an endpoint at p. Since F is one-one, this is impossible.

One generally uses S^∞ with the lift of $dx^2 + dy^2$ to define a Riemann surface on which $\log z$ is single valued for $z = x + iy$. Whenever such cutting any pasting methods are used to produce a surface S for which the lift of $dx^2 + dy^2$ to S defines a Riemann surface on which a particular multivalued holomorphic function of $z = x + iy$ is single valued, the lift of $dxdy$ in place of $dx^2 + dy^2$ defines a Lorentz surface instead. (Here we omit from S any branch points, such as the origin for $\sqrt{z}$.) This provides a number of interesting Lorentz surfaces, which are usually not simply connected.

To construct our next family of simply connected Lorentz surfaces, start with two copies R_1 and R_2 of the grid box $(-1, 1) \times (-1, 1)$ in the x, y-plane, cut along the line segment $\{y \equiv 0, -1 < x \leq 0\}$. Reattach $\{y = 0, -1 < x < 0\}$ to the upper edge of the cut on R_1, discarding just the origin. Attach the upper edge of the cut on R_1 to the lower edge of the cut on R_2 to obtain a surface Σ. There is a natural projection of Σ to the x, y-plane, and local coordinates x, y on Σ (which are proper h-null for the lift h of the metric $dxdy$ to Σ) can be used to define grid boxes on Σ. One argues as above that $\mathcal{L}_\Sigma = (\Sigma, [h])$ is not conformally homeomorphic to any subset of $E^{2,1}$. Now reattach $\{y = 0, -1 < x < 0\}$ to the upper edge of the cut on $R_2 \subset \Sigma$ to obtain $\hat{\Sigma}$. Eventually, $\hat{\Sigma}$ will be used to construct a family of examples which prove Theorem 2.

First, however, let Q be the union of the disc $(x+4)^2+(y+1)^2 < 4$ with the grid box $(-3,0)\times(-2,0)$. On the circle C given by $(x+4)^2+(y+1)^2 = 4$, let Γ be the half closed arc $C \cap \{x > -4\} \cap \{y \geq 0\}$. For any p on Γ, let $R(p)$ be the grid box inscribed in C with a vertex at p. Then any grid box in Q with *exactly* four of its boundary points on the boundary of Q must be an $R(p)$ for some p on Γ. Now pick a sequence of points $\{p_k\}$ on Γ whose ordinates are strictly increasing, with $p_1 = (\sqrt{3}-4, 0)$ and $p_k \to (-4, 1)$ as $k \to \infty$. For each r with $0 < r < 1$, take a binary expansion $r = .r_1r_2\cdots$, so that $r_k = 0$ or $r_k = 1$ for each $k = 1, 2, \ldots$. In case $r_k = 0$, let the symbol q_k^r denote the point p_k. In case $r_k = 1$, let the symbol q_k^r denote a fixed closed subarc of Γ containing p_k but disjoint from $q_{k-1}^r \cup \{p_{k+1}\}$. For any $k = 1, 2, \ldots$, let I_k^r be the line segment joining the highest point on q_k^r to the lowest point on q_{k+1}^r. Form an arc Γ_r which joins the endpoints of Γ by using the I_k^r's to connect the q_k^r's. Replace the arc Γ on C by the arc Γ_r to obtain a Jordan curve C_r, and let Q_r be the union of the interior of C_r with the grid box $(-3,0)\times(-2,0)$. By construction, $Q_r \subset Q$. Let $\mathcal{L}_r = (Q_r, [dxdy])$ in the following result. (See the examples $\mathcal{L}_a$ in §3.2 of [7].)

Lemma 4. $\mathcal{L}_r \sim_0 \mathcal{L}_{r'}$ *if and only if* $r = r'$.

Proof. Suppose $F : \mathcal{L}_r \to \mathcal{L}_{r'}$ is a conformal homeomorphism. Since

$$I = \{y \equiv -1\} \cap Q_r = \{y \equiv -1\} \cap Q_{r'}$$

is the only maximal horizontal line segment in Q_r (resp. $Q_{r'}$) which is not the edge of a grid box in Q_r (resp. $Q_{r'}$), F takes I onto itself. Similarly, since

$$J = \{x \equiv -4\} \cap Q_r = \{x \equiv -4\} \cap Q_{r'}$$

is the only maximal vertical line segment in Q_r (resp. $Q_{r'}$) which is not the edge of a grid box in Q_r (resp. $Q_{r'}$), F takes J onto itself. The grid box $(\sqrt{3}-4, 0)\times(-2,0)$ is maximal among grid boxes in Q_r (resp. $Q_{r'}$) with three edges contained in the boundary of Q_r (resp. $Q_{r'}$). Thus F carries $(\sqrt{3}-4,0)\times(-2,0)$ onto itself, and must respect the sense of motion to the right on I, and therefore of motion upward on J. If F takes the point $(x,-1)$

to $(\alpha(x), -1)$ for $-6 < x < 0$, use the values $\alpha(-6) = -6$ and $\alpha(0) = 0$ to define a strictly increasing continuous function $\alpha : [-6, 0] \to [-6, 0]$. Similarly, if F takes the point $(-4, y)$ to $(-4, \beta(y))$ for $-3 < y < 1$, use the values $\beta(-3) = -3$ and $\beta(1) = 1$ to define a strictly increasing continuous function $\beta : [-3, 1] \to [-3, 1]$. Thus F extends to a homeomorphism $\bar{F} : \bar{Q}_r \to \bar{Q}_{r'}$ which takes any point (x, y) to $(\alpha(x), \beta(y))$.

Because each point p on $\Gamma \cap \Gamma_r$ (resp. on $\Gamma \cap \Gamma_{r'}$) corresponds to a grid box $R(p)$ with a vertex at p, and since no other grid boxes in Q_r (resp. $Q_{r'}$) have exactly four boundary points on the boundary of Q_r (resp. $Q_{r'}$), F takes each grid box $R(p)$ for p on $\Gamma \cap \Gamma_r$ onto a grid box $R(p')$ for p' on $\Gamma \cap \Gamma_{r'}$. This implies that $\bar{F}$ establishes a one-one correspondence between the collection of sets q_k^r on Γ_r and the collection of sets $q_k^{r'}$ on $\Gamma_{r'}$, taking a point (resp. arc) q_k^r onto a point (resp. arc) $q_{k'}^{r'}$. But since $\bar{F}$ respects the order of ordinate values, $\bar{F}$ must take q_k^r onto $q_k^{r'}$ for each $k = 1, 2, \ldots$, making q_k^r a point (resp. an arc) if and only if $q_k^{r'}$ is a point (resp. an arc). In case $r \neq r'$, there is at least one value of $k = 1, 2, \ldots$ for which $r_k \neq r'_k$, so that q_k^r is a point and $q_k^{r'}$ an interval, or vice versa. This proves the lemma. □

Now, for each r in $(0, 1)$, attach Q_r to $\hat{\Sigma}$ along the interval $\{y \equiv 0, -1 < x < 0\}$ on the upper edge of the cut on $R_2 \subset \hat{\Sigma}$ to obtain a surface $S(r)$. There is a natural projection of $S(r)$ into the x, y-plane, and local coordinates x, y on $S(r)$ which are proper h-null for the lift h of the metric $dxdy$ to $S(r)$ can be used to define grid boxes on $S(r)$. If $\mathcal{L}(r) = (S(r), [h])$, any conformal homeomorphism $F : \mathcal{L}(r) \to E^{2,1}$ restricts to a conformal homeomorphism $F : \mathcal{L}_\Sigma \to E^{2,1}$, which is impossible. Any conformal homeomorphism $F : \mathcal{L}(r) \to \mathcal{L}(r')$ restricts to a conformal homeomorphism $F : \mathcal{L}_r \to \mathcal{L}_{r'}$ for the Lorentz surfaces $\mathcal{L}_r$ and $\mathcal{L}_{r'}$ in Lemma 4, so that $r = r'$. This gives the following result.

Theorem 2. *There are uncountably many C^0 conformally distinct simply connected Lorentz surfaces, each C^0 conformally distinct from every subset of $E^{2,1}$.*

To construct our next family of examples, translate Q_r for each r in $(0, 1)$ one unit to the right and one unit upward. Then remove the line segment $\{y \equiv 0, 0 \leq x < 1\}$ to obtain a surface $\hat{Q}_r$. The open sets

$$U_1 = (\hat{Q}_r \cap \{y > 0\}) \cup (\hat{Q}_r \cap \{x < 0\})$$

and

$$U_2 = (-1, 1) \times (-1, 0)$$

cover $\hat{Q}_r$, and the chart maps $\chi_1 : U_1 \to x, y$-plane and $\chi_2 : U_2 \to x, y$-plane given by $\chi_1(x, y) = (x_1, y_1)$ and $\chi_2(x, y) = (x_2, y_2)$ with $x_1 = x$, $y_1 = y$ and $x_2 = x^3$, $y_2 = y$ are C^∞ related on $U_1 \cap U_2 = (-1, 0) \times (-1, 0)$. Thus χ_1 and χ_2 determine a C^∞ atlas on $\hat{Q}_r$, defining a new surface Q_r^*. Take a C^∞ partition of unity $\{\phi_1, \phi_2\}$ on Q_r^* subordinate to the covering $\{U_1, U_2\}$, and use the C^∞ metric

$$h^* = \phi_1 dx_1 dy_1 + \phi_2 dx_2 dy_2$$

on Q_r^*. Here the proper h^*-null coordinates x_1, y_1 and x_2, y_2 on Q_r^* determine the same grid boxes on Q_r^* as are identified by x, y on $\hat{Q}_r$. If $\mathcal{L}_r^* = (Q_r^*, [h^*])$ and $\hat{\mathcal{L}}^r = (\hat{Q}_r, [dxdy])$, the identity map shows that $\mathcal{L}_r^* \sim_0 \hat{\mathcal{L}}^r$. However, $\mathcal{L}_r^*$ is not C^1 conformally equivalent to any subset of $E^{2,1}$.

To see this, suppose $F : \mathcal{L}_r^* \to E^{2,1}$ is a C^1 conformal diffeomorphism between $\mathcal{L}_r^*$ and $F(\mathcal{L}_r^*)$. Define the surface $Q^* = Q_r^* \cap \{x > -1\}$ which does not depend on the choice of r. Use on Q^* the non-standard C^∞ structure induced by the C^∞ structure on Q_r^*. If $\mathcal{L}^* = (Q^*, [h^*])$, then F restricts to a C^1 conformal diffeomorphism between $\mathcal{L}^*$ and $F(\mathcal{L}^*)$. The following result from [8] says this is impossible.

Lemma 5. $\mathcal{L}^* = (Q^*, [h^*])$ *is not C^1 conformally diffeomorphic to any subset of $E^{2,1}$.*

Proof. Suppose $F : \mathcal{L}^* \to E^{2,1}$ is a C^1 conformal diffeomorphism between $\mathcal{L}^*$ and $F(\mathcal{L}^*)$, so that F takes horizontal (resp. vertical) line segments onto horizontal (resp. vertical) line segments. Compose F with a 180° rotation if necessary, so that F respects motion to the right (resp. upward) on horizontal (resp. vertical) line segments. Let γ^+ (resp. γ^-) be the portion of $y \equiv 1/2$ (resp. $y \equiv -1/2$) in $\mathcal{L}^*$. Let γ_0^+ (resp. γ_0^-) be the portion of γ^+ (resp. γ^-) in $\{x \leq 0\}$. Then F takes points on γ_0^+ and γ_0^- with matching abscissas to points in $E^{2,1}$ with matching abscissas. Suppose $F(\gamma^+)$ is no longer than $F(\gamma_-)$. The map taking each point on $F(\gamma^+)$ to the point directly beneath it on $F(\gamma^-)$ is a C^∞ diffeomorphism from $F(\gamma^+)$ onto the portion of $F(\gamma^-)$ beneath it, which must contain $F(\gamma_0^-)$ as a proper subset. This map pulls back under F^{-1} to a C^1 diffeomorphism f from γ^+ onto a portion of γ^- containing γ_0^- as a proper subset. If $f : \gamma^+ \to \gamma^-$ takes the point $(x, 1/2)$ on γ^+ to the point $(\sigma(x), -1/2)$ on γ^- for $-1 < x < 1$, then $\sigma(x) \equiv x$ for $-1 < x \leq 0$ and (because the C^∞ structure on γ^- is determined by a chart map taking x to x^3), the function $\sigma^3(x)$ must be C^1 with positive first derivative for $-1 < x < 1$. But when $-1 < x < 0$, $\sigma(x) = x$ and

$$\frac{d}{dx}(\sigma^3) = 3\sigma^2\frac{d\sigma}{dx} = 3\sigma^2$$

goes to zero as $x \to 0^-$, a contradiction. The argument is much the same if $F(\gamma^+)$ is longer than $F(\gamma^-)$. □

For each r in $(0,1)$, we have shown that $\mathcal{L}_r^*$ is not C^1 conformally homeomorphic to any subset of $E^{2,1}$. One argues that $\mathcal{L}_r^* \sim_0 \mathcal{L}_{r'}^*$ if and only if $r = r'$ just as we argued above that $\mathcal{L}_r \sim_0 \mathcal{L}_{r'}$ if and only if $r = r'$. This justifies the following result.

Theorem 3. *There are uncountably many C^0 conformally distinct simply connected Lorentz surfaces, each C^0 conformally equivalent to a subset of $E^{2,1}$, and each C^1 conformally distinct from every subset of $E^{2,1}$.*

Suppose now that R^r is a rectangular region in the x,y-plane enclosed by lines of slope ± 1, whose spacelike edges each have length $r > 0$, and whose timelike edges each have length 1. For each R^r, define the Lorentz surface $\mathcal{L}^r = (R^r, [dxdy])$. Comparison with Lemma 1 makes the following result from [6] all the more surprising.

Theorem 4. $\mathcal{L}^r \sim_0 \mathcal{L}^{r'}$ *if and only if* $r = r'$.

Proof. We sketch the argument from [6]. For each $r > 0$, position R^r with its bottom vertex at the origin. Define a sequence $\{P_k^r\}$ of points on the boundary dR^r of R^r as follows. Take $P_0^r = (0,0)$ and let P_1^r be the top endpoint of the maximal vertical line segment in R^r whose bottom endpoint is at P_0^r. Suppose $k \geq 2$. If P_{k-1}^r is a vertex of R^r, take $P_k^r = P_{k-2}^r$. If P_{k-1}^r is not a vertex of R^r, one maximal null line interval in R^r with an endpoint at P_{k-1}^r ends at P_{k-2}^r, and the other ends at a point different from P_{k-2}^r, which we take as P_k^r.

Suppose there is a conformal homeomorphism $F : R^r \to R^{r'}$. Then F takes horizontal (resp. vertical) line segments onto horizontal (resp. vertical) line segments, and grid boxes onto grid boxes. Moreover, a point on dR^r (resp. $dR^{r'}$) is the vertex of a grid box in R^r (resp. $R^{r'}$) if and only if it is *not* a vertex of R^r (resp. $R^{r'}$). It follows that F extends to a homeomorphism $\bar{F} : \bar{R}^r \to \bar{R}^{r'}$ taking vertices of R^r to vertices of $R^{r'}$ and endpoints of a maximal horizontal (resp. vertical) line segment in R^r onto the endpoints of a maximal horizontal (resp. vertical) line segment in $R^{r'}$. Assume that $\bar{F}$ takes $(0,0)$ to $(0,0)$. (Otherwise, $\bar{F}$ takes $(0,0)$ to the top vertex of $R^{r'}$, and F can be replaced by its composition with 180° rotation of $R^{r'}$ about its center.) Then $\bar{F}$ carries the vertices and edges of R^r to the corresponding vertices and edges of $R^{r'}$. In addition, $\bar{F}$ must take P_k^r to $P_k^{r'}$ for each $k = 0, 1, \ldots$. But if $r \neq r'$, there is a $k_0 = 1, 2, \ldots$ for which $P_{k_0}^r$ and $P_{k_0}^{r'}$ are neither at corresponding vertices nor at corresponding sides of R^r and $R^{r'}$. This contradiction shows that $r = r'$. □

Theorem 4 shows that there are uncountably many C^0 conformally distinct rectangular regions in $E^{2,1}$. Even though every $\mathcal{L}^r$ in Theorem 4 is a bounded, convex, Lorentz subsurface of $E^{2,1}$ which is symmetric with respect to a spacelike line and a timelike line, the following result is proved in [6].

A Finiteness Theorem. *There are exactly* 21 C^0 *conformally distinct bounded convex Lorentz subsurfaces of* $E^{2,1}$ *which are symmetric with respect to a null line.*

For a fuller understanding of simply connected Lorentz surfaces, see [2], [3], [7] and [10].

References

[1] L.V. Ahlfors, *Lectures on Quasiconformal Mappings*, Wadsworth Inc., 1987.

[2] R. Kulkarni, "An analogue of the Riemann mapping theorem for Lorentz metrics," *Proc. Royal Soc. Lond.* **A401** (1985), 117–130.

[3] F. Luo and R. Stong, "Conformal embedding of a disc with a Lorentz metric into the plane," preprint.

[4] T.K. Milnor, "A conformal analog of Bernstein's theorem for timelike surfaces in Minkowski 3-space," *Contemp. Math.* **64** (1987), 123–130.

[5] ___________, "Entire timelike minimal surfaces in $E^{3,1}$," *Mich. Math. J.* **37** (1990), 163–177.

[6] R.W. Smyth, "Uncountably many C^0 conformally distinct Lorentz surfaces and a finiteness theorem," *Proc. AMS*, to appear.

[7] ___________, "Characterization of Lorentz surfaces via the conformal boundary," Thesis, Rutgers U., 1995.

[8] R.W. Smyth and T. Weinstein, "Conformally homeomorphic Lorentz surfaces need not be conformally diffeomorphic," *Proc. AMS* **123** (1995), 3499–3506.
[9] M. Spivak, *A Comprehensive Introduction to Differential Geometry IV*, Publish or Perish, 2nd edition, 1979.
[10] T. Weinstein, *An Introduction to Lorentz Surfaces*, de Gruyter Expositions in Mathematic **22** (1996).

Robert Smyth
Department of Mathematics
Georgian Court College
Lakewood, NJ 08701

Tilla Weinstein
Department of Mathematics
Rutgers University
New Brunswick, NJ 08903

Received April 1995

On a Theorem of Milnor and Thom

Nolan R. Wallach

1. Introduction

In [M1],[T], Milnor and Thom (independently) proved an estimate on the sum of the Betti numbers (relative to an arbitrary field of coefficients) of the set of zeros of polynomials of degree at most $k > 0$ in $\mathbb{R}^n$, essentially $C_n k^n$ (Milnor's estimate is $k(2k-1)^{n-1}$, Thom's is essentially twice Milnor's). In particular, this result gives a quantitative version of Whitney's (earlier) theorem [W], that says that the number of connected components is finite.

Most applications of the Theorem of Milnor and Thom are to the implied estimate on the number of connected components. For example, in [B] the estimate was used to determine lower bounds for the complexity of certain algebraic computation trees. One purpose of this article is to give a proof (following Milnor's methods) of the estimate on the number of connected components that uses only advanced calculus, elementary topology and Sard's theorem (the special cases of Sard's theorem that are used will also be sketched in this article) that should be accessible to mathematicians and computer scientists who are not experts in algebraic topology. Another is that the proof of Lemma 1 in [M1], left quite a bit to the reader. The first two sections of this article are devoted to an an elementary proof of this lemma (see Theorem 3.4). We also give a less elemenatary proof in section 7 that gives a quantitative upper bound for the number of irreducible components of a variety over an algebraically closed field in terms of the degrees of a defining set of equations. This result may be of independent interest. Sections 7,8,9 involve more algebraic geometry and constitute whatever is new in this paper.

In [M1], Milnor indicates that he has no examples where $C_n \neq 1$. This suggests the problem of proving (or disproving) the contention that we can take $C_n = 1$. In section 8 we prove that if $n = 2$ the answer (for the number of connected components) is affirmative. In section 9 we give an affirmative answer for non-singular hypersurfaces for the sum of the Betti numbers. This result gives a sharper upper bound for the sum of the Betti numbers of a set of the form $\mathbb{R}^n - X$ where X is the zero set of a of polynomials

---

Research partially supported by an NSF Summer Grant.

with real coefficients.

This article is an outgowth of lectures that the author gave on real algebraic geometry during a three quarter course in algebraic geometry at the University of California, San Diego. Beside graduate students in mathematics there were also regular participants from computer science and economics. We would like to thank R. Paturi for suggesting the Milnor-Thom theorem as a topic in the course and for his lectures on complexity related to [B].

Finally, this article is dedicated to the memory of my friend Joe D'Atri. His untimely death has left a void in the differential geometry community. His interests in and out of mathematics enriched all of our lives.

2. Generic finite varieties

Let k denote an algebraically closed field and let V denote an n-dimensional vector space over k. We use the notation $\mathcal{P}^r(V)$ for the space of polynomial functions on V that are homogeneous of degree r. We set

$$W_{m_1,\dots,m_n} = \mathcal{P}^{m_1}(V) \times \cdots \times \mathcal{P}^{m_n}(V).$$

We look upon $W_{m_1,\dots,m_n}$ as an $N = \sum \binom{m_n+n-1}{n-1}$ dimensional vector space over k. We fix $m_i > 0$, $i = 1,\dots,n$ and set $W = W_{m_1,\dots,m_n}$. If $g \in W$ then we look upon g as both a polynomial map of k^n to k^n and an ordered set of polynomials. Set $\mathcal{P}(V)$ equal to the algebra of polynomials on V. If $g \in W$ then set I_g equal to the ideal generated by the entries of g.

We note that I_g is a graded subspace of $\mathcal{P}(V)$. Thus $R_g = \mathcal{P}(V)/I_g$ inherits a natural grade. Set R_g^j equal to the j-th homogeneous component and $h_g(t) = \sum_j t^j \dim R_g^j$ (thought of as a formal power series). We set

$$h(t) = \prod_{j=1}^{n}(1 + t + \dots + t^{m_j-1}).$$

The following result is no doubt well known.

Proposition 2.1. *The set, $\Omega_{m_1,\dots,m_n}$, of all $g \in W$ such that $h_g(t) = h(t)$ is non-empty and Zariski open in W.*

We will need some notation before we give our (elementary) proof. Let Z be an n-dimensional vector space over k and let $z_1, \dots, z_n$ be a basis of Z. We grade Z by setting $\deg(z_i) = m_i$. Then $Z = \oplus Z^p$ with Z^p the span of the z_i with $\deg(z_i) = p$. We grade $\mathcal{P}(V) \otimes Z$ by setting $(\mathcal{P}(V) \otimes Z)^j =$

$\sum \mathcal{P}^{j-i}(V) \otimes Z^i$. We define

$$\partial(g) : \mathcal{P}(V) \otimes Z \longrightarrow \mathcal{P}(V)$$

by

$$\partial(g)(f \otimes z_i) = f g_i$$

(here $g = (g_1, ..., g_n)$). Then $\partial(g)(\mathcal{P}(V) \otimes Z)^j \subset \mathcal{P}^j(V)$ and $\partial(g)(\mathcal{P}(V) \otimes Z) = I_g$.

Define $h_j \in \mathbb{Z}$ by

$$h(t) = \sum_j h_j t^j.$$

Set $d = m_1 + ... + m_n - n$. Set $p_j = \binom{j+n-1}{n-1} - h_j$. It is easy to see that $p_j \geq 0$. For each $1 \leq j \leq d+1$ choose bases of $(\mathcal{P}(V) \otimes Z)^j$ and $\mathcal{P}^j(V)$ and if $p_j > 0$ let $\Phi_{j,i}(g)$ be an enumeration of the $p_j \times p_j$ minors of the restriction of $\partial(g)$ to $(\mathcal{P}(V) \otimes Z)^j$. We set

$$\Omega^o = \{g \in W | \text{if } p_j > 0 \text{ there exists } i \text{ s}\Phi_{j,i}(g) \neq 0\}.$$

It is clear that Ω^o is a Zariski open subset. We also note that $g = (x_1^{m_1}, ..., x_n^{m_n}) \in \Omega^o$. So Ω^o is non-empty.

We note that I_g is a graded subspace of $\mathcal{P}(V)$. Thus $R_g = \mathcal{P}(V)/I_g$ inherits a natural grade. Set R_g^j equal to the j-th homogeneous component and $h_g(t) = \sum_j t^j \dim R_g^j$ (thought of as a formal power series).

Lemma 2.2. *If $g \in \Omega^o$ then $h_g = h$.*

Proof. Fix $g \in \Omega^o$. Since $h_{d+1} = 0$,

$$\partial(g)(\mathcal{P}(V) \otimes Z)^{d+1} = \mathcal{P}(V)^{d+1}.$$

Also, if $0 \leq j \leq d$ then $\dim R_g^j \leq \dim \mathcal{P}^j(V) - p_j = h_j$. Let U be a graded subspace of $\mathcal{P}(V)$ such that $U \oplus I_g = \mathcal{P}(V)$. It is easy to see that $\dim U \cap \mathcal{P}^j(V) = \dim R_g^j$ and that if $u_1, ..., u_p$ is a basis of U then $\sum k[g_1, ..., g_n]u_i = \mathcal{P}(V)$. Let $w_1, ..., w_n$ be indeterminates and grade $k[w_1, ..., w_n]$ by setting $\deg w_i = m_i$. Then we have a graded surjection, Ψ, of $k[w_1, ..., w_n] \otimes U$ to $\mathcal{P}(V)$ given by $f[w_1, ..., w_n] \otimes u \mapsto f[g_1, ..., g_n]u$. Let $Y = \ker \Psi$. Then Y is graded and we have an identity of formal power series

$$\frac{h_g(t)}{\prod_{j=1}^n (1 - t^{m_j})} - \sum t^j \dim Y^j = \frac{1}{(1-t)^n}.$$

Since

$$\frac{h(t)}{\prod_{j=1}^{n}(1-t^{m_j})} = \frac{1}{(1-t)^n},$$

we conclude that $\dim R_g^j = h_j$. That is $h_g(t) = h(t)$. This completes the proof of the Lemma.

If $g \in W$ and if $h_g(t) = h(t)$ then it is easily seen that $\dim \partial(g)(\mathcal{P}(V)\otimes Z)^j = p_j$ for $1 \le j \le d+1$. Hence $g \in \Omega^o$. Thus $\Omega^o = \Omega$.

In order to drop the condition of homogeneity we must recall some elementary facts about the relationship between filtrations and gradings. Let I be an ideal in $\mathcal{P}(V)$. Set $R = \mathcal{P}(V)/I$. We put $\mathcal{P}_j(V) = \sum_{i\le j}\mathcal{P}^j(V)$. Let π denote the natural projection of $\mathcal{P}(V)$ onto R. Set $R_j = \pi(\mathcal{P}_j(V))$. Then $R_j \subset R_{j+1}$ and $\cup R_j = R$. Put (a usual) $Gr^j R = R_j/R_{j-1}$ $(R_{-1} = 0)$ and $GrR = \oplus Gr^j R$. We use the obvious addition and multiplication on GrR to make it into a graded algebra over k.

If $f \in \mathcal{P}_j(V)$, $f \notin \mathcal{P}_{j-1}(V)$ then $f = f_0 + \ldots + f_j$ with $f_i \in \mathcal{P}^i(V)$ and $f_j \ne 0$. We put $f_{top} = f_j$. Denote by I_{top} the linear span of the f_{top} for $f \in I$, $f \ne 0$. Then it is easy to see that I_{top} is a graded ideal and that GrR is isomorphic with $\mathcal{P}(V)/I_{top}$ as a graded algebra over k.

With this formalism in place we can state the main result of this section.

Proposition 2.3. *Let $f_1, \ldots, f_n \in \mathcal{P}(V)$ set $g_i = (f_i)_{top}$ and assume that $\deg g_i = m_i > 0$. Let I be the ideal $\sum \mathcal{P}(V)f_i$. If $g = (g_1, \ldots, g_n) \in \Omega_{m_1,\ldots,m_n}$ the algebraic set $X = \{p \in V | f_i(p) = 0, i = 1, \ldots, n\}$ has at most $m_1 \cdots m_n$ elements.*

Proof. Let J be the radical of the ideal I. Then the nullstellensatz implies that the algebra of regular functions on X is $\bar{R} = \mathcal{P}(V)/J$. Let $R = \mathcal{P}(V)/I$. Then it is clear that $\dim Gr^j\bar{R} \le \dim Gr^j R = \dim(\mathcal{P}(V)/I_{top})^j$. Now, $I_{top} \supset I_g$. Thus $\dim(\mathcal{P}(V)/I_{top})^j \le \dim R_g^j = h_j$ (all notation is as above). Thus $\dim \bar{R} \le \sum h_j = h(1) = m_1 \cdots m_n$. The nullstellensatz implies that the elements of $\bar{R}$ separate the points of X. The proposition now follows.

3. Some observations about isolated elements varieties over $\mathbb{C}$

If $f : S^k \to S^k$ is a continuous map then $\deg f$ is defined to be the action of f on $H^k(S^k, \mathbb{Z}) \cong \mathbb{Z}$. Set $\omega = \sum_{i=1}^{k+1}(-1)^{i+1}x_i dx_1 \wedge \cdots \wedge dx_{i-1} \wedge dx \wedge \cdots \wedge dx_{k+1}$ and $C_k = \int_{S^k}\omega$. If f is smooth then

$$\deg f = C_k^{-1}\int_{S^k} f^*\omega.$$

For our purposes this definition of $\deg f$ is sufficient although it is only obvious that this integral representation of $\deg f$ yields a real number. What is obvious is that if $f : [0,1] \times S^k \to S^k$ is smooth and if $f_0(x) = f(0,x)$, $f_1(x) = f(1,x)$ then $\deg f_0 = \deg f_1$. This follows from Stokes theorem.

The following result is taken from [M2,Lemma B.1,p.111].

Lemma 3.1. *If $f : \mathbb{C}^n \to \mathbb{C}^n$ is a polynomial map and $p \in \mathbb{C}^n$ is such that*

(1) $f(p) = 0$.

(2) $\det Df(p) \neq 0$ *(i.e.* $\det \left[\frac{\partial f_i}{\partial x_j}(p)\right] \neq 0$*).*

Then there exists $r > 0$ such that if $0 < \|x\| \leq r$ then $f(x+p) \neq 0$. Let $0 < s \leq r$ and set $\Phi_s(x) = \frac{f(p+sx)}{\|f(p+sx)\|}$ for $x \in S^{2n-1}$. Then $\deg \Phi_s = 1$ for $0 < s \leq r$.

Proof. We first note that if $0 < s < r$ and we set $h(t,x) = \Phi_{s+t(r-s)}$ then $h_0 = \Phi_s$ and $h_1 = \Phi_r$. Thus $\deg \Phi_s = \deg \Phi_r$ for $0 < r < s$. Set $A = Df(p)$ then $f(p+x) = Ax + E(x)$ and if $\|x\| \leq r$ then $\|E(x)\| \leq C\|x\|^2$ with $C > 0$ fixed. Since $\det A \neq 0$ there exists s with $0 < s < r$ such that if $\|x\| = s$ then $\|E(x)\| < \frac{1}{2}\|Ax\|$. Set $g(t,x) = \frac{sAx+tE(sx)}{\|sAx+tE(sx)\|}$ for $x \in S^{2n-1}$. Then $g_1 = \Phi_s$ and $g_0(x) = \frac{Ax}{\|Ax\|}$. Thus $\deg \Phi_s = \deg g_0$. Finally $GL(n,\mathbb{C})$ is connected thus there exists a smooth curve $\sigma(t)$ in $GL(n,\mathbb{C})$ such that $\sigma(0) = I$ and $\sigma(1) = A$. Set $u(t,x) = \frac{\sigma(t)x}{\|\sigma(t)x\|}$. Then $u_0(x) = x$, $x \in S^{2n-1}$ and $u_1 = g_0$. The Lemma follows.

Lemma 3.2. *Let $f : \mathbb{C}^n \to \mathbb{C}^n$ and let $p \in \mathbb{C}^n$. Suppose that there exists $r > 0$ such that $f(x) \neq 0$ for $\|x - p\| \leq r$. If $0 < s \leq r$ then define Φ_s as in Lemma 3.1. Then $\deg \Phi_s = 0$.*

Proof. Set $g(t,x) = \Phi_{tr}(x)$. Then $g_0(x) = \frac{f(p)}{\|f(p)\|}$ for 1 $x \in S^{2n-1}$ and $g_1 = \Phi_r$. Clearly $\deg g_0 = 0$.

Here is an immediate implication:

Proposition 3.3. *Let $f : \mathbb{C}^n \to \mathbb{C}^n$ be a polynomial mapping. Assume that $p \in \mathbb{C}^n$ and $f(x) \neq 0$ for $\|x - p\| = r$. Let Φ_r be defined as in Lemma 3.1. If $\deg \Phi_r \neq 0$ then there exists $x \in \mathbb{C}^n$ such that $\|x - p\| < r$ with $f(x) = 0$.*

We now combine this with the observations in the previous section.

Theorem 3.4. *Let $f : \mathbb{C}^n \mapsto \mathbb{C}^n$ be a polynomial map. Let $\deg f_i = m_i$. Then there are at most $m_1 m_2 \cdots m_n$ elements $p \in \mathbb{C}^n$ such that $f(p) = 0$ and $\det Df(p) \neq 0$.*

Proof. We will use standard multindex notation. That is, if $I = (i_1, ..., i_n)$, $i_j \in \mathbb{N} = \{0, 1, 2, ...\}$ then $|I| = i_1 + ... + i_n$ and $x^I = x_1^{i_1} \cdots x_n^{i_n}$. We may assume $m_i > 0$ for all i (otherwise we are discussing the empty set). Let $f(x) = \sum a_I x^I$ with $a_I \in \mathbb{C}^n$. Given $m > 0$, $m \in \mathbb{Z}$ ther exists $g^m = (g_1^m, ..., g_n^m)$ such that g_i^m is a polynomial on $\mathbb{C}^n$, $(g^m)_{top} \in \Omega_{m_1,...,m_n}$ and $g^m = \sum_I b_I^m x^I$ with $\|a_I - b_I^m\| < \frac{1}{m}$. Let $p_1, ..., p_s$ be distinct elements with $f(p_i) = 0$ and $\det Df(p_i) \neq 0$. Then there exists $r > 0$ such that the sets $B_{p_i}(r) = \{x \in \mathbb{C}^n | \|x - p_i\| \leq r\}$ are disjoint and if $f(p) = 0$ with $p \in B_{p_i}(r)$ then $p = p_i$. If $x \in \mathbb{C}^n$ then

$$\|f(x) - g^m(x)\| \leq \sum_I \|a_I - b_I^m\| |x|^I.$$

Hence there exists a constant $C > 0$ such that if $x \in \cup B_{p_i}(r)$ then $\|f(x) - g^m(x)\| < \frac{C}{m}$. In particular this implies that if m is sufficiently large and if $\|x - p_i\| = r$ then $\|f(x)\| > \frac{1}{2}\|f(x) - g^m(x)\|$. Set

$$h(t, x) = \frac{f(p_i + rx) + t(g^m(p_i + rx) - f(p_i + rx))}{\|f(p_i + rx) + t(g^m(p_i + rx) - f(p_i + rx))\|}.$$

Then $h_0(x) = \Phi_r(x)$ (for f) and $h_1(x) = \frac{g^m(p_i+rx)}{\|g^m(p_i+rx)\|}$ for $\|x\| = 1$. Thus $\deg h_1 = 1$ by Lemma 3.1. Hence Proposition 3.4 implies that there exists $q_i \in B_{p_i}(r)$ such that $g^m(q_i) = 0$. Since the $B_{p_i}(r)$ are mutually disjoint, Proposition 2.3 implies that $s \leq m_1 \cdots m_n$. This completes the proof.

We note that Lefschetz has shown that if p is an isolated 0 of a polynomial map $f : \mathbb{C}^n \to \mathbb{C}^n$ and if $r > 0$ is such $f(x) \neq 0$ for $0 < \|x - p\| \leq r$ then if Φ_s is as in Lemma 3.1 then $\deg \Phi_s \geq 1$ for $0 < s \leq r$(cf. [M2,p.114]). The proof of Theorem combined with this result implies the following refinement:

Theorem 3.4′. *Let f be as in Theorem 3.4. Then there are at most $m_1 \cdots m_n$ isolated zeros of f.*

In section 7 we will give an algebraic proof of a sharpening of this result.

4. On Milnor's Theorem 1

Let f be a polynomial of degree k with real coefficients in n variables. We set $X = X(f) = \{x \in \mathbb{R}^n | f(x) = 0\}$.

Lemma 4.0. *If X is compact and $n \geq 2$ then k is even.*

Proof. Assume that k is odd. We show that f has arbitrarily large zeros. Let $f = g + h$ with g homogeneous of degree k and $\deg h < k$. Let $b \in \mathbb{R}^n$

be such that $g(b) \neq 0$. If $a \in \mathbb{R}^n$ and $\langle a, b\rangle = 0$ then $\varphi(t) = f(a + tb) = t^k g(b) + u(t)$ with $\deg u < k$. Thus φ is a polynomial of degree k in t. Since k is odd, φ must have a real zero, ξ. Now, $\|a+\xi b\| \geq \|a\|$ and a is arbitrary subject to $\langle a, b\rangle = 0$, the lemma follows.

We now assume that X is compact non-empty and $n \geq 2$ (so k is even). We also assume that if $x \in X$ then $df_x \neq 0$. Thus X is a smooth manifold of dimension $n - 1$. If $\omega \in S^{n-1}$ then we set $h_\omega(x) = \langle x, \omega\rangle$ for $x \in X$. If $p \in X$ is a critical point for h_ω then ω must be orthogonal to the tangent space of X at p. Thus ω must be a multiple of $N(p) = \left(\frac{\partial f}{\partial x_1}(p), \ldots, \frac{\partial f}{\partial x_n}(p)\right)$. If $x \in \mathbb{R}^n, x \neq 0$, then we write $[x]$ for the corresponding one dimensional subspace with basis x. That is, $[x] \in \mathbb{P}^{n-1}(\mathbb{R})$ (real projective space of dimension $n - 1$). Set $\pi(x) = [N(x)]$, $x \in X$. Then π is a smooth mapping from X to $\mathbb{P}^{n-1}(\mathbb{R})$. $p \in X$ is a critical point of h_ω if and only if $\pi(p) = [\omega]$. Since X is compact, this implies that π is surjective.

Sard's theorem implies that the set of crtical values (i.e. the set of $\pi(p)$ such that $d\pi_p$ is not bijective) has dense complement in $\mathbb{P}^{n-1}(\mathbb{R})$(see also Lemma 6.4). We make an orthogonal change of variables and we assume that $[e_n]$ is not a critical value. (Here $e_1 = (1, 0, \ldots, 0), \ldots, e_n = (0, 0, \ldots, 1)$). Set $g = (f, \frac{\partial f}{\partial x_1}, \ldots, \frac{\partial f}{\partial x_{n-1}})$. Then g is a polynomial map of $\mathbb{R}^n$ to $\mathbb{R}^n$.

Lemma 4.1. *The assumptions are as above. Let $h = h_{e_n}$. Then the set of critical points of h is precisely the set of points $p \in \mathbb{R}^n$ such that $g(p) = 0$ and at such a p, $\det Dg(p) \neq 0$. Finally, each of the critical points in X of h is non-degenerate.*

Before we prove this Lemma we indicate how Milnor uses it.

Theorem 4.2. *Assume that $n > 1$. Let f be a polynomial with real coefficients in n variables. Assume that*

1. $X = \{x \in \mathbb{R}^n | f(x) = 0\}$ is compact.

2. If $x \in X$ then $df_x \neq 0$.

Then the number of connected components of X is at most $\frac{k(k-1)^{n-1}}{2}$.

Note. We using the Morse inequalities ([M3,I.5]) one can see that in fact one has the sum of the Betti numbers of X is at most $k(k - 1)^{n-1}$.

Proof. We may use the assumptions and notation in Lemma 4.1. Then in light of Lemma 4.1, Theorem 3.4 implies that h has at most $k(k - 1)^n$ critical points in X. On each connected component of X, h must have a maximum and a minimum. Since, h has only a finite number of critical points there must be at least 2 in each connected component. The Morse inequalities imply the note above.

We will now prove the lemma. If $p \in X$ is a critical point for h then $\pi(p) = [e_n]$. Thus $g(p) = 0$ and $\frac{\partial f}{\partial x_n}(p) \neq 0$. Write $p = (p', p_n)$ then the imp theorem implies that there exists a neighborhood, U of p' in $\mathbb{R}^{n-1}$ and a smooth function φ on U with $\varphi(p') = p_n$ and $f(y, \varphi(y)) = 0$ for $y \in U$. If $1 \leq i \leq n-1$ then

$$0 = \frac{\partial f}{\partial y_i}(y, \varphi(y)) = \frac{\partial f}{\partial x_i}(y, \varphi(y)) + \frac{\partial \varphi}{\partial y_i}(y) \frac{\partial f}{\partial x_n}(y, \varphi(y)). \tag{1}$$

Our assumption implies that p' is a critical point of φ. Thus if we differentiate the above equation relative to y_j with $1 \leq j \leq n-1$ and evaluate at $y = p'$ we have

$$\frac{\partial^2 f}{\partial x_i \partial x_j}(p) = -\frac{\partial f}{\partial x_n}(p) \frac{\partial^2 \varphi}{\partial y_i \partial y_j}(p'). \tag{2}$$

We now use the assumption that $[e_n]$ is a regular value of π. If $q \in X$ is close to p then $\frac{\partial f}{\partial x_n}(q) \neq 0$. Thus

$$\pi(q) = [u_1(q), ..., u_{n-1}(q), 1]$$

with $u_i(q) = \frac{\partial f}{\partial x_i}(q) / \frac{\partial f}{\partial x_n}(q)$, $i = 1, ..., n-1$. We calculate $\frac{\partial u_i}{\partial y_j}(y, \varphi(y))$ at $y = p'$ using the chain the fact that $\frac{\partial \varphi}{\partial y_i}(p') = 0$ (see (1) above) fi $= 1, ..., n-1$ and find that if $1 \leq i, j \leq n-1$ then

$$\frac{\partial u_i}{\partial y_j}(p', \varphi(p')) = \frac{\frac{\partial^2 f}{\partial x_i \partial x_j}(p)}{\frac{\partial f}{\partial x_n}(p)}.$$

The assumption that $[e_n]$ is a regular value of π means that $\det \left[\frac{\partial u_i}{\partial y_j}(p', \varphi(p'))\right] \neq 0$. This combined with (2) above implies that p is a non-degenerate critical point of h. Also since $\pi(p) = [e_n]$ $\det \left[\frac{\partial g_i}{\partial x_j}(p)\right]_{1 \leq i,j \leq n} = \frac{\partial f}{\partial x_n}(p) \det \left[\frac{\partial^2 f}{\partial x_i \partial x_j}(p)\right]_{1 \leq i,j \leq n-1}$. The lemma now follows.

5. On Milnor's Theorem 2

We now show how Milnor derives his main theorem from Theorem 4.2. Set $B(r) = \{x \in \mathbb{R}^n | \|x\| \leq r\}$. Let $f_1, ..., f_m$ be polynomials with real coefficients of degree at most k. Following Milnor we set for $\epsilon > 0$

$$u_\epsilon(x) = f_1(x)^2 + ... + f_m(x)^2 + \epsilon^2 \|x\|^2$$

for $\epsilon > 0$. Set $K(\epsilon,\delta) = \{x \in \mathbb{R}^n | u_\epsilon(x) \leq \delta^2\}$ for $\delta > 0$ then $K(\epsilon,\delta) \subset B(\frac{\delta}{\epsilon})$ hence compact. Set$X = \{x \in \mathbb{R}^n | f_i(x) = 0, i = 1,..,m\}$. If $r \leq \frac{\delta}{\epsilon}$ then $X \cap B(r) \subset K(\epsilon,\delta)$. We assume that X is non-empty. Hence there exists $r_o > 0$ such that if $r \geq r_o$ then $B(r) \cap X \neq \emptyset$. Fix $r \geq r_o$. Sard's theorem (cf. also Lemma 6.2) implies that there exists a sequence $\{\epsilon_i\}$ with $\epsilon_i \geq \epsilon_{i+1} > 0$ and $\delta_i > 0$ such that $\frac{\delta_i}{\epsilon_i} \geq \frac{\delta_{i+1}}{\epsilon_{i+1}}$, $\lim_{i\to\infty} \frac{\delta_i}{\epsilon_i} = r$ and δ_i is a regular value of u_ϵ.

We note

Lemma 5.1. *$K(\epsilon_i,\delta_i) \supset K(\epsilon_{i+1},\delta_{i+1})$ and $\cap K(\epsilon_i,\delta_i) = X \cap B(r)$.*

Proof. We note that $\frac{\delta_{i+1}}{\epsilon_{i+1}} \leq \frac{\delta_i}{\epsilon_i}$ implies that $\delta_{i+1} \leq \frac{\delta_i \epsilon_{i+1}}{\epsilon_i} \leq \delta_i$ since $\epsilon_i \geq \epsilon_{i+1}$. Thus writing the inequality $u_{\epsilon_{i+1}}(x) \leq \delta_{i+1}$ as

$$\frac{f_1(x)^2 + ... + f_m(x)^2}{\delta_{i+1}} + \frac{\epsilon_{i+1}\|x\|^2}{\delta_{i+1}} \leq 1.$$

The asserted inclusions now follow from

$$\frac{f_1(x)^2 + ... + f_m(x)^2}{\delta_i} + \frac{\epsilon_i\|x\|^2}{\delta_i} \leq \frac{f_1(x)^2 + ... + f_m(x)^2}{\delta_{i+1}} + \frac{\epsilon_{i+1}\|x\|^2}{\delta_{i+1}}.$$

The above form of the definition of $K(\epsilon_i,\delta_i)$ also implies the assertion about the intersection.

Set $\partial K(\epsilon,\delta) = \{x \in \mathbb{R}^n | u_\epsilon(x) = \delta\}$. For each i, $\partial K(\epsilon_i,\delta_i)$ and $u_{\epsilon_i} - \delta_i$ satisfy (1) and (2) of Theorem 4.2. Thus we have

I. For each i, the number of connected components of $\partial K(\epsilon_i,\delta_i)$ is at most $k(2k-1)^{n-1}$.

Lemma 5.2. *Let $u : \mathbb{R}^n \to \mathbb{R}$ be a continuous map. If $r \in \mathbb{R}$ then the number of connected components of $u^{-1}((-\infty,r])$ is less than or equal to the number of connected components of $u^{-1}(r)$.*

Proof. We may assume that $Y \neq \emptyset$. Let $Y = u^{-1}((-\infty,r])$, $Z = u^{-1}(r)$ then $Y - Z$ is open in $\mathbb{R}^n$ ($Y - Z = \{x \in \mathbb{R}^n | u(x) < r\}$). Suppose that W is a connected component of Y and $W \cap Z = \emptyset$. There exists an open subset U of $\mathbb{R}^n$ such that $U \cap Y = W$. Thus $W = U \cap (Y - Z)$ which is open in $\mathbb{R}^n$. Hence W is open and closed in $\mathbb{R}^n$ hence empty. This is a contradiction. Thus every connected component of Y has a non-empty intersection with Z. This implies the Lemma.

Note. In the case when $u = u_{\epsilon_i}$ and $r = \delta_i$, Milnor shows (using Alexander duality) the sum of the Betti numbers of $K(\epsilon_i,\delta_i)$ is less than or equal to half the sum of the Betti numbers of $K(\epsilon_i,\delta_i)$.

If Z is a closed subset of $\mathbb{R}^n$ set $b_0(Z)$ equal to the number of connected components of Z. In light of I Lemma 5.2 implies that

II. $b_0(K(\epsilon_i,\delta_i)) \leq k(2k-1)^{n-1}$.

Lemma 5.3. *Let C_i be compact subsets of $\mathbb{R}^n$ with $C_i \supset C_{i+1}$ and $\cap C_i = C$. If $b_0(C_i) \leq d$ then $b_0(C) \leq d$.*

Proof. We may assume that $C \neq \emptyset$. Let $C = Y_1 \cup Y_2 \cup \cdots \cup Y_s$ be the decomposition of C into connected components. Let $C_{i,j}$, $j = 1,...,d_i \leq d$ be the connected components of C_i. Then each Y_j is contained in a unique connected component $C_{i,l(i,j)}$ of C_i. We note that $C_{i,l(i,j)} \supset C_{i+1,l(i+1,j)}$. We assert that for each j there exists $r(j)$ such that if $i \geq r(j)$ then $C_{i,l(i,j)} \cap C = Y_j$. This will clearly prove the lemma. Suppose not then there is an infinite sequence $r_1 < r_2 < ...$ and $z_k \in C_{r_k,l(r_k,j)} - Y_j$, $z_k \in Y$. Taking a subsequence, if necessary, we may assume $\lim_{k\to\infty} z_k = z_o$ and $z_o \in Y$. But $z_0 \in \cap C_{r_k,l(r_k,j)} = V_j$ which is connected. Since $V_j \supset Y_j$, $V_j = Y_j$ and we have a contradiction.

This lemma combined with II and Lemma 5.1 implies

III. $b_0(B(r) \cap X) \leq k(2k-1)^{n-1}$.

Note. Using standard properties of Čech cohomology (commuting with infinite decreasing sequences of compact spaces) Milnor has the same estimate as in III for the sum of the Betti numbers.

We can now prove the main theorem.

Theorem 5.4. *Let $f_1,...,f_m$ be polynomials in n variables with real coefficients. Let $X = \{x \in \mathbb{R}^n | f_i(x) = 0,\ i = 1,...,m\}$. Then the number of connected components of X is less than or equal to $k(2k-1)^{n-1}$.*

Proof. We may assume that $X \neq \emptyset$. Let $X_1,...,X_s$ be connected components of X. Let $r > 0$ be so large that $B(r) \cap X_i \neq \emptyset$ for $i = 1,...,s$. Then $B(r) \cap X_i$ is a union of $p_i > 0$ connected components. Since $\cup_i B(r) \cap X_i$ is open and closed in $B(r) \cap X$ this implies $s \leq p_1 + ... + p_s \leq b_o(B(r) \cap X) \leq k(2k-1)^{n-1}$. The theorem follows.

Note. Milnor in fact gives the same inequality for the sum of the Betti numbers. Here he uses a more sophisticated argument (in fact two). We recommend that the reader consult the original paper of Milnor. We also note that [T,Lemme 3.,p.260] also impies this assertion.

We also observe that the argument above implies that if $f_p(x) = \prod_{j=1}^{k}(x_p - j)$, $p = 1,...,n$, if $r > k$ and (in the notation above) if i is sufficiently large then $\partial K(\epsilon_i, r_i)$ is a union of at least k^n connected smooth manifolds.

In [M1], Milnor notes that he can find no examples with the sum of the Betti numbers greater than k^n. This suggests the following

Problem. Can we replace $k(2k-1)^{n-1}$ in Theorem 5.4 with k^n?

In section 8 we will give an affirmative answer to this question in the case when $n = 2$. In section 9 we will give an affirmative answer for the sum of the Betti numbers for a non-singular hypersurface.

6. Some further results

Let $f_1, ..., f_m, g_1, ..., g_q$ be polynomials in n variables with real coefficients. Let $X = \{x \in \mathbb{R}^n | f_i(x) = 0, i = 1, ..., m\}$, $Y = \{x \in \mathbb{R}^n | g_j(x) = 0, j = 1, ..., q\}$. We will now show how the result of Milnor and Thom applies to $X - Y = \{x \in X | x \notin Y\}$. We first observe that if $x \notin Y$ then some $g_i(x) \neq 0$. This since the polynomials g_i have real coefficients this is the same as $h(x) = \sum g_j(x)^2 \neq 0$ (if $m = 1$ set $h = g_1$). Thus $X - Y$ is homeomorphic with

$$\{(x,t) \in \mathbb{R}^{n+1} | f_i(x) = 0, i = 1, ..., m, th(x) = 1\}.$$

We can now apply the result to the variety given in this way.

Theorem 6.1. *Let $k = \max_{i,j}\{\deg f_i, 2\deg g_j + 1\}$ if $q > 1$ and $k = \max\{\deg g_1 + 1, \deg f_i\}$ if $q = 1$. Then the sum of the Betti numbers of $X - Y$ is less than or equal to $k(2k-1)^n$.*

This in particular applies to the situation $\mathbb{R}^n - Y$. An important example of this is the case $g = \prod_{i<j}(x_i - x_j)$. Then $\deg g = \frac{n(n-1)}{2}$. Let $Y = \{x \in \mathbb{R}^n | g(x) = 0\}$. Then it standard that $\mathbb{R}^n - Y$ is a union of $n!$ non-empty convex subsets of $\mathbb{R}^n$. The estimate of Theorem 6.1 is

$$(\frac{n(n-1)}{2} + 1)(n(n-1)+1)^n \sim n^{2n+2}/2.$$

We will come back to this example in section 9.

We also record algebraic variants of Sard's theorem that are used in the proofs of Milnor's theorems. Let f be a polynomial in n variables with complex coeffiecients (for this any field of characteristic 0 will do). Let $\Sigma(f) = \{x \in \mathbb{C}^n | df_x = 0\}$. The following results will use a bit more algebraic geometry than the rest of the exposition.

Lemma 6.2. *The set $f(\Sigma(f))$ is finite.*

Proof. Let Y be an irreducible component of $\Sigma(f)$. We show that $f(Y)$ is a point in $\mathbb{C}$. Since Y is irreducible the Zariski closure of $f(Y)$ is irreducible as a subvariety of $\mathbb{C}$. Thus the Zariski closure of $f(Y)$ is either a point or all of $\mathbb{C}$. Suppose that we are in the latter situation. Let Y^o be the set of simple points of Y. Then Y^o is a non-singular quasi-affine variety and the Zariski closure of Y^o is Y. Thus $f(Y^o)$ has Zariski interior in $\mathbb{C}$. But f is constant on each connected component of Y^o in the classical topology of Y^o (the subspace topology of Y^o in $\mathbb{C}^n$ with the Euclidian metric topology). Since Y^o has only a countable number of connected components, we have a contradiction.

A similar elementary argument using algebraic geometry proves the following result.

Lemma 6.3. *Let X be an irreducible smooth n-dimensional affine variety over $\mathbb{C}$ and let $f : X \to Y$ where $Y = \mathbb{C}^n$ or $Y = \mathbb{P}^n$. Let $\Sigma = \{x \in X | df_x$ is not surjective$\}$ then the closure of $f(\Sigma)$ in the Zariski topology of Y has dimension at most $n-1$.*

Corollary 6.4. *Let φ be a polynomial in n indeterminates with real coefficients and let $X = \{x \in \mathbb{C}^n | \varphi(x) = 0, d\varphi_x \neq 0\}$. Let $f : X \to \mathbb{P}^{n-1}$ be a regular map such that $f(X \cap \mathbb{R}^n) \subset \mathbb{P}^{n-1}(\mathbb{R})$. Let $\Sigma = \{x \in X | df_x$ is not surjective$\}$ then there exists a non-zero, homogeneous polynomial, u, with real coefficients such that $f(\Sigma \cap \mathbb{R}^n) \subset \{x \in \mathbb{P}^{n-1}(\mathbb{R}) | u(x) = 0\}$.*

7. An estimate on the number of irreducible components

In this section K will denote an algebraically closed field. We will be using a bit more algebraic geometry that was needed in the earlier sections. Let $\mathbb{P}^n$ denote the n dimensional projective space over K and $\mathbb{A}^n$ the n dimensional affine space. This section will be devoted to the proof of the following result.

Theorem 7.1. *Let X be (Zariski) closed in $\mathbb{A}^n$ (resp. $\mathbb{P}^n$) given as the zero locus of polynomials $f_1, ..., f_m$ (resp. homogeneous) of degree at most k. Then the number of irreducible components of X is at most k^n.*

We first note that the affine case follows from the projective case. Indeed, by adding a variable x_0 we can homogenize $f_1, ..., f_m$ to be homogeneous of degree k. Let Y be the corresponding projective variety. Let $Y = \cup_{i=1}^d Y_i$ be an irredundent decomposition into irreducible components. Then $X = \cup_{i=1}^d Y_i \cap \mathbb{A}^n$. Now throw away the redundent terms (using the fact that $\mathbb{A}^n \cap Y_i$ is open in Y_i and hence irreducible as an affine variety).

We now concentrate on the projective case. We may assume that all of the f_i are homogeneous of degree k in variables $x_0, x_1, ..., x_n$. If $X \subset \mathbb{P}^n$ is closed and irreducible then $\deg X$ is the leading coefficient of $(\dim X)!h(t)$ with h the Hilbert polynomial of the homogeneous coordinate ring of X. In the projective case we will prove the following sharper result

Theorem 7.1′. *Let $f_1, ..., f_m$ be homogeneous polynomials of degree k and let X be the zero locus of $\{f_1, ..., f_m\}$ in $\mathbb{P}^n$. Let $X = X_1 \cup \cdots \cup X_d$ be an irredundent decomposition of X into irreducible components then*

$$\sum_i \deg X_i \leq k^n.$$

The following simple lemma will be used in the proof of the theorem.

Lemma 7.2. *Let $X \subset \mathbb{P}^n$ be closed. Let V be a subspace of $K[x_0, ..., x_n]$ consisting of homogeneous elements of degree k. Then we can label the irreducible components of X as $X_1, ..., X_d$ with $V_{|X_i} = 0$ for $i \leq s$ and there exists $f \in V$ such that $f_{|X_i} \neq 0$ for $i > s$.*

Proof. Order the index set $\{1, ..., d\}$ by inclusion. Let S be a maximal subset subject to the condition $V_{|X_i} = 0$ for $i \in S$. Set $S^c = \{i \notin S | 1 \leq i \leq d\}$. Assume that for each $f \in V$, $f \neq 0$, there exists $i \in V^c$ such that $f_{|X_i} = 0$. Let for $i \in S^c$, $V_i = \{f \in V | f_{|V_i} = 0\}$. Then $V = \cup_{i \in S^c} V_i$. This implies that there exists $i \in S^c$ such that $V_{|X_i} = 0$. This contradicts the maximality of S. Thus there exists $f \in V$ such that $f_{|X_i} \neq 0$ for $i \in S^c$. This completes the proof of the Lemma.

We now prove Theorem 7.1′. Obviously, we may assume $n \geq 2$. We prove the result by induction on k. If $k = 1$ then the result is obvious. Assume the result for $1, ..., k-1$. Let $\mathcal{P}^k$ denote the space of $f \in K[x_0, ..., x_n]$ that are homogeneous of degree k. Set $V = \{f \in \mathcal{P}^k | f_{|X} = 0\}$. Then $X = \{x \in \mathbb{P}^n | V(x) = 0\}$. Choose $g_1 \in V$, $g_1 \neq 0$. Let $Y_1 = \{x \in \mathbb{P}^n | g_1(x) = 0\}$. Apply Lemma 7.2 to Y_1 and find that $Y_1 = X_1 \cup X_2 \cup \cdots \cup X_{s_1} \cup Z_1 \cup \cdots \cup Z_{t_1}$ an irredundent decomposition into irreducible components such that $V_{|X_i} = 0$ for $i = 1, ..., s_1$ and if $t_1 > 0$ then there exists $g_2 \in V$ such that $g_2(Z_i) \neq 0$ for $i = 1, ..., t_1$. If $t_1 = 0$ then $\{X_1, ..., X_{s_1}\}$ is the set of irreducible components of X. Since g has degree k, $\sum_{i \leq s_1} \deg X_i \leq k$ and the result is proved in this case. Now assume that $t_1 > 0$. If $s_1 > 0$ then there must be a non-trivial irreducible factor, h, of g_1 that divides every element of V. Then $X = X(h) \cup X(V/h)$ (here if S is a set of homogenous polynomials then $X(S) = \{x \in \mathbb{P}^n | f(x) = 0, x \in S\}$). The inductive hypothesis applies to $X(h)$ and to $X(V/h)$. If $a > 0$ and $b > 0$ that $a^n + b^n < (a+b)^n$. So the result follows if $s_1 > 0$. Thus we may assume that $X = \cup Z_i$. Thus $Y_2 = \{z \in Y_1 | g_2(z) = 0\}$. Let

$Y_2 = X_1 \cup \cdots \cup X_{s_2}$ be an irredundent decomposition into irreducible components. We note that Bezout's theorem (cf. [H;Theorem I.7.7,p. 53]) implies that there are integers $c_i > 0$ such that

$$\sum c_i \deg(X_i) \leq \deg g_1 \deg g_2 = k^2.$$

Now $\dim X_i = n - 2$. Thus if $n = 2$ then Y_2 is a finite set with at most k^2 elements. Since $X \subset Y_2$, the Theorem 7.1 is now completely proved for $n = 2$. We thus assume $n > 2$. If $Y_2 = X$ we are also done. Otherwise, we can write $Y_2 = X_1 \cup X_2 \cup \cdots X_{p_2} \cup Z_1 \cup \cdots \cup Z_{q_2}$ (irreducible decomposition into irreducibles) with $V_{|X_i} = 0$ and there exists $g_3 \in V$ with $g_3(Z_i) \neq 0$. Set

$$Y_3 = X_1 \cup \cdots \cup X_{p_2} \cup \cup_i Z_i \cap X(g_3)$$
$$\{z \in \mathbb{P}^n | g_i(z) = 0, i = 1, 2, 3\}.$$

Let $Z_i \cap X(g_3) = \cup_{j=1}^{q_{ij}} X_{ij}$ be an irredundent decomposion into irreducible components. Then applying Bezout's theorem there exist $c_{ij} > 0$, c_{ij} integers such that

$$\sum_j c_{ij} \deg X_{ij} \leq \deg Z_i \deg g_3.$$

Since $k > 1$ we see that we can write

$$Y_3 = X_1 \cup X_2 \cup \cdots \cup X_{p_2} \cup X_{p_2+1} \cup \cdots \cup X_{p_3}$$

with X_i irreducible and such that there exist positive integers d_{2i} so that

$$\sum_i d_{2i} \deg X_i \leq k^3.$$

Also, $\dim X_i = n - 3$ for $i > p_2$. If $n = 3$ we can argue as in the case of $n = 2$ to complete the proof of the theorem. So assume $n > 3$. Then either $Y_3 = X$ or we can continue the argument to find $g_4 \in V$. Obviously, we can continue this process. If the process continues to n stages then we are done as in the case of $n = 2$. If it stops in $r < n$ steps then we have an upper bound on the sum of the degrees of the irreducible components of the form k^r. The theorem now follows.

8. A sharp result in the case $n = 2$

The purpose of this section is to prove

Theorem 8.1. *$f_1, ..., f_m$ be polynomials with real coefficients in 2 variables of degree at most k. Then $X = \{x \in \mathbb{R}^2 | f_i(x) = 0, i = 1, 2\}$ has at most k^2 connected components.*

We will prove this result by induction on k. If $k = 0, 1$ then the result is obvious. So assume the theorem for degrees at most $k - 1$. Let V denote the span over $\mathbb{C}$ of $f_1, ..., f_m$. If S is a set of polynomials (with real or complex coefficients) set $X_{\mathbb{R}}(S) = \{x \in \mathbb{R}^2 | f(x) = 0, f \in S\}$. Let $g \in V$ be an element of minimal degree. If for each $h \in V$ there exists a non-constant irreducible factor, u, of g such that u divides h. Then there exists a non-constant factor, u, of g that divides every element of V. Then $X = X_{\mathbb{R}}(u) \cup X_{\mathbb{R}}(V/u)$. Thus if $\deg u = r < k$ then $b_0(X) \leq b_0(X_{\mathbb{R}}(u)) \cup b_o(X_{\mathbb{R}}(V/u)) \leq r^2 + (k - r)^2$, by the inductive hypothesis. So we may assume that $\deg u = k$. But then $u = g$ and hence $X = X_{\mathbb{R}}(g)$. Suppose that there exists h in V such that h and g are relatively prime. Then (notation as in section 7). $\dim X(g, h) = 0$ and $X \subset X(g, h)$. $X(g, h)$ has at most k^2 elements by Theorem 7.1. Thus we are left with the case when $X = X_{\mathbb{R}}(g)$ with g irreducible over $\mathbb{C}$. Assume that g is not a multiple of a polynomial with real coefficients. Let $g(x) = u(x) + iv(x)$ with u, v polynomials with real coefficients. If u and v have a non-trivial (complex) factor w in common then w divides g which is contrary to our assumption. Thus u and v are relatively prime over $\mathbb{C}$. Since $X = X_{\mathbb{R}}(u, v) \subset X(u, v)$ which is fini Theorem 7.1 implies the result in this case. We are thus left with the case when g has real coefficients and is irreducible over $\mathbb{C}$.

If $a, b \in \mathbb{R}$ then set

$$u_{a,b}(x, y) = (x - a)\frac{\partial f}{\partial y}(x, y) - (y - b)\frac{\partial f}{\partial x}(x, y).$$

Suppose that for every $(a, b) \in \mathbb{R}^2 - X$, g divides $u_{a,b}$. Then $u_{a,b}(X(g)) = 0$ for all $a, b \in \mathbb{R}^2$. Differentiating this identity implies that $X(g)$ has no simple points. We may thus choose $(a, b) \in \mathbb{R}^2 - X$ such that g and $u_{a,b}$ are relatively prime over $\mathbb{C}$. Fix such a $u = u_{a,b}$. Let $X_1, ..., X_l$ be the connected components of X. We assume that if $1 \leq i \leq q$ and if $p \in X_i$ then at least one of $\frac{\partial f}{\partial x}(p)$, $\frac{\partial f}{\partial y}(p)$ is non-zero. We also assume that if $i > q$ then X_i contains a point where both of the partials are equal to 0. Set $\varphi(x, y) = \frac{1}{2}((x - a)^2 + (y - b)^2)$. Since each X_i is closed, φ must attain a minimum in X_i. If $1 \leq i \leq q$ then at such a minimum, p, $0 = d\varphi_p \wedge df_p = u(p)dx \wedge dy$. We therefore see that for each $1 \leq i \leq l$ there exists an element $p \in X_i$ such that $p \in X(g, u)$. But $X(g, u)$ has at most k^2 elements by Theorem 7.1.

9. A better estimate for smooth algebraic hypersurfaces

Let f be a polynomial with real coefficients in n indeterminates. Set $X = \{x \in \mathbb{R}^n | f(x) = 0\}$. The purpose of this section is to prove

Theorem 9.1. *If for exery $x \in X$, $df_x \neq 0$ then $b_0(X) \leq (\deg f)^n$.*

We note that if $k = \deg f$ then the Milnor-Thom theorem would give the inequality $k(2k-1)^{n-1}$.

The proof of Theorem 9.1 is based on the following simple result.

Lemma 9.2. *Let V be a finite dimensional subspace of the polynomials in n variables over $\mathbb{C}$. Let $X(V) = \{x \in \mathbb{C}^n | V(x) = 0\}$. If $x \in X(V)$ and if $\dim\{dg_x | g \in V\} = n$ then there is a Zariski open subset U of $\mathbb{C}^n$ such that $U \cap X(V) = \{x\}$.*

Proof. Let $f_1, ..., f_n \in V$ be such that

$$df_1 \wedge \cdots \wedge df_n = \varphi dx_1 \wedge \cdots \wedge dx_n$$

is non-zero at x. So $\varphi(x) \neq 0$. Let $U_1 = \{y \in \mathbb{C}^n | \varphi(y) \neq 0\}$. Then $U \cap X(f_1, ..., f_n)$ is isomorphic with the variety $Y = \{(y,t) | f_i(y) = 0, i = 1, ..., n, \varphi(y)t = 1\}$. Set $u_i(y,t) = f_i(y)$, $i = 1, ..., n$ and $u_{n+1}(y,t) = \varphi(y)t - 1$.

$$du_1 \wedge \cdots \wedge du_{n+1} = \varphi^2 dy_1 \wedge \cdots \wedge dy_n \wedge dt.$$

This implies that $\dim Y = 0$. So Y is finite. Set $Y - \{x\} = F$ and $U = U_1 - F$. Then

$$\{x\} \subset U \cap X(V) \subset U \cap X(f_1, ..., f_n) = \{x\}.$$

This completes the proof of the lemma.

We now prove Theorem 9.1. Clearly, we may assume that $n \geq 2$. Let $a \in \mathbb{R}^n$ be such that if $\alpha(x) = \frac{1}{2}\sum_i (x_i - a_i)^2$ then α has non-degenerate critical points in X (such an a exists by the Lemma of Andrioti-Frankel (cf. [M3,Theorem 6.6,p.36]). If $1 \leq i < j \leq n$ we set $\psi_{ij}(x) = (x_i - a_i)\frac{\partial f}{\partial x_j}(x) - (x_j - a_j)\frac{\partial f}{\partial x_i}(x)$. Then

$$d\alpha \wedge df = \sum_{i<j} \psi_{ij} dx_i \wedge dx_j.$$

Thus $y \in X$ is a crtical point for α if and only if $f(x) = \psi_{ij}(x) = 0$ for $1 \leq i,j \leq n$. Thus if V is the complex span of $\{f\} \cup \{\psi_{ij} | i < j\}$. Then the set of critical points of α is $X(V) \cap \mathbb{R}^n$.

Let $p \in X$ be a critical point for α. After relabeling coordinates we may assume $\frac{\partial f}{\partial x_n}(p) \neq 0$. Set $p = (p', p_n)$. We can find a neighborhood U of p' in $\mathbb{R}^{n-1}$ and a smooth function $\varphi : U \to \mathbb{R}$ such that $\varphi(p') = p_n$ and $f(y, \varphi(y)) = 0$ for all $y \in U$. The condition that p is a non-degenerate critial point is just that

$$\Delta(p') = \det\left[\frac{\partial^2 \alpha}{\partial y_i \partial y_j}(y, \varphi(y))\right]_{y=p'} \neq 0.$$

Assume that $i \leq n-1$ then

$$\frac{\partial \alpha}{\partial y_i}(y, \varphi(y)) = (y_i - a_i) + (\varphi(y) - a_n)\frac{\partial \varphi}{\partial y_i}(y).$$

Also

$$\frac{\partial \varphi}{\partial y_i}(y) = -\left(\frac{\partial f}{\partial x_n}(y, \varphi(y))\right)^{-1} \frac{\partial f}{\partial x_i}(y, \varphi(y)).$$

So it follows that

$$\frac{\partial \alpha}{\partial y_i}(y, \varphi(y)) = \left(\frac{\partial f}{\partial x_n}(y, \varphi(y))\right)^{-1} \psi_{i,n}(y, \varphi(y)).$$

Set $u_i = \psi_{i,n}$, $i = 1, \ldots, n-1$ and $u_n = f$. Then a direct calculation (similar to the one in section 4) shows that

$$\det\left[\frac{\partial u_i}{\partial x_j}(p)\right] = \left(\frac{\partial f}{\partial x_n}(p)\right)^{n+1} \Delta(p').$$

We can now apply Lemma 9.2 to conclude that each critical point of α in X is an irreducible component of $X(V)$. Theorem 7.1 now implies that α has at most k^n ($k = \deg f$) critical points.

Since each connected component of X is closed α must have a minimum on each connected component. Thus each connected component contains at least one critical point. This completes the proof of Theorem 9.1.

We note that the function α in the proof above is proper. We may thus apply [M3, Theorem 3.5,p.20] to deduce

Theorem 9.1′. *Let f, X be as in Theorem 9.1. Then the sum of the Betti numbers of X is at most k^n ($k = \deg f$).*

We note that if f is a non-zero polynomial in n indeterminates with real coefficients then $U = \{x \in \mathbb{R}^n | f(x) \neq 0\}$ is isomorphic with the smooth hypersurface $Y = \{(x,t)|f(x)t = 1\}$ in $\mathbb{R}^{n+1}$. Thus Theorem 9.1′ applies and we have

Corollary 9.2. *Let f and U be as above and assume that* $\deg f = k$. *Then the sum of the Betti numbers of U is at most* $(k+1)^{n+1}$.

In the example of section 6, this theorem improves the estimate by a factor of 2^{-n}.

References

[B] M. Ben-Or, Lower bounds for algebraic computation trees, in *Proc. 15th ACM Symposium on the Theory of Computing*, May 1983, 80-86.

[H] Robin Hartshorne, *Algebraic Geometry*, Graduate Texts in Mathematics, Vol. 52, Springer-Verlag, New York, 1977.

[M1] J. Milnor, On the Betti numbers of real varieties, *Proc. Amer. Math. Soc.*, **15** (1964), 275-280.

[M2] J. Milnor, *Singular Points of Complex Hypersurfaces*, Annals of Math Studies, Study 61, Princeton, 1968.

[M3] J. Milnor, *Morse Theory*, Annals of Math Studies, Study 51, 1963.

[T] R. Thom, Sur l'homologie des variétés algébriques réelles, in *Differential and Combinatorial Topology*, Princeton University Press, 1965, 255-265.

[W] H. Whitney, Elementary structure of real algebraic varieties,*Ann. of Math.*, **66** (1957),545-546.

Department of Mathematics
University of California at San Diego
La Jolla, CA 92093

Received June 1995

Riemannian Exponential Maps and Decompositions of Reductive Lie Groups

Joseph A. Wolf and Roger Zierau

ABSTRACT. Let X be a complete connected riemannian manifold, Y a closed submanifold, and $\mathbb{N}_{Y,X} \to Y$ the normal bundle of Y in X. Then the exponential map $\exp_{Y,X} : \mathbb{N}_{Y,X} \to X$ is surjective. When X is a riemannian symmetric space $X = G/K$, G reductive, this extends a number of decomposition theorems of the form $G = H \cdot \exp_G(\mathfrak{s} \cap \mathfrak{r}) \cdot K$, and when Y is totally geodesic in X it extends a number of "Euler angle type" formulae of the form $G = HAK$. The principal new features here are that H can be any reductive subgroup of G and the symmetric space X may have compact and/or euclidean factors. There are also some consequences for pseudo–riemannian manifolds and for open G–orbits on complex flag manifolds $G_{\mathbb{C}}/Q$. The papers [11] and [12] use the result with compact factors, and [3] uses the pseudo–riemannian result.

1. Riemannian Exponential Map

Let X be a complete connected riemannian manifold. Fix a closed submanifold $Y \subset X$. The **normal bundle** $\mathbb{N}_{Y,X} \to Y$ is the sub–bundle of the restriction $\mathbb{T}(X)|_Y$ of the tangent bundle of X, whose fibre over $y \in Y$ is the orthocomplement $T_y(Y)^\perp \subset T_y(X)$ of the tangent space to Y at y in the tangent space to X at y. The **exponential map** $\exp_{Y,X} : \mathbb{N}_{Y,X} \to X$ is just the corresponding restriction of the usual riemannian exponential map $\exp_X : \mathbb{T}(X) \to X$. In this note we will see that the rather easy theorem

Theorem 1.1. *The exponential map* $\exp_{Y,X} : \mathbb{N}_{Y,X} \to X$ *is surjective.*

has a number of interesting consequences for the structure of real reductive Lie groups. Some of these consequences were known through rather delicate results of Mostow [4]. Others are new and are needed in [3], [11], and [12].

The case of Theorem 1.1 where Y is a single point $\{y\}$, is part of the classical Hopf–Rinow Theorem: every point $x \in X$ can be joined to y by a geodesic. Our argument relies on that case. The case where X has sectional curvature $\leqq 0$ and $Y \subset X$ is a totally geodesic submanifold, was studied by Hermann [2]; there $\exp_{Y,X} : \mathbb{N}_{Y,X} \to X$ is a covering map.

Research partially supported by N.S.F. Grants DMS 93 21285 (JAW) and DMS 93 03224 (RZ). The second author thanks the MSRI for hospitality during the fall of 1994.

Proof. Let $x \in X$. Choose $w \in Y$ and let $m = d(x,w)$ where $d(\cdot,\cdot)$ denotes riemannian distance. Then $E = \{v \in Y \mid d(x,v) \leqq m\}$ is compact, so we have $y \in E$ minimizing the distance from x to any point of Y. Now the minimizing geodesic arc from x to y has tangent vector at y that is orthogonal to $T_y(Y)$ inside $T_y(X)$. In other words, there is a tangent vector $\xi \in T_y(Y)^{\perp}$ such that $\exp_{Y,X}(\xi) = x$. We have proved that $x \in \exp_{Y,X}(\mathbb{N}_{Y,X})$. □

2. Reductive Group Decomposition

In order to extract some structural results on Lie groups from Theorem 1.1 we fix

$$\begin{aligned} &G : \text{reductive Lie group,} \\ &\theta : \text{involutive automorphism of } G, \\ &K : \text{open subgroup of the fixed point set } G^{\theta} \\ &\qquad \text{with } X = G/K \text{connected, and} \\ &ds^2 : G\text{–invariant } \theta\text{–invariant riemannian metric on } X = G/K. \end{aligned} \tag{2.1}$$

Let $\mathfrak{g}$ denote the Lie algebra of G. In (2.1) there is no restriction on how the center of $\mathfrak{g}$ is allocated between the ± 1 eigenspaces of θ. Compare [8]. In any case, (X, ds^2) is a connected riemannian symmetric space. The usual case is when G is a connected semisimple Lie group with no compact factors, θ is a Cartan involution of G, and $K = G^{\theta}$. Here however X could have compact or euclidean factors, in particular could be compact. Now fix

$$H : \text{closed } \theta\text{–invariant subgroup of } G \tag{2.2}$$

and denote

$$\begin{aligned} &Y = H(x_0) \subset X \text{ where } x_0 = 1K, \\ &\qquad \text{identity coset in } G/K \text{ and base point in } X. \end{aligned} \tag{2.3}$$

In view of (2.1), a subalgebra $\mathfrak{h} \subset \mathfrak{g}$ of the Lie algebra of G is reductive in $\mathfrak{g}$ if and only if some conjugate $\mathrm{Ad}(g)\mathfrak{h}$ is θ–invariant. See [6, §12.1] for the case where θ is a Cartan involution; the general case follows. Let $\mathfrak{h}$ be the Lie algebra of H. Then (2.2) is essentially (up to conjugacy of θ in the group of automorphisms of G) equivalent to the condition that H be a reductive subgroup of G.

Decompose the Lie algebras $\mathfrak{g}$ and $\mathfrak{h}$ into ± 1 eigenspaces of θ,

$$\mathfrak{g} = \mathfrak{k} + \mathfrak{s} \text{ and } \mathfrak{h} = (\mathfrak{k} \cap \mathfrak{h}) + (\mathfrak{s} \cap \mathfrak{h}) \tag{2.4}$$

where $\mathfrak{k}$ is both the +1 eigenspace of θ and the Lie algebra of K. In view of (2.2),

$$\mathfrak{g} = \mathfrak{h} + \mathfrak{r} \qquad (2.5a)$$

where

$$\mathrm{Ad}(H)\mathfrak{r} = \mathfrak{r}\ , \mathfrak{k} = (\mathfrak{k} \cap \mathfrak{h}) + (\mathfrak{k} \cap \mathfrak{r})\ ,\ \text{and } \mathfrak{s} = (\mathfrak{s} \cap \mathfrak{h}) + (\mathfrak{s} \cap \mathfrak{r}). \qquad (2.5b)$$

If β denotes the positive definite bilinear form on $\mathfrak{s}$ that corresponds to ds^2 then we may assume that the decomposition (2.5) of $\mathfrak{s}$ is an orthogonal direct sum.

The tangent space $T_{gx_0}(X)$ is represented by $\mathrm{Ad}(g)\mathfrak{s}$ for $g \in G$. The subspace $T_{hx_0}(Y) \subset T_{hx_0}(X)$ is represented by $\mathrm{Ad}(h)(\mathfrak{s} \cap \mathfrak{h})$ whenever $h \in H$, and the normal space $T_{hx_0}(Y)^{\perp}$ is represented by $\mathrm{Ad}(h)(\mathfrak{s} \cap \mathfrak{r})$. Since X is a riemannian symmetric space, the riemannian and Lie group exponential maps are related by $\exp_X(g_*\xi) = \exp_G(\mathrm{Ad}(g)\xi)K = \exp_G(\mathrm{Ad}(g)\xi)x_0$ whenever $g \in G$ and $\xi \in \mathfrak{s} = T_{x_0}(X)$. Thus

Lemma 2.6. *Let $h \in H$. Then the exponential map $\exp_{Y,X} : \mathbb{N}_{Y,X} \to X$ is given on the fibre $T_{hx_0}(Y)^{\perp}$ at hx_0 by*

$$\begin{aligned}\exp_{Y,X}(\mathrm{Ad}(h)\xi) &= \exp_G(\mathrm{Ad}(h)\xi)hx_0 \\ &= h\exp_G(\xi)x_0 \ \textit{for}\ \xi \in (\mathfrak{s} \cap \mathfrak{r}).\end{aligned}$$

Theorem 1.1 and Lemma 2.6 combine to give the first statement of Theorem 2.7 below, and the second statement follows from the first by $g \mapsto g^{-1}$.

Theorem 2.7. *$G = H \cdot \exp_G(\mathfrak{s} \cap \mathfrak{r}) \cdot K$ in the sense that $\phi : (h, \xi, k) \mapsto h\ \exp_G(\xi)\ k$ is a real analytic map of $H \times (\mathfrak{s} \cap \mathfrak{r}) \times K$ onto G. Similarly $G = K \cdot \exp_G(\mathfrak{s} \cap \mathfrak{r}) \cdot H$.*
{Of course ϕ cannot be injective: if $\ell \in H \cap K$ then $\phi(\ell, 0, 1) = \phi(1, 0, \ell)$.}

3. Pseudo–Riemannian Exponential Map

As G is reductive and $X = G/K$ is riemannian symmetric, the riemannian metric ds^2 comes from a nondegenerate $\mathrm{Ad}(G)$–invariant symmetric bilinear form (again call it β) on $\mathfrak{g}$. The restriction of β to $\mathfrak{r}$ is nondegenerate because H is reductive in G. Now we have a pseudo–riemannian manifold

$$D = G/H \text{ with metric } d\sigma^2 \text{ defined by } \beta|_{\mathfrak{r}}\ . \qquad (3.1)$$

$(D, d\sigma^2)$ has a compact totally geodesic submanifold

$$E = K(d_0) \subset D \text{ where } d_0 = 1H \in G/H \text{ is the base point in } D. \qquad (3.2)$$

This situation is especially interesting when D is an open G–orbit on a complex flag manifold $G_{\mathbb{C}}/Q$; then E is a maximal compact subvariety and its $G_{\mathbb{C}}$–translates inside D carry a lot of geometric and analytic information on both G and D. Compare [3], [7], [8], [9], [10] and [12].

As before, we have the normal bundle $\mathbb{N}_{E,D} \to D$, sub–bundle of the restriction $\mathbb{T}(D)|_E$ of the tangent bundle of D, whose fibre over $d \in D$ is the orthocomplement $T_d(E)^\perp \subset T_d(D)$ of the tangent space to E at d in the tangent space to D at d. Here it is important to notice that $T_d(E)$ is a $d\sigma^2$–nondegenerate subspace of $T_d(D)$. The exponential map $\exp_{E,D} : \mathbb{N}_{E,D} \to D$ again is just the corresponding restriction of the usual exponential map $\exp_D : \mathbb{T}(D) \to D$. As in Lemma 2.6,

Lemma 3.3. *Let $k \in K$. Then the exponential map $\exp_{E,D} : \mathbb{N}_{E,D} \to D$ is given on the fibre $T_{kd_0}(E)^\perp$ at kd_0 by*
$$\exp_{E,D}(\mathrm{Ad}(k)\xi) = \exp_G(\mathrm{Ad}(k)\xi)kd_0 = k\exp_G(\xi)d_0 \text{ for } \xi \in (\mathfrak{s} \cap \mathfrak{r}).$$

Lemma 3.3 combines with the second statement of Theorem 2.7 to yield

Theorem 3.4. *The exponential map $\exp_{E,D} : \mathbb{N}_{E,D} \to D$ is surjective.*

4. Symmetric Space Case and Euler Angle Decompositions

Now consider the case where $D = G/H$ is a pseudo–riemannian symmetric space. In other words, there is an involutive automorphism τ of G such that H is an open subgroup of the fixed point set G^τ. Then τ and θ commute because $\theta(H) = H$, and $\mathfrak{r}$ is the -1 eigenspace of τ on $\mathfrak{g}$.

Decompose the Lie algebra $\mathfrak{g}$ into ± 1 eigenspaces of θ and τ,

$$\mathfrak{g} = \mathfrak{k} + \mathfrak{s} = \mathfrak{h} + \mathfrak{r} = (\mathfrak{k} \cap \mathfrak{h}) + (\mathfrak{k} \cap \mathfrak{r}) + (\mathfrak{s} \cap \mathfrak{h}) + (\mathfrak{s} \cap \mathfrak{r}). \tag{4.1}$$

Let $L \subset G^{\tau\theta}$ be the identity component of the fixed point set of $\tau\theta$. Its Lie algebra $\mathfrak{l} = (\mathfrak{k} \cap \mathfrak{h}) + (\mathfrak{s} \cap \mathfrak{r})$ and $L(x_0) \cong L/(K \cap L)$ is riemannian symmetric. Denote

$$\mathfrak{a} : \text{ maximal abelian subspace of } \mathfrak{s} \cap \mathfrak{r} \text{ and } A = \exp_G(\mathfrak{a}) \tag{4.2}$$

Then it is standard that $\mathfrak{a}$ is unique up to $(K \cap L)$–conjugacy and $L = (K \cap L)A(K \cap L)$. But $K \cap L$ is a maximal compactly embedded subgroup of L, hence connected because L is connected, so $(K \cap L) \subset (K \cap H)$. As $\exp_G(\mathfrak{s} \cap \mathfrak{r}) \subset L$, this combines with Theorem 2.7 to yield

Theorem 4.3. $G = HAK = KAH$ *as in* Theorem 2.7.

In case $K = H$ this is the classical "Cartan decomposition", generalizing the Euler angle decomposition of $SO(3)$. In case G is a connected semisimple group of noncompact type and with finite center, decompositions of this sort derive from results of Mostow [4] and have been used extensively in representation theory. See [1] and [5]. When G is compact, the decomposition seems to be new.

References

[1] M. Flensted–Jensen, *Discrete series for semisimple symmetric spaces*, Annals of Math. **111** (1980), 253–311.

[2] R. Hermann, *Homogeneous Riemannian manifolds of non-positive sectional curvature*, Proc. Koninkl. Nederl. Akad. Wet. Ser. A **66** (1963), 47–56.

[3] C. Leslie, *Geometry of open orbits in complex flag manifolds*, thesis in preparation.

[4] G. D. Mostow, *Some new decomposition theorems for semisimple Lie groups. In "Lie Groups and Lie Algebras"*, Mem. Amer. Math. Soc. **14** (1955), 31–54.

[5] J. Rawnsley, W. Schmid and J. A. Wolf, *Singular unitary representations and indefinite harmonic theory*, J. Functional Analysis **51** (1983), 1–114.

[6] J. A. Wolf, *Spaces of Constant Curvature, Fifth Edition*, Publish or Perish, 1984.

[7] ———, *The action of a real semisimple Lie group on a complex manifold,* I*: Orbit structure and holomorphic arc components*, Bull. Amer. Math. Soc. **75** (1969), 1121–1237.

[8] ———, *The action of a real semisimple Lie group on a complex manifold,* II*: Unitary representations on partially holomorphic cohomology spaces*, Memoirs Amer. Math. Soc. **138** (1974).

[9] ———, *The Stein condition for cycle spaces of open orbits on complex flag manifolds*, Annals of Math. **136** (1992), 541–555.

[10] ———, *Exhaustion functions and cohomology vanishing theorems for open orbits on complex flag manifolds*, Mathematical Research Letters **2** (1995), 179–191.

[11] J. A. Wolf and R. Zierau, *Cayley transforms and orbit structure in complex flag manifolds*, to appear.

[12] ———, *Linear cycle spaces in flag domains*, in preparation.

Department of Mathematics
University of California
Berkeley, California 94720

Department of Mathematics
Oklahoma State University
Stillwater, Oklahoma 74074

`jawolf@math.berkeley.edu`

`zierau@math.okstate.edu`

Received June 1995; revised November 1995

Weakly Symmetric Spaces

Wolfgang Ziller *

A Riemannian manifold M is called weakly symmetric, if for any two points p and q in M, there exists an isometry f of M which interchanges p and q. An equivalent condition is that for every geodesic $\gamma(t)$ in M, there exists an isometry f which reverses the geodesic, i.e. $f(\gamma(t)) = \gamma(-t)$. A Riemannian symmetric space is clearly weakly symmetric. These manifolds were first studied by A.Selberg [S] who showed that for a weakly symmetric space, the algebra of isometry invariant differential operators is commutative.

We will present a number of new examples of such manifolds in this paper. But before we do so, let us mention the connection of this property with several other geometric properties. A Riemannian manifold is called a D'Atri space if the geodesic symmetries around each point (which are always defined locally) are volume-preserving up to sign. This condition was first discussed by Joe D'Atri [D], who he showed that every naturally reductive homogeneous space is a D'Atri space. One easily shows that the geodesic symmetry is volume-preserving iff every small geodesic sphere has the same Gauss Kronecker curvature at antipodal points. Hence a weakly symmetric space must be a D'Atri space (see [BV]).

Another well known consequence of being naturally reductive is that all geodesics are images of one-parameter groups of isometries, i.e., that every geodesic $\gamma(t)$ is of the form $\gamma(t) = \exp(tX) \circ \gamma(0)$ for some Killing vector field X. Manifolds with this property are called g.o. spaces (geodesic orbit spaces). In [KV1] it was shown that a g.o. space is also a D'Atri space and in [BKV] it was shown that a weakly symmetric space is a g.o. space. See [G2] for some new results on g.o. spaces and the survey article [KPV] for a discussion of D'Atri spaces, g.o. spaces, commutative spaces, and weakly symmetric spaces.

Many of the examples in this paper are not naturally reductive. Hence they also give new examples of g.o. spaces, D'Atri spaces, and commutative spaces. Previously known examples of weakly symmetric spaces which are not symmetric [BV] are the distance spheres in rank 1 symmetric spaces and tubes around some totally geodesic submanifolds in rank 1 symmetric spaces as well as horospheres in non-compact rank 1 symmetric spaces. We will start with an alternative description of why these spaces are weakly

* Partly supported by a grant from the National Science Foundation.

symmetric and then go on to construct new examples. These new examples will be metrics on $\mathbb{C}P^{2n+1}$, the symmetric space SO(2n)/U(n), irreducible metrics on $S^7 \times S^7$, and infinitely many 3-dimensional families of irreducible weakly symmetric metrics on $S^2 \times S^{2n+1}$. Furthermore we exhibit two weakly symmetric spaces which are homeomorphic but not diffeomorphic and some new examples of left invariant metrics on nilpotent Lie groups which are weakly symmetric. We will also classify all Riemmanian nilmanifolds of Heisenberg type which are weakly symmetric.

We first make an obvious reformulation of the definition of being weakly symmetric. If H_p is the full isotropy group at p, i.e. the set of all isometries f with $f(p) = p$, then we need that for every $v \in T_pM$, there exists a $f \in H_p$ with $d(f)_p(v) = -v$. A weakly symmetric space is clearly homogeneous and since for a homogeneous space the isotropy groups at different points are conjugate, we need to require this condition only at one point. Hence we can equivalently start with a homogeneous space $M = G/H_p$ with isotropy representation χ (i.e. $\chi(g) = d(g)_p$ for $g \in H_p$) and require that

$$(*) \qquad \text{for every } v \in T_pM \text{ there exists a } g \in H_p \text{ with } \chi(g)(v) = -v.$$

Hence, if a homogeneous space G/H satisfies condition (*), every G-invariant metric on G/H is weakly symmetric.

Example 1. Distance Spheres and horospheres. To illustrate this process, let us look at the transitive action of U(n) on $M = S^{2n-1}(1) \subset \mathbb{C}^n$. If $p = (1, 0, .., 0)$, then $H_p = \mathrm{U(n-1)}$. T_pM can be viewed as $\mathbb{C}^n \oplus \mathbb{R}$ where $\mathbb{R}$ is spanned by ip and the isotropy representation is given by $\chi(A)(v, a) = (Av, a)$. There exists a further isometry C of M given by conjugation on each coordinate. It also lies in H_p and $\chi(C)(v, a) = (\bar{v}, -a)$. Hence if $G = \mathrm{U(n)} \cup \mathrm{C} \circ \mathrm{U(n)}$, then $H_p = \mathrm{U(n-1)} \cup \mathrm{C} \circ \mathrm{U(n-1)}$ and H_p clearly satisfies the above property (*). Indeed, we can first send (v, a) to $(\bar{v}, -a)$ and since U(n) acts transitively on the unit sphere in $\mathbb{C}^n$, we can then choose an element in U(n) which sends $\bar{v}$ to $-v$. Hence any G-invariant metric on M is weakly symmetric. If h_1 and h_2 are the usual inner products on $\mathbb{C}^n$ and $\mathbb{R}$, then any G-invariant metric is of the form $ah_1 + bh_2$ where a and b are arbitrary positive constants. Hence we obtain a 2-parameter family of weakly symmetric metrics or a one-parameter family up to scaling. The distance spheres in complex projective space $\mathbb{C}P^n$ and in complex hyperbolic space $\mathbb{C}H^n$ clearly belong to this family since the isotropy group of a point in each space together with the isotropy representation agrees with the above G. In fact, up to scaling, these distance spheres describe all G-invariant metrics which are not isometric to the round sphere metric, see e.g. [Z2].

This generalizes to the other rank 1 symmetric spaces. For the distance spheres in $\mathbb{H}P^n$ and $\mathbb{H}H^n$ we look at the isotropy representation of both symmetric spaces which induces the transitive action of Sp(n) $\times$ Sp(1) on $M = S^{4n-1}(1) \subset \mathbb{H}^n$ given by $(A,q)(v) = A(v)q^{-1}$. The isotropy group of $p = (1,0,..,0)$ is equal to $\mathrm{Sp(n-1)} \times \triangle\mathrm{Sp}(1)$ (Sp(1) embedded diagonally). T_pM can be identified with $\mathbb{H}^{n-1} \oplus \mathrm{Im}(\mathbb{H})$ and the isotropy action becomes $\chi(A,q)(v,w) = (A(v), qwq^{-1})$. This representation satisfies (*) since Sp(n$-$1) also acts transitively on the unit sphere in $\mathbb{H}^{n-1}$ and $w \to qwq^{-1}$ is the standard action of $\mathrm{Sp}(1)/\mathbb{Z}_2 = \mathrm{SO}(3)$ on $\mathbb{R}^3$. Again, there exists, up to scaling, a one-parameter family of weakly symmetric metrics which agrees with the distance spheres in $\mathbb{H}P^n$ and $\mathbb{H}H^n$ (together with the round sphere metric). Notice though that if we write M as Sp(n)/Sp(n $-$ 1), we obtain a much larger class of Sp(n)-invariant metrics, but they are not weakly symmetric anymore, unless they belong to the previous class.

The Cayley plane exhibits a somewhat different phenomena. Here we look at $M = S^{15} = \mathrm{Spin}(9)/\mathrm{Spin}(7)$ with tangent space $T_pM = \mathbb{R}^7 \oplus \mathbb{R}^8$ and isotropy representation $\chi(A)(v,w) = (\rho_7(A)(v), \Delta_7(A)(w))$ where ρ_7 is the standard representation of Spin(7) on $\mathbb{R}^7$ (via the two-fold cover of Spin(7) onto SO(7)) and Δ_7 is the spin representation of Spin(7) on $\mathbb{R}^8$. Both representations act transitively on the unit sphere and hence each has property (*). But in general the direct sum of two representations that have property (*) does not have to have property (*) again. To see that the direct sum in this case still has this property for $(v,w) \in \mathbb{R}^7 \oplus \mathbb{R}^8$, we first choose an element g in Spin(7) with $\rho_7(g)(v) = -v$. We then consider the isotropy group of v in Spin(7) under the action of ρ_7. This isotropy group is equal Spin(6) = SU(4) and the restriction of the action of Spin(7) on $\mathbb{R}^8$ to SU(4) becomes the standard action of SU(4) on $\mathbb{R}^8$ which is still transitive on spheres. Hence we can find an element $h \in$ SU(4) with $\Delta_7(h) \circ \Delta_7(g)(w) = -w$ and hence hg has the desired property. Again we obtain a one-parameter family, up to scaling, of weakly symmetric manifolds which consist of the distance spheres in CayP^2 and CayH^2 and the round sphere metric.

In [Z1], [Z2], and [WZ1],p.625, it is shown that the first and second class of metrics are all naturally reductive. For the third class it is shown that the distance spheres in CayP^2 and CayH^2 are not naturally reductive, with one exception, namely the distance sphere in CayP^2 whose radius is 2/3 the distance to the cut locus, which corresponds up to scaling to the metric on S^{15} induced by the Killing form on Spin(9) and is hence naturally reductive. Therefore among the one-parameter family of metrics on S^{15}

there are precisely two which are naturally reductive, the above metric and the round sphere metric.

An alternative description of the above three families of metrics is in terms of the Hopf fibrations. If we start with one of the 3 Hopf fibrations:

$$S^1 \to S^{2n+1} \to \mathbb{C}P^n$$
$$S^3 \to S^{4n+3} \to \mathbb{H}P^n$$
$$S^7 \to S^{15} \to S^8,$$

then the projection of the total space to the base becomes a Riemannian submersion if the metric on total space and fibre are spheres of radius 1. We can then scale the fibre and base to get a two parameter family of Riemannian submersions which agree with the above two-parameter families of metrics. In each case, the group G then consists of the full group of isometries of the total space $S^n(1)$ which takes fibres to fibres (see e.g. [GWZ]).

In all three cases we can also look at the horospheres in the non-compact symmetric space of rank 1. They are again weakly symmetric since horospheres through a point can be viewed as limits of distance spheres through the same point (and with the same tangent space). The isotropy group and isotropy action for the distance spheres and horospheres is therefore the same, although the group G acting transitively is different in each case. In fact, there exists a nilpotent Lie group N that acts simply transitively on the horospheres, the Heisenberg group in the case of $\mathbb{C}H^n$ and the quaternionic and Cayley analogue of the Heisenberg group in case of $\mathbb{H}H^n$ and CayH^2. Hence the metric can also be regarded as a left invariant metric on these nipotent Lie groups N. The isotropy groups H_p that we discussed for the distance spheres are the full isotropy groups at $e \in N$ for these left invariant metrics and the full isometry group G is the semidirect product of N with H. (See also the discussion in Example 6.) We therefore obtain, up to scaling, a one-parameter family of weakly symmetric metrics on these homogeneous spaces which agree with the one-parameter family of metrics one obtains on the horospheres. But notice that these metrics are, up to isometry, just a multiple of each other, and hence the metric is unique up to scaling. It follows immediately from [G1], Theorem 4.8, that the horospheres in $\mathbb{C}H^n$ and $\mathbb{H}H^n$ are naturally reductive and the horospheres in CayH^2 are not naturally reductive.

There also exists another non-compact version of these examples which can be regarded as the dual to the compact examples (and this construction will also carry over to all further examples). Let us first consider U(n)/U(n − 1) and the inclusions U(n − 1) ⊂ U(n − 1)U(1) ⊂ U(n).

Whenever $H \subset K \subset G$, then the isotropy representation of G/H is the direct sum of the isotropy representation of K/H and the restriction of the isotropy representation of G/K to H. Hence, if we consider $\mathrm{U(n-1)} \subset \mathrm{U(n-1)U(1)} \subset \mathrm{U(n-1,1)}$, then $\mathrm{U(n-1,1)/U(n-1)} = \mathrm{SU(n-1,1)/SU(n-1)}$ has the same isotropy representation as $\mathrm{U(n)/U(n-1)}$ since the complex projective space $\mathrm{U(n)/U(n-1)U(1)}$ and the complex hyperbolic space $\mathrm{U(n-1,1)/U(n-1)U(1)}$ have the same isotropy representation. Furthermore, there exists an automorphism of $\mathrm{U(n-1,1)}$ which induces complex conjugation on $\mathrm{U(n-1)U(1)}$ and hence $\mathrm{SU(n-1,1)/SU(n-1)}$ again has a one-parameter family of weakly symmetric metrics. In the case of $n = 2$ this gives a weakly symmetric metric on $\mathrm{SU(1,1)} = \mathrm{SL}(2,\mathbb{R})$, which was actually the first non-symmetric example discovered by Selberg [S]. $\mathrm{SU(n-1,1)/SU(n-1)}$ can also be regarded as a tube of radius r around the totally geodesic submanifold $\mathbb{C}H^{n-1} \subset \mathbb{C}H^n$ since $\mathrm{SU(n-1,1)}$ acts transitively on such tubes.

Similarly from the inclusions $\mathrm{Sp(n-1)} \subset \mathrm{Sp(n-1)Sp(1)} \subset \mathrm{Sp(n)}$ and $\mathrm{Sp(n-1)} \subset \mathrm{Sp(n-1)Sp(1)} \subset \mathrm{Sp(n-1,1)}$, it follows that $\mathrm{Sp(n-1,1)/Sp(n-1)}$ carries, up to scaling, a one-parameter family of weakly symmetric metrics which agree with the tubes of radius r around $\mathbb{H}H^{n-1} \subset \mathbb{H}H^n$. And from $\mathrm{Spin}(7) \subset \mathrm{Spin}(8) \subset \mathrm{Spin}(9)$ and $\mathrm{Spin}(7) \subset \mathrm{Spin}(8) \subset \mathrm{Spin}(8,1)$ it follows that $\mathrm{Spin}(8,1)/\mathrm{Spin}(7)$ carries a one-parameter family of weakly symmetric metrics which agree with the tubes of radius r around $\mathrm{Cay}H^1 \subset \mathrm{Cay}H^2$. It follows immediately from [G1], Theorem 5.2, that the tubes around $\mathbb{C}H^{n-1} \subset \mathbb{C}H^n$ and $\mathbb{H}H^{n-1} \subset \mathbb{H}H^n$ are naturally reductive and the tubes around $\mathrm{Cay}H^1 \subset \mathrm{Cay}H^2$ are not.

Example 2. $\mathbb{C}\mathbf{P}^{2n+1}$. All of the examples described so far can be found in [BV], using a different method. To construct some new examples, we first look at the Hopf fibration

$$S^2 \to \mathbb{C}P^{2n+1} \to \mathbb{H}P^n$$

which sends complex lines in $\mathbb{C}^{2n+2} = \mathbb{H}^{n+1}$ into the corresponding quaternionic lines. We again obtain a two-parameter family of metrics by scaling fibre and base. The group G of isometries of $M = \mathbb{C}P^{2n+1}$ which takes fibres to fibres is now $\mathrm{Sp(n+1)} \subset \mathrm{SU(2n+2)}$ with isotropy group $H = \mathrm{Sp(n)U(1)} \subset \mathrm{Sp(n)Sp(1)}$ and the Hopf fibration becoming the homogeneous fibration

$$\mathrm{Sp(1)/U(1)} \to \mathrm{Sp(n+1)/Sp(n)U(1)} \to \mathrm{Sp(n+1)/Sp(n)Sp(1)}.$$

To describe the isotropy action, we observe that it preserves vertical and horizontal space, hence $T_pM = \mathbb{R}^2 \oplus \mathbb{H}^n$ and H acts on R^2 via the rotation of U(1) on $\mathbb{R}^2$. In fact, if $e^{i\theta} \in$ U(1), then it acts under the isotropy representation of Sp(1)/U(1) by multiplication with $e^{2i\theta}$ on $\mathbb{R}^2 = \mathbb{C}$. Furthermore, H acts on $\mathbb{H}^n$ as the restriction of the action of Sp(n)Sp(1) (the isotropy action of Sp(n + 1)/Sp(n)Sp(1)) to Sp(n)U(1) and hence as $(A, z)(v) = A(v)\bar{z}$. We therefore first use $e^{i\pi/2}$ which induces -id on $\mathbb{R}^2$ and right multiplication with $-i$ on $\mathbb{H}^n$ and then choose an element in Sp(n) which takes $v(-i) \in \mathbb{H}^n$ to $-v$ which we can do since Sp(n) acts transitively on the unit sphere in $\mathbb{H}^n$. Hence H satisfies (*) and we obtain, up to scaling, a one-parameter family of weakly symmetric metrics on $\mathbb{C}P^{2n+1}$. Only two of these are naturally reductive (see [Z2]) : the one induced by the Killing form of Sp(n + 1) and the symmetric metric on $\mathbb{C}P^{2n+1}$ (This follows from [DZ], p.5, and the fact that the only groups acting transitively on $\mathbb{C}P^{2n+1}$ are SU(2n + 2) and Sp(n + 1)). Since these metrics, together with the previously described metrics on spheres, are precisely the homogeneous metrics on spheres and projective spaces [Z2], this shows that most homogeneous metrics on compact rank 1 symmetric spaces are weakly symmetric.

There also exists a non-compact dual to this example. From the inclusions Sp(n)U(1) $\subset$ Sp(n)Sp(1) $\subset$ Sp(n, 1) and the fact that the quaternionic hyperbolic space Sp(n, 1)/Sp(n)Sp(1) has the same isotropy representation as the quaternionic projective space Sp(n + 1)/Sp(n)Sp(1), it follows that Sp(n, 1)/Sp(n)U(1) and Sp(n + 1)/Sp(n)U(1) have the same isotropy representation and hence Sp(n, 1)/Sp(n)U(1) also has a one-parameter family of weakly symmetric metrics. [G1] again implies that these metrics are not naturally reductive.

Example 3. SO(2n)/U(n) . The next example are homogeneous metrics on the symmetric space M = SO(2n)/U(n) which can be described as the set of orthogonal complex structures on $\mathbb{R}^{2n}$. The subgroup SO(2n − 1) $\subset$ SO(2n) still acts transitively on M with isotropy group U(n − 1) (see [K1] for details) and from the inclusions U(n − 1) $\subset$ SO(2n − 2) $\subset$ SO(2n − 1) we obtain the homogeneous fibration:

$$\mathrm{SO(2n-2)/U(n-1)} \to \mathrm{M} \to \mathrm{SO(2n-1)/SO(2n-2)} = \mathrm{S}^{2n-2}$$

The isotropy representation preserves vertical and horizontal space and is easily seen to be $\Lambda^2\mu_{n-1} \oplus \mu_{n-1}$ where μ_{n-1} is the standard representation of U(n−1) on $\mathbb{C}^{n-1}$. Both representations contain -id (induced by -id in the second summand and by $e^{i\pi/2}$ id in the first summand) and hence satisfy (*). To see that the direct sum satisfies (*) also, let $(v, w) \in \Lambda^2\mathbb{C}^n \oplus \mathbb{C}^n$ and choose an orthonormal basis $e_1, ..., e_{n-1}$ in $\mathbb{C}^{n-1}$ with e_1 parallel to w and $v = a_1e_1 \wedge e_2 + a_2e_2 \wedge e_3 + ... + a_{n-2}e_{n-2} \wedge e_{n-1}$ for some constants a_i.

Then consider the element g in U(n − 1) which takes $e_1, e_3, e_5, \ldots$ into their negatives and fixes $e_2, e_4, \ldots$. It satisfies $\Lambda^2(g)(v) = -v$ and $g(w) = -w$ and hence the isotropy representation satisfies(*). We obtain, up to scaling, a one-parameterone-parameterone-parameter family of weakly symmetric metrics on $M = \mathrm{SO(2n)/U(n)} = \mathrm{SO(2n-1)/U(n-1)}$ only two of which are naturally reductive, the one induced by the Killing form of SO(2n) and the one induced by the Killing form of SO(2n − 1). We also again have a non-compact dual $M = \mathrm{SO(2n-2,1)/U(n-1)}$ which has the same isotropy representation and hence is weakly symmetric, but not naturally reductive by [G1]. For $n = 3$, $\mathrm{SO(6)/U(3)} = \mathbb{C}\mathrm{P}^3$ and the example in this case agrees with the previous one. In [KV2] and [G2] this special case of SO(5)/U(2) and SO(4, 1)/U(2) was observed to be a g.o. space. For $n = 4$, SO(8)/U(4) can also be regarded as the Grassmannian of oriented 2-planes in $\mathbb{R}^8$.

Example 4. $\mathbf{S^7 \times S^7}$. Our next example is $M = S^7 \times S^7 = \mathrm{Spin}(8)/\mathrm{G}_2$ (see e.g. [K2] for details). The isotropy representation is $\pi_7 \oplus \pi_7$ where π_7 is the seven-

dimensional representation of G_2 via automorphisms on the imaginary Cayley numbers. It is well known that G_2 acts transitively on two-frames. Hence if we want to send (v, w) to $(-v, -w)$, we can choose the element of G_2 which induces a rotation by π in the two-plane spanned by v and w (If v and w are parallel it is even simpler to find the desired element of G_2). Hence all Spin(8) invariant metrics are weakly symmetric. Since the two representations are equivalent, there exists a three-parameter family of such metrics, (or a two-parameter family up to scaling). A two-dimensional subfamily consists of the product metrics of constant curvature metrics on each S^7, which are of course symmetric, and one easily shows that all other metrics are irreducible as Riemannian manifolds. One also easily shows that the only metrics which are naturally reductive are the product metrics and the metric induced by the Killing form of Spin(8). Hence most of these weakly symmetric metrics are not naturally reductive. Of course $S^7 \times S^7$ also carries product metrics of the weakly symmetric metrics on S^7 discussed earlier, which are hence also weakly symmetric, but not isometric to the ones discussed here. We also get a non-compact dual $\mathrm{Spin}(7,1)/\mathrm{G}_2$ from the inclusions $G_2 \subset \mathrm{Spin}(7) \subset \mathrm{Spin}(7,1)$ and [G1] implies that this example is not naturally reductive.

Example 5. Principle Torus Bundles. An interesting example consists of principle torus bundles over products of complex projective spaces (see [WZ2] for details). Let $G = \mathrm{SU}(n_1+1) \times \ldots \times \mathrm{SU}(n_k+1)$ and $H = \mathrm{SU}(n_1) \times \ldots \times \mathrm{SU}(n_k) \times \mathrm{T}^r$ where T^r is any subtorus of $T^k = \mathrm{U}(1) \times \ldots \times \mathrm{U}(1)$ and the i-th U(1) is the normalizer of $\mathrm{SU}(n_i)$ in $\mathrm{SU}(n_i+1)$. Now we consider $M = G/H$. Since $H \subset \mathrm{U}(n_1)\mathrm{U}(n_2)\ldots\mathrm{U}(n_k) \subset \mathrm{G}$ we can view M as a principal torus bundle over $\mathbb{C}P^{n_1}\times\ldots\times\mathbb{C}P^{n_k}$ where the fibre has dimension $k-r$. The tangent space T_pM splits up into $\mathfrak{m}_1\oplus\ldots\oplus\mathfrak{m}_k\oplus\mathfrak{n}$ where $\mathfrak{m}_i$ is the perpendicular complement of $\mathrm{SU(n_i)U(1)}$ in $\mathrm{SU(n_i+1)}$ and $\mathfrak{n}$ is the perpendicular complement of T^r in T^k. The isotropy representation of H acts trivially on $\mathfrak{n}$ and on $\mathfrak{m}_i=\mathbb{C}^{n_i}$ the ith factor $\mathrm{SU(n_i)}$ acts via its tautological representation and all other factors act trivially. In addition we have the automorphism of G which is complex conjugation of the matrices in each component. It leaves

H invariant and hence induces a map of G/H which fixes p and whose derivative at p is given by complex conjugation on $\mathfrak{m}_i$ and by $-id$ on $\mathfrak{n}$. If $n_i>1$, $\mathrm{SU(n_i)}$ acts transitively on the unit sphere in $\mathfrak{m}_i$. But if $n_i=1$, we need to construct some further isometries. Notice that the torus T^k is contained in the normalizer of H and hence right translation by elements in T^k are well defined and conjugation by such elements induce diffeomorphisms of G/H which fix p and whose derivative at p acts by multiplication with a unit complex number on $\mathfrak{m}_i$ and trivially on $\mathfrak{n}$. Hence they act by isometries for any G-invariant metric. If $n_i=1$, these maps induce rotations which act transitively on the unit sphere in $\mathfrak{m}_i$. Hence H, together with these maps, satisfies (*) and every G-invariant metric is weakly symmetric. A G-invariant metric is a multiple of the Killing form of G on $\mathfrak{m}_i$ and an arbitrary inner product on $\mathfrak{n}$. (Special cases of these examples were also considered in [GV] from a different viewpoint.)

These metrics can be viewed as the total space of a Riemannian submersion onto $\mathbb{C}P^{n_1}\times\ldots\times\mathbb{C}P^{n_k}$ with totally geodesic flat fibres and it immediately follows from [DZ] p. 9 and 66 that all metrics are naturally reductive with respect to the transitive action of $G\times T^{k-r}$.

These examples exhibit very interesting topological properties (see [WZ2] for details). Consider first the case of $n_1=n_2=1$ and $k=2, r=1$ which consists of circle bundles over $S^2\times S^2$. Here $M_{p,q}=S^3\times S^3/S^1$ where the circle is embedded with slope p/q in the maximal torus $S^1\times S^1\subset S^3\times S^3$. Each homogeneous space has a three-parameter family of weakly symmetric and naturally reductive metrics, and as manifolds they are all diffeomorphic to $S^3\times S^2$. This gives rise to infinitely many transitive actions of $S^3\times S^3$ on $S^3\times S^2$ and one shows that they are all distinct as topological actions. Hence the invariant Riemannian metrics are all non-isometric which implies that $S^3\times S^2$ carries infinitely many three-dimensional families of weakly symmetric metrics. For $p/q=0$, these are the symmetric product metrics on $S^3\times S^2$, but for $p/q\neq 0$ they are all irreducible as Riemannian manifolds. For $p/q=1$, these are the natural metrics on the unit tangent bundles $T_1S^3=\mathrm{SO(4)/SO(2)}$.

By choosing $n_1=1$ and $n_2=m>1$, $k=2, r=1$, and using the results in [WZ2], one obtains the same conclusion for the manifold $S^2 \times S^{2m+1}$. In the case of $n_1 = 1$ and $n_2 = 2$ and $k = 2, r = 1$, the seven-dimensional manifolds G/H were classified up to diffeomorphism and homeomorphism in [KS]. Suprisingly, there are two such manifolds which are homeomorphic but not diffeomorphic. Hence there exist weakly symmetric spaces which are homeomorphic but not diffeomorphic, a phenomena which is impossible for symmetric spaces.

One special case, that is of independent interest, is the case where $k = 2, r = 1$ and the circle U(1) is embedded diagonally in $T^2 = \mathrm{U}(1) \times \mathrm{U}(1)$. The manifold G/H can then also be viewed as a tube around the linear projective subspace $\mathbb{C}P^{n_1}$ in $\mathbb{C}P^{n_1+n_2+1}$ since G acts transitively on this tube with isotropy group H. Hence these tubes are all weakly symmetric.

The latter case easily generalizes to symplectic groups and shows that tubes around symplectic subspaces in $\mathbb{H}P^n$ are weakly symmetric. But the more general version of bundles over products of projective spaces does not generalize to the quaternionic case.

These principal torus bundles also have dual examples due to the inclusions $H \subset \mathrm{U}(n_1)\mathrm{U}(n_2)...\mathrm{U}(n_k) \subset \mathrm{SU}(n_1, 1) \times ... \times \mathrm{SU}(n_k, 1)$ and the fact that there also exists an automorphism on $\mathrm{SU}(n_i, 1)$ which induces complex conjugation on $\mathfrak{m}_i$ and -id on the Lie algebra of U(1). Hence every invariant metric on $\mathrm{SU}(n_1, 1)...\mathrm{SU}(n_k, 1)/\mathrm{H}$ is weakly symmetric. The special case of $k = 2, r = 1$ and U(1) embedded diagonally corresponds to the tubes around $\mathbb{C}H^k \subset \mathbb{C}H^n$. One can also mix up the compact and non-compact examples by considering the inclusions $H \subset \mathrm{U}(n_1)\mathrm{U}(n_2)...\mathrm{U}(n_k) \subset \mathrm{G}_1 \times ... \times \mathrm{G}_k = \mathrm{G}$ where G_i can be either $\mathrm{SU}(n_i + 1)$ or $\mathrm{SU}(n_i, 1)$. We can also let G_i be the semidirect product of the $2n_i + 1$-dimensional real Heisenberg group with $\mathrm{U}(n_i)$ since, as we observed earlier, the Heisenberg group has the same isotropy action as $\mathrm{U}(n_i + 1)/\mathrm{U}(n_i)$. In all cases G/H is weakly symmetric and naturally reductive by [G1].

Example 6. Nilpotent Lie Groups. We now examine weakly symmetric metrics on nilpotent Lie groups N. It follows from [G2] that N must be 2-step nilpotent. We will use the notation and results in [G2]. Let $\mathfrak{n}$ be the Lie algebra of N with a given inner product, $\mathfrak{z} = [\mathfrak{n}, \mathfrak{n}]$ and $\mathfrak{a} = \mathfrak{z}^\perp$. The equation $< j(z)u, v > = < z, [u, v] >$ defines a linear map $j : \mathfrak{z} \to \mathfrak{so}(\mathfrak{a})$. If we have another $\mathfrak{n}', \mathfrak{z}', j'$ then N' it is isometric to N iff there exist isometries $A : \mathfrak{a} \to \mathfrak{a}'$ and $B : \mathfrak{z} \to \mathfrak{z}'$ with $j'(B(z)) = Aj(z)A^{-1}$. We can hence normalize so that $\mathfrak{a} = \mathbb{R}^n$ with its usual inner product and we can identify $\mathfrak{z}$ with $j(\mathfrak{z}) \subset \mathfrak{so}(n)$. Then the embedding $\mathfrak{z} \subset \mathfrak{so}(n)$ and the metric on $\mathfrak{z}$ determine the Riemannian nilmanifold N. The Lie groups of Heisenberg type are those with $j(z)^2 = -|z|^2 id$. In [G1], [G2] it was proved that:

a) N is naturally reductive iff $\mathfrak{z}$ is a subalgebra of $\mathfrak{so}(n)$ and the metric on $\mathfrak{z}$ is biinvariant, i.e., Int($\mathfrak{z}$) invariant.

b) N is a g.o. manifold iff $\mathfrak{z}$ is a linear subspace that satisfies the transitive normalizer condition: For each $z \in \mathfrak{z}, x \in \mathfrak{a} = \mathbb{R}^n$ there exists an A in the normalizer of $\mathfrak{z} \subset \mathfrak{so}(n)$ with $[A, z] = 0$ and $A(x) = z(x)$. Furthermore the metric on $\mathfrak{z}$ must be such that the above A can be chosen so that $adA_{|\mathfrak{z}}$ is skew-symmetric.

c) The isometry group of N consists of the left translations by elements of N, which is normal in the full isometry group, and the isotropy group at $e \in N$ consisting of those $(A, B) \in (O(\mathfrak{a}), O(\mathfrak{z})) \subset O(\mathfrak{n})$ such that A normalizes $\mathfrak{z}$ and $Ad(A)_{|\mathfrak{z}} = B$. Hence N is weakly symmetric iff this isotropy group satisfies (*).

Let us first look at the case of dim $\mathfrak{z} = 1$. The line $\mathfrak{z} \subset \mathfrak{so}(n)$ can be assumed to lie in the maximal torus of $\mathfrak{so}(n)$ and we can assume that $n = 2k$ with $\mathbb{R}^n = \mathbb{C}^k$ and for $z \in \mathfrak{z}, |z| = 1$ we have $j(z)(z_1, ..., z_k) = (a_1 i z_1, ..., a_k i z_k)$ with a_i real and $a_1 \geq ... \geq |a_k|$. Notice that the Lie group N is always isomorphic to the Heisenberg group but that the isometry class of the metric is uniquely determined by $a_1, ..., a_k$. N is of Heisenberg type iff $a_1 = ... = a_k = 1$ and this is the case of the horosphere in $\mathbb{C}H^{n+1}$ discussed earlier. In the three-dimensional case, the metric is actually unique up to scaling due to the automorphism λid on $\mathfrak{a}$ and $\lambda^2 id$ on $\mathfrak{z}$. All these metrics are clearly naturally reductive. The isotropy group contains $A(v) = \bar{v}$ with $B = Ad(A)|\mathfrak{z} = -id$ and the maximal torus $A(z_1, ..., z_k) = (e^{i\theta_1} z_1, ..., e^{i\theta_1} z_1)$ with $B = id$ (and further elements if some of the a_i are equal, up to U(n) if all a_i are equal). Hence (*) is satisfied and we obain a k-parameter family of weakly symmetric metrics on the Heisenberg group.

A similar example is the case where dim $\mathfrak{z} = 2$ and $\mathfrak{z}$ is a subalgebra of $\mathfrak{so}(n)$. $\mathfrak{z}$ must then be abelian and all metrics are naturally reductive. We can again assume n even and $j(e_1)(z_1, ..., z_k) = (a_1 i z_1, ..., a_k i z_k)$, and $j(e_2)(z_1, ..., z_k) = (b_1 i z_1, ..., b_k i z_k)$ for an orthonormal basis e_1, e_2 of $\mathfrak{z}$. One still has the same elements in the isotropy group as in the previous example, and hence we obtain a $2k$-dimensional family of weakly symmetric metrics.

We next look at the case where $\mathfrak{z} = \text{Im}(\mathbb{H})$, $\mathfrak{a} = \mathbb{H}^a \oplus \mathbb{H}^b = \mathbb{R}^{4a+4b}$ with $j(q)(v, w) = (qv, wq)$. If we choose on $\text{Im}(\mathbb{H}) = \mathbb{R}^3$ the usual inner product, we obtain the Riemannian nilmanifolds of Heisenberg type with dim $\mathfrak{z} = 3$ ($a = n, b = 0$ corresponds to the horospheres in $\mathbb{H}H^{n+1}$ discussed earlier), but we can also choose an arbitrary inner product on $\mathfrak{z}$. Since $\mathfrak{z}$ is a subalgebra, we obtain a copy of Sp(1) in the normalizer: $A_r(v, w) = (rv, wr^{-1})$ induces via conjugation $B(q) = rqr^{-1}$ on $\mathfrak{z}$ (which achieves any element of SO(3)). Using these automorphisms, we can assume that the metric on $\mathfrak{z}$ is diagonal with entries x, y, z. The metric is naturally reductive iff $x = y = z$, which, up to scaling, is isometric to the quaternionic Heisenberg group. The normalizer of $\mathfrak{z} \subset so(n)$ is equal to Sp(1) × Sp(a) × Sp(b) where $r \in$ Sp(1) acts as above and $(R, S) \in$ Sp(a) × Sp(b) acts via $(v, w) \to (vR, Sw)$ which commutes with $\mathfrak{z}$ and hence induces $B = id$ on $\mathfrak{z}$. We claim that the metric is weakly symmetric iff $x = y = z$. Indeed in that case we can first choose r such that $rqr^{-1} = -q$ and then R, S such that $R(rv) = -v, S(wr^{-1}) = -w$. Furthermore, if x, y, z are not all the same, this destroys some of the isometries in Sp(1) which are all needed to prove (*). But notice that all of these metrics are g.o. spaces. Indeed for $q \in \mathfrak{z}, (v, w) \in \mathfrak{a}$ we can choose $A = (R, S)$ with $(R, S)(v, w) = j(q)(v, w)$ and since $Ad(A)_{|\mathfrak{z}}$ is trivial, no restrictions on the metric on $\mathfrak{z}$ are necessary.

A similar case is the generalized Heisenberg Lie algebra with dim $\mathfrak{z} = 2$. Here $\mathfrak{z}$ is spanned by $i, j \in \mathrm{Im}(\mathbb{H})$, $\mathfrak{a} = \mathbb{H}^n$, and $j(q)(v) = qv$. The only element from the above Sp(1) which remains in the normalizer of $\mathfrak{z}$ is A_q with $q = k$ which induces -id on $\mathfrak{z}$. The rest of the normalizer is still Sp(n) acting via $v \to vA, q \to q$. Hence the isotropy group clearly satisfies (*). Notice that there is no restriction on the metric on $\mathfrak{z}$ but that we can also not normalize it as in the previous example. Hence we obtain a three-parameter family of weakly symmetric metrics on N, only one of which is of Heisenberg type. These metrics are not naturally reductive since $\mathfrak{z}$ is not a subalgebra.

The Lie algebras we considered so far include the Lie algebras of Heisenberg type with dim $\mathfrak{z} \leq 3$ (but with metrics not all of which are of Heisenberg type). Notice that it follows easily from [G1] that a Riemannian nilmanifold of Heisenberg type is naturally reductive iff dim $\mathfrak{z} = 1$ or 3. One can also easily determine the other Riemannian nilmanifolds of Heisenberg type which are weakly symmetric (see also [BRV]). For this we use the fact that a weakly symmetric space is also a commutative space and the classification of Ricci [Ri] of manifolds of Heisenberg type which are commutative spaces. They are the ones with dim $\mathfrak{z} = 1, 2, 3$ discussed above and dim $\mathfrak{z} = 5, 6, 7$, $\mathfrak{z} \subset \mathrm{Im(Cay)}$ with $\mathfrak{a} =$ Cay and $j(q)(v) = qv$ and dim $\mathfrak{z} = 7$, $\mathfrak{z} = \mathrm{Im(Cay)}$ with $\mathfrak{a} = \mathrm{Cay} \oplus \mathrm{Cay}$ and $j(q)(v, w) = (qv, qw)$. We will now show that each of these examples is also weakly symmetric.

In [R1] the isotropy group of all Heisenberg type Lie groups was determined. It always contains Spin(m) with $m = \dim \mathfrak{z}$ which acts on $\mathfrak{z}$ via the two -fold cover onto SO($\mathfrak{z}$) and on $\mathfrak{a}$ via the spin representation Δ_m. (Indeed, this follows immediately from the fact that $\mathfrak{a}$ is a Clifford module over $\mathfrak{z}$.) The isotropy group is essentially the direct product of this group with the group of automorphisms of $\mathfrak{a}$ as a Clifford module over $\mathfrak{z}$ together with the id on $\mathfrak{z}$. We will only need the action of Spin(m) in the above examples. If dim $\mathfrak{z} = 5, 6, 7$, this representation is $\rho_k \oplus \Delta_k$, $k = 5, 6, 7$ acting on $R^k \oplus R^8$ and for dim $\mathfrak{z} = 7$, $\mathfrak{a} = \text{Cay} \oplus \text{Cay}$ it is $\rho_7 \oplus \Delta_7 \oplus \Delta_7$. We will now show that each representation has property (*). For $\rho_7 \oplus \Delta_7$ we already did this in the case of the distance spheres in the Cayley plane. Using the same method we consider the action of $\text{Spin}(6) = \text{SU}(4)$ via $\rho_6 \oplus \Delta_6$ on $\mathbb{R}^6 \oplus \mathbb{R}^8$. Here Δ_6 is the standard action of SU(4) on $\mathbb{R}^8 = \mathbb{C}^4$. We again first use ρ_6 to reverse a vector on $\mathbb{R}^6$ and then consider the isotropy group of this vector, which is $\text{Spin}(5) = \text{Sp}(2)$. The action of SU(4) on $\mathbb{C}^4$ restricts to the standard action of Sp(2) on $\mathbb{C}^4 = \mathbb{H}^2$ which acts transitively on spheres and hence $\rho_6 \oplus \Delta_6$ satisfies (*). Next we consider the action of $\text{Spin}(5) = \text{Sp}(2)$ via $\rho_5 \oplus \Delta_5$ on $\mathbb{R}^5 \oplus \mathbb{R}^8$ and Δ_5 is the standard action of Sp(2) on $\mathbb{R}^8 = \mathbb{H}^2$. The subgroup of Sp(2) fixing a vector in $\mathbb{R}^5$ is $\text{Spin}(4) = \text{Sp}(1) \times \text{Sp}(1)$ acting on $\mathbb{H}^2$ via multiplication on each coordinate. Hence we need to also know how an element that reverses a vector in $\mathbb{R}^5$ acts on $\mathbb{H}^2$. But i id $\in \text{Sp}(2)$ under the two fold cover onto SO(5) goes into the element which is -id on a two plane and id on the perpendicular three-plane. It hence reverses one vector in $\mathbb{R}^5$ and multiplies each coordinate in $\mathbb{H}^2$ with i. We can now use the action of $\text{Sp}(1) \times \text{Sp}(1)$ to take (iv, iw) into $(-v, -w)$. For the representation $\rho_7 \oplus \Delta_7 \oplus \Delta_7$ we observe that the action of SU(4), the isotropy group of a vector in $\mathbb{R}^7$, is equal to $\mu_4 \oplus \mu_4$ on $\mathfrak{a}$, where μ_4 is the standard action of SU(4) on $\mathbb{C}^4$. But given two vectors in $\mathbb{C}^4$, there clearly exists an element in SU(4) that reverses both of them and hence the representation satisfies (*).

Thus a Lie algebra of Heisenberg type is weakly symmetric iff it is a commutative space. In [R1] the Lie algebras of Heisenberg type were classified which are g.o. spaces. Besides the above examples, they contain one more case which is hence g.o. but not weakly symmetric.

We give one more example of a weakly symmetric metric on a nilpotent group, which is very different from the groups of Heisenberg type. It is the case where $\mathfrak{z} = \mathfrak{so}(n)$ and the metric on $\mathfrak{so}(n)$ is biinvariant. The Lie group can be viewed as $\mathbb{R}^n \oplus \Lambda^2\mathbb{R}^n$ with the only nonvanishing Lie bracket given by $[v, w] = v \wedge w$. The full isotropy group is equal to $\text{O}(n)$ which acts on $\mathbb{R}^n \oplus \Lambda^2\mathbb{R}^n$ in the canonical fashion. (Notice that under the identification $\mathfrak{so}(n) = \Lambda^2\mathbb{R}^n$ the adjoint representation becomes the natural action of $\text{O}(n)$ on $\Lambda^2\mathbb{R}^n$.) Now the same proof as in Example 3 shows that this representation satisfies (*) and hence we obtain a two-parameter family of weakly symmetric metrics.

Example 7. General Case. In the previous examples we have seen many cases where the group G is two-step nilpotent, semisimple compact, or semisimple non-compact. The general case can in some sense be reduced to these subcases as it was done in [G2] for g.o. spaces. In fact, since weakly symmetric spaces are g.o. spaces, the results in [G2] can be applied.

In particular, if G is nilpotent, then it must be two-step nilpotent. In general G can be decomposed into a two-step nilpotent part, a semisimple non-compact part and a semisimple compact part, and the orbit of each are totally geodesic submanifolds which one easily shows are again weakly symmetric. Nevertheless these three parts can still be put together in interesting ways as the last case in Example 5 shows.

Finally, I would like to thank the editors for allowing me to contribute to this volume honoring Joseph D'Atri. Early in my career we spent a year together at the Institute for Advanced Study in Princeton which I fondly remember for the many discussions we had and the joint work that came out of it.

Bibliography

[BV] J. Berndt and L. Vanhecke, Geometry of weakly symmetric spaces, Preprint 1995.

[BKV] J. Berndt, O.Kowalski and L. Vanhecke, Geodesics in weakly symmetric spaces, Preprint 1995.

[BRV] J. Berndt, F. Ricci and L. Vanhecke, Weakly symmetric groups of Heisenberg type, Preprint 1995.

[D] J. E. D'Atri, Geodesic spheres and symmetries in naturally reductive spaces, *Michigan Math. J.* **22** (1975), 71–76.

[DZ] J. E. D'Atri and W. Ziller, Naturally reductive metrics and Einstein metrics on compact Lie groups, *Mem. Amer. Math. Soc.* **18**, No. 215 , (1979).

[G1] C. Gordon, Naturally reductive homogeneous Riemannian manifolds, *Can. J. Math.* **37** (1985), 467–487.

[G2] C. Gordon, Homogeneous Riemannian manifolds whose geodesics are orbits, see article in this volume.

[GV] J.C. Gonzalez-Davila and L. Vanhecke, New examples of weakly symmetric spaces, Preprint 1995.

[GWZ] H. Gluck, F. Warner and W. Ziller, The geometry of the Hopf fibrations, *L'Ens. Math.* **32** (1986), 173–198.

[K1] M. Kerr, Some new homogenous Einstein metrics on symmetric spaces, Preprint 1994.

[K2] M. Kerr, Homogeneous Einstein metrics on products of symmetric spaces, Preprint 1995.

[KPV] O. Kowalski, F. Prüfer and L. Vanhecke, D'Atri spaces, see article in this volume.

[KV1] O. Kowalski and L. Vanhecke, A generalization of a theorem on naturally reductive homogeneous spaces, *Proc. Amer. Math. Soc.* **91**, (1984), 433-435.

[KV2] O. Kowalski and L. Vanhecke, Riemannian manifolds with homogeneous geodesics, *Boll. Un. Math. Ital. B (7)* **5** (1991), 189-246.

[KS] M. Kreck and S. Stolz, A diffeomorphism classification of 7-dimensional homogeneous Einstein manifolds with $SU(2) \times SU(2) \times U(1)$ symmetry, *Ann. Math.* **127** (1988), 373–388.

[Ri] F. Ricci, Commutative algebras of invariant functions on groups of Heisenberg type, *J. London Math. Soc.* **32** (1985), 265–271.

[R1] C. Riehm, The automorphism group of a composition of quadratic forms, *Trans. A.M.S.* **269** (1982), 403–414.

[R2] C. Riehm, Explicit spin representations and Lie algebras of Heisenberg type, *J. London Math. Soc.* **29** (1984), 49–62.

[S] A. Selberg, Harmonic analysis and discontinuous groups in weakly symmetric Riemannian spaces with applications to Dirichlet series, *J. Indian Math. Soc.* **20**(1956), 47 –87.

[WZ1] M. Wang and W. Ziller, On normal homogeneous Einstein manifolds, *Ann. Scient. Ec. Norm. Sup.* **18** (1985), 563–633.

[WZ2] M. Wang and W. Ziller, Einstein metrics on principal torus bundles, *J. Diff. Geom.* **31** (1990), 215–248.

[Z1] W. Ziller The Jacobi equation on naturally reductive compact Riemannian homogeneous spaces, *Comm. Math. Helv.* **52** (1977), 573–590.

[Z2] W. Ziller, Homogeneous Einstein metrics on spheres and projective spaces, *Math. Ann.* **259** (1982), 351–358.

University of Pennsylvania
Philadelphia, Pa 19104
wziller@math.upenn.edu

Received December 1995

Progress in Nonlinear Differential Equations and Their Applications

Editor
Haim Brezis
Département de Mathématiques
Université P. et M. Curie
4, Place Jussieu
75252 Paris Cedex 05
France
and
Department of Mathematics
Rutgers University
New Brunswick, NJ 08903
U.S.A.

Progress in Nonlinear Differential Equations and Their Applications is a book series that lies at the interface of pure and applied mathematics. Many differential equations are motivated by problems arising in such diversified fields as Mechanics, Physics, Differential Geometry, Engineering, Control Theory, Biology, and Economics. This series is open to both the theoretical and applied aspects, hopefully stimulating a fruitful interaction between the two sides. It will publish monographs, polished notes arising from lectures and seminars, graduate level texts, and proceedings of focused and refereed conferences.

We encourage preparation of manuscripts in some form of TeX for delivery in camera-ready copy, which leads to rapid publication, or in electronic form for interfacing with laser printers or typesetters.

Proposals should be sent directly to the editor or to: Birkhäuser Boston, 675 Massachusetts Avenue, Cambridge, MA 02139

PNLDE 5 Composite Media and Homogenization Theory
Gianni Dal Maso and Gian Fausto Dell'Antonio, editors

PNLDE 6 Infinite Dimensional Morse Theory and Multiple Solution Problems
Kung-ching Chang

PNLDE 7 Nonlinear Differential Equations and their Equilibrium States, 3
N.G. Lloyd, W.M. Ni, L.A. Peletier, J. Serrin, editors

PNLDE 8 Introduction to Γ–Convergence
Gianni Dal Maso

PNLDE 9 Differential Inclusions in Nonsmooth Mechanical Problems: Shocks and Dry Frictions
Manuel D. P. Monteiro Marques

PNLDE 10 Periodic Solutions of Singular Lagrangian Systems
Antonio Ambrosetti and Vittorio Coti Zelati

PNLDE 11 Nonlinear Waves and Weak Turbulence with Applications in Oceanography and Condensed Matter Physics
N. Fitzmaurice, D. Gurarie, F. McCaugham, and W. A. Woyczynski, editors

PNLDE 12 Semester on Dynamical Systems: Euler International Math Institute, St. Petersburg, 1991
Kuksin, Lazutkin, and Poeschel, editors

PNLDE 13 Ginzburg-Landau Vortices
F. Bethuel, H. Brezis, and F. Hélein

PNLDE 14 Variational Methods for Image Segmentation
Jean-Michel Morel and Sergio Solomini

PNLDE 15 Topological Nonlinear Analysis: Degree, Singularity, and Variations
Michele Matzeu and Alfonso Vignoli, editors

PNLDE 16 Analytic Semigroups and Optimal Regularity in Parabolic Problems
A. Lunardi

PNLDE 17 Blowup for Nonlinear Hyperbolic Equations
Serge Alinhac

PNLDE 18 The Heat Kernel Lefschetz Fixed Point Formula for the Spin-c Dirac Operator
J. J. Duistermaat

PNLDE 19 Chaos and Nonlinear Dynamical Systems
I. Hoveijn

PNLDE 20 Topics in Geometry: In Memory of Joseph D'Atri
Simon Gindikin, editor

PNLDE 21 Partial Differential Equations and Mathematical Physics: The Danish-Swedish Analysis Seminar, 1995
Lars Hörmander and Anders Melin, editors

PNLDE 22 Partial Differential Equations and Functional Analysis: In Memory of Pierre Grisvard
J. Cea, D. Chenais, G. Geymonat, and J.-L. Lions, editors

PNLDE 23 Geometry of Harmonic Maps
Yuanlong Xin